IET ENERGY ENGINEERING SERIES 127

Lightning Electromagnetics

Other volumes in this series:

Volume 1	**Power Circuit Breaker Theory and Design** C.H. Flurscheim (Editor)
Volume 4	**Industrial Microwave Heating** A.C. Metaxas and R.J. Meredith
Volume 7	**Insulators for High Voltages** J.S.T. Looms
Volume 8	**Variable Frequency AC Motor Drive Systems** D. Finney
Volume 10	**SF_6 Switchgear** H.M. Ryan and G.R. Jones
Volume 11	**Conduction and Induction Heating** E.J. Davies
Volume 13	**Statistical Techniques for High Voltage Engineering** W. Hauschild and W. Mosch
Volume 14	**Uninterruptible Power Supplies** J. Platts and J.D. St Aubyn (Editors)
Volume 15	**Digital Protection for Power Systems** A.T. Johns and S.K. Salman
Volume 16	**Electricity Economics and Planning** T.W. Berrie
Volume 18	**Vacuum Switchgear** A. Greenwood
Volume 19	**Electrical Safety: a guide to causes and prevention of hazards** J. Maxwell Adams
Volume 21	**Electricity Distribution Network Design, 2nd Edition** E. Lakervi and E.J. Holmes
Volume 22	**Artificial Intelligence Techniques in Power Systems** K. Warwick, A.O. Ekwue and R. Aggarwal (Editors)
Volume 24	**Power System Commissioning and Maintenance Practice** K. Harker
Volume 25	**Engineers' Handbook of Industrial Microwave Heating** R.J. Meredith
Volume 26	**Small Electric Motors** H. Moczala *et al.*
Volume 27	**AC-DC Power System Analysis** J. Arrillaga and B.C. Smith
Volume 29	**High Voltage Direct Current Transmission, 2nd Edition** J. Arrillaga
Volume 30	**Flexible AC Transmission Systems (FACTS)** Y-H. Song (Editor)
Volume 31	**Embedded generation** N. Jenkins *et al.*
Volume 32	**High Voltage Engineering and Testing, 2nd Edition** H.M. Ryan (Editor)
Volume 33	**Overvoltage Protection of Low-Voltage Systems, Revised Edition** P. Hasse
Volume 36	**Voltage Quality in Electrical Power Systems** J. Schlabbach *et al.*
Volume 37	**Electrical Steels for Rotating Machines** P. Beckley
Volume 38	**The Electric Car: Development and future of battery, hybrid and fuel-cell cars** M. Westbrook
Volume 39	**Power Systems Electromagnetic Transients Simulation** J. Arrillaga and N. Watson
Volume 40	**Advances in High Voltage Engineering** M. Haddad and D. Warne
Volume 41	**Electrical Operation of Electrostatic Precipitators** K. Parker
Volume 43	**Thermal Power Plant Simulation and Control** D. Flynn
Volume 44	**Economic Evaluation of Projects in the Electricity Supply Industry** H. Khatib
Volume 45	**Propulsion Systems for Hybrid Vehicles** J. Miller
Volume 46	**Distribution Switchgear** S. Stewart
Volume 47	**Protection of Electricity Distribution Networks, 2nd Edition** J. Gers and E. Holmes
Volume 48	**Wood Pole Overhead Lines** B. Wareing
Volume 49	**Electric Fuses, 3rd Edition** A. Wright and G. Newbery
Volume 50	**Wind Power Integration: Connection and system operational aspects** B. Fox *et al.*
Volume 51	**Short Circuit Currents** J. Schlabbach
Volume 52	**Nuclear Power** J. Wood
Volume 53	**Condition Assessment of High Voltage Insulation in Power System Equipment** R.E. James and Q. Su
Volume 55	**Local Energy: Distributed generation of heat and power** J. Wood
Volume 56	**Condition Monitoring of Rotating Electrical Machines** P. Tavner, L. Ran, J. Penman and H. Sedding
Volume 57	**The Control Techniques Drives and Controls Handbook, 2nd Edition** B. Drury
Volume 58	**Lightning Protection** V. Cooray (Editor)
Volume 59	**Ultracapacitor Applications** J.M. Miller
Volume 62	**Lightning Electromagnetics** V. Cooray

Volume 63 **Energy Storage for Power Systems, 2nd Edition** A. Ter-Gazarian
Volume 65 **Protection of Electricity Distribution Networks, 3rd Edition** J. Gers
Volume 66 **High Voltage Engineering Testing, 3rd Edition** H. Ryan (Editor)
Volume 67 **Multicore Simulation of Power System Transients** F.M. Uriate
Volume 68 **Distribution System Analysis and Automation** J. Gers
Volume 69 **The Lightening Flash, 2nd Edition** V. Cooray (Editor)
Volume 70 **Economic Evaluation of Projects in the Electricity Supply Industry, 3rd Edition** H. Khatib
Volume 72 **Control Circuits in Power Electronics: Practical issues in design and implementation** M. Castilla (Editor)
Volume 73 **Wide Area Monitoring, Protection and Control Systems: The enabler for Smarter Grids** A. Vaccaro and A. Zobaa (Editors)
Volume 74 **Power Electronic Converters and Systems: Frontiers and applications** A.M. Trzynadlowski (Editor)
Volume 75 **Power Distribution Automation** B. Das (Editor)
Volume 76 **Power System Stability: Modelling, analysis and control** A.A. Sallam and B. Om P. Malik
Volume 78 **Numerical Analysis of Power System Transients and Dynamics** A. Ametani (Editor)
Volume 79 **Vehicle-to-Grid: Linking electric vehicles to the smart grid** J. Lu and J. Hossain (Editors)
Volume 81 **Cyber-Physical-Social Systems and Constructs in Electric Power Engineering** S. Suryanarayanan, R. Roche and T.M. Hansen (Editors)
Volume 82 **Periodic Control of Power Electronic Converters** F. Blaabjerg, K. Zhou, D. Wang and Y. Yang
Volume 86 **Advances in Power System Modelling, Control and Stability Analysis** F. Milano (Editor)
Volume 87 **Cogeneration: Technologies, Optimisation and Implentation** C.A. Frangopoulos (Editor)
Volume 88 **Smarter Energy: from Smart Metering to the Smart Grid** H. Sun, N. Hatziargyriou, H.V. Poor, L. Carpanini and M.A. Sánchez Fornié (Editors)
Volume 89 **Hydrogen Production, Separation and Purification for Energy** A. Basile, F. Dalena, J. Tong and T.N. Veziroğlu (Editors)
Volume 90 **Clean Energy Microgrids** S. Obara and J. Morel (Editors)
Volume 91 **Fuzzy Logic Control in Energy Systems with Design Applications in Matlab/Simulink®** İ.H. Altaş
Volume 92 **Power Quality in Future Electrical Power Systems** A.F. Zobaa and S.H.E.A. Aleem (Editors)
Volume 93 **Cogeneration and District Energy Systems: Modelling, Analysis and Optimization** M.A. Rosen and S. Koohi-Fayegh
Volume 94 **Introduction to the Smart Grid: Concepts, technologies and evolution** S.K. Salman
Volume 95 **Communication, Control and Security Challenges for the Smart Grid** S.M. Muyeen and S. Rahman (Editors)
Volume 96 **Industrial Power Systems with Distributed and Embedded Generation** R. Belu
Volume 97 **Synchronized Phasor Measurements for Smart Grids** M.J.B. Reddy and D.K. Mohanta (Editors)
Volume 98 **Large Scale Grid Integration of Renewable Energy Sources** A. Moreno-Munoz (Editor)
Volume 100 **Modeling and Dynamic Behaviour of Hydropower Plants** N. Kishor and J. Fraile-Ardanuy (Editors)
Volume 101 **Methane and Hydrogen for Energy Storage** R. Carriveau and D.S-K. Ting
Volume 104 **Power Transformer Condition Monitoring and Diagnosis** A. Abu-Siada (Editor)
Volume 106 **Surface Passivation of Industrial Crystalline Silicon Solar Cells** J. John (Editor)
Volume 107 **Bifacial Photovoltaics: Technology, applications and economics** J. Libal and R. Kopecek (Editors)

Volume 108 **Fault Diagnosis of Induction Motors** J. Faiz, V. Ghorbanian and G. Joksimović
Volume 109 **Cooling of Rotating Electrical Machines:** Fundamentals, modelling, testing and design D. Staton, E. Chong, S. Pickering and A. Boglietti
Volume 110 **High Voltage Power Network Construction** K. Harker
Volume 111 **Energy Storage at Different Voltage Levels: Technology, integration, and market aspects** A.F. Zobaa, P.F. Ribeiro, S.H.A. Aleem and S.N. Afifi (Editors)
Volume 112 **Wireless Power Transfer: Theory, Technology and Application** N. Shinohara
Volume 114 **Lightning-Induced Effects in Electrical and Telecommunication Systems** Y. Baba and V.A. Rakov
Volume 115 **DC Distribution Systems and Microgrids** T. Dragičević, F. Blaabjerg and P. Wheeler
Volume 116 **Modelling and Simulation of HVDC Transmission** M. Han (Editor)
Volume 117 **Structural Control and Fault Detection of Wind Turbine Systems** H.R. Karimi
Volume 118 **Modelling and Simulation of Complex Power Systems** A. Monti and A. Benigni
Volume 119 **Thermal Power Plant Control and Instrumentation: The control of boilers and HRSGs, 2nd Edition** D. Lindsley, J. Grist and D. Parker
Volume 120 **Fault Diagnosis for Robust Inverter Power Drives** A. Ginart (Editor)
Volume 121 **Monitoring and Control using Synchrophasors in Power Systems with Renewables** I. Kamwa and C. Lu (Editors)
Volume 123 **Power Systems Electromagnetic Transients Simulation, 2nd Edition** N. Watson and J. Arrillaga
Volume 124 **Power Market Transformation** B. Murray
Volume 125 **Wind Energy Modeling and Simulation Volume 1: Atmosphere and plant** P. Veers (Editor)
Volume 126 **Diagnosis and Fault Tolerance of Electrical Machines, Power Electronics and Drives** A.J.M. Cardoso
Volume 128 **Characterization of Wide Bandgap Power Semiconductor Devices** F. Wang, Z. Zhang and E.A. Jones
Volume 129 **Renewable Energy from the Oceans: From wave, tidal and gradient systems to offshore wind and solar** D. Coiro and T. Sant (Editors)
Volume 130 **Wind and Solar Based Energy Systems for Communities** R. Carriveau and D.S-K. Ting (Editors)
Volume 131 **Metaheuristic Optimization in Power Engineering** J. Radosavljević
Volume 132 **Power Line Communication Systems for Smart Grids** I.R.S Casella and A. Anpalagan
Volume 134 **Hydrogen Passivation and Laser Doping for Silicon Solar Cells** B. Hallam and C. Chan (Editors)
Volume 139 **Variability, Scalability and Stability of Microgrids** S.M. Muyeen, S.M. Islam and F. Blaabjerg (Editors)
Volume 142 **Wind Turbine System Design: Volume 1: Nacelles, drive trains and verification** J. Wenske (Editor)
Volume 143 **Medium Voltage DC System Architectures** B. Grainger and R.D. Doncker (Editors)
Volume 145 **Condition Monitoring of Rotating Electrical Machines** P. Tavner, L. Ran and C. Crabtree
Volume 146 **Energy Storage for Power Systems, 3rd Edition** A.G. Ter-Gazarian
Volume 147 **Distribution Systems Analysis and Automation 2nd Edition** J. Gers
Volume 151 **SiC Power Module Design: Performance, robustness and reliability** A. Castellazzi and A. Irace (Editors)
Volume 152 **Power Electronic Devices: Applications, failure mechanisms and reliability** F. Iannuzzo (Editor)
Volume 153 **Signal Processing for Fault Detection and Diagnosis in Electric Machines and Systems** M. Benbouzid (Editor)
Volume 155 **Energy Generation and Efficiency Technologies for Green Residential Buildings** D. Ting and R. Carriveau (Editors)
Volume 156 **Lithium-ion Batteries Enabled by Silicon Anodes** C. Ban and K. Xu (Editors)

Volume 157 **Electrical Steels, 2 Volumes** A. Moses, K. Jenkins, Philip Anderson and H. Stanbury
Volume 158 **Advanced Dielectric Materials for Electrostatic Capacitors** Q. Li (Editor)
Volume 159 **Transforming the Grid Towards Fully Renewable Energy** O. Probst, S. Castellanos and R. Palacios (Editors)
Volume 160 **Microgrids for Rural Areas: Research and case studies** R.K. Chauhan, K. Chauhan and S.N. Singh (Editors)
Volume 161 **Artificial Intelligence for Smarter Power Systems: Fuzzy Logic and Neural Networks** M.G. Simoes
Volume 165 **Digital Protection for Power Systems 2nd Edition** Salman K Salman
Volume 166 **Advanced Characterization of Thin Film Solar Cells** N. Haegel and M. Al-Jassim (Editors)
Volume 167 **Power Grids with Renewable Energy Storage, integration and digitalization** A.A. Sallam and B. OM P. Malik
Volume 169 **Small Wind and Hydrokinetic Turbines** P. Clausen, J. Whale and D. Wood (Editors)
Volume 170 **Reliability of Power Electronics Converters for Solar Photovoltaic Applications** F. Blaabjerg, A.l Haque, H. Wang, Z. Abdin Jaffery and Y. Yang (Editors)
Volume 171 **Utility-scale Wind Turbines and Wind Farms** A. Vasel-Be-Hagh and D.S.-K. Ting
Volume 172 **Lighting interaction with Power Systems, 2 volumes** A. Piantini (Editor)
Volume 174 **Silicon Solar Cell Metallization and Module Technology** T. Dullweber (Editor)
Volume 175 **n-Type Crystalline Silicon Photovoltaics: Technology, applications and economics** D. Munoz and R. Kopecek (Editors)
Volume 178 **Integrated Motor Drives** Xu Deng and B. Mecrow (Editors)
Volume 180 **Protection of Electricity Distribution Networks, 4th Edition** J. Gers and E. Holmes
Volume 182 **Surge Protection for Low Voltage Systems** A. Rousseau (Editor)
Volume 184 **Compressed Air Energy Storage: Types, systems and applications** D. Ting and J. Stagner
Volume 186 **Synchronous Reluctance Machines: Analysis, optimization and applications** N. Bianchi, C. Babetto and G. Bacco
Volume 191 **Electric Fuses: Fundamentals and new applications 4th Edition** N. Nurse, A. Wright and P.G. Newbery
Volume 193 **Overhead Electric Power Lines: Theory and practice** S. Chattopadhyay and A. Das
Volume 194 **Offshore Wind Power Reliability, availability and maintenance, 2nd edition** P. Tavner
Volume 196 **Cyber Security for Microgrids** S. Sahoo, F. Blaajberg and T. Dragicevic
Volume 198 **Battery Management Systems and Inductive Balancing** A. Van den Bossche and A. Farzan Moghaddam
Volume 199 **Model Predictive Control for Microgrids: From power electronic converters to energy management** J. Hu, J.M. Guerrero and S. Islam
Volume 204 **Electromagnetic Transients in Large HV Cable Networks: Modeling and calculations** Ametani, Xue, Ohno and Khalilnezhad
Volume 208 **Nanogrids and Picogrids and their Integration with Electric Vehicles** S. Chattopadhyay
Volume 210 **Superconducting Magnetic Energy Storage in Power Grids** M.H. Ali
Volume 211 **Blockchain Technology for Smart Grids: Implementation, management and security** H.L. Gururaj, K.V. Ravi, F. Flammini, H. Lin, B. Goutham, K.B.R. Sunil and C. Sivapragash
Volume 212 **Battery State Estimation: Methods and Models** S. Wang
Volume 215 **Industrial Demand Response: Methods, best practices, case studies, and applications** H.H. Alhelou, A. Moreno-Muñoz and P. Siano (Editors)
Volume 213 **Wide Area Monitoring of Interconnected Power Systems 2nd Edition** A.R. Messina

Volume 217 **Advances in Power System Modelling, Control and Stability Analysis 2nd Edition** F. Milano (Editor)
Volume 225 **Fusion-Fission Hybrid Nuclear Reactors: For enhanced nuclear fuel utilization and radioactive waste reduction** W.M. Stacey
Volume 228 **Digital Technologies for Solar Photovoltaic Systems: From general to rural and remote installations** S. Motahhir (Editor)
Volume 238 **AI for Status Monitoring of Utility Scale Batteries** S. Wang, K. Liu, Y. Wang, D. Stroe, C. Fernandez and J.M. Guerrero
Volume 905 **Power system protection, 4 volumes**

Lightning Electromagnetics

Volume 1: Return stroke modelling and electromagnetic radiation
2nd Edition

Edited by
Vernon Cooray, Farhad Rachidi and Marcos Rubinstein

The Institution of Engineering and Technology

Published by The Institution of Engineering and Technology, London, United Kingdom

The Institution of Engineering and Technology is registered as a Charity in England & Wales (no. 211014) and Scotland (no. SC038698).

First published 2012
2nd Edition published 2022

The Institution of Engineering and Technology
Futures Place
Kings Way, Stevenage
Hertfordshire SG1 2UA, United Kingdom

www.theiet.org

British Library Cataloguing in Publication Data
A catalogue record for this product is available from the British Library

ISBN 978-1-78561-539-9 (Volume 1 hardback)
ISBN 978-1-78561-540-5 (Volume 1 pdf)
ISBN 978-1-78561-541-2 (Volume 2 hardback)
ISBN 978-1-78561-542-9 (Volume 2 pdf)
ISBN 978-1-78561-543-6 (2 Volume set hardback)

Typeset in India by MPS Limited

Cover image: Sam's Photography/Moment via Getty Images

Contents

About the editors

Vernon Cooray is a professor emeritus at the Department of Electrical Engineering of Uppsala University, Sweden, he is a fellow of the IEEE and recipient of the Berger award, and he was also in charge of the HV Laboratory at Uppsala University. He has authored and co-authored about 550 scientific publications and edited and authored five books, served as keynote speaker and session convener at various international conferences, on journal boards, and as president of ICLP.

Farhad Rachidi is a professor at Ecole Polytechnique Federale de Lausanne (EPFL), Switzerland. Fellow of IEEE and recipient of the Blondel medal and Karl Berger Award, he is the head of the EMC Laboratory at the EPFL. His research focus includes lightning electromagnetics, and EMP interaction with transmission lines. He has served on key journal boards and as chairman or convener to key events and working groups, and has published about 300 papers in peer-reviewed journals.

Marcos Rubinstein is a professor at the University of Applied Sciences and Arts Western Switzerland, Yverdon-les-Bains, where he leads the Applied Electromagnetics Group and serves as the chairman of the International Project on Electromagnetic Radiation from Lightning to Tall Structures. He is an IEEE fellow and a member of the Institute for Information and Communication Technologies. Professor Rubinstein's work has been recognized with numerous awards, including the prestigious Karl Berger Award. He is also the author or co-author of over 300 scientific publications in technical and scientific journals and conferences.

Acknowledgements

We wish to thank all our colleagues who have spent a good deal of their free time writing the chapters of this book.

We wish to express our sincere thanks to Ms. Olivia Wilkins and Ms. Nikki Tarplett from the IET publishers and Mr. N. Srinivasan from MPS Limited for their outstanding support throughout the publishing project. Despite unexpected delays in our submissions, they remained patient and accommodating, always willing to listen to our suggestions and provide valuable feedback. We are truly grateful for their professionalism and dedication.

Chapter 1

Basic electromagnetic theory – a summary

Vernon Cooray[1], Pasan Hettiarachchi[1] and Gerald Cooray[2,3]

1.1 Introduction

The goal of this chapter is to provide a summary of the basic concepts of electromagnetic theory as a complement to the subject matter, most of which is related to electromagnetism, discussed in this book. The chapter covers only the concepts which are necessary to understand the electromagnetics of lightning flashes. A detailed description of the electromagnetic theory can be found in [1–3].

1.2 The nomenclature

I – Current, units: Amperes (A)
J – Current density, units: Amperes/meter^2 (A/m^2)
E – Electric field intensity; units: volts/meter (V/m); in this chapter, this quantity is referred to as E-field
D – Electric flux density, units: Coulombs/meter^2 (C/m^2); in this chapter, this quantity is referred to as D-field
H – Magnetic field intensity, units: Amperes/meter (A/m); in this chapter, this quantity is referred to as H-field
B – Magnetic flux density, units: Tesla (T) or Webers/meter^2 (Wb/m^2); in this chapter, this quantity is referred to as B-field
σ – Conductivity, units: Siemens/meter (S/m)
ϕ – Scalar potential, units: volts
A – Vector potential, units: Tesla meter (Tm) or Webers/meter (Wb/m)
ε_o – Electrical permittivity of vacuum, units: Farads/meter (F/m)
μ_o – Magnetic permeability of vacuum, units: Henrys/meter (H/m)
ε_r – Relative dielectric constant, a ratio; no units
μ_r – Relative magnetic permeability, a ratio; no units

[1]Department of Electrical Engineering, Uppsala University, Sweden
[2]Wellcome Trust Centre for Neuroimaging, Institute of Neurology, University College London, UK
[3]Department of Clinical Neuro Science, Karolinska Institute, Stockholm, Sweden

In this chapter, we consider homogeneous (no spatial variation of properties), isotropic (properties are rotation invariant), and linear mediums. If the medium is isotropic and linear, then one can write

$$\mathbf{D} = \varepsilon_0 \varepsilon_r \mathbf{E} \tag{1.1}$$

In a similar manner, **B** and **H** in an isotropic and linear medium is related by

$$\mathbf{B} = \mu_0 \mu_r \mathbf{H} \tag{1.2}$$

1.3 Coordinate systems

Electromagnetic theory is frequently described either in Cartesian, Spherical, or Cylindrical coordinate systems. These systems together with the relevant three coordinates are shown in Figure 1.1. Unit vectors (vectors whose magnitude is unity; see the next section for definition) associated with these coordinates are denoted in this chapter by $\mathbf{a_x}$, $\mathbf{a_y}$, and $\mathbf{a_z}$ in Cartesian coordinates, $\mathbf{a_z}$, $\mathbf{a_\rho}$, and $\mathbf{a_\varphi}$ in Cylindrical coordinates and $\mathbf{a_r}$, $\mathbf{a_\theta}$, and $\mathbf{a_\varphi}$ in Spherical coordinate system. In this chapter, we will be using mainly the Cartesian coordinate system.

1.4 Important vector relationships

In electromagnetic theory, electric and magnetic fields are represented by vectors. The most important vector relationships are referred to in the following. As mentioned above, in this chapter unit vectors in the direction of x, y, and z-axis are denoted by $\mathbf{a}_x$, $\mathbf{a}_y$, and $\mathbf{a}_z$. Thus, any three-dimensional vector, say **P**, is defined by the equation

$$\mathbf{P} = p_x \mathbf{a_x} + p_y \mathbf{a_y} + p_z \mathbf{a_z} \tag{1.3}$$

where p_x, p_y, and p_z are the magnitudes of the component of the vector in x, y, and z directions. The magnitude of this vector is

$$|P| = \sqrt{p_x^2 + p_y^2 + p_z^2} \tag{1.4}$$

A unit vector in the direction of **P** is then given by

$$\mathbf{a}_P = \frac{\mathbf{P}}{\sqrt{p_x^2 + p_y^2 + p_z^2}} \tag{1.5}$$

Note that the magnitude of $\mathbf{a}_p$ i.e. $|\mathbf{a}_p| = 1$.

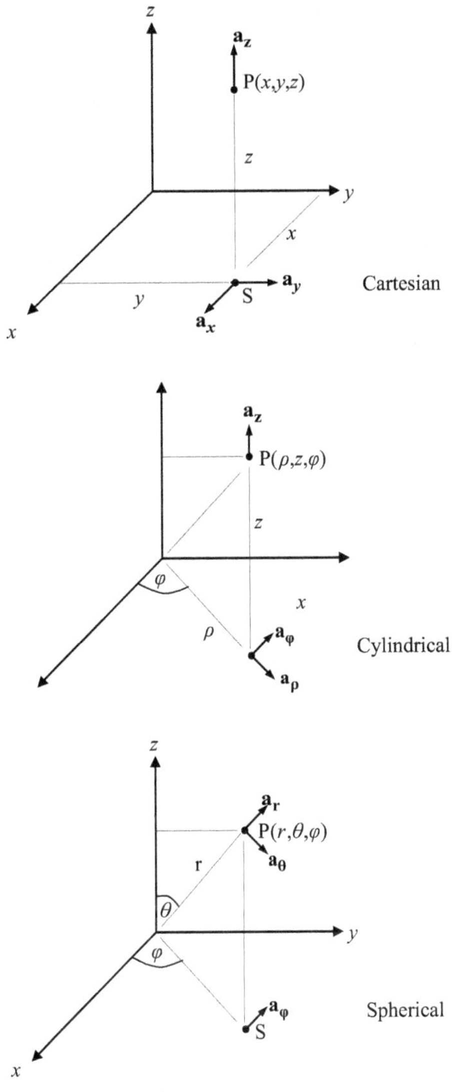

Figure 1.1 Pictorial definition of unit vectors and coordinates in different coordinate systems

1.4.1 The scalar product of vectors

The scalar product of two vectors is given by

$$\mathbf{P} \cdot \mathbf{Q} = |P||Q|\cos\ \theta_{pq} \tag{1.6}$$

where θ_{PQ} is the angle between them. Thus it follows that

$$\mathbf{P} \cdot \mathbf{Q} = \mathbf{Q} \cdot \mathbf{P} \tag{1.7}$$

Let us now resolve the two vectors into its components in the x, y, and z directions and perform the scalar product. The result would be the following,

$$\mathbf{P} \cdot \mathbf{Q} = (p_x\mathbf{a_x} + p_y\mathbf{a_y} + p_z\mathbf{a_z}) \cdot (q_x\mathbf{a_x} + q_y\mathbf{a_y} + q_z\mathbf{a_z}) \tag{1.8}$$

Performing the scalar product term by term and noting that $\mathbf{a}_x \cdot \mathbf{a}_y = 0$, $\mathbf{a}_x \cdot \mathbf{a}_z = 0$ and $\mathbf{a}_y \cdot \mathbf{a}_z = 0$ (because the angle between these vectors is equal to 90°) one obtains

$$\mathbf{P} \cdot \mathbf{Q} = p_xq_x + p_yq_y + p_zq_z \tag{1.9}$$

Note also that

$$\mathbf{P} \cdot \mathbf{P} = |P|^2 \tag{1.10}$$

One can use the scalar product to find the component of a vector in a given direction. For example, the component of vector **P** in the direction specified by the unit vector **a** is

$$\mathbf{P} \cdot \mathbf{a} = |P||a|\cos\,\theta_{Pa} = |P|\cos\,\theta_{Pa} \tag{1.11}$$

where θ_{Pa} is the smaller of the two angles between **P** and **a**.

1.4.2 The vector product of two vectors

The vector product of two vectors is defined by the equation

$$\mathbf{P} \times \mathbf{Q} = |P||Q|\sin\theta_{PQ}\ \mathbf{a}_n \tag{1.12}$$

In the above equation, θ_{PQ} is the small angle between **P** and **Q** and $\mathbf{a}_n$ is a vector which is perpendicular to both **P** and **Q**. The direction of $\mathbf{a}_n$ is given by the right handed screw rule. Namely, rotate a right handed screw from **P** to **Q** around the angle θ_{PQ} the direction of motion of the screw defines the positive direction of $\mathbf{a}_n$.

Let us now take the vector product of the two vectors resolved into its components in the direction of x, y, and z. We can then write

$$\mathbf{P} \times \mathbf{Q} = (p_x\mathbf{a_x} + p_y\mathbf{a_y} + p_z\mathbf{a_z}) \times (q_x\mathbf{a_x} + q_y\mathbf{a_y} + q_z\mathbf{a_z}) \tag{1.13}$$

After taking the vector product of each individual term, we find

$$\begin{aligned}\mathbf{P} \times \mathbf{Q} = {} & p_xq_x(\mathbf{a}_x \times \mathbf{a}_x) + p_xq_y(\mathbf{a}_x \times \mathbf{a}_y) + p_xq_z(\mathbf{a}_x \times \mathbf{a}_z) + p_yq_x(\mathbf{a}_y \times \mathbf{a}_x) \\ & + p_yq_y(\mathbf{a}_y \times \mathbf{a}_y) + p_yq_z(\mathbf{a}_y \times \mathbf{a}_z) + p_zq_x(\mathbf{a}_z \times \mathbf{a}_x) \\ & + p_zq_y(\mathbf{a}_z \times \mathbf{a}_y) + p_zq_z(\mathbf{a}_z \times \mathbf{a}_z)\end{aligned} \tag{1.14}$$

Noting that $\mathbf{a}_x \times \mathbf{a}_x = 0$, $\mathbf{a}_y \times \mathbf{a}_y = 0$ and $\mathbf{a}_z \times \mathbf{a}_z = 0$ (because the angle between them is zero) one obtains

$$\mathbf{P} \times \mathbf{Q} = (p_y q_z - p_z q_y)\mathbf{a}_x + (p_z q_x - p_x q_z)\mathbf{a}_y + (p_x q_y - p_y q_x)\mathbf{a}_z \tag{1.15}$$

This can be obtained by the determinant of the following matrix:

$$\mathbf{P} \times \mathbf{Q} = \begin{vmatrix} \mathbf{a}_x & \mathbf{a}_y & \mathbf{a}_z \\ p_x & p_y & p_z \\ q_x & q_y & q_z \end{vmatrix} \tag{1.16}$$

1.4.3 *Vector field*

In a vector field, every point in space is associated with the field under consideration. The field can be either electric or magnetic in our case. Since the field is defined by a vector, we have to define the value of this vector in every point in space. The field at any general point (x, y, z) is given by

$$\mathbf{F} = f_x(x, y, z)\mathbf{a}_x + f_y(x, y, z)\mathbf{a}_y + f_z(x, y, z)\mathbf{a}_z \tag{1.17}$$

This is called a vector field. As one can see **F** defines the field at every point in space and how it varies from one point to another depends on the functions f_x, f_y, and f_z.

1.4.4 *The Nabla operator and its operations*

The Nabla operator is defined in Cartesian coordinates by

$$\nabla = \mathbf{a_x}\frac{\partial}{\partial x} + \mathbf{a_y}\frac{\partial}{\partial y} + \mathbf{a_z}\frac{\partial}{\partial z} \tag{1.18}$$

In cylindrical and spherical coordinates, it is given by

$$\nabla = \mathbf{a}_\rho \frac{\partial}{\partial \rho} + \mathbf{a}_\varphi \frac{1}{\rho}\frac{\partial}{\partial \varphi} + \mathbf{a_z}\frac{\partial}{\partial z} \tag{1.19}$$

and

$$\nabla = \mathbf{a_r}\frac{\partial}{\partial r} + \mathbf{a}_\theta \frac{1}{r}\frac{\partial}{\partial \theta} + \mathbf{a}_\varphi \frac{1}{r \sin\theta}\frac{\partial}{\partial \varphi} \tag{1.20}$$

respectively. The Nabla operator can operate either on a scalar function or on a vector function. In the later case, it could operate either as a scalar product or a vector product.

1.4.4.1 The Gradient of a scalar function

Let us consider a scalar function $S(x,y,z)$. Consider the operation

$$\nabla S(x, y, z) = \mathbf{a}_x \frac{\partial}{\partial x}S(x, y, z) + \mathbf{a}_y \frac{\partial}{\partial y}S(x, y, z) + \mathbf{a}_z \frac{\partial}{\partial z}S(x, y, z) \tag{1.21}$$

Observe that when the Nabla operator operates on a scalar function it gives rise to a vector. The resulting vector is called the Gradient of the scalar function $S(x, y, z)$ and is denoted by Grad S. We can also define the same scalar function either in cylindrical or spherical coordinate systems. In cylindrical coordinates Grad S is given by

$$\nabla S(\rho,\varphi,z) = \mathbf{a}_\rho \frac{\partial}{\partial \rho} S(\rho,\varphi,z) + \mathbf{a}_\varphi \frac{1}{\rho}\frac{\partial}{\partial \varphi} S(\rho,\varphi,z) + \mathbf{a}_z \frac{\partial}{\partial z} S(\rho,\varphi,z) \tag{1.22}$$

and in the spherical coordinates, it is given by

$$\nabla S(r,\theta,\varphi) = \mathbf{a}_r \frac{\partial}{\partial r} S(r,\theta,\varphi) + \mathbf{a}_\theta \frac{1}{r}\frac{\partial}{\partial \theta} S(r,\theta,\varphi) + \mathbf{a}_\varphi \frac{1}{r\sin\theta}\frac{\partial}{\partial \varphi} S(r,\theta,\varphi) \tag{1.23}$$

1.4.4.2 The divergence of a vector field

The scalar product of the Nabla vector with a vector field **F** is defined as the divergence of that vector field. In Cartesian coordinates, it becomes

$$\nabla \cdot \mathbf{F} = \frac{\partial}{\partial x} f_x(x,y,z) + \frac{\partial}{\partial y} f_y(x,y,z) + \frac{\partial}{\partial z} f_z(x,y,z) \tag{1.24}$$

Note that this operation results in a scalar function and it is denoted by Div **F**.

Now if we take the divergence of the vector field given by (4.19) (i.e. Grad S), we obtain

$$\nabla \cdot \nabla S(x,y,z) = \frac{\partial^2}{\partial x^2} S(x,y,z) + \frac{\partial^2}{\partial y^2} S(x,y,z) + \frac{\partial^2}{\partial z^2} S(x,y,z) \tag{1.25}$$

This is denoted by $\nabla^2 S(x,y,z)$. This is called the Laplacian of the scalar function S. One can also perform the operation $(\nabla \cdot \nabla)\mathbf{F}$. This results in

$$\nabla^2 \mathbf{F}(x,y,z) = \frac{\partial^2}{\partial x^2} \mathbf{F}(x,y,z) + \frac{\partial^2}{\partial y^2} \mathbf{F}(x,y,z) + \frac{\partial^2}{\partial z^2} \mathbf{F}(x,y,z) \tag{1.26}$$

This is called the Laplacian of the vector field **F**. The divergence and the Laplacian in cylindrical coordinates are given by

$$\nabla \cdot \mathbf{F} = \frac{1}{\rho}\frac{\partial}{\partial \rho} \rho f_\rho(\rho,\varphi,z) + \frac{1}{\rho}\frac{\partial}{\partial \varphi} f_\varphi(\rho,\varphi,z) + \frac{\partial}{\partial z} f_z(\rho,\varphi,z) \tag{1.27}$$

$$\nabla \cdot \nabla S(\rho,\varphi,z) = \frac{1}{\rho}\frac{\partial}{\partial \rho}\left\{\rho \frac{\partial S(\rho,\varphi,z)}{\partial \rho}\right\} + \frac{1}{\rho^2}\frac{\partial^2}{\partial \varphi^2} S(\rho,\varphi,z) + \frac{\partial^2}{\partial z^2} S(\rho,\varphi,z) \tag{1.28}$$

and in spherical coordinates they are given by

$$\nabla \cdot \mathbf{F} = \frac{1}{r^2}\frac{\partial}{\partial r}r^2 f_r(r,\theta,\varphi) + \frac{1}{r\sin\theta}\frac{\partial}{\partial\theta}\sin\theta f_\theta(r,\theta,\varphi) + \frac{1}{r\sin\theta}\frac{\partial}{\partial\varphi}f_\varphi(r,\theta,\varphi) \tag{1.29}$$

$$\nabla\cdot\nabla S(r,\theta,\varphi) = \frac{1}{r^2}\frac{\partial}{\partial r}\left\{r^2\frac{\partial S(r,\theta,\varphi)}{\partial r}\right\} + \frac{1}{r^2\sin\theta}\frac{\partial}{\partial\theta}\left\{\sin\theta\frac{\partial S(r,\theta,\varphi)}{\partial\theta}\right\} + \frac{1}{r^2\sin^2\theta}\frac{\partial^2 S(r,\theta,\varphi)}{\partial\varphi^2} \tag{1.30}$$

In the above equations f_ρ, f_φ, f_z and f_r, f_θ, f_φ are, respectively, the components of F in cylindrical and spherical coordinate systems.

1.4.4.3 The Curl of a vector field

The vector product of the Nabla operator with another vector or a vector field, say **F**, is called the Curl of that vector field. It is denoted by Curl **F**. Using the technique described previously to obtain the vector product, Curl **F** in Cartesian coordinates is given by

$$\text{Curl } \mathbf{F} = \nabla\times\mathbf{F} = \begin{vmatrix} \mathbf{a}_x & \mathbf{a}_y & \mathbf{a}_z \\ \dfrac{\partial}{\partial x} & \dfrac{\partial}{\partial y} & \dfrac{\partial}{\partial z} \\ f_x & f_y & f_z \end{vmatrix} \tag{1.31}$$

After expanding the above, we obtain

$$\text{Curl } \mathbf{F} = \mathbf{a}_x\left[\frac{\partial f_z}{\partial y} - \frac{\partial f_y}{\partial z}\right] + \mathbf{a}_y\left[\frac{\partial f_x}{\partial z} - \frac{\partial f_z}{\partial x}\right] + \mathbf{a}_z\left[\frac{\partial f_y}{\partial x} - \frac{\partial f_x}{\partial y}\right] \tag{1.32}$$

In cylindrical coordinates, it is given by

$$\text{Curl } \mathbf{F} = \nabla\times\mathbf{F} = \begin{vmatrix} \dfrac{\mathbf{a}_\rho}{\rho} & \mathbf{a}_\varphi & \dfrac{\mathbf{a}_z}{\rho} \\ \dfrac{\partial}{\partial \rho} & \dfrac{\partial}{\partial \varphi} & \dfrac{\partial}{\partial z} \\ f_\rho & \rho f_\varphi & f_z \end{vmatrix} \tag{1.33}$$

after expanding the matrix, we obtain

$$\text{Curl } \mathbf{F} = \mathbf{a}_\rho\left[\frac{1}{\rho}\frac{\partial f_z}{\partial \varphi} - \frac{\partial f_\varphi}{\partial z}\right] + \mathbf{a}_\varphi\left[\frac{\partial f_\rho}{\partial z} - \frac{\partial f_z}{\partial \rho}\right] + \mathbf{a}_z\frac{1}{\rho}\left[\frac{\partial \rho f_\varphi}{\partial \rho} - \frac{\partial f_\rho}{\partial \varphi}\right] \tag{1.34}$$

In spherical coordinates, it is given by

$$\text{Curl } \mathbf{F} = \nabla \times \mathbf{F} = \frac{1}{r^2 \sin\theta} \begin{vmatrix} \mathbf{a_r} & r\mathbf{a_\theta} & r\sin\theta \mathbf{a_\varphi} \\ \frac{\partial}{\partial r} & \frac{\partial}{\partial \theta} & \frac{\partial}{\partial \varphi} \\ f_r & rf_\theta & r\sin\theta f_\varphi \end{vmatrix} \tag{1.35}$$

Expanding the matrix, we obtain

$$\text{Curl } \mathbf{F} = \mathbf{a}_r \frac{1}{r^2 \sin\theta}\left[\frac{\partial(f_\varphi r \sin\theta)}{\partial\theta} - \frac{\partial(rf_\theta)}{\partial\varphi}\right] + \mathbf{a}_\theta \frac{1}{r}\left[\frac{1}{\sin\theta}\frac{\partial f_r}{\partial\varphi} - \frac{\partial(rf_\varphi)}{\partial r}\right] + \mathbf{a}_\varphi \frac{1}{r}\left[\frac{\partial(rf_\theta)}{\partial r} - \frac{\partial f_r}{\partial\theta}\right] \tag{1.36}$$

1.4.5 Important vector identities

In handling the equations of electromagnetic theory the following three vector identities will be used in several places in this chapter. They are

$$\nabla \times (\nabla \times \mathbf{F}) = \text{Grad}(\text{Div } \mathbf{F}) - \nabla^2 \mathbf{F} \tag{1.37}$$

$$\text{Div } (\mathbf{F} \times \mathbf{G}) = -\mathbf{F} \cdot (\text{Curl } \mathbf{G}) + \mathbf{G} \cdot (\text{Curl } \mathbf{F}) \tag{1.38}$$

$$\text{Grad } (S) \times \mathbf{F} = \text{Curl } (S\mathbf{F}) - S \text{ Curl } \mathbf{F} \tag{1.39}$$

Note that in the above equation S is a scalar function.

1.4.6 Relationship between the Curl of a vector field and the line integral of that vector field around a closed path

Consider an infinitesimally small area ds in a vector field. This area can be defined as a vector **ds** with magnitude ds. The direction of the vector is defined as the direction of the normal to the area. Now, consider the line integral of vector field **F** performed around the outer boundary of the area ds. This line integral is denoted by

$$\text{Line integral} = \oint_l \mathbf{F} \cdot \mathbf{dl} \tag{1.40}$$

The circle on the integral denotes that the path is closed and the symbol l stands for line integral. It can be shown that as ds goes to zero the ratio $\frac{\oint \mathbf{F} \cdot \mathbf{dl}}{|ds|}$ approaches the component of **Curl F** in the direction of **ds**.

$$\oint \mathbf{F} \cdot \mathbf{dl} \rightarrow \text{Curl } \mathbf{F} \cdot \mathbf{ds} \text{ as } |ds| \rightarrow 0 \tag{1.41}$$

The positive direction of **ds** is given by the right handed screw rule as follows. Rotate a right handed screw in the circular direction in which the line integral is performed. The direction of motion of the screw defines the positive direction of **ds**. The above relationship shows that Curl $\mathbf{F} \cdot \mathbf{ds}$ gives the value of the line integral of **F** performed along the periphery of the infinitesimal area **ds**.

1.4.7 The flux of a vector field through a surface

Let us denote a surface by S. The flux of **F** going through this area is given by the integration

$$\text{Flux} = \int_S \mathbf{F} \cdot \mathbf{ds} \tag{1.42}$$

The subscript S on the integration sign indicates that it is a surface integral. What was done above is to divide the surface into infinitesimal areas **ds** and summing up the contribution of $\mathbf{F} \cdot \mathbf{ds}$ (which is the flux of **F** through the infinitesimal area) from all the infinitesimal areas on the surface S.

1.4.8 Relationship between the divergence of a vector field and the flux of that vector field through a closed surface

Consider a small volume dv located at P. This volume is bounded by a closed surface S. The flux of **F** coming out of the closed surface is

$$\text{Flux of } \mathbf{F} \text{ coming out from a closed surface S} = \oint_S \mathbf{F} \cdot \mathbf{ds} \tag{1.43}$$

where **ds** is an infinitesimal area on the closed surface S. The circle on the integral sign indicates that the surface is closed. The positive direction of **ds** is defined as the outward normal to the closed surface at the point of interest. One can show that as the volume dv approaches zero the quantity $\frac{\oint_s \mathbf{F} \cdot \mathbf{ds}}{dv}$ approaches the divergence of the vector field at point P. That is

$$\frac{\oint_s \mathbf{F} \cdot \mathbf{ds}}{\Delta v} \to \text{Div } \mathbf{F} \text{ } as \text{ } \Delta v \to 0 \tag{1.44}$$

In other words, if one considers an infinitesimal volume dv in a vector field, the quantity div **F** dv gives the flux of **F** out of the infinitesimal volume.

1.4.9 Divergence theorem

The divergence theorem connects a volume integral of a vector field to a surface integral over the surface that bounds the volume. According to the divergence theorem

$$\int_v \text{div } \mathbf{F} \, dv = \oint_s \mathbf{F} \cdot \mathbf{ds} \tag{1.45}$$

The symbol v on the integration sign indicates that it is a volume integral. Let us divide the volume into a large number of infinitesimal volumes. Now, calculate the divergence of **F** at the location of each of these infinitesimal volumes and multiply that with the value of the infinitesimal volume dv. According to the above theorem the sum of these products is equal to the flux of the vector field coming out of the surface bounding the volume. This theorem can be interpreted physically as follows. Observe that according to (1.44), the quantity Div $\mathbf{F}\,dv$ is the flux of **F** coming out of the infinitesimal volume dv. Thus the left-hand side of the (1.45) sums up the total flux of **F** generated inside the volume. This quantity must be equal to the total flux of **F** leaving the volume through the closed surface bounding this volume. This is exactly what the right-hand side of the equation says.

Using the divergence theorem one can convert a volume integral to a surface integral. This is a useful theorem in many branches of physics including electromagnetic theory.

1.4.10 Stokes theorem

Stokes theorem provides a relationship between a line integral of a vector field around a closed path to a surface integral of the Curl of that vector field over a surface bounded by the closed path. Note that there is infinite number of surfaces bounded by that path and the relationship given below is valid for any of those surfaces. According to Stokes theorem

$$\int_s \text{Curl}\,\mathbf{F}\cdot\mathbf{ds} = \oint_l \mathbf{F}\cdot\mathbf{dl} \tag{1.46}$$

The positive direction of **ds** in calculating the flux of Curl **F** in the above equation is given by the right handed screw rule with the circular direction of rotation of the screw is given by the direction in which the line integral is performed.

Note that according to (1.41), Curl $\mathbf{F}\cdot\mathbf{ds}$ is nothing but the integral of F performed around the outer boundary of the infinitesimal area ds. As you sum up the contribution from each of the infinitesimal surface areas, the contributions from adjacent areas cancel out (note that the integral is performed in the same circular direction in each of the areas) leaving behind the line integral of **F** around the boundary of the surface under consideration. This is exactly what the right-hand side of the integral says.

1.5 Static electric fields

1.5.1 Coulomb's law

This law specifies the force between two charges. Consider two charges q_1 and q_s located in vacuum as shown in Figure 1.2. The location of the charges q_1 and q_s (at point P) are defined by the vectors $\mathbf{R_1}$ and $\mathbf{R}_s$. The vector $\mathbf{R}_{1s}$ defines the vector joining the two charges as shown in Figure 1.2. According to Coulomb's law, the

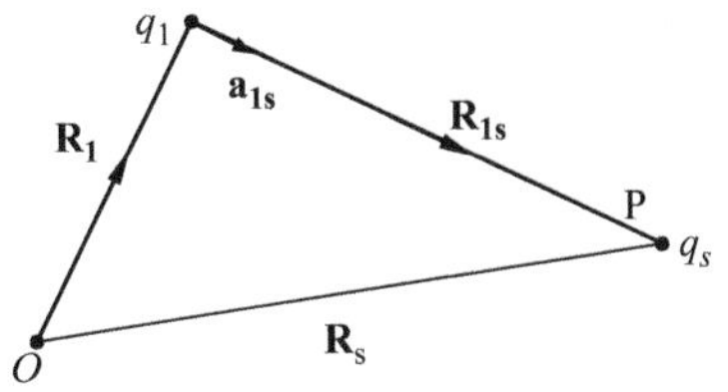

Figure 1.2 Definition of vectors used in defining Coulomb's law (1.47)

force on the test charge q_s caused by the charge q_1 is given by

$$\mathbf{F} = \frac{q_1 q_s}{4\pi\varepsilon_0 |R_{1s}|^2}\mathbf{a}_{1s} \tag{1.47}$$

$$\mathbf{a}_{1s} = \frac{\mathbf{R}_{1s}}{|\mathbf{R}_{1s}|} = \frac{\mathbf{R}_s - \mathbf{R}_1}{|\mathbf{R}_s - \mathbf{R}_1|} \tag{1.48}$$

Note that the force is directed along the line joining the two charges. The E-field, **E**, at point P (shown in Figure 1.2) due to charge q_1 is defined as the force per unit charge at point P as the magnitude of the test charge goes to zero. The last statement is necessary to guarantee that the test charge q_s will not disturb the electric field. Thus, the E-field at point P (in vacuum) produced by the charge q_1 is given by

$$\mathbf{E} = \frac{q_1 q_s}{4\pi\varepsilon_0 q_s |R_{1s}|^2}\mathbf{a}_{1s} \ as \ q_s \to 0 \tag{1.49}$$

In the case of several charges, the electric field produced by all the charges is the vector sum of the electric fields produced by individual charges. This is called the principle of superposition.

Consider a charge distribution in space defined by the volume charge density ρ which may vary from one point to another in space. Consider a small volume located at the point defined by the vector $\mathbf{r}_o$ (see Figure 1.3). The charge density at this point is $\rho(r_o)$ The E-field produced by this element at point of observation P is given by

$$dE(r) = \frac{\rho_v(r_o)\cdot dv}{4\pi\varepsilon_0|\mathbf{r} - \mathbf{r}_o|^2}\cdot\frac{\mathbf{r} - \mathbf{r}_o}{|\mathbf{r} - \mathbf{r}_o|} \tag{1.50}$$

Then the total E-field at point P is given by

$$\mathbf{E}(r) = \int \frac{\rho_v(r_o)\cdot dv}{4\pi\varepsilon_0|\mathbf{r} - \mathbf{r}_o|^2}\cdot\frac{\mathbf{r} - \mathbf{r}_o}{|\mathbf{r} - \mathbf{r}_o|} \tag{1.51}$$

In writing down the above equation, we have assumed that the charges and the point of observation were located in vacuum. If they are located in a homogeneous and isotropic dielectric medium with relative dielectric constant ε_r, the ε_o in the above equations has to be replaced by $\varepsilon_r\varepsilon_o$.

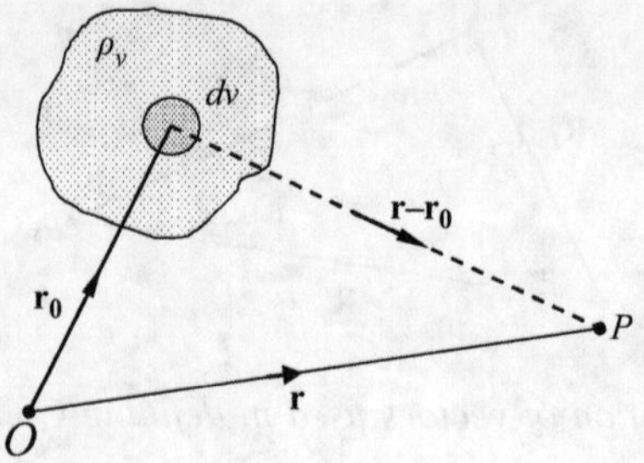

Figure 1.3 Vector notation used in defining the electric field at point P due to a volume charge distribution [(1.50) and (1.51)]

1.5.2 Electric field produced by static charges is a conservative field

The nature of the electric field produced by static charges is such that the work done on a point charge while moving it around a closed path located in such an electric field is equal to zero. Since the work done W in moving a charge q around a closed path is just the charge times the line integral of the electric field around the closed path, one can write

$$W = -q \oint_l \mathbf{E} \cdot \mathbf{dl} \tag{1.52}$$

Since $W = 0$ one can conclude that

$$\oint_l \mathbf{E} \cdot \mathbf{dl} = 0 \tag{1.53}$$

A field that satisfies the above condition is called a conservative field. One can also infer from the above condition that the work done in moving a charge from one point to another point in an electric field is independent of that path. Note that (1.53) provides us with information concerning static electric fields in an integral form. It gives you information concerning the spatial variation of the static electric field. But, one can combine this equation together with Stokes theorem to obtain information concerning the electric field at a given point. Now, according to the Stokes theorem, we can write

$$\oint_l \mathbf{E} \cdot \mathbf{dl} = \int_s (\mathrm{Curl}\mathbf{E}) \cdot \mathbf{ds} \tag{1.54}$$

Since $\oint_l \mathbf{E} \cdot \mathbf{dl} = 0$, this indicates that

$$\int_s (\mathrm{Curl}\mathbf{E}) \cdot \mathbf{ds} = 0 \tag{1.55}$$

The above is true for any surface. Thus, it implies that

$$\text{Curl } \mathbf{E} = 0 \tag{1.56}$$

for the electric fields produced by static charges. Equation (1.56) provides us with information concerning the static electric field at a given point. However, both (1.53) and (1.55) convey the same information.

1.5.3 Gauss's law

Consider a closed volume in vacuum where there is an electric field. This volume is bounded by a closed surface *S*. The total flux of the E-field coming out of this closed surface is

$$\text{Flux out of the closed surface} = \oint_s \mathbf{E} \cdot \mathbf{ds} \tag{1.57}$$

In the above integration, the positive direction of **ds** is defined as the outward normal to the surface at the point of interest. According to Gauss's law, the total flux of the E-field coming out of a closed surface is equal to the total charge located inside the closed surface divided by ε_o. That is

$$\oint_s \mathbf{E} \cdot \mathbf{ds} = \frac{Q}{\varepsilon_o} \tag{1.58}$$

The same equation can be written as

$$\oint_s \mathbf{D} \cdot \mathbf{ds} = Q \tag{1.59}$$

where **D** is the electric flux density. In the case that the volume under consideration is inside a dielectric medium with relative dielectric constant ε_r, (1.59) becomes

$$\oint_s \mathbf{E} \cdot \mathbf{ds} = \frac{Q}{\varepsilon_r \varepsilon_o} \tag{1.60}$$

However, (1.59) remains the same inside a dielectric medium. Now, the total charge inside a volume can be obtained by integrating the volume charge density ρ_v over the whole volume. That is

$$Q = \int_v \rho_v dv \tag{1.61}$$

With this expression for the total charge in the volume, (1.58) can be written as

$$\oint_s \mathbf{E} \cdot \mathbf{ds} = \frac{1}{\varepsilon_o} \int_v \rho_v dv \tag{1.62}$$

Now, according to the divergence theorem

$$\oint_s \mathbf{E} \cdot \mathbf{ds} = \int_v \text{Div } \mathbf{E} \, dv \tag{1.63}$$

substituting for the left-hand side from 1.62 we obtain

$$\int_v \text{Div}\mathbf{E} \, dv = \frac{1}{\varepsilon_o} \int_v \rho_v dv \tag{1.64}$$

Since this is true for any volume, we can conclude that in vacuum (or in air)

$$\text{Div } \mathbf{E} = \frac{\rho_v}{\varepsilon_o} \tag{1.65}$$

If the region is inside a dielectric medium, then ε_o has to be replaced by $\varepsilon_o \varepsilon_r$.

1.5.4 Electric scalar potential

The work done by an external agent against the electric forces to move a point positive charge from point P1 to P2 in an electric field is given by

$$W = -\int_{P1}^{P2} q\mathbf{E} \cdot \mathbf{dl} \tag{1.66}$$

We have seen earlier that for the E-field produced by static charges

$$\text{Curl } \mathbf{E} = 0 \tag{1.67}$$

This indicates that we can express the electric field as the gradient of a scalar function. Let us denote this scalar function by ϕ. Let us express the E-field as

$$\mathbf{E} = -\text{Grad } \phi \tag{1.68}$$

This is in agreement with (1.67) because Curl (Grad ϕ) = 0. The function ϕ, which is a scalar function of location, is called the scalar potential. Substituting this expression for the E-field in (1.66), we obtain the work done by an external agent to move the charge q from P1 to P2 as

$$W = \int_{P1}^{P2} q\text{Grad } \phi(r) \cdot \mathbf{dl} \tag{1.69}$$

Note that by denoting the scalar potential as $\phi(r)$ we are indicating that scalar potential depends on the location under consideration. The above equation can be written as

$$W = q\int_{P1}^{P2} d\phi(r) = q\{\phi(r_2) - \phi(r_1)\} \tag{1.70}$$

In the above equation, r_1 and r_2 represents the location of the points P1 and P2, respectively. The quantity W/q is called the potential difference between the two

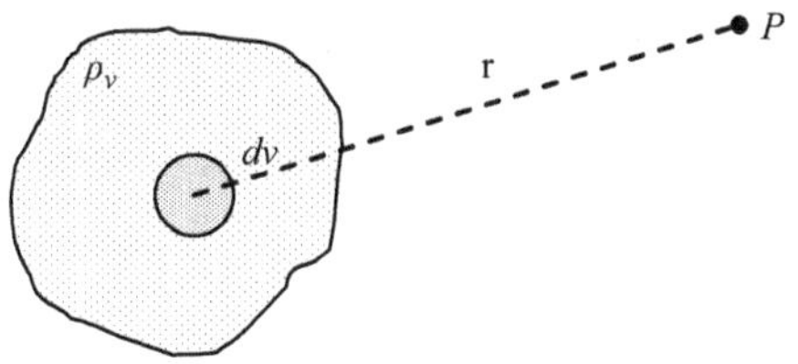

Figure 1.4 Geometry pertinent to the definition of potential at point P due to a volume charge distribution (1.74)

points P2 and P1. If the point P1 is at infinity where we define the scalar potential to be zero, then

$$\frac{W}{q} = -\int_{\infty}^{P2} \mathbf{E} \cdot \mathbf{dl} = \int_{\infty}^{P2} d\phi(r) = \phi(r_2) \tag{1.71}$$

In the above equation, W/q is the work done by an external agent against the electrical forces to move a unit point charge from infinity to the position P2. The function $\phi(r_2)$ is called the potential at point P2.

Consider a point charge Q located at the origin O. The E-field produced by this point charge at any point P is given by

$$\mathbf{E} = \frac{Q}{4\pi\varepsilon_0}\frac{1}{r^2}\mathbf{a_r} \tag{1.72}$$

where $\mathbf{a_r}$ is a unit vector in the direction of OP. The potential at point P due to the point charge Q is given by

$$\phi = \frac{Q}{4\pi\varepsilon_o r} \tag{1.73}$$

where the two quantities (i.e. $\mathbf{E}$ and ϕ) are connected by $\mathbf{E} = -\text{Grad}\phi$. Note also that the unit of potential is Joules per Coulomb. Using (1.61), the potential due to a charge distribution (see Figure 1.4) can be written as

$$\phi = \int_v \frac{\rho_v dv}{4\pi\varepsilon r} \tag{1.74}$$

In the above equation, ρ_v is the volume charge density.

1.5.5 Poisson and Laplace equations

Let us write $\mathbf{E} = -\text{Grad}\phi$ separated into three components in Cartesian coordinates. That is

$$\mathbf{E} = -\frac{\partial\phi}{\partial x}\mathbf{a_x} - \frac{\partial\phi}{\partial y}\mathbf{a}_y - \frac{\partial\phi}{\partial z}\mathbf{a}_z \tag{1.75}$$

Taking the divergence of the above, we obtain

$$\mathrm{Div}\mathbf{E} = -\frac{\partial^2 \phi}{\partial x^2} - \frac{\partial^2 \phi}{\partial y^2} - \frac{\partial^2 \phi}{\partial z^2} \tag{1.76}$$

Since Div $\mathbf{E} = \frac{\rho_v}{\varepsilon_o}$ in vacuum
we find

$$\frac{\partial^2 \phi}{\partial x^2} + \frac{\partial^2 \phi}{\partial y^2} + \frac{\partial^2 \phi}{\partial z^2} = \frac{-\rho_v}{\varepsilon_0} \tag{1.77}$$

This equation is called Poisson's equation. The solution of electro static problems involves solving this equation with the given boundary conditions. Another important feature of this equation is that its solution is unique. That means, for a given set of boundary conditions only one solution exists to Poisson's equation. Thus, if one can find a solution of (1.77) that satisfies the specified boundary conditions by any means (even including guessing) then that solution is the correct one. In charge free region, this equation reduces to

$$\frac{\partial^2 \phi}{\partial x^2} + \frac{\partial^2 \phi}{\partial y^2} + \frac{\partial^2 \phi}{\partial z^2} = 0 \tag{1.78}$$

Equation (1.78) is called the Laplace equation.

1.5.6 *Concept of images*

A problem that occurs frequently in lightning research is to calculate the electric field produced by a charge located over a conducting ground plane. The concept of images can be used to calculate the electric field of charges located near conducting boundaries. In the theory a conductor having a certain potential is replaced by one or several image charges in such a way that the conductor surface is replaced by an equipotential surface at the same potential. This concept is illustrated in Figure 1.5.

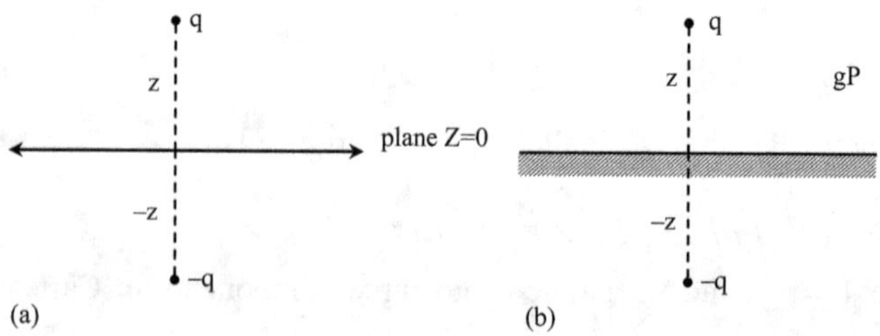

Figure 1.5 In image theory a conductor having a certain potential (in this case zero potential) is replaced by one or several image charges in such a way that the conductor surface is replaced by an equipotential surface at the same potential

In this transformation, the boundary conditions are conserved and the electric field obtained outside the conductor is the correct electric field. In order to illustrate the technique, consider the electric field produced by two point charges one positive and the other negative. The charges are located on the z-axis at the locations z and $-z$. In this charge distribution, the plane $z = 0$ is an equipotential plane with potential equal to zero. That means, this plane can be replaced by a conductor with zero potential without changing the field configuration in the upper half space (i.e. $z > 0$). Thus, the electric field due to a point charge located over a conducting plane at zero potential (in dealing with electrostatic problems the ground is usually considered as a conducting plane at zero potential) can be obtained by replacing the ground plane by a negative (opposite) charge at the image position. In reality, the negative mirror image represents the effects of induced negative charges on the conducting plane due to the influence of the positive charge located above it. The distribution of the induced negative surface charge density on the surface can be obtained using the electrostatic boundary conditions at the conducting surface (see the next section).

1.5.7 Electrostatic boundary conditions

In order to satisfy the laws of electricity in different media, the electric field has to satisfy certain boundary conditions at a boundary separating two media. These conditions are known as boundary conditions. They can be derived by using the fact that the electrostatic field in both media should satisfy (1.53) and (1.60). Consider a boundary that separates two mediums, say 1 and 2, with relative dielectric constant ε_{r1} and ε_{r2}, respectively (see Figure 1.6). Let us resolve the electric field in the two media into components parallel (tangential) and perpendicular (normal) to the surface. Let us denote the E-field parallel to the surface in the two media by E_{t1} and E_{t2}. The components perpendicular to the surface are denoted by E_{n1} and E_{n2}. The corresponding D-fields are D_{t1}, D_{t2}, D_{n1}, and D_{n2}.

Consider a closed path as shown in Figure 1.6a. Applying (1.53) for this closed path and reducing the thickness of the path Δl to zero one finds

$$E_{t1}\Delta W - E_{t2}\Delta W = 0 \tag{1.79}$$

Thus the boundary condition satisfied by the tangential E-fields in the two media is

$$E_{t1} = E_{t2} \tag{1.80}$$

That is, the E-field component parallel to the surface is continuous across the boundary. In terms of D-fields (1.80) can be written as

$$\frac{D_{t1}}{D_{t2}} = \frac{\varepsilon_{1r}}{\varepsilon_{2r}} \tag{1.81}$$

Now consider a closed volume in the shape of a pill box as shown in Figure 1.6b. Applying the (1.60) (Gauss law) to this pill box and reducing the

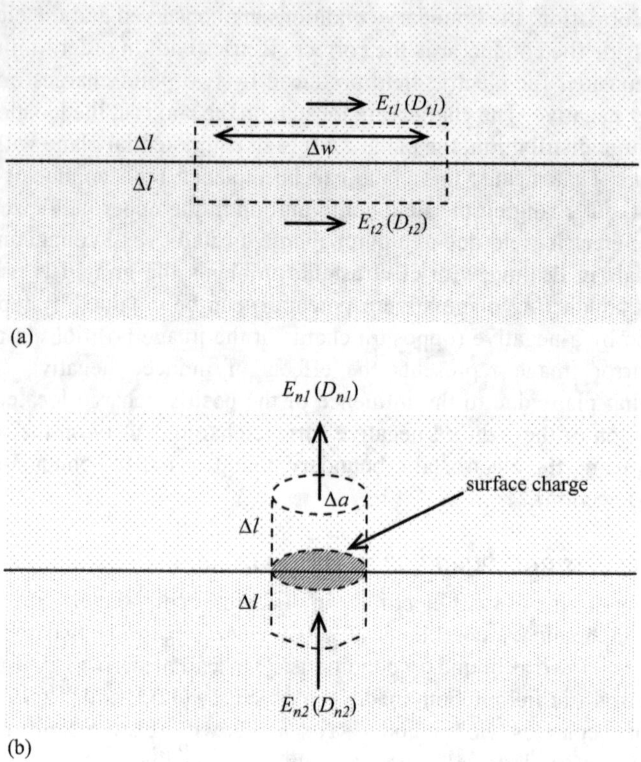

Figure 1.6 (a) The contour along which the line integral of electric field is evaluated to obtain the electrostatic boundary conditions satisfied by the tangential electric field components. (b) The surface bounding the volume into which Gauss's law is applied to obtain the electrostatic boundary conditions satisfied by the normal electric field components.

height of the pill box Δl to zero we obtain

$$D_{n1}\Delta a - D_{n2}\Delta a = \Delta Q = \rho_s \Delta a \tag{1.82}$$

where ρ_s is the surface charge density on the surface and Δa is the area of cross section of the pill box. Thus the boundary condition satisfied by the normal component of D-field is

$$D_{n1} - D_{n2} = \rho_s \tag{1.83}$$

If the surface charge is zero (i.e. $\rho_s = 0$) then the normal component of the D-field is continuous across the boundary. In terms of E-fields, the boundary condition is

$$\varepsilon_o \varepsilon_{r1} E_{n1} - \varepsilon_o \varepsilon_{r2} E_{n2} = \rho_s \tag{1.84}$$

When the charge density is zero, the above equation reduces to

$$\varepsilon_{r1}E_{n1} = \varepsilon_{r2}E_{n2} \tag{1.85}$$

Now assume that one of the mediums under consideration is a vacuum (or air) and the other a conductor. A conductor contains free electrons and the moment one attempts to apply an electric field inside the conductor the free electrons will displace inside the conductor and create an opposite electric field so as to make the electric field inside the conductor zero. This process of relaxation (or removal) of the electric field takes some time and this time, as we will see later in Section 1.6.3, depends on the conductivity and the dielectric constant of the conducting medium. In a perfect conductor the relaxation time is zero and therefore the electric field inside is zero at all times. If the conductivity is finite, then the relaxation time is not zero and it may take some time before the field is neutralized. However, it becomes zero under static conditions. Since the electric field is zero inside a perfect conductor, the boundary condition (1.80) shows that the tangential component of the E-field is zero on its surface. This is the case since the tangential component of the E-field is zero inside the conductor. Thus the E-field at the surface of a perfect conductor is perpendicular to the surface. Since the component of the E-field normal to the surface inside the conductor is zero the normal component of the E-field at the surface is given by

$$\varepsilon_o E_n = \rho_s \tag{1.86}$$

where ρ_s is the surface charge density on the conductor.

1.6 Electric currents, charge conservation, and static magnetic fields

Magnetic fields are generated by electric currents and therefore, before we proceed further let us consider how to represent the electric current in a given medium in terms of the properties of the charged particles that exist in the medium.

1.6.1 Electric current

Electric currents are created by the movement of charges. Consider a medium which contains mobile electric charges. The polarity of the charges can be either positive or negative. They can also be either ions or electrons. For our purpose at hand assume that the polarity of the mobile charge in the medium is positive. Let the density of the charged particles in the medium (i.e. number of charged particles per unit volume) be n_+. We also assume that the particles are singly charged. That is, the charge on them is equal to one electronic charge. If an electric field is applied to this medium, the charged particles will experience a force and they start accelerating in the direction of the electric field. However, due to the collisions with other particles in the medium, they will reach a steady velocity that depends on the background electric field and the size and the mass of the charged particles.

This steady velocity is called drift velocity and it is a function of the electric field. The relationship between the drift velocity and the E-field is given by

$$\mathbf{V_{d+}} = \mu_+ \mathbf{E} \tag{1.87}$$

The parameter μ_+ is called the mobility of the positive charged particles under consideration. The movement of the charge gives rise to a flow of electric current in the medium. The current density **J** in the direction of the electric field (i.e. the current flowing through a unit cross section located perpendicular to the electric field) is given by

$$\mathbf{J} = e\mathbf{V_d}n_+ \tag{1.88}$$

where e is the electronic charge. If the medium contained charged particles of both polarities all of which are singly charged then

$$\mathbf{J} = e\mathbf{V_{d+}}n_+ + e\mathbf{V_{d-}}n_- \tag{1.89}$$

where the subscript with negative sign indicates quantities related to the negative charge particles. The above equation can be written as

$$\mathbf{J} = e\mu_+\mathbf{E}n_+ + e\mu_-\mathbf{E}n_- \tag{1.90}$$

This can be written as

$$\mathbf{J} = \sigma\mathbf{E} \tag{1.91}$$

where

$$\sigma = e\mu_+n_+ + e\mu_-n_- \tag{1.92}$$

The parameter σ is called the conductivity of the medium. In general, a medium may contain charged particles of both polarities and of different masses and charges. In evaluating the conductivity, one has to consider the contribution of each of these particles to the current flow and to the conductivity.

1.6.2 Conservation of electric charge

It is a law of nature that the electric charge cannot be created or destroyed. It can only move from one place to another in the form of electric currents. The conservation of charge provides restrictions on the electric current that can be present in a given region. Consider a closed volume in a region where there is a current flow. Let S be a closed surface enclosing this volume. Let the total current coming out of the volume be denoted by I. Let **J** be the current density at any given point on the surface. Then the total current coming out of the surface is

$$I = \oint_s \mathbf{J} \cdot \mathbf{ds} \tag{1.93}$$

Since the charge cannot be destroyed, the outward flow of current from the volume should reduce the charge inside the volume at a rate given by

$$I = \oint_s \mathbf{J} \cdot \mathbf{ds} = -\frac{dQ}{dt} \tag{1.94}$$

where Q is the charge in the volume at a given time. Now, the charge inside the volume at a given instant can be written as an integral of the volume charge density. Thus, we can write

$$\frac{dQ}{dt} = \frac{d}{dt}\left(\int_v \rho_v dv\right) \tag{1.95}$$

Substituting this in (1.94), we obtain

$$\oint_s \mathbf{J} \cdot \mathbf{ds} = -\frac{d}{dt}\left(\int_v \rho_v dv\right) \tag{1.96}$$

Now according to the divergence theorem, we can write

$$\oint_s \mathbf{J} \cdot \mathbf{ds} = \int_v \mathrm{div}\mathbf{J} dv \tag{1.97}$$

Substituting this in (1.96) we obtain

$$\int_v \mathrm{div}\mathbf{J} dv = -\left(\int_v \frac{d\rho_v}{dt} dv\right) \tag{1.98}$$

Since the above relationship is true for any volume, we conclude that

$$\mathrm{div}\mathbf{J} = -\frac{\partial \rho_v}{\partial t} \tag{1.99}$$

This equation is an expression, in point form, of the conservation of electric charge.

1.6.3 Re-distribution of excess charge placed inside a conducting body

Consider an isotropic and homogeneous conductor with relative dielectric constant of ε_r and conductivity σ. Assume that at time equal to zero an excess charge is placed inside the conductor with charge density $\rho_v(r, 0)$. This charge will generate an electric field inside the conductor generating a current that will redistribute this charge and displace it to the surface of the conductor (recall that one can have electric charges on the surface of conductors) and making the electric field inside the conductor zero. Let us evaluate how fast this process will take place.

From the equation of charge conservation, we have

$$\text{div}\mathbf{J}(r,t) = -\frac{\partial \rho_v(r,t)}{\partial t} \tag{1.100}$$

where $\rho_v(r,t)$ is the charge density at any time inside the conductor. Substituting for $\mathbf{J}$ in the above equation from $\mathbf{J} = \sigma\mathbf{E}(r,t)$ we obtain

$$\sigma \text{div}\mathbf{E}(r,t) = -\frac{\partial \rho_v(r,t)}{\partial t} \tag{1.101}$$

We also know from Gauss law that

$$\text{div}\mathbf{E}(r,t) = \frac{\rho_v(r,t)}{\varepsilon_o \varepsilon_r} \tag{1.102}$$

Substituting this in the previous equation, we obtain

$$\rho_v(r,t) = -\frac{\varepsilon_o \varepsilon_r}{\sigma}\frac{\partial \rho_v(r,t)}{\partial t} \tag{1.103}$$

The solution of (1.103) is

$$\rho_v(r,t) = \rho_v(r,0)e^{-\frac{\sigma t}{\varepsilon_o \varepsilon_r}} \tag{1.104}$$

The above expression for the variation of charge density inside the conductor shows that the charge inside the conductor will decreases exponentially in time. The quantity $\frac{\varepsilon_o \varepsilon_r}{\sigma}$ is called the relaxation time of the conductor. In the same way, if we create an electric field inside a conductor it will decrease to zero exponentially with a time constant equal to the relaxation time. In a perfect conductor, the conductivity is infinite and thus the relaxation time is zero. Thus the electric field inside a perfect conductor is zero under all circumstances.

1.6.4 Magnetic field produced by a current element – Biot–Savarts law

Consider a current element *dl* through which a current I is flowing (see Figure 1.7). The current element is located at point O. The current element can be represented by a vector **dl** with magnitude *dl* and direction specified by the direction of current flow. According to Biot–Savarts law, the B-field produced by this current element at point P is given by

$$dB = \mu_o I \cdot \frac{\mathbf{dl} \times \mathbf{a_r}}{4\pi |R|^2} \tag{1.105}$$

In the above equation $\mathbf{a_r}$ is a unit vector in the direction of OP; see Figure 1.7. Note that the direction of the magnetic field is perpendicular to both the current element **dl** and the vector joining the current element and the point of observation (i.e. $\mathbf{a_r}$). The magnitude of the magnetic field is proportional to the current in the current element and its strength decreases with $1/R^2$.

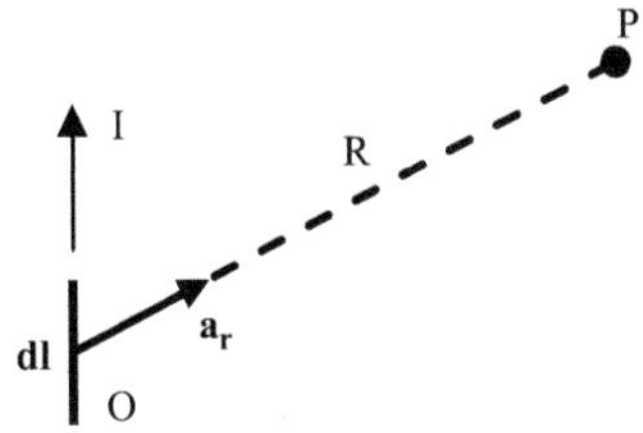

Figure 1.7 The geometry relevant to the evaluation of magnetic field at point P produced by a current element dl

The magnetic field produced by a conductor of any shape can be calculated by dividing the conductor into elementary sections and summing up the contribution to the B-field from each element using the above equation.

1.6.5 Gauss's law for magnetic fields

One interesting fact of nature is the observation that magnetic poles, the sources of magnetic fields, always come in pairs. This means that the net flux of the magnetic field coming out from a closed surface has to be zero because there are no net magnetic charge (or poles) inside any given volume. Thus, for any closed volume, one can write

$$\oint_s \mathbf{B} \cdot \mathbf{ds} = 0 \tag{1.106}$$

The above says that the total flux of B-field coming out of a closed surface is equal to zero. Using divergence theorem, one can write

$$\oint_s \mathbf{B} \cdot \mathbf{ds} = \int_v \mathrm{Div}\mathbf{B}\, dv \tag{1.107}$$

Substituting this in (1.106), we obtain

$$\int_v \mathrm{Div}\mathbf{B}\, dv = 0 \tag{1.108}$$

Since this is true for any closed volume, one can conclude that

$$\mathrm{Div}\mathbf{B} = 0 \tag{1.109}$$

1.6.6 Amperes law

Amperes law relates the integral of the B-field or the H-field around a closed path to the electric current passing through it. Consider a closed path of any shape. Let S be any surface bounded by this closed path. As we have indicated previously there are infinite number of surfaces satisfying this condition. The net current, *I*, passing through the closed path is given by

$$I = \int_s \mathbf{J} \cdot \mathbf{ds} \tag{1.110}$$

In the above equation, the net current is calculated by first considering a surface S bounded by the closed loop and taking the flux of current density through it. Note that in the above equation **J.ds** gives the current passing through the element **ds**. According to Ampere's law,

$$\oint_l \mathbf{H} \cdot \mathbf{dl} = \int_s \mathbf{J} \cdot \mathbf{ds} \tag{1.111}$$

In the above equation, the left-hand side defines the integral of the H-field around a closed path and the right-hand side defines the electric current passing thorough the same closed path. The parameter **ds** defines a surface element on the surface and its positive direction is defined by the tight hand screw law. That is, the positive direction of **ds** is the direction of motion of a right handed screw when rotated in the circular direction in which the line integral is performed. Now, using Stokes theorem, one can write

$$\oint_l H \cdot \mathbf{dl} = \int_s (\mathrm{Curl}\mathbf{H}) \cdot \mathbf{ds} \tag{1.112}$$

Combining (1.111) and (1.112), we obtain

$$\int_s \mathbf{J} \cdot \mathbf{ds} = \int_s (\mathrm{Curl}\mathbf{H}) \cdot \mathbf{ds} \tag{1.113}$$

Since this is true for any surface, we can conclude that

$$\mathrm{Curl}\mathbf{H} = \mathbf{J} \tag{1.114}$$

In terms of B-field, this can be written as

$$\mathrm{Curl}\mathbf{B} = \mu_0 \mathbf{J} \tag{1.115}$$

1.6.7 Boundary conditions for the static magnetic field

Similar to electric fields, the magnetic field should also satisfy boundary conditions at the boundary between two media. These boundary conditions can be obtained easily by applying Gauss law for the magnetic fields and Amperes law at the boundary. Consider a boundary that separates two mediums 1 and 2 with relative magnetic permeabilities μ_{r1} and μ_{r2} respectively (Figure 1.8a and b). Let us resolve the magnetic field in the two media into components parallel (tangential) and perpendicular (normal) to the surface. Let us denote the H-field parallel to the surface in the two media by H_{t1} and H_{t2}. The components perpendicular to the surface are denoted by H_{n1} and H_{n2}. The corresponding B-fields are B_{t1}, B_{t2}, B_{n1} and B_{n2}.

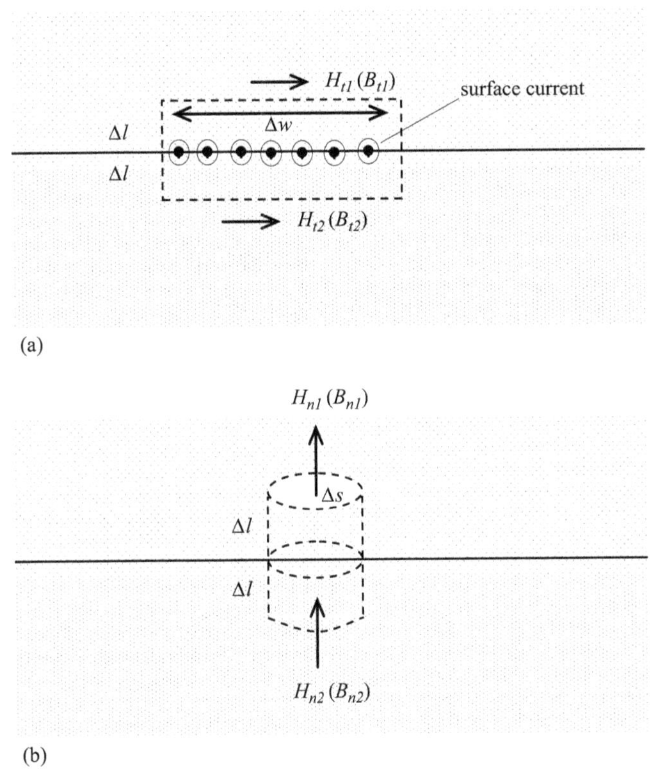

Figure 1.8 (a) The contour along which the line integral of magnetic field is evaluated to obtain the magnetostatic boundary conditions satisfied by the tangential magnetic field components. (b) The surface bounding the volume into which Gauss's law is applied to obtain the magnetostatic boundary conditions satisfied by the normal magnetic field components

Consider a closed path as shown in Figure 1.8a. Applying Amperes law for this closed path and reducing the thickness Δl of the path to zero, one finds

$$H_{t1} - H_{t2} = J_s \tag{1.116}$$

where J_s is the surface current density per unit length flowing perpendicular to the closed path. If the surface current on the surface is zero, the above reduces to

$$H_{t1} = H_{t2} \tag{1.117}$$

This tells us that if there are no surface currents, the H-field parallel to the surface is continuous across the boundary. The boundary conditions to be satisfied by the parallel components of B-fields can be written as

$$\frac{B_{t1}}{\mu_o \mu_{1r}} - \frac{B_{t2}}{\mu_o \mu_{2r}} = J_s \tag{1.118}$$

$$\frac{B_{t1}}{B_{t2}} = \frac{\mu_{1r}}{\mu_{2r}} \text{ (when is equal to zero)} \tag{1.119}$$

Now consider a closed volume in the shape of a pill box as shown in Figure 1.8b. Applying (1.106) (Gauss law for magnetic fields) to this pill box and reducing the height of the pill box to zero, we obtain

$$B_{n1}\Delta s - B_{n2}\Delta s = 0 \tag{1.120}$$

Thus the boundary condition satisfied by the normal component of the B-field is

$$B_{n1} = B_{n2} \tag{1.121}$$

The boundary condition to be satisfied by the normal component of the H-field is then given by

$$\mu_{1r} H_{n1} = \mu_{2r} H_{n2} \tag{1.122}$$

1.6.8 Vector potential

The fact that the magnetic field is divergence free (i.e. Div $\mathbf{B} = 0$) makes it possible to define a vector function $\mathbf{A}$ with the characteristics that

$$\mathbf{B} = \text{Curl } \mathbf{A} \tag{1.123}$$

This satisfies the condition that Div $\mathbf{B} = 0$ because the divergence of a Curl of a function is zero. The function $\mathbf{A}$ defined above is called the vector potential. Let us now find the relationship between the vector potential and the current giving rise to the magnetic field. Let us start with the Ampere's law in point form:

$$\text{Curl}\mathbf{B} = \mu_0 \mathbf{J} \tag{1.124}$$

Substituting $\mathbf{B} = \text{Curl } \mathbf{A}$ in the above equation, one obtains

$$\text{Curl}(\text{Curl}\mathbf{A}) = \mu_0 \mathbf{J} \tag{1.125}$$

Expanding Curl(Curl$\mathbf{A}$) using the vector identity given in (1.37), we obtain

$$\text{Grad}(\text{Div}\mathbf{A}) - \nabla^2 \mathbf{A} = \mu_0 \mathbf{J} \tag{1.126}$$

Now, any vector field is not completely defined until both its Curl and divergence is specified. We have already defined the Curl of the function $\mathbf{A}$ but we are free to select the value of the function div$\mathbf{A}$. Let us select div$\mathbf{A} = 0$. This selection is called Coulomb's Gauge. With this selection, we find

$$\nabla^2 \mathbf{A} = -\mu_0 \mathbf{J} \tag{1.127}$$

The relationship given in (1.127) is a vector equation and it can be separated into its components as follows:

$$\begin{aligned}\nabla^2 A_x &= -\mu_0 J_x \\ \nabla^2 A_y &= -\mu_0 J_y \\ \nabla^2 A_z &= -\mu_0 J_z\end{aligned} \tag{1.128}$$

Note that A_x, A_y, A_z and J_x, J_y, J_z are scalar quantities. The above equation gives the relationship between the vector potential and the current generating the magnetic field. Let us now consider the solution of this equation.

1.6.8.1 Vector potential of a current distribution

Consider a current distribution in space. See Figure 1.9. Let us specify the current density at a given point by $\mathbf{J}(r')$ where r' defines the coordinates of the current source i.e. related to coordinates x', y', and z'. According to Biot–Savarts Law, the magnetic field at point P is given by

$$\mathbf{B} = \frac{\mu_o}{4\pi} \int_v \mathbf{J}(r') \times \left(\frac{\mathbf{a_r}}{r^2}\right) dv \tag{1.129}$$

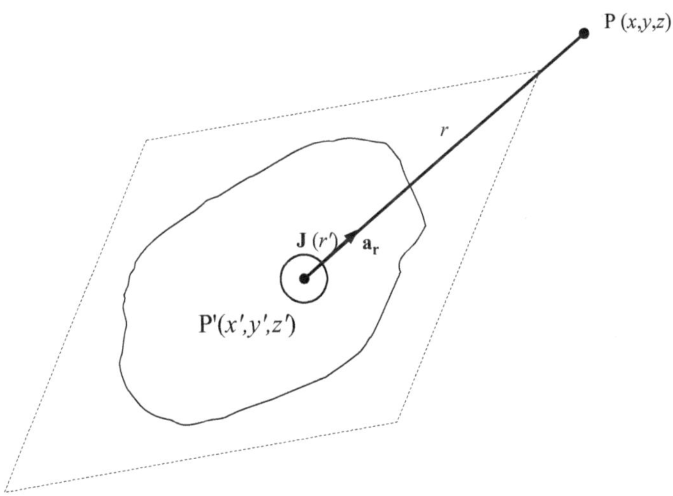

Figure 1.9 The geometry relevant to the derivation of vector potential at point P due to a current distribution

where in the above equation $\mathbf{a_r}$ is a unit vector directed in the direction of r (see Figure 1.9). The above equation can be written as

$$\mathbf{B} = \frac{\mu_o}{4\pi}\int_v \nabla\left(\frac{1}{r}\right) \times \mathbf{J}(r')dv \tag{1.130}$$

Note that it is possible to write (1.129) as (1.130) because ∇ contains derivatives with respect to x, y, and z (coordinates of point of the field point) whereas $\mathbf{J}$ is only a function of source coordinates, x', y', and z'. Now using the vector identity given by (1.39), we obtain

$$\nabla\left(\frac{1}{r}\right) \times \mathbf{J}(r') = \nabla \times \left(\frac{1}{r}\mathbf{J}(r')\right) - \left(\frac{1}{r}\right)\nabla \times \mathbf{J}(r') \tag{1.131}$$

The second term on the right-hand side is equal to zero because ∇ contains derivatives with respect to x, y, and z (coordinates of the point of observation) whereas $\mathbf{J}$ is not a function of x, y, and z. Substituting this in the previous equation, we obtain

$$\mathbf{B} = \frac{\mu_o}{4\pi}\int_v \mathrm{Curl}\left\{\frac{\mathbf{J}(r')}{r}\right\}dv \tag{1.132}$$

Changing the order of differentiation and integration, we find

$$\mathbf{B} = \frac{\mu_o}{4\pi}\mathrm{Curl}\int_v \left\{\frac{\mathbf{J}(r')}{r}\right\}dv \tag{1.133}$$

Comparing this with (1.123), we conclude that

$$\mathbf{A} = \frac{\mu_0}{4\pi}\int_v \frac{\mathbf{J}(r')}{r}dv \tag{1.134}$$

The x, y, and z components of this vector relationship are

$$\begin{aligned} A_x &= \int_v \frac{J_x(r')\mu_0}{4\pi r}dv \\ A_y &= \int_v \frac{J_y(r')\mu_0}{4\pi r}dv \\ A_z &= \int_v \frac{J_z(r')\mu_0}{4\pi r}dv \end{aligned} \tag{1.135}$$

1.6.8.2 Vector potential due to a current element dl

The problem under consideration is shown in Figure 1.10. Consider a current element $\mathbf{dl}$. The magnitude or the length of this element dl i.e. $dl = |\mathbf{dl}|$. The direction of the vector $\mathbf{dl}$ is in the direction of current flow. Assume that the current is flowing uniformly across the cross section of the element. The current density in

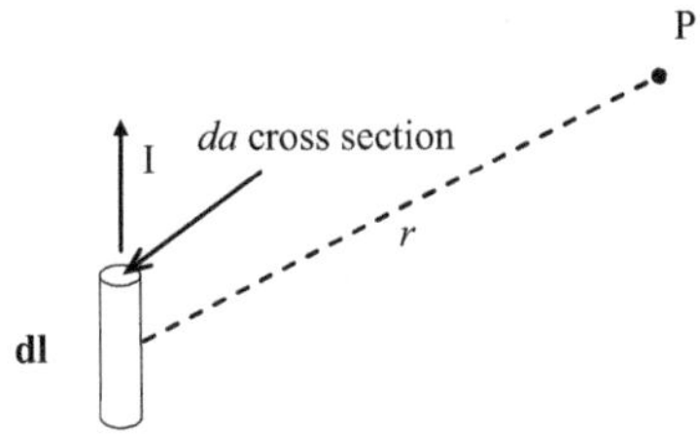

Figure 1.10 The geometry necessary to calculate the vector potential at point P due to a current element of length dl *and of cross section* da

the element is given by **J**. The vector potential due to the current element at a distance r when $r \gg dl$ is

$$\mathbf{A} = \frac{\mu_0}{4\pi}\frac{\mathbf{J}}{r}(dl \cdot da) \tag{1.136}$$

where da is the cross-sectional area of the element. Note that the quantity $dl.da$ is the volume of the current element. Since the current density multiplied by the cross-sectional area gives the total current flowing through the element, we can write

$$\mathbf{A} = \frac{\mu_0}{4\pi}\frac{\mathbf{dl}\,I}{r} \tag{1.137}$$

The above equation gives the vector potential at a given point due to a current element of length dl.

1.6.9 Force on a charged particle

Consider a region where there is both an E- and a B-field. Assume that at any given instant a charged particle of charge q is at point (x,y,z) and it is moving with velocity $\mathbf{v}$. The particle will experience a force both due to E and B fields. This force is given by,

$$\mathbf{F} = q\mathbf{E}(x, y, z) + \mathbf{v} \times \mathbf{B}(x, y, z) \tag{1.138}$$

This force is called the Lorentz force. If the particle is stationary with respect to the B-field it will experience a force only due to the E-field.

1.7 Energy density of an electric field

Let us evaluate the energy density of an electric field in vacuum. This we can do by creating an electric field in a given region in vacuum and evaluating the energy that was spent in creating that electric field. Consider a plane parallel capacitor in vacuum charged to potential V. The energy of the capacitor is stored in its electric field. If one neglects the end effects of the capacitor, the electric field is uniform between the two plates. The total energy spent in charging the capacitor is

$$W = \frac{1}{2} qV \tag{1.139}$$

where q is the charge on the capacitor. Neglecting the end effects, the E-field inside the capacitor is given by $E = \frac{V}{d}$ where d is the separation between the plates. Applying Gauss law and recalling that the E-field inside a conductor (in our case the plates of the capacitor) is zero, we find that

$$E = \frac{q}{A} \frac{1}{\varepsilon_0} \tag{1.140}$$

where A is the area of the plates of the capacitor. Substituting for q and V in terms of E, we find that

$$W = \frac{1}{2} \varepsilon_0 E^2 (Ad) \tag{1.141}$$

Since Ad is the volume of the capacitor, the energy per unit volume is given by

$$\frac{W}{Ad} = \frac{1}{2} \varepsilon_o E^2 \tag{1.142}$$

Since the energy of the capacitor is stored in the electric field, one can conclude that the energy density, EE_{den} of an E-field in vacuum is given by

$$EE_{den} = \frac{1}{2} \varepsilon_o E^2 \tag{1.143}$$

1.8 Electrodynamics – time varying electric and magnetic fields

So far we have dealt with static electric and magnetic fields. Now, let us consider situations in which the electric and magnetic fields vary in time. Such situations occur when the charge densities (the sources of electric field) and current densities (sources of magnetic fields) vary in time. The variation of these quantities with time can take place in any arbitrary manner but for many applications it is convenient to assume that the variations are sinusoidal. Indeed, any time variation can be represented as a sum of *sine* and *cosine* variations using Fourier analysis.

Consider an E-field which is varying as a cosine function. It is given by

$$E = E_{x0} \cos \omega t \tag{1.144}$$

The quantity ω is called the angular frequency (measured in radians per second) and it is related to the frequency f (measured in cycles per second) by the relationship $\omega = 2\pi f$.

Using the identity

$$e^{j\omega t} = \cos\, \omega t + j \sin \omega t \tag{1.145}$$

One can write the E-field as the real part of $E_{x0}e^{j\omega t}$, that is

$$E = \mathrm{Re}\left[E_{x0}e^{j\omega t}\right] \tag{1.146}$$

An E-field varying as a sine function can be written as the imaginary part of $E_{x0}e^{j\omega t}$ i.e. $\mathrm{Im}[E_{x0}e^{j\omega t}]$. The advantage of this technique is that all the operations that need to be done on the E-field (or any other field of interest) can be done on $E_{x0}e^{j\omega t}$ and the final solution can be obtained by taking the real part of the resulting expression if the electric field was changing as a cosine function. For example, consider the operation of taking the time derivative of the electric field. The result is

$$\frac{dE}{dt} = -E_{x0}\omega \sin\, \omega t \tag{1.147}$$

On the other hand, one can perform the operation directly on $E = E_{x0}e^{j\omega t}$ which results in

$$\frac{dE}{dt} = E_{x0}j\omega e^{j\omega t} = E_{x0}[j\omega \cos\, \omega t - \omega \sin \omega t] \tag{1.148}$$

The real part of this is $-E_{x0}\omega \sin \omega t$ which is the solution we need. So, instead of working with sine or cosine functions one can work with $E = E_{x0}e^{j\omega t}$ and extract the final answer either as the real or the imaginary part of the final expression depending on whether the starting point was $E = \mathrm{Re}[e^{j\omega t}]$ or $E = \mathrm{Im}[e^{j\omega t}]$. One advantage of this technique is that it simplify all the equations by removing the time derivatives.

1.8.1 Faraday's law

The essence of Faraday's law is that it defines and quantifies the natural law that a changing magnetic field gives rise to an electric field. Consider a closed path in a region where there is a changing magnetic field. According to Faraday's law the E-field **E** generated by this changing magnetic field is such that

$$\oint_l \mathbf{E} \cdot \mathbf{dl} = -\frac{d\psi}{dt} \tag{1.149}$$

In the above equation, the left-hand side is the line integral of the electric field taken along the closed path and the right-hand side is equal to the negative rate of change of magnetic flux ψ passing through the closed path. The magnetic flux passing through the closed path can be calculated as

$$\psi = \int_s \mathbf{B} \cdot \mathbf{ds} \tag{1.150}$$

where, as indicated previously, the surface integral is carried out over a surface that is bounded by the closed path. The positive direction of **ds** is the direction of motion of a right-handed screw when rotated in the circular direction in which the line integral is performed. Since the electromotive force, *emf*, generated around the closed path under consideration is given by the line integral of the E-field along that path, one can write

$$emf = -\frac{d\psi}{dt} \tag{1.151}$$

If the closed path is replaced by a conducting wire, the *emf* given above will generate a current flow in the wire. The minus sign tells us that the direction of the induced current is such that it opposes the change in the magnetic field that gives rise to the *emf*. This phenomenon is described as Lenz's law.

Now, using Stokes theorem, one can write

$$\int_s (\text{Curl}\mathbf{E}) \cdot \mathbf{ds} = \oint_l \mathbf{E} \cdot \mathbf{dl} \tag{1.152}$$

Combining (1.149), (1.150), and (1.152), one can write

$$\int_s (\text{Curl}\mathbf{E}) \cdot \mathbf{ds} = -\frac{d}{dt}\int_s \mathbf{B} \cdot \mathbf{ds} \tag{1.153}$$

If the path of the line integral is fixed in space in time, one can change the order of differentiation and integration. That gives rise to

$$\int_s (\text{Curl}\mathbf{E}) \cdot \mathbf{ds} = -\int_s \frac{\partial \mathbf{B}}{\partial t} \cdot \mathbf{ds} \tag{1.154}$$

Since the above equation is true for any surface, one can conclude that

$$\text{Curl}\mathbf{E} = -\frac{\partial \mathbf{B}}{\partial t} \tag{1.155}$$

The above equation is the point form description of Faraday's law.

1.8.2 Maxwell's modification of Ampere's law – the displacement current term

After scrutinizing the experimentally derived laws available to him, Maxwell realized that the Amperes law as given in (1.111) or (1.114) cannot be complete. For example, take the Divergence of both sides of (1.114). We will get

$$\text{Div}(\text{Curl}\mathbf{H}) = \text{Div}\mathbf{J} \tag{1.156}$$

Since divergence of a curl of a function is always zero, the above equation implies that Div$\mathbf{J}$ = 0. This cannot be true in general because

$$\text{Div}\mathbf{J} = -\frac{\partial \rho_v}{\partial t} \neq 0 \tag{1.157}$$

This shows that a term is missing in (1.111) and (1.114). Let us assume for the moment that the missing term is **F**. The goal is to find an expression for this function in terms of known quantities. With this addition (1.114) can be written as

$$\mathrm{Curl}\mathbf{H} = \mathbf{J} + \mathbf{F} \tag{1.158}$$

Taking the divergence of both sides, we obtain

$$\mathrm{Div}(\mathrm{Curl}\mathbf{H}) = \mathrm{Div}\mathbf{J} + \mathrm{Div}\mathbf{F} \tag{1.159}$$

From this, we obtain (since $\mathrm{Div}(\mathrm{Curl}\mathbf{H}) = 0$)

$$\mathrm{Div}\mathbf{F} = -\mathrm{Div}\mathbf{J} \tag{1.160}$$

However, from charge conservation, we have already seen that

$$\mathrm{Div}\mathbf{J} = -\frac{\partial \rho_v}{\partial t} \tag{1.161}$$

where ρ_v is the volume charge density. At the same time, from Gauss's law, we know that

$$\rho_v = \mathrm{Div}\mathbf{D} \tag{1.162}$$

Combining (1.160), (1.161), and (1.162), we obtain

$$\mathrm{Div}\mathbf{F} = \mathrm{Div}\left(\frac{\partial \mathbf{D}}{\partial t}\right) \tag{1.163}$$

From this, we conclude that the function **F** is given by

$$\mathbf{F} = \frac{\partial \mathbf{D}}{\partial t} \tag{1.164}$$

Thus the corrected form of Ampere's law is

$$\mathrm{Curl}\mathbf{H} = \mathbf{J} + \frac{\partial \mathbf{D}}{\partial t} \tag{1.165}$$

Maxwell coined the term 'displacement current density' for the new term that appears in the right-hand side of this equation. Note that the first term on the right-hand side is the conduction current density. The most important conclusion that one can reach from this equation is that a changing electric field will give rise to a magnetic field.

1.8.3 Energy density in a magnetic field

Consider a region in the form of a long tube in vacuum (see Figure 1.11). The cross-sectional area of the tube is A. Let us setup a uniform B-field in this region. This can be done by creating circular current around this region as shown in Figure 1.11. The energy stored in the B-field can be obtain by the energy spent in creating it.

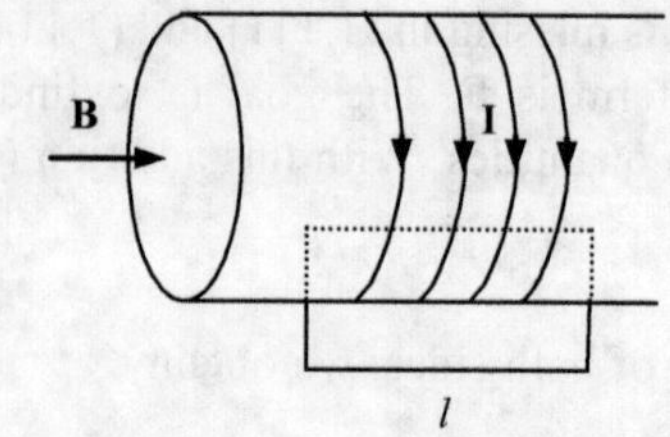

Figure 1.11 The spatial arrangement of currents that can provide a uniform magnetic field in a volume in the form of a long tube

We start with zero current and increase it in some arbitrary manner until the final B-field in the volume is B_0. At any instant, the circular current per unit length flowing around the cylinder is $i(t)$ and the corresponding B-field at that time is $B_t(t)$. Applying the Ampere's law along the closed path shown in Figure 1.1, we obtain (note that B is zero outside the cylinder)

$$B_t(t) \cdot l = \mu_0 \cdot i(t) \cdot l \tag{1.166}$$

Now, as the magnetic field is increasing it will generate an electric field, according to Faradays law, that opposes the increase in current. Consider a time interval $t \rightarrow t + dt$. The work done, dW, by the current per unit length opposing the effect of this electric field during this time interval is

$$dW = i(t) \cdot A \cdot \frac{dB_t(t)}{dt} dt \tag{1.167}$$

The total work done per by the current per unit length in establishing the final B-field B

$$W = \int_0^{t_0} i(t) \cdot A \cdot \frac{dB_t(t)}{dt} dt \tag{1.168}$$

where t_o is the time taken to establish the final B-field. Substituting for $i(t)$ from (1.166), we obtain

$$W = \int_0^{t_0} \frac{B_t(t)}{\mu_0} \cdot A \cdot \frac{dB_t(t)}{dt} dt \tag{1.169}$$

This can be written as

$$W = \int_0^{t_0} \frac{A}{2\mu_0} \cdot \frac{dB_t^2(t)}{dt} dt \tag{1.170}$$

After performing the integration, we obtain

$$W = \frac{1}{2\mu_0} \left[B_t^2(t)\right]_0^{t_0} A \tag{1.171}$$

Since the B-field at $t = 0$ is zero and its value at $t = t_o$ is B, we obtain

$$W = \frac{1}{2\mu_0} B_0{}^2 A \tag{1.172}$$

Recall that this is the energy spent in establishing the B-field per unit length of tube. This energy is now stored in the B-field. Since the volume per unit length of the tube is equal to A (the cross-sectional area times unity), the energy per unit volume of the B-field or the energy density of the B-filed, EB_{den}, is

$$EB_{den} = \frac{W}{A} = \frac{1}{2\mu_0} B^2 \tag{1.173}$$

1.9 Summary of the laws of electricity

The laws of electricity that we have analyzed so far can be summarized as

Integral form		Point form	
$\oint \mathbf{E} \cdot \mathbf{dl} = -\int_s \frac{\partial \mathbf{B}}{\partial t} \cdot \mathbf{ds}$	(1.174a)	$\text{Curl}\mathbf{E} = -\frac{\partial \mathbf{B}}{\partial t}$	(1.174b)
$\oint \mathbf{H} \cdot \mathbf{dl} = \int_s \mathbf{J} \cdot \mathbf{ds} + \int_s \frac{\partial \mathbf{D}}{\partial t} \cdot \mathbf{ds}$	(1.175a)	$\text{Curl}\mathbf{H} = \mathbf{J} + \frac{\partial \mathbf{D}}{\partial t}$	(1.175b)
$\oint \mathbf{D} \cdot \mathbf{ds} = \int_v \rho_v dv$	(1.176a)	$\text{Div}\mathbf{D} = \rho_v$	(1.176b)
$\oint \mathbf{B} \cdot \mathbf{ds} = 0$	(1.177a)	$\text{Div}\mathbf{B} = 0$	(1.177b)

These equations are known as Maxwell's equations. In the case of sinusoidally varying fields (i.e. fields varying as $e^{j\omega t}$) (1.174b) and (1.175b) can be written as

$$\text{Curl}\mathbf{E} = -j\omega\mathbf{B} \tag{1.178}$$

$$\text{Curl}\mathbf{H} = \mathbf{J} + j\omega\mathbf{D} \tag{1.179}$$

In a medium of relative dielectric constant ε_r and conductivity σ (9.6) can be written as

$$\text{Curl}\mathbf{H} = \mathbf{J} + j\omega\varepsilon_o\varepsilon_r\mathbf{E} \tag{1.180}$$

Substituting $\sigma\mathbf{E}$ for $\mathbf{J}$, we obtain

$$\text{Curl}\mathbf{H} = \mathbf{E}\{\sigma + j\omega\varepsilon_o\varepsilon_r\} \tag{1.181}$$

Note that in a dielectric medium where $\mathbf{J} = 0$ we have $\mathrm{Curl}\mathbf{H} = j\omega\varepsilon_o\varepsilon_r\mathbf{E}$. Comparison of this with (1.181) shows us that a conducting medium can be considered as a medium with a complex dielectric constant given by $\left\{\frac{\sigma}{j\omega\varepsilon_0} + \varepsilon_r\right\}$.

1.10 Wave equation

Any wave, either sound, electromagnetic, or any other type satisfies the wave equation. The wave equation in three dimensions can be written as

$$\frac{\partial^2 W}{\partial t^2} = v^2\left\{\frac{\partial^2 W}{\partial x^2} + \frac{\partial^2 W}{\partial y^2} + \frac{\partial^2 W}{\partial z^2}\right\} \tag{1.182}$$

In this equation, W is the displacement associated with the wave (in the case of a transverse wave propagating along a string W is the displacement of the string from its equilibrium position) and v is the speed of propagation of the wave. In one dimension, for example a wave moving in z direction, it reduces to

$$\frac{\partial^2 W}{\partial t^2} = v^2\frac{\partial^2 W}{\partial z^2} \tag{1.183}$$

The solution of this equation for a wave moving along the positive z-axis takes the form,

$$W(t,z) = f\left(t - \frac{z}{v}\right) \tag{1.184}$$

For example, one-dimensional sinusoidal wave, which is a solution of (1.183), can be represented by

$$W(t,z) = W_o \sin\omega\left\{t - \frac{z}{v}\right\} \tag{1.185a}$$

or

$$W(t,z) = W_o \sin\{\omega t - kz\} \tag{1.185b}$$

In the above representation of the solutions, ω is the angular frequency which is related to the frequency f by $\omega = 2\pi f$. The wavelength of the wave λ is given by $\lambda = v/f$. The parameter $k = \omega/v$ is called the wave number of the wave.

1.11 Maxwell's prediction of electromagnetic waves

Consider Maxwell's equations in vacuum. We assume that electric and magnetic fields vary as $e^{j\omega t}$. The space is free of charges and currents except at the sources of electromagnetic fields. We have

$$\mathrm{Curl}\mathbf{E} = -j\omega\mu_0\mathbf{H} \tag{1.186}$$

Taking the Curl of both sides of this equation, we obtain

$$\text{Curl}(\text{Curl}\mathbf{E}) = -j\omega\mu_o\text{Curl}\mathbf{H} \tag{1.187}$$

Using vector identity given in (1.37) and noting that in a region of space where there are no charges div $\mathbf{E} = 0$ we obtain

$$\text{Curl}(\text{Curl}\mathbf{E}) = -\nabla^2\mathbf{E} \tag{1.188}$$

Substituting for the left-hand side from (1.187), we obtain

$$\nabla^2\mathbf{E} = j\omega\mu_o\text{Curl}\mathbf{H} \tag{1.189}$$

However, we also know that in a region in vacuum where there are no current sources (i.e. $\mathbf{J} = 0$)

$$\text{Curl}\mathbf{H} = j\omega\varepsilon_0\mathbf{E} \tag{1.190}$$

Substituting for Curl**H** from (1.190) in (1.189), we obtain

$$\nabla^2\mathbf{E} = -\mu_0\varepsilon_0\omega^2\mathbf{E} \tag{1.191}$$

Since **E** has time dependence $e^{j\omega t}$ the above equation can be written as

$$\frac{\partial^2\mathbf{E}}{\partial t^2} = \frac{1}{\mu_o\varepsilon_o}\nabla^2\mathbf{E} \tag{1.192}$$

If this is resolved into three components, we obtain

$$\frac{\partial^2 E_x}{\partial t^2} = \frac{1}{\mu_0\varepsilon_0}\left\{\frac{\partial^2 E_x}{\partial x^2} + \frac{\partial^2 E_x}{\partial y^2} + \frac{\partial^2 E_x}{\partial z^2}\right\} \tag{1.193a}$$

$$\frac{\partial^2 E_y}{\partial t^2} = \frac{1}{\mu_0\varepsilon_0}\left\{\frac{\partial^2 E_y}{\partial x^2} + \frac{\partial^2 E_y}{\partial y^2} + \frac{\partial^2 E_y}{\partial z^2}\right\} \tag{1.193b}$$

$$\frac{\partial^2 E_z}{\partial t^2} = \frac{1}{\mu_0\varepsilon_0}\left\{\frac{\partial^2 E_z}{\partial x^2} + \frac{\partial^2 E_z}{\partial y^2} + \frac{\partial^2 E_z}{\partial z^2}\right\} \tag{1.193c}$$

This shows that each component of **E** satisfies the wave equation. The speed of propagation of the wave v is given by $\frac{1}{\sqrt{\mu_0\varepsilon_0}}$. Now let us consider the magnetic field. We start with the Maxwell's equation

$$\text{Curl}\mathbf{H} = j\omega\varepsilon_o\mathbf{E} \tag{1.194}$$

Taking the Curl of both sides of this equation, we obtain

$$\text{Curl}(\text{Curl}\mathbf{H}) = j\omega\varepsilon_o\text{Curl}\mathbf{E} \tag{1.195}$$

Using the vector identity given in (1.37) and noting that Div $\mathbf{H} = 0$, we obtain

$$\text{Curl}(\text{Curl}\mathbf{H}) = -\nabla^2\mathbf{H} \tag{1.196}$$

Substituting to the left-hand side from (1.195), we obtain

$$\nabla^2 \mathbf{H} = -j\omega\varepsilon_o \text{Curl}\mathbf{E} \tag{1.197}$$

But from Maxwell's equations, we have Curl$\mathbf{E} = -j\omega\mu_0\mathbf{H}$. Substituting for Curl $\mathbf{E}$ in (1.197) and recalling that $\mathbf{H}$ has time dependence $e^{j\omega t}$, we get

$$\frac{\partial^2 \mathbf{H}}{\partial t^2} = \frac{1}{\mu_o\varepsilon_o}\nabla^2\mathbf{H} \tag{1.198}$$

The above equation shows that each component of $\mathbf{H}$ also satisfies the wave equation. The speed of propagation of the wave is again equal to $\frac{1}{\sqrt{\mu_0\varepsilon_0}}$.

The above analysis shows that Maxwell's equations predict the existence of electromagnetic waves which move in vacuum with speed $\frac{1}{\sqrt{\mu_0\varepsilon_0}}$. This speed is indeed equal to the speed of light in vacuum showing that light is an electromagnetic wave.

1.12 Plane wave solution

1.12.1 The electric field of the plane wave

Consider a plane electromagnetic wave moving in the *z*-direction. For a plane electromagnetic wave moving in the *z* direction, the electric field is independent of *x* and *y*. Without losing the generality, we can assume that the electric field associated with this wave is in *x*-direction. We denote it by E_x. As mentioned previously in the plane wave under consideration E_x does not vary in *x* and *y* directions. Thus, the wave equation for the electric field is

$$\frac{\partial^2 E_x}{\partial z^2} = \mu_0\varepsilon_0\frac{\partial E_x}{\partial t^2} \tag{1.199}$$

We also assume, as before, that the time variation of E_x has the form $e^{j\omega t}$. Thus the solution of the above wave equation takes the form

$$E_x = E_{x0}e^{j\omega\left(t-\frac{z}{v}\right)} \tag{1.200}$$

or

$$E_x = E_{x0}e^{j\omega t}e^{-jkz} \tag{1.201}$$

The solution can also be written in terms of sine or cosine functions. For example, it can be written as

$$E_x = E_{x0}\cos(\omega t - kz) \tag{1.202}$$

or

$$E_x = E_{x0}\cos\left\{\omega\left(t - \frac{z}{v}\right)\right\} \tag{1.203}$$

Since $v = 1/\sqrt{\mu_0\varepsilon_0}$, we can write

$$E_x = E_{x0}\cos\{\omega(t - \sqrt{\mu_0\varepsilon_0}z)\} \tag{1.204}$$

1.12.2 The magnetic field of the plane wave

Recall that we are considering a plane wave moving in z-direction. The electric field is in the x direction and it does not vary in x or y direction (because it is a plane wave).

Now consider the equation

$$\mathrm{Curl}\mathbf{E} = -j\omega\mu_0\mathbf{H} \tag{1.205}$$

Separating this into components, we obtain

$$\begin{aligned}(\mathrm{Curl}\mathbf{E})_x &= -j\omega\mu_0\mathbf{H_x}\\ (\mathrm{Curl}\mathbf{E})_y &= -j\omega\mu_0\mathbf{H_y}\\ (\mathrm{Curl}\mathbf{E})_z &= -j\omega\mu_0\mathbf{H_z}\end{aligned} \tag{1.206}$$

Now, expanding Curl**E** into its components in the x, y, and z directions, we obtain

$$\mathrm{Curl}\mathbf{E} = \left\{\frac{\partial E_z}{\partial y} - \frac{\partial E_y}{\partial z}\right\}\mathbf{a_x} - \left\{\frac{\partial E_z}{\partial x} - \frac{\partial E_x}{\partial z}\right\}\mathbf{a_y} + \left\{\frac{\partial E_y}{\partial x} - \frac{\partial E_x}{\partial y}\right\}\mathbf{a_z} \tag{1.207}$$

Since $E_z = 0$ and $E_y = 0$ (E-field is in the x direction) and all terms operated by $\frac{\partial}{\partial x}$ and $\frac{\partial}{\partial y}$ are zero (no variation in x and y direction), we obtain

$$(\mathrm{Curl}\mathbf{E})_x = 0 \tag{1.208}$$

$$(\mathrm{Curl}\mathbf{E})_z = 0 \tag{1.209}$$

$$(\mathrm{Curl}\mathbf{E})_y = \frac{\partial E_x}{\partial z} \tag{1.210}$$

Substituting for $(\mathrm{Curl}\mathbf{E})_y$ from (1.206), we obtain

$$\frac{\partial E_x}{\partial z} = -j\omega\mu_0 H_y \tag{1.211}$$

substituting for E_x from (1.201), we obtain an expression for H-field as

$$H_y = \left(-\frac{1}{j\omega\mu_0}\right)(-jk)E_{x0}e^{-jkz}e^{j\omega t} \tag{1.212}$$

Substituting $k = \omega\sqrt{\mu_0\varepsilon_0}$ for the term inside the bracket in the above equation, we obtain

$$H_y = \sqrt{\frac{\varepsilon_0}{\mu_0}}E_{x0}e^{-jkz}e^{j\omega t} \tag{1.213}$$

Thus, the E and H and B-field components of the plane wave moving in the z-direction are given by

$$E_x = E_{x0}\cos(\omega t - kz) \tag{1.214}$$

$$H_y = \sqrt{\frac{\varepsilon_0}{\mu_0}}E_{x0}\cos(\omega t - kz) \tag{1.215}$$

$$B_y = \sqrt{\mu_0\varepsilon_o}\,E_{x0}\cos(\omega t - kz) \tag{1.216}$$

The above shows that if the electric field component of a plane wave moving in the z-direction is directed in the x-direction then the magnetic field is directed in the y-direction. That is, the magnetic field is perpendicular to the electric field. In fact, the direction of propagation of the wave is given by the direction of $\mathbf{E} \times \mathbf{H}$ (or $\mathbf{E} \times \mathbf{B}$) where E and H (or B) are the electric and magnetic field components of the wave. In fact in a plane wave, **E**, **H** (or **B**) and the direction of propagation (in the above case x, y, and z, respectively) form an orthogonal set. In a plane wave in vacuum (moving in the z-direction), the ratio of E-field and H-field components is given by

$$\frac{E_x}{H_y} = \sqrt{\frac{\mu_0}{\varepsilon_0}} \tag{1.217}$$

The parameter $\sqrt{\mu_o/\varepsilon_o}$ is called the intrinsic impedance of vacuum (or free space) and is denoted usually by η_0. The ratio of the E-field and B-field components of the plane wave is given by

$$\frac{E_x}{B_y} = \frac{1}{\sqrt{\mu_0\varepsilon_0}} = c \tag{1.218}$$

where c is the speed of light in vacuum.

1.12.3 Energy transported by a plane wave – Poynting's theorem

Consider the divergence of quantity $\mathbf{E} \times \mathbf{H}$ in a plane wave. We obtain (from (1.38))

$$\mathrm{Div}(\mathbf{E} \times \mathbf{H}) = -\mathbf{E}\cdot(\mathrm{Curl}\mathbf{H}) + \mathbf{H}\cdot(\mathrm{Curl}\mathbf{E}) \tag{1.219}$$

Substituting for Curl**H** and Curl**E** from Maxwell's equations, we obtain

$$\mathrm{Div}(\mathbf{E} \times \mathbf{H}) = -\mathbf{E}\cdot\varepsilon_0\frac{\partial \mathbf{E}}{\partial t} - \mathbf{H}\cdot\mu_0\frac{\partial \mathbf{H}}{\partial t} \tag{1.220}$$

The above can be written as

$$\mathrm{Div}(\mathbf{E} \times \mathbf{H}) = -\frac{\partial}{\partial t}\left\{\frac{\varepsilon_o E^2}{2} + \frac{\mu_o H^2}{2}\right\} \tag{1.221}$$

Consider a closed volume in a region in vacuum where a plane wave is propagating. This volume is bounded by a surface S. Now take the integral of the quantity Div($\mathbf{E} \times \mathbf{H}$) in this closed volume. We obtain

$$\int_v \mathrm{Div}(\mathbf{E} \times \mathbf{H})dv = -\frac{\partial}{\partial t}\int_v \left\{ \frac{\varepsilon_o E^2}{2} + \frac{\mu_o H^2}{2} \right\} dv \tag{1.222}$$

Note that the quantity inside the bracket on the right-hand side of the above equation is the energy densities of the E- and B-fields [(1.143) and (1.173)]. Thus the right-hand side of this equation gives the rate of change of the energy or the energy lost from the closed volume. Now, converting the left-hand side using the divergence theorem, we obtain

$$\oint_s (\mathbf{E} \times \mathbf{H}) \cdot \mathbf{ds} = -\frac{\partial}{\partial t}\int_v \left\{ \frac{\varepsilon_o E^2}{2} + \frac{\mu_o H^2}{2} \right\} dv \tag{1.223}$$

Since the right-hand side gives the rate of energy dissipation from the closed volume, the left-hand side tells us that the energy moving out from a unit area in a direction perpendicular to that unit area in a unit time is $\mathbf{E} \times \mathbf{H}$. Note that the direction of $\mathbf{E} \times \mathbf{H}$ is the direction of propagation of the wave. Thus the energy passing through a unit area per unit time perpendicular to the direction of propagation of the wave is $\mathbf{E} \times \mathbf{H}$. This is called Poynting's theorem and $\mathbf{E} \times \mathbf{H}$ is called the Poynting vector. In the example considered previously for a plane wave moving in the z-direction the E- and H-field components are given by (1.214) and (1.215). Using these in the Poynting vector, we find the power flow P_z in the z-direction as

$$P_z = \frac{{E_{x0}}^2}{\eta_0}\cos^2(\omega t - kz) \tag{1.224}$$

The units of P_z is Watts per meter^2 (W/m^2).

1.13 Maxwell's equations and plane waves in different media (summary)

1.13.1 Vacuum

In vacuum, we have $\varepsilon_r = 1; \mu_r = 1; \mathbf{J} = 0$. We also assume that the fields are varying as $e^{j\omega t}$. Thus the corresponding Maxwell's equations are

$$\mathrm{Div}\,\mathbf{E} = \frac{\rho}{\varepsilon_o} \tag{1.225}$$

$$\mathrm{Div}\,\mathbf{B} = 0 \tag{1.226}$$

$$\mathrm{Curl}\mathbf{E} = -j\omega\mathbf{B} \tag{1.227}$$

$$\mathrm{Curl}\mathbf{B} = j\omega\varepsilon_o\mu_0\mathbf{E} \tag{1.228}$$

The E-field (assumed to be in the x-direction) associated with a plane moving in the z-direction is given by

$$E_x = E_{x0} e^{j\omega t} e^{-j\omega z\sqrt{\varepsilon_0\mu_0}} \tag{1.229a}$$

or

$$E_x = E_{x0} \sin\left\{\omega\left(t - z\sqrt{\varepsilon_0\mu_0}\right)\right\} \tag{1.229b}$$

The H-field is given by (this is derived in Section 1.12.2)

$$H_y = \sqrt{\frac{\varepsilon_o}{\mu_o}} E_{x0} e^{j\omega t} e^{-j\omega z\sqrt{\varepsilon_0\mu_0}} \tag{1.230a}$$

or

$$H_y = \sqrt{\frac{\varepsilon_o}{\mu_o}} E_{x0} \sin\left\{\omega\left(t - z\sqrt{\varepsilon_0\mu_0}\right)\right\} \tag{1.230b}$$

The speed of propagation of the wave, v, is given by $v = 1/\sqrt{\varepsilon_0\mu_0}$.

1.13.2 Isotropic and linear dielectric and magnetic media

Since the conductivity of the medium is zero the conducting current $\mathbf{J} = 0$. Maxwell's equations take the form

$$\mathrm{Div}\mathbf{E} = \frac{\rho}{\varepsilon_r\varepsilon_o} \tag{1.231}$$

$$\mathrm{Div}\mathbf{B} = 0 \tag{1.232}$$

$$\mathrm{Curl}\mathbf{E} = -j\omega\mathbf{B} \tag{1.233}$$

$$\mathrm{Curl}\mathbf{B} = j\omega\mu_0\varepsilon_0\mu_r\mu_o\mathbf{E} \tag{1.234}$$

The E-field (assumed to be in the x-direction) associated with a plane moving in the z-direction is given by

$$E_x = E_{x0} e^{j\omega t} e^{-j\omega z\sqrt{\varepsilon_0\varepsilon_r\mu_r\mu_0}} \tag{1.235a}$$

or

$$E_x = E_{x0} \sin\left\{\omega\left(t - z\sqrt{\varepsilon_0\varepsilon_r\mu_r\mu_0}\right)\right\} \tag{1.235b}$$

The H-field is given by (following the same procedure as in Section 1.12.2)

$$H_y = \sqrt{\frac{\varepsilon_r\varepsilon_o}{\mu_r\mu_o}} E_{x0} e^{j\omega t} e^{-j\omega z\sqrt{\varepsilon_0\mu_0}} \tag{1.236a}$$

or

$$H_y = \sqrt{\frac{\varepsilon_r\varepsilon_o}{\mu_r\mu_o}} E_{x0} \sin\left\{\omega\left(t - z\sqrt{\varepsilon_0\mu_0}\right)\right\} \tag{1.236b}$$

The speed of propagation of the wave, v, is given by $v = 1/\sqrt{\varepsilon_0\mu_r\mu_0\varepsilon_r}$.

1.13.3 Conducting media

In a conducting media, an electric field will give rise to a current. Thus, we have to assume that $\mathbf{J} \neq 0$. Maxwell's equations take the form

$$\text{Div}\,\mathbf{E} = \frac{\rho}{\varepsilon_r \varepsilon_o} \tag{1.237}$$

$$\text{Div}\,\mathbf{B} = 0 \tag{1.238}$$

$$\text{Curl}\mathbf{E} = -j\omega\mathbf{B} \tag{1.239}$$

$$\text{Curl}\mathbf{B} = \mu_r\mu_o\mathbf{J} + j\omega\mu_0\mu_r\varepsilon_o\varepsilon_r\mathbf{E} \tag{1.240}$$

The last equation can be written as (substituting $\sigma\mathbf{E}$ for $\mathbf{J}$)

$$\text{Curl}\mathbf{B} = j\omega\mathbf{E}\left\{\frac{\mu_r\mu_o\sigma}{j\omega} + \mu_0\mu_r\varepsilon_o\varepsilon_r\right\} \tag{1.241}$$

In a highly conducting media, the conductivity σ is very large with respect to $\omega\varepsilon_o\varepsilon_r$ (i.e. we assume that the conduction current is much larger than the displacement current) resulting in

$$\text{Curl}\mathbf{B} = j\omega\mathbf{E}\left\{\frac{\mu_r\mu_o\sigma}{j\omega}\right\} \tag{1.242}$$

In this case, the electric field associated with the plane wave moving in the z-direction becomes

$$E_x = E_{x0}e^{j\omega t}e^{-j\omega z\sqrt{\frac{\mu_r\mu_0\sigma}{j\omega}}} \tag{1.243}$$

This can be written as

$$E_x = E_{x0}e^{j\omega t}e^{-j\omega z(a+jb))} \tag{1.244}$$

with

$$a = \sqrt{\frac{\mu_o\mu_r\sigma}{2\omega}} \tag{1.245}$$

$$b = -\sqrt{\frac{\mu_o\mu_r\sigma}{2\omega}} \tag{1.246}$$

Substituting for a and b, we obtain

$$E_x = E_{x0}e^{j\omega\left\{t - z\sqrt{\frac{\mu_r\mu_0\sigma}{2\omega}}\right\}}e^{-z\sqrt{\frac{\mu_r\mu_0\sigma\omega}{2}}} \tag{1.247}$$

The above equation can be written as

$$E_x = E_{x0}e^{j\omega\left\{t - z\sqrt{\frac{\mu_r\mu_0\sigma}{2\omega}}\right\}}e^{-\frac{z}{\delta}} \tag{1.248}$$

Starting with Maxwell' equation, Curl$\mathbf{E} = -j\omega\mu_0\mathbf{H}$ and following the same procedure used in Section 1.12.2, we obtain the H-field of the plane wave as

$$H_y = \frac{E_{x0}}{\sqrt{\frac{\omega\mu_0\mu_r}{\sigma}}} e^{j\omega\left[t - z\sqrt{\frac{\mu_0\mu_r\sigma}{2\omega}}\right]} e^{-\frac{z}{\delta}} e^{-j\frac{\pi}{4}} \tag{1.249}$$

Note first that there is a phase difference between the E-field and the H-field. Moreover, the above equations show that as the electromagnetic wave moves in the z-direction its amplitude will decreases exponentially. The parameter δ is called the skin-depth of the medium. Thus, the skin depth is given by

$$\delta = \sqrt{\frac{2}{\mu_r\mu_o\sigma\omega}} \tag{1.250}$$

In terms of frequency of the wave, f, the skin depth is given by

$$\delta = \sqrt{\frac{1}{\mu_r\mu_o\pi\sigma f}} \tag{1.251}$$

The speed of propagation of the wave, v, becomes

$$v = \sqrt{\frac{2\omega}{\mu_0\mu_r\sigma}} \tag{1.252}$$

This shows that the speed of propagation of the wave is a function of ω. Thus waves of different frequencies travel at different speeds in conducting media.

1.14 Retarded potentials

Recall that the scalar and vector potentials of static charge and current distributions are given by

$$\phi = \frac{1}{4\pi\varepsilon_0}\int\frac{\rho_v dv}{r} \tag{1.253}$$

$$\mathbf{A} = \frac{\mu}{2\pi}\int\frac{\mathbf{J}(r)dv}{r} \tag{1.254}$$

Recall also that under static conditions, ϕ and $\mathbf{A}$ satisfy the equations

$$\nabla^2\phi = -\frac{\rho}{\varepsilon_0} \tag{1.255}$$

$$\nabla^2\mathbf{A} = -\mu_0\mathbf{J} \tag{1.256}$$

Now let us consider the case in which both **J** and ρ_v depend on time. We start with Maxwell's equation

$$\text{Curl}\mathbf{B} = \mu_0\mathbf{J} + \mu_0\frac{\partial \mathbf{D}}{\partial t} \tag{1.257}$$

This can be written as

$$\text{Curl}(\text{Curl}\mathbf{A}) = \mu_0\mathbf{J} + \mu_0\varepsilon_0\frac{\partial \mathbf{E}}{\partial t} \tag{1.258}$$

In the presence of time varying fields, the electric field has two components: one due to static charges and the other due to changing magnetic fields. The contribution due to static charges is given by $-\text{Grad}\phi$. The E-field component produced by the changing magnetic field, say $\mathbf{E}_m$ is given by

$$\text{Curl}\mathbf{E}_m = -\frac{\partial \mathbf{B}}{\partial t} \tag{1.259}$$

substituting $\mathbf{B} = \text{Curl}\mathbf{A}$, we find that

$$\mathbf{E}_m = -\frac{\partial \mathbf{A}}{\partial t} \tag{1.260}$$

Thus the total E-field is given by

$$\mathbf{E} = -\frac{\partial \mathbf{A}}{\partial t} - \text{Grad}\phi \tag{1.261}$$

Expanding the left-hand side of (1.258) and substituting for **E** from (1.261), we obtain

$$\text{Grad}(\text{Div}\mathbf{A}) - \nabla^2\mathbf{A} = \mu_0\mathbf{J} - \mu_0\varepsilon_0\frac{\partial^2 \mathbf{A}}{\partial t^2} - \mu_0\varepsilon_0\frac{\partial}{\partial t}\text{grad}\phi \tag{1.262}$$

Now we have already defined Curl **A** but we have the freedom to select the function Div **A**. Let us select it as

$$\text{Div}\mathbf{A} = -\mu_0\varepsilon_0\frac{\partial \phi}{\partial t} \tag{1.263}$$

This is called the Lorentz Gauge. Substituting this in (1.262), we obtain

$$-\mu_0\varepsilon_0\frac{\partial}{\partial t}\text{Grad}\phi - \nabla^2\mathbf{A} = \mu_0\mathbf{J} - \mu_0\varepsilon_0\frac{\partial^2 \mathbf{A}}{\partial t^2} - \mu_0\varepsilon_0\frac{\partial}{\partial t}\text{Grad}\phi \tag{1.264}$$

This reduces to

$$\nabla^2\mathbf{A} - \mu_0\varepsilon_0\frac{\partial^2 \mathbf{A}}{\partial t^2} = -\mu_0\mathbf{J} \tag{1.265}$$

Now let us consider the scalar potential. We start with (1.261). Taking the divergence of both sides of this equation, we obtain

$$\mathrm{Div}\mathbf{E} = -\nabla^2\phi - \frac{\partial}{\partial t}\mathrm{Div}\mathbf{A} \tag{1.266}$$

However, in vacuum $\mathrm{Div}\mathbf{E} = \rho/\varepsilon_o$ and substituting this in the above, we obtain

$$\nabla^2\phi + \frac{\partial}{\partial t}\mathrm{Div}\mathbf{A} = -\frac{\rho}{\varepsilon_o} \tag{1.267}$$

Now substituting for Div **A** from the Lorentz condition [i.e. (1.263)], we obtain

$$\nabla^2\phi - \mu_0\varepsilon_0\frac{\partial^2\phi}{\partial t^2} = -\frac{\rho}{\varepsilon} \tag{1.268}$$

Note that both (1.265) and (1.268) reduce to (1.256) and (1.255), respectively under static conditions.

In the case of time varying charges and currents, the field at a given time is produced by charges and currents at some earlier time t_r where $t_r = t - \frac{r}{c}$ and r is the distance to the location of the charge or the current that give rise to the field. So the general solution for ϕ and A are

$$\phi = \frac{1}{4\pi\varepsilon_0}\int\frac{\rho_v(r, t - r/c)}{r}dv \tag{1.269}$$

$$\mathbf{A} = \frac{\mu_0}{4\pi}\int\frac{\mathbf{J}(r, t - r/c)}{r}dv \tag{1.270}$$

In the above equation, the time $t - r/c$ is called the retarded time and ϕ and A as given by (1.269) and (1.270) are called retarded potentials.

1.15 Electromagnetic fields of a current element – electric dipole

Consider a current element of length dl where the current is directed along the z-axis. The geometry is shown in Figure 1.12. The current in the element is given by

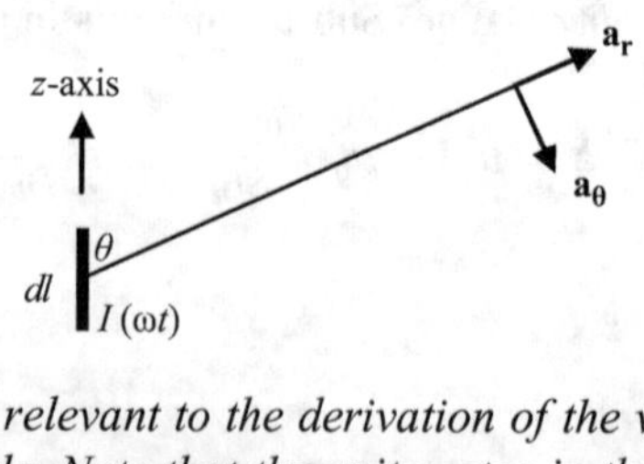

Figure 1.12 Geometry relevant to the derivation of the vector potential due to a short dipole. Note that the unit vector in the azimuthal direction is given by $\mathbf{a_r} \times \mathbf{a_\theta}$

$I = I_0 e^{j\omega t}$. Under the conditions $r >> dl$ and $\lambda >> dl$, where λ is the wavelength, the vector and scalar potential of the current element at point P will reduce to

$$\mathbf{A} = \frac{\mu_0}{4\pi} I_o \, dl \frac{e^{j(\omega t - k_0 r)}}{r} \mathbf{a_z} \tag{1.271}$$

$$\phi = \frac{I_o \, dl \cos\,\theta e^{j(\omega t - k_0 r)}}{4\pi\varepsilon_0 c} \left\{ \frac{1}{r} + \frac{c}{j\omega r^2} \right\} \tag{1.272}$$

In the above equations, $k_0 = \omega / c$. Now using the relationships $\mathbf{E} = -j\omega\mathbf{A} -$ Gradϕ and $\mathbf{B} =$ Curl $\mathbf{A}$, we obtain

$$\mathbf{E_r} = \frac{I_0 l e^{j\omega t} e^{-jk_o r}}{2\pi\varepsilon_0} \cos\,\theta \left[\frac{1}{cr^2} + \frac{1}{j\omega r^3} \right] \mathbf{a_r} \tag{1.273}$$

$$\mathbf{E_\theta} = \frac{I_0 l e^{j\omega t} e^{-jk_o r}}{4\pi\varepsilon_0} \sin\theta \left[\frac{j\omega}{c^2 r} + \frac{1}{cr^2} + \frac{1}{j\omega r^3} \right] \mathbf{a_\theta} \tag{1.274}$$

$$\mathbf{B_\varphi} = \frac{\mu_0 I_0 l e^{j\omega t} e^{-jk_o r}}{4\pi} \sin\theta \left[\frac{j\omega}{cr} + \frac{1}{r^2} \right] \mathbf{a_\varphi} \tag{1.275}$$

Note that the E-field has three components one varying as $1/r$ (called the radiation field), the other varying as $1/r^2$ (called the induction field) and the third component varying as $1/r^3$ (called the static field). When the distance to the point of observation is such that $k_o r >> 1$, the contributions from the static and induction fields can be neglected leaving behind only the radiation fields. Thus in the radiation field zone $\mathbf{E_r}$ is negligible and the electric and magnetic fields reduce to

$$\mathbf{E_\theta} = \frac{I_0 l e^{j\omega t} e^{-jk_o r}}{4\pi\varepsilon_0 c^2} \sin\theta \left[\frac{j\omega}{r} \right] \mathbf{a_\theta} \tag{1.276}$$

$$\mathbf{B_\varphi} = \frac{I_0 l e^{j\omega t} e^{-jk_o r}}{4\pi\varepsilon_0 c^3} \sin\theta \left[\frac{j\omega}{r} \right] \mathbf{a_\varphi} \tag{1.277}$$

Observe that, except for the radiation fields, the other terms do not carry any physical meaning or they are not directly connected to the physical processes at the source that generate these fields (see Chapter 2 of this volume). These equations (or their counterpart in time domain) are frequently used in calculating the electromagnetic fields of lightning flashes. In these calculations the lightning channel is divided in a series of infinitesimal current elements and the total field is obtained as a sum of the fields generated by these elements. The field due to each current element in frequency domain is given by the above equations.

1.16 Electromagnetic fields of a lightning return stroke

Consider a lightning return stroke located over ground which is assumed to be a perfect conductor. We assume that the lightning channel is straight and vertical. Let us divide this channel into a large number of elements. Consider a current element

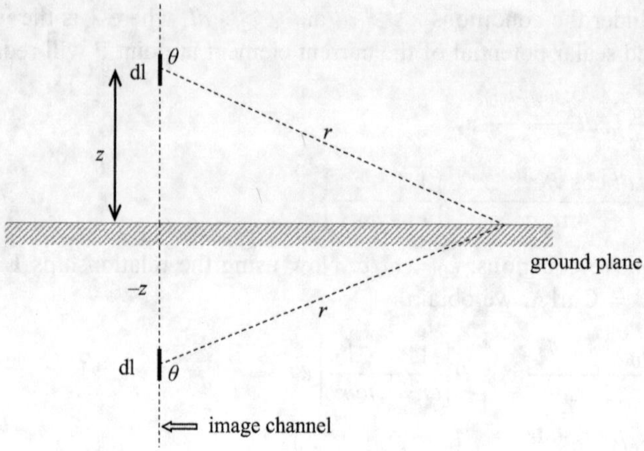

Figure 1.13 Geometry relevant to the calculation of electromagnetic fields due to a lightning channel located above a perfectly conducting ground plane

of length dz located at a height z above a perfectly conducting ground plane at zero potential (Figure 1.13). This plane represents the Earth's surface. The goal is to write down an expression for the electric field at the surface of the conductor. The effect of the ground plane can be taken into account by replacing it by an image current element located at a distance h below the ground plane (Figure 1.13). Note that when a positive current flows upwards along the current element a negative current flows downwards in the image. Writing down the E-field components due to the two current elements using (1.273) to (1.274) and summing up the contributions we obtain the component of the E-field perpendicular to the ground (along the z-axis) as

$$dE_z = \frac{I_0 dz e^{j(\omega t - \beta r)}}{2\pi\varepsilon_0}\left[-\frac{\sin^2\theta\, j\omega}{c^2 r} + (2 - 3\sin^2\theta)\frac{1}{cr^2} + \frac{1}{j\omega r^3}(2 - 3\sin^2\theta)\right] \quad (1.278)$$

The E-field component parallel to the surface or the horizontal E-field is zero. The B-field at the surface is given by

$$dB_\varphi = \frac{\mu_0 I_0 dz e^{j\omega t} e^{-jk_o r}}{2\pi}\sin\theta\left[\frac{j\omega}{cr} + \frac{1}{r^2}\right] \quad (1.279)$$

Transforming these equations into time domain, one obtains

$$de_z(t) = \frac{dz}{2\pi\varepsilon_0}\left[-\frac{\sin^2\theta}{c^2 r}\frac{dI(t - r/c)}{dt} + (2 - 3\sin^2\theta)\frac{I(t - r/c)}{cr^2}\right.$$
$$\left. + \frac{1}{r^3}(2 - 3\sin^2\theta)\int_{r/c}^{t} I(\tau - r/c)d\tau\right] \quad (1.280)$$

$$db_{\varphi}(t) = \frac{\mu_0 dz}{2\pi}\left[\frac{\sin\theta}{cr}\frac{dI(t-r/c)}{dt} + \frac{\sin\theta}{r^2}I(t-r/c)\right] \quad (1.281)$$

The electromagnetic field generated by a lightning return stroke can be obtained by first dividing the lightning channel into infinitesimal elements and then summing the contribution from each element taking into account the time delays properly. That is, the total E- and B-fields due to a lightning flash are given by

$$\begin{aligned} e_z(t) = \int_0^H \frac{dz}{2\pi\varepsilon_0}\Bigg[&-\frac{\sin^2\theta}{c^2 r}\frac{\partial I(z,t-r/c)}{\partial t} + \frac{(2-3\sin^2\theta)}{cr^2}I(z,t-r/c) \\ &+\frac{1}{r^3}(2-3\sin^2\theta)dz\int_{r/c}^{t} I(z,\tau-r/c)d\tau\Bigg] \end{aligned} \quad (1.282)$$

$$B_{\varphi}(t) = \int_0^H \frac{\mu_0 dz}{2\pi}\left[\frac{\sin\ \theta}{cr}\frac{\partial I(z,t-r/c)}{\partial t} + \frac{\sin\ \theta}{r^2}I(z,t-r/c)\right] \quad (1.283)$$

Following the nomenclature adopted from the dipole fields, the terms that vary as $1/r$ are called the radiation field, the terms that vary as $1/r^2$ are called the induction field and the terms that vary as $1/r^3$ are called static field. However, as mentioned earlier, except for the radiation fields, the other terms do not carry any physical meaning or they are not directly connected to the physical processes at the source that generate these fields (see Chapter 2 of this volume).

When the distance is very large compared to the dimension of the channel only the terms which vary as $1/r$ will contribute to the field. These radiation fields are given by (note also that under these conditions $\sin\theta = 1$)

$$e_{z,rad}(t) = -\int_0^H \frac{dz}{2\pi\varepsilon_0}\left[\frac{1}{c^2 r}\frac{\partial I(z,t-r/c)}{\partial t}\right] \quad (1.284)$$

$$B_{\varphi,rad}(t) = \int_0^H \frac{\mu_0 dz}{2\pi}\left[\frac{1}{cr}\frac{\partial I(z,t-r/c)}{\partial t}\right] \quad (1.285)$$

Note that these fields satisfy the condition

$$E_{z,rad}(t) = c\ B_{\varphi,ard}(t) \quad (1.286)$$

where c is the speed of light in vacuum.

References

[1] W. K. H. Panofsky and M. Phillips, *Classical Electricity and Magnetism*, Addison-Wesley, New York, NY, 1962.

[2] Kraus, J. D. and K. R. Carver, *Electromagnetics (second edition)*, McGraw-Hill, Inc., USA, 1973.

[3] J. D. Jackson, *Classical Electrodynamics*, Wiley, New York, NY, 1999.

Chapter 2

Application of electromagnetic fields of accelerating charges to obtain the electromagnetic fields of engineering return stroke models

Gerald Cooray[1,2] *and Vernon Cooray*[3]

2.1 Introduction

In publication [1], Cooray and Cooray demonstrated that electromagnetic fields from accelerating charges can be utilized to evaluate the electromagnetic fields from lightning return strokes. In that publication, they have documented in details how to utilize these equations to calculate electromagnetic fields of return strokes. The applications of this technique were further explored in [2,3] and the technique was generalized in [4] for calculating electromagnetic fields from arbitrary distributions of electric charge and currents. More recently, the technique was used to calculate the electromagnetic fields in the dart leader channel [5].

The basics of this technique is summarized in this chapter by applying it to evaluate the electromagnetic fields generated by three types of engineering return stroke models, namely, current propagation (CP), current generation (CG), and current dissipation (CD) (see Chapter 3 of this volume). But, first, let us consider the electromagnetic fields of accelerating electric charges.

2.2 Electromagnetic fields of a moving charge

The theory of electromagnetic fields generated by accelerating charges is described in any standard text book on electromagnetic theory, and it suffices to quote the results directly [1,6]. The geometry relevant to the problem under consideration is depicted in Figure 2.1. In this diagram, $\mathbf{a}_r$ and $\mathbf{a}_\theta$ are unit vectors one in the direction of increasing r and the other in the angular direction of increasing θ. The unit vector $\mathbf{a}_\varphi$ is directed along the vector $\mathbf{a}_r \times \mathbf{a}_\theta$. A charged particle is moving

[1]Wellcome Trust Centre for Neuroimaging, Institute of Neurology, University College London, UK
[2]Department of Clinical Neuro Science, Karolinska Institute, Stockholm, Sweden
[3]Department of Electrical Engineering, Uppsala University, Sweden

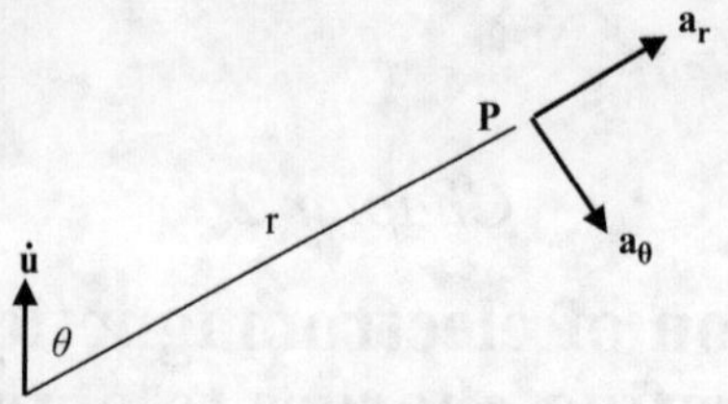

Figure 2.1 Definition of the parameters that appear in (2.1) and (2.2)

with speed **u** and acceleration **u̇** along the direction z defined by the unit vector $\mathbf{a_z}$. We assume that the direction of **u** does not change with time; that is, both **u** and **u̇** are acting in the same direction. The electric field produced by this charge at point P is given by

$$\mathbf{E} = \frac{q}{4\pi\varepsilon_o r^2}\frac{1}{\left(1-\frac{u}{c}\cos\theta\right)^3}\left(\mathbf{a_r}-\frac{u}{c}\mathbf{a_z}\right)\left(1-\frac{u^2}{c^2}\right)+\frac{q\sin\theta}{4\pi\varepsilon_o c^2 r}\frac{\dot{u}}{\left(1-\frac{u}{c}\cos\theta\right)^2}\mathbf{a_\theta} \tag{2.1}$$

$$\mathbf{B} = \frac{q}{4\pi\varepsilon_o c^2 r^2}\frac{u\sin\theta}{\left(1-\frac{u}{c}\cos\theta\right)^3}\left(1-\frac{u^2}{c^2}\right)\mathbf{a_\varphi}+\frac{q\sin\theta}{4\pi\varepsilon_o c^3 r}\frac{\dot{u}}{\left(1-\frac{u}{c}\cos\theta\right)^2}\mathbf{a_\varphi} \tag{2.2}$$

Note that the expressions for **E** and **B** both consist of two terms. The second term, which depends on the acceleration of the charge, is the radiation field, and the first term, which is actually the modification of the Coulombs law for moving charges, is called the velocity field. Note that the velocity field becomes zero when the speed of propagation of the charge is equal to the speed of light.

2.3 Electromagnetic fields of a propagating current pulse

Consider the geometry shown in the top diagram of Figure 2.2. A pulse of current originates at point S1 and travels along the z-axis with constant speed u and without any attenuation or dispersion. At the initiation of the current, charges will be accelerated from rest to a speed u. Once they attain this speed, they travel with constant velocity along the z-axis. The acceleration of charge at S1 generates a radiation field and the uniform propagation of charge along the z-axis generates a velocity field. Recently, using (2.1) and (2.2), Cooray and Cooray [1] derived expressions for the radiation field produced by the acceleration of charge at S1 and for the velocity field produced by the uniform motion. According to their results, the expressions for the electric and magnetic radiation fields generated by the charge acceleration are given respectively by

$$\mathbf{e}_{rad}(t) = \frac{i(t-r/c)u\sin\theta}{4\pi\varepsilon_o c^2 r}\frac{1}{\left[1-\frac{u\cos\theta}{c}\right]}\mathbf{a_\theta} \tag{2.3}$$

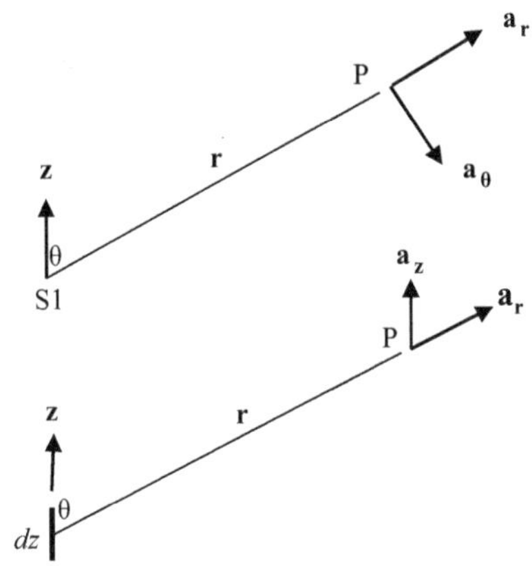

Figure 2.2 Geometry relevant to the parameters in (2.3) and (2.4) (top diagram) and 5 and 6 (bottom diagram)

$$\mathbf{b}_{rad}(t) = \frac{i(t - r/c)u \sin\theta}{4\pi\varepsilon_o c^3 r} \frac{1}{\left[1 - \frac{u \cos\theta}{c}\right]} \mathbf{a}_\varphi \tag{2.4}$$

In these equations, $i(t)$ is the temporal variation of the current emanating from S1. Now, consider a spatial element of length dz through which a current pulse $i(t)$ is moving with speed u (see the diagram at the bottom of Figure 2.2). The velocity fields generated by this element at point P are given by [1]

$$\mathbf{de}_{vel} = \frac{i(t - r/c)dz}{4\pi\varepsilon_o r^2 \left[1 - \frac{u}{c}\cos\theta\right]^2} \left[1 - \frac{u^2}{c^2}\right] \left[\frac{\mathbf{a_r}}{u} - \frac{\mathbf{a_z}}{c}\right] \tag{2.5}$$

$$\mathbf{db}_{vel} = \frac{i(t - r/c)\sin\theta dz}{4\pi\varepsilon_o r^2 c^2 \left[1 - \frac{u}{c}\cos\theta\right]^2} \left[1 - \frac{u^2}{c^2}\right] \mathbf{a}_\varphi \tag{2.6}$$

In the next section, these equations are used to derive the electric and magnetic fields of a current channel of length l for the case of a current pulse that propagates with constant velocity.

2.4 Electromagnetic fields generated by a current pulse propagating from one point in space to another along a straight line with uniform velocity and without attenuation

The geometry under consideration is shown in Figure 2.3. A current pulse $i(t)$ originates at point S1 and travels without attenuation or dispersion towards S2. At S2, the current is terminated. The total electric field at point P, generated by this process has five components. They are as follows: (i) the radiation field generated from S1 as the charge accelerates when the current is initiated, (ii) the radiation

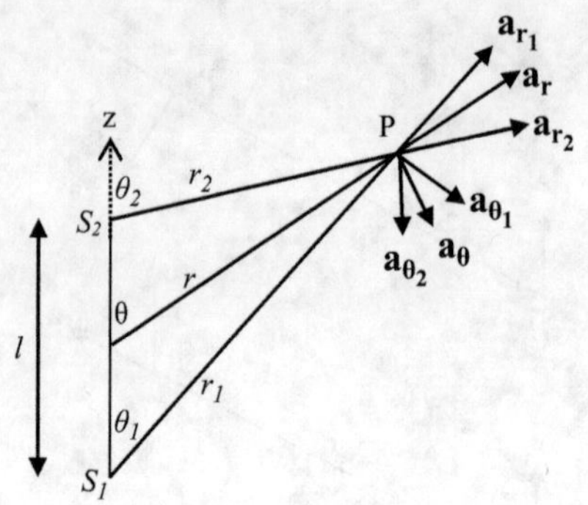

Figure 2.3 Geometry used in deriving the electromagnetic fields of a current channel

field generated from S2 during the charge deceleration as the current is terminated, (iii) the electrostatic field generated by the negative charge accumulated at S1 when the positive charge travels towards S2, (iv) the electrostatic field generated by the accumulation of positive charge at S2, and (v) the velocity field generated as the current pulse moves from S1 to S2. The magnetic field generated by the element consists of three terms, namely, two radiation fields generated at S1 and S2, and the velocity field generated as the current propagates along the path. Let us now write down the expressions for these field components.

2.4.1 The electric radiation field generated from S1

Let us assume that the current pulse leaving S1 can be represented by $i(t)$. The radiation field at point P is given by (2.3), and with the geometry under consideration here, one can rewrite this as

$$\mathbf{e}_{rad,S_1} = \frac{i(t - r_1/c)u \sin\theta_1}{4\pi\varepsilon_o c^2 r_1} \frac{1}{\left[1 - \frac{u \cos\theta_1}{c}\right]} \mathbf{a}_{\theta_1} \quad t > r_1/c \tag{2.7}$$

2.4.2 The electric radiation field generated from S2

The radiation field generated from S2 as the charges decelerate is given by

$$\mathbf{e}_{rad,S_2} = -\frac{i(t - l/u - r_2/c)u \sin\theta_2}{4\pi\varepsilon_o c^2 r_2} \frac{1}{\left[1 - \frac{u \cos\theta_2}{c}\right]} \mathbf{a}_{\theta_2} \quad t > l/u + r_2/c \tag{2.8}$$

2.4.3 The static field generated by the accumulation of charge at S1

The charge accumulation at S1 is equal to the integral of the current, and the field component generated by the charges is given by

$$\mathbf{e}_{stat,S_1} = -\frac{\int_{r_1/c}^{t} i(\tau - r_1/c)d\tau}{4\pi\varepsilon_o r_1^2} \mathbf{a}_{r_1} \quad t > r_1/c \tag{2.9}$$

2.4.4 The static field generated by the accumulation of positive charge at S2

The component of the static field generated by the accumulation of positive charge at S2 is given by

$$\mathbf{e}_{stat,S_2} = \frac{\int_{l/u+r_2/c}^{t} i(\tau - l/u - r_2/c)d\tau}{4\pi\varepsilon_o r_2^2} \mathbf{a}_{\mathbf{r}_2} \quad t > l/u + r_2/c \tag{2.10}$$

2.4.5 The velocity field generated as the current pulse propagates along the channel element

The component attributable to the velocity field generated as the current pulse propagates along the channel element can be written directly using (2.5). We first write down the expression for the velocity field generated by a channel element of infinitesimal length $d\xi$ located at a distance ξ from S1. In order to obtain the total velocity field we will integrate this expression from $\xi = 0$ equal to $\xi = l$. The result is

$$\mathbf{e}_{vel} = \int_0^l \frac{i(t - \xi/u - r/c)\left\{1 - \frac{u^2}{c^2}\right\}}{4\pi\varepsilon_o r^2 \left[1 - \frac{u}{c}\cos\theta\right]^2} \left[\frac{\mathbf{a_r}}{u} - \frac{\mathbf{a_z}}{c}\right] d\xi \tag{2.11}$$

Note, though, that in writing down this equation, we have assumed that the current pulse does not vary as it travels along the element l.

2.4.6 Magnetic radiation field generated from S1

The magnetic radiation field generated from S1 is given by

$$\mathbf{b}_{rad,S_1} = \frac{i(t - r_1/c)u \sin\theta_1}{4\pi\varepsilon_o c^3 r_1} \frac{1}{\left[1 - \frac{u\cos\theta_1}{c}\right]} \mathbf{a}_\varphi \quad t > r_1/c \tag{2.12}$$

Note that the magnetic field is in the azimuthal direction.

2.4.7 Magnetic radiation field generated from S2

The magnetic radiation field generated from S2 is given by

$$\mathbf{b}_{rad,S_2} = -\frac{i(t - l/u - r_2/c)u \sin\theta_2}{4\pi\varepsilon_o c^3 r_2} \frac{1}{\left[1 - \frac{u\cos\theta_2}{c}\right]} \mathbf{a}_\varphi \quad t > l/u + r_2/c \tag{2.13}$$

2.4.8 Magnetic velocity field generated as the current pulse propagate along the channel element

The velocity field generated as the current pulse propagate along the channel element is given by

$$\mathbf{b}_{vel} = \int_0^l \frac{i(t - \xi/u - r/c)\left\{1 - \frac{u^2}{c^2}\right\}\sin\theta}{4\pi\varepsilon_o r^2 c^2 \left[1 - \frac{u}{c}\cos\theta\right]^2} \mathbf{a}_\varphi d\xi \tag{2.14}$$

Again, note that, in writing down this equation, it was assumed that the current pulse does not vary as it travels along the straight path from S1 to S2.

The field components given by (2.7)–(2.14) provide a complete description of the electric and magnetic fields generated by the current pulse propagating with uniform velocity and without attenuation.

2.5 Effect of change in current on the radiation field

Consider the case of a current pulse that propagates along a channel with uniform speed u. The current pulse changes its amplitude as it propagates along the channel. Let us represent the current at any height along the channel by $i(z,t)$. Consider a channel element dz located at height z. The geometry is shown in Figure 2.4. Let us represent the change in the current amplitude as the current pulse moves through the channel element by $di(z,t)$. One can show that $di(z,t)$ is given by [4]

$$di(z,t) = \left\{\frac{\partial i(z,t)}{\partial z} + \frac{1}{u}\frac{\partial i(z,t)}{\partial t}\right\}dz \qquad (2.15)$$

The radiation field caused by this change in current amplitude while passing through the element dz is

$$\mathbf{de}_{rad}(t) = \frac{u \sin\theta}{4\pi\varepsilon_o c^2 r}\left\{\frac{\partial i(z,t)}{\partial z} + \frac{1}{u}\frac{\partial i(z,t)}{\partial t}\right\}dz \frac{1}{\left[1 - \frac{u\cos\theta}{c}\right]}\,\mathbf{a}_\theta \qquad (2.16)$$

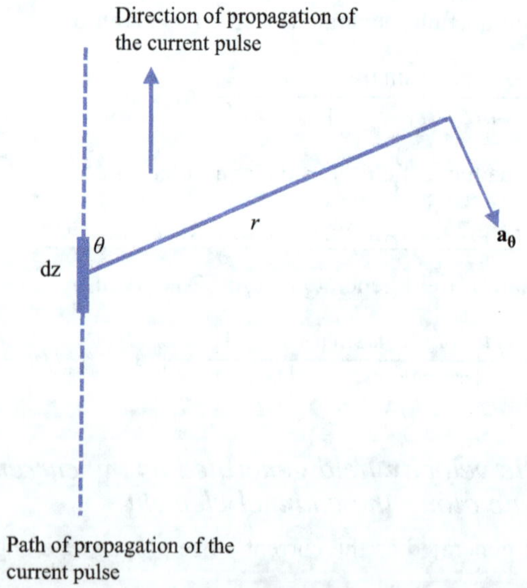

Figure 2.4 The geometry relevant to the derivation of field expressions for the change in current amplitude and return stroke speed

2.6 Effect of change in speed on the radiation field

Consider a current pulse propagating along a channel with variable speed denoted by $u(z)$. The geometry of interest is again given in Figure 2.4. As the current pulse propagates across a channel element of length dz located at height z, the change in the speed of propagation of the current pulse is

$$du = \frac{\partial u(z)}{\partial z} dz \tag{2.17}$$

The radiation field generated due to this change in the speed is

$$\mathbf{de}_{rad}(t) = \frac{i(z,t)\sin\theta\, du(z)}{4\pi\varepsilon_o c^2 r \quad dz} dz \frac{1}{\left[1 - \frac{u(z)\cos\,\theta}{c}\right]} \mathbf{a}_\theta \tag{2.18}$$

2.7 Electromagnetic fields of return strokes simulated by different models

In order to calculate the electromagnetic fields from return strokes, it is necessary to know the spatial and temporal variation of the return stroke current. There are models in the literature that attempt to simulate the return strokes. The most successful of them are the engineering return stroke models. Detailed information concerning these return stroke models are given in Chapter 3 of this volume. Here we will consider different categories of return stroke models and derive the electromagnetic fields pertinent to them using the charge acceleration equations. In the analysis we keep the return stroke channel straight and the return stroke speed constant. The results given below however can be easily generalized to variable return stroke speeds.

2.7.1 Electromagnetic fields of modified transmission line model

In the modified transmission line model (MTL), a current pulse injected at the channel base, say $i_b(t)$, travels with uniform speed u along the channel. As the current propagates upwards it undergoes attenuation (a full description of the model is given in Chapter 3). The attenuation of the current as a function of height is denoted by the function $A(z)$.

The current at any given point along the channel of the MTL model is given by

$$i(z,t) = A(z) i_b(t - z/u) \quad t > z/u \tag{2.19}$$

The electromagnetic fields generated by this model return stroke can be divided into Radiation, Velocity and Static fields. Radiation fields are generated at the initiation of the current at the channel base and due to the attenuation of the current along the channel. Velocity fields are generated by the propagation of the current along the channel and the static fields are generated by the stationary charge resulting from the charge deposition by the return stroke. For the MTL models one obtains

$$\left\{ \frac{\partial i(z,t)}{\partial z} + \frac{1}{u}\frac{\partial i(z,t)}{\partial t} \right\} = \frac{dA(z)}{dz} i_b(t - z/u) \tag{2.20}$$

Furthermore, the amount of charge deposited by the return stroke on the channel element dz is given by

$$Q(z,t) = -\int_{z/u}^{t} i_b(\tau - z/u)\frac{dA(z)}{dz} d\tau \quad t > z/u \tag{2.21}$$

Note that we have inserted a negative sign in the above equation to make the deposited charge positive when the return stroke current amplitude decreases with height. Now, we are in a position to write down the expressions for the components of the electric field at ground level at a horizontal distance D from the strike point. The relevant geometry is shown in Figure 2.5. Observe that we treat electric fields directed along the positive z-axis (i.e., coming out of the ground plane) as positive. The radiation field at the point of observation generated during the initiation of the channel base current is given by

$$\mathbf{e}_{z,ini} = -\frac{i_b(t - D/c)u}{2\pi\varepsilon_o c^2 D} \tag{2.22}$$

Consider a channel element dz located at a height z from ground level. The radiation field at ground level generated due to the attenuation of the current with height is given by [from (2.16) and (2.20)]

$$\mathbf{de}_{z,atte}(t) = -\frac{u \sin^2\theta}{4\pi\varepsilon_o c^2 r}\frac{dA(z)}{dz} i_b(t - z/u - r/c)dz \frac{1}{\left[1 - \frac{u \cos\theta}{c}\right]} \tag{2.23}$$

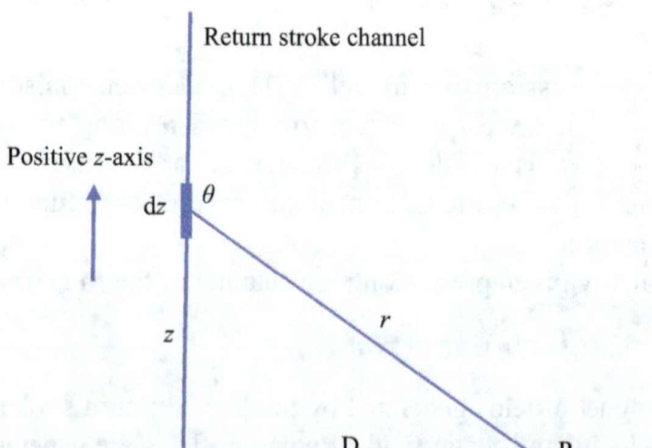

Figure 2.5 Geometry relevant to the derivation of field expressions for the MTL and CD type models

The total radiation field generated by the attenuation of the current is

$$\mathbf{e}_{z,atte}(t) = -\int_0^{l(t)} \frac{u\sin^2\theta}{4\pi\varepsilon_o c^2 r}\frac{dA(z)}{dz} i_b(t - z/u - r/c)\frac{dz}{\left[1 - \frac{u\cos\theta}{c}\right]} \tag{2.24}$$

In the above equation, $l(t)$ is the length of the return stroke channel as seen by an observe located at the point of observation. Observe that we have first obtained the radiation field generated by the change of current when the return stroke propagates along the channel element and then integrated the results to obtain the total radiation field generated by current attenuation. In a similar manner, the velocity field and the static field generated by the return stroke are given by, respectively

$$\mathbf{e}_{z,vel} = \int_0^{l(t)} \frac{i_b(t - \xi/u - r/c)A(z)u\left\{1 - \frac{u^2}{c^2}\right\}}{2\pi\varepsilon_o r^2\left[1 - \frac{u}{c}\cos\theta\right]^2}\left[\frac{\cos\theta}{u} - \frac{1}{c}\right]dl \tag{2.25}$$

$$\mathbf{e}_{z,stat} = \int_0^{l(t)} \frac{Q(z, t - z/u - r/c)\cos\theta}{2\pi\varepsilon_o r^2} dz \tag{2.26}$$

The total electric field generated by the MTL model is then given by

$$E_{z,MTL}(t) = e_{z,ini}(t) + e_{z,atte}(t) + e_{z,vel}(t) + e_{z,stat}(t) \tag{2.27}$$

Observe that, unlike the dipole procedure, the derivative of the current does not appear inside the integrant. This makes it easier to evaluate the electromagnetic fields specially if there is a current discontinuity at the front of the return stroke.

Now, in the case of the transmission line model (see Chapter 3) $A(z) = 1$ and the radiation field generated by the attenuation of the current as given by (2.24) goes to zero. Furthermore, there is no deposition of charge along the channel (i.e. $Q = 0$). In this case, the total electric field reduces to

$$E_{z,TL}(t) = e_{z,ini}(t) + e_{z,vel}(t) \tag{2.28}$$

Observe that in the case of dipole techniques (given in Chapter 1) one has to solve three integrals to obtain the total electromagnetic field of the transmission line model whereas here the results can be obtained by numerically solving one integral. If the speed of propagation of the current pulse in the case of TL model is equal to the speed of light in free space, the total electric field reduces to (with $u = c$)

$$E_{z,TL}(t) = e_{z,ini}(t) \tag{2.29}$$

In this case, the electric field at any point in space can be obtained analytically without resolving to numerical integration. Indeed, this special case can be easily implemented in the calculation of induced overvoltage's in power lines because the electromagnetic field components can be obtained analytically for points located both at and above the ground [7].

2.7.2 Electromagnetic fields of CG type model

CG type return stroke models hypothesize that the return stroke current at any given point of the channel is generated by the corona currents resulting from the neutralization of the corona charge located on the leader channel. According to this model, as the return stroke front travels upwards it will release corona currents from the corona sheath and these currents travel along the core of the return stroke channel to ground with the speed of light. For example, consider a channel element dz located at height z from ground level. As the return stroke current front reaches this channel element it will release a corona current which will travel to ground along the core of the channel with speed of light. The channel base current is created by the cumulative sum of the corona currents generated from all the channel elements located along the channel. In this model, radiation fields are generated during the initiation of corona currents at each height and also when the corona current is absorbed into ground when it reaches the bottom of the channel. The velocity field associated with the model is zero because the corona currents are moving down with the speed of light. During the generation of the corona currents, positive charge will be deposited on the corona sheath and these charges will give rise to a static field. Let us represent the corona current per unit length released by the corona sheath at each height by $i_c(z,t)$ and the channel base current by $i_b(t)$. The charge per unit length deposited by the corona current per unit length as a function to time is given by $Q(z,t) = dz\int_0^t i_c(z,\tau)d\tau$. Let us now write down the expressions for these components of the electromagnetic field. Consider an infinitesimal channel length of dz. The relevant geometry is shown in Figure 2.6. Observe that the definition of the angle is θ different to that of Figure 2.5. The radiation field generated by the initiation of the corona current in this channel

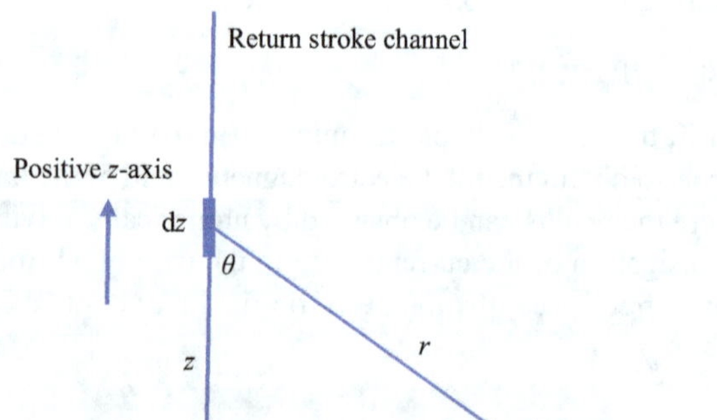

Figure 2.6 Geometry relevant to the derivation of field expressions for the MTL and CG type models. Note that the location of the angle θ is different to that of Figure 2.5

element is

$$de_{z,cor-ini} = -\frac{i_c(z,t-r/c)\sin^2\theta}{2\pi\varepsilon_o cr(1-\cos\theta)}dz \tag{2.30}$$

The total radiation field generated by the initiation of corona currents along the channel is given by

$$E_{z,cor-ini} = -\int_0^{l(t)} \frac{i_c(z,t-r/c)\sin^2\theta}{2\pi\varepsilon_o cr(1-\cos\theta)}dz \tag{2.31}$$

The static field generated by the charge deposited on the channel element is given by

$$de_{z,stat} = -\frac{Q(z,t-r/c)\cos\theta dz}{4\pi\varepsilon_0 r^2} \tag{2.32}$$

Thus, the total static field generated by the return stroke is

$$E_{z,stat} = -\int_0^{l(t)} \frac{Q(z,t-r/c)\cos\theta}{4\pi\varepsilon_o r^2}dz \tag{2.33}$$

Finally, we have to consider the radiation field generated by the absorption or termination of the corona currents at ground level. However, using the fact that the channel base current is generated by the cumulative effect of the corona currents reaching the ground level, the radiation field generated during the termination of the corona current can be written directly as

$$E_{z,cor-term} = -\frac{I_b(t-D/c)}{2\pi\varepsilon_o cD} \tag{2.34}$$

Thus, the total electric field produced by the CG type model is then given by

$$E(t) = E_{z,cor-ini}(t) + E_{z,cor-term}(t) + E_{z,stat}(t) \tag{2.35}$$

2.7.3 CD type models

The CD models can be considered as a form of time reversed CG model. In this model type, a current is injected at the channel base and this current travels upwards carrying positive charge along the leader channel with the speed of light. This current is neutralized by corona currents carrying negative charge that propagates upwards also with the same speed. In this respect all the current pulses are propagating upwards with the speed of light. However, the interaction of the upward moving corona currents and the injected currents interact in such a way that the corona currents eat into the front of the injected current thus making the net current in the return stroke channel to propagate upwards with a speed less than the speed of light. This is the speed of the return stroke front. Let us denote this speed by u. The corona currents are generated by any channel element when the return stroke front reaches that current element. Let us denote the injected current at the channel base by $i_b(t)$ and the corona current generated per unit length by a channel element located at height z by $i_c(z,t)$. The relevant geometry is given in Figure 2.5. In this model, radiation fields are generated during the injection of the current at the channel base and by the initiation of corona currents at each height. As in the CG type models, the velocity field associated with the model is zero because the corona currents are moving upwards with the speed of light. During the generation of the corona currents, positive charge will be deposited on the corona sheath and these charges will give rise to a static field. The charge per unit length deposited by the corona current per unit length as a function to time is given by $Q(t) = dz \int_0^t i_c(z,\tau)d\tau$. Let us now write down the expressions for these components of the electromagnetic field. The radiation field generated during the injection of the channel base current into the channel is given by

$$E_{z,injected} = -\frac{I_b(t - D/c)}{2\pi\varepsilon_o cD} \tag{2.36}$$

Consider an infinitesimal channel length of dz located at height z. The initiation of corona current at this channel element gives rise to a radiation field given by

$$de_{z,cor-ini} = \frac{i_c(z, t - r/c)\sin^2\theta}{2\pi\varepsilon_o cr(1 - \cos\theta)}dz \tag{2.37}$$

The total radiation field generated by the initiation of corona currents along the whole channel is then given by

$$E_{z,cor-ini} = \int_0^{l(t)} \frac{i_c(z, t - r/c)\sin^2\theta}{2\pi\varepsilon_o cr(1 - \cos\theta)}dz \tag{2.38}$$

The static field generated by the static charges deposited along the channel is given by

$$E_{z,stat} = \int_0^{L(t)} \frac{Q(z, t - r/c)\cos\theta}{4\pi\varepsilon_o r^2}dz \tag{2.39}$$

Finally the total electric field generated by the return stroke simulated by a CD type model is given by

$$E(t) = E_{z,cor-ini}(t) + E_{z,injected}(t) + E_{z,stat}(t) \tag{2.40}$$

Observe that the electric field generated by the CD model can be obtained directly from the field equations of the CG model by changing the speed of propagation of the corona currents from c to $-c$, or vice versa (see also Chapter 3 of this volume).

2.8 Concluding remarks

In this chapter, we have described how to calculate the electromagnetic fields generated by return strokes simulated by different types of engineering return stroke models, namely CP, CG, and CD using accelerating charge equations. The same total fields of course can also be obtained using dipole technique. Acceleration charge technique has several advantages over the dipole technique. First, note that there is no need to perform an integration of the current derivative over the length of the channel. Second, in the dipole technique in order to obtain the charge one has to integrate the total current at any given level which in turn has to be obtained numerically by using model parameters. However, in the charge acceleration equation technique, the charge necessary for the static field in most cases can be obtained analytically and thus one can skip additional numerical integrations in mathematical routines. Thus, there are many cases where the charge acceleration equations provide results that can be realized with less computations. Furthermore, the charge acceleration technique provides a direct connection between the field components and the physical process that leads to the electric field.

References

[1] Cooray, V. and Cooray, G., The electromagnetic fields of an accelerating charge: Applications in lightning return stroke models, *Trans. IEEE (EMC)*, 2010, *52*, 944–955.

[2] Cooray, V. and Cooray, G., Electromagnetic fields of accelerating charges: Applications in lightning protection, *Electric Power Systems Research*, 2017, *145*, 234–247.

[3] Cooray, V., Cooray, G., Rubinstein, M., and Rachidi, F., On the apparent non-uniqueness of the electromagnetic field components of return strokes revisited, *Atmosphere,* 2021, *12*, 1319. https://doi.org/10.3390/atmos12101319

[4] Cooray, V., Cooray, G., Rubinstein, M., and Rachidi, F., Generalized electric field equations of a time-varying current distribution based on the electromagnetic fields of moving and accelerating charges, *Atmosphere,* 2019, *10*, 367. https://doi.org/10.3390/atmos10070367

[5] Kereszy, I., Rakov, V., Czumbil, L., *et al.*, Energetic radiation from subsequent-stroke leaders: The role of reduced air density in decayed lightning channels, *Applied Science,* 2022, *12*, 7520. https://doi.org/10.3390/app12157520

[6] Pannofsky, W. K. H. and Phillips, M., *Classical Electricity and Magnetism*, Addison-Wesley, Reading, MA, 1962.

[7] Cooray, V., Rubinstein, M., and Rachidi, F., Field-to-transmission line coupling models with special attention to Cooray–Rubinstein approximation, *IEEE Transactions (EMC)*, 2020, *63*(2), 484–493.

Chapter 3
Basic features of engineering return stroke models

Vernon Cooray[1]

3.1 Introduction

From the point of view of an electrical engineer, the return stroke is the most important event in a lightning flash; it is the return stroke that causes most of the destruction and disturbance in electrical and telecommunication networks. In their attempts to provide protection, engineers seek the aid of return stroke models for three reasons: first, they would like to characterize and quantify the electromagnetic fields produced by return strokes at various distances to provide them with the input for mathematical routines that analyze the transient voltages and currents induced in electrical networks by these fields. This calls for return stroke models that are capable of generating electromagnetic fields similar to those created by natural return strokes. Second, their profession demands detailed knowledge of the effects of direct injection of lightning current. In a real situation, this direct injection will be superimposed on currents and voltages induced by electromagnetic fields in the system under consideration. This necessitates the use of return stroke models that are capable of generating channel base currents similar to those in nature. Finally, in order to evaluate the level of threat posed by lightning, engineers require statistical distributions of peak currents and peak current derivatives in lightning flashes. Even though the characteristics of return stroke currents can be obtained through measurements at towers equipped with current measuring devices or by utilizing rocket triggering techniques, gathering statistically significant data samples in different regions and under different weather conditions is an exceptionally difficult enterprise. Accurate return stroke models can simplify this task to a large extent by providing the connection between the electromagnetic fields and the currents so that the latter can be extracted from the measured electromagnetic fields.

In the case of return strokes, a model is a mathematical formulate that is capable of predicting the temporal and spatial variation of the return stroke current, the variation of return stroke speed, the temporal and spatial characteristics of optical

[1]Department of Electrical Engineering, Uppsala University, Sweden

radiation, the features of electromagnetic fields at different distances, and the signature of thunder. From the point of view of an engineer, the lightning parameters of particular interest are the return stroke current and its electromagnetic fields, whence most of the return stroke models available today, especially the engineering models, are constructed to predict either one or both of these features. On the basis of the concepts and aims of return stroke models, they can be separated into four main groups, namely, (1) the electro-thermodynamic models, (2) the transmission line or LCR models, (3) antenna models, and (4) engineering models [1,2]. Here we concentrate on the engineering models.

Even though the name "engineering models" may give the impression that these models have down played the physics completely, all these models are build using a valid physical concept as a foundation. However, the physics is somewhat neglected when selecting the input parameters necessary to build up the model. In many of these models, input parameters are not evaluated using fundamental principles. Some of these parameters are obtained from available experimental data and the other parameters are selected, sometimes without any physical foundation, in such a way that the model predictions agree with experimental data. One problem with this procedure is the lack of uniqueness in the way in which the input parameters could be combined to generate the required result. The basic concept on which all the engineering models are built on, whether it is stated as such or not, can be traced back to the physics of transmission lines. Using the basic physics of transmission line theory, three types of engineering models can be constructed. These are the Current Propagation Models, Current Generation Models, and Current Dissipation Models. At a later stage in this chapter, we will show how these different types of return stroke models are related to each other.

In this chapter, we will describe the basic features of the three types of engineering return stroke models and how to obtain the temporal and spatial variation of the current in these models that is necessary in the calculation of electromagnetic fields.

Some parts of this chapter were adapted from the book chapter titled "Return stroke models with special attention to engineering applications" that appeared in The lightning Flash, Second Edition, Edited by V. Cooray and published by IET in 2014.

3.2 Current propagation models (CP models)

3.2.1 Basic concept

Consider an uniform and lossless transmission line. A current pulse injected into this line will propagate along the line with uniform speed without any change in the amplitude of the waveshape. The transmission line does not interfere with the current (of course this is not true in the case of a transmission line going into corona). It will only provide a path for the propagation of the current pulse from one location to another. This is the basis of the current propagation models. In these models it assumed that the return stroke is a current pulse originating at ground level and propagating from ground to cloud along the leader channel. As the current propagates upwards, its amplitude could decrease and the waveshape of the current could get modified due to dispersion. Furthermore, the speed of the return stroke could also change as it propagates upwards. The transmission line model introduced by Uman and McLain [3] assumes that the current pulse propagates without attenuation (it is worth mentioning here that this model is a special case of the more general model introduced previously by Dennis and Pierce [4]). In the model introduced by Nucci *et al.* [5], the current amplitude decreases linearly with height and in the model introduced by Rakov and Dulzon [6] the current amplitude decreases linearly with height. Cooray and Orville [5] and Cooray *et al.* [7] introduced return stroke models where the current not only attenuates but also disperses as it propagates along the leader channel. These modification to the original transmission line model are called modified transmission line models or MTL models. In the next section, an expression for the return stroke current pertinent to the most general representation of CP-type models is presented.

3.2.2 Most general description

Consider a CP model where both the current amplitude and the current wave shape change with height. Moreover, the return stroke speed is also be allowed to change as the return stroke propagates upwards. The current at any given height z and at time t of such a model can be represented by the general expression

$$I(z,t) = A(z)F(z, t - z/v_{av}(z)) \quad t > z/v_{av}(z) \tag{3.1}$$

In the above equation, $A(z)$ is the function that describes the attenuation of the amplitude of the return stroke current and $v_{av}(z)$ is the average return stroke speed over the channel section that spans from ground to height z. That is

$$v_{av}(z) = z \Big/ \int_0^z \frac{d\xi}{v(\xi)} \tag{3.2}$$

Note that $v(z)$ is the return stroke speed at height z. In (3.1), the function $F(z,t)$ describes the wave shape of the current at height z. One can define the function $F(z,t)$ as follows:

$$F(z,t) = \int_0^t I_b(\tau)\Phi(z, t-\tau)d\tau \tag{3.3}$$

where $I_b(t)$ is the channel base current and $\Phi(z,t)$ is a function that describes how a Dirac Delta function injected from the channel base is being modified due to dispersion with height. However, this operation itself leads to the deposition of charge along the channel and this could be avoided by normalizing this function to unity. That is

$$\int_{-\infty}^{+\infty} \Phi(z,t)dt = 1 \tag{3.4}$$

Equation (3.1), together with (3.2)–(3.4), can be used to represent any CP model that incorporate both current attenuation and dispersion. Observe that in the case of the MTL models introduced by Nucci *et al.* [5] and Rakov and Dulzon [6], $\Phi(z,t) = \delta(t)$ and in the case of the transmission line model, in addition to the above condition $A(z) = 1$.

3.3 Current generation models (CG models)

3.3.1 Basic concept

If a transmission line goes into corona, the corona currents released at each line element will give rise to currents propagating along the line and an observer will be able to measure a current appearing at the base of the line. A similar scenario is used in the current generation models to describe the creation of the return stroke current. In these models, the leader channel is treated as a charged transmission line and the return stroke current is generated by a wave of ground potential that travels along it from ground to cloud. The arrival of the wave front (i.e. return stroke front) at a given point on the leader channel changes its potential from cloud potential to ground potential causing the release of bound charge on the central core and on the corona sheath giving rise to the current in the channel (this is called the corona current in the literature). These models postulate that as the return stroke front propagates upwards the charge stored on the leader channel collapses into the highly conducting core of the return stroke channel. Accordingly, each point on the leader channel can be treated as a current source which is turned on by the arrival of the return stroke front at that point. The corona current injected by these sources into the highly conducting return stroke channel core travels to ground with a speed denoted by v_c. As we will see later, in most of the return stroke models it is assumed that $v_c = c$ where c is the speed of light.

The basic concept of CG models was first introduced by Wagner [8]. He assumed that the neutralization of the corona sheath takes a finite time and therefore the corona current can be represented by a decaying exponential function. The decay time constant associated with this function is called the *corona decay time constant*. Wagner assumed however that the speed of propagation of the corona current down the return stroke channel is infinite. Lin *et al.* [9] introduced a model in which both CG and CP concepts are incorporated in the same model. In the portion of the current described by CG concept, the corona current is represented by a double exponential function. The speed of propagation of corona current down the channel is assumed to be the same as the speed of light. A modified form of this model is introduced by Master *et al.* [10] but in this modification the CG description remained intact. Heidler [11] constructed a model based on this principle in which the channel base current and the return stroke speed are assumed as input parameters. Furthermore, it was assumed that the neutralization of the corona sheath is instantaneous and hence the corona current generated by a given channel section can be represented by a Dirac Delta function. The speed of propagation of the corona current down the return stroke channel is assumed to be equal to the speed of light. This model gives rise to a current discontinuity at the return stroke front which, according to the author's understanding is not physically reasonable. Hubert [12] constructed a current generation model rather similar to that of the Wagner's model with the exception that the downward speed of propagation of the corona current is equal to the speed of light. He utilized this model to reproduce experimental data (both current and electromagnetic fields) obtained from triggered lightning. Cooray [13,14] introduced a model in which the distribution of the charge deposited by the return stroke (i.e. sum of the positive charge necessary to neutralize the negative charge on the leader and the positive charge induced on the channel due to the action of the background electric field) and the decay time constant of the corona current are taken as input parameters with the model predicting the channel base current and return stroke speed. Moreover, he took into consideration that the process of neutralization of the corona sheath takes a finite time in reality and, as a consequence, the corona current was represented by an exponential function with a finite duration. This is the first model in which the decay time constant of the corona current (and hence the duration of the corona current) is assumed to increase with height. Since the leader channel contains a hot core surrounded by a corona sheath, he also divided the corona current into two parts, one fast and the other slow. The fast one was associated with the neutralization of the core and the slow one with the neutralization of the corona sheath. Furthermore, by treating the dart leader as an arc and assuming that the electric field at the return stroke front is equal to the electric field that exists in this arc channel, he manages to derive the speed of the return stroke. Diendorfer and Uman [15] introduced a model in which the channel base current, return stroke speed and the corona decay time constant were assumed as input parameters. They also divided the corona current into two parts one fast and the other slow. Thottappillil *et al.* [16] and Thottappillil and Uman [17] modified this model to include variable return stroke speed and a corona decay time constant that varies with height. Cooray [18] developed the ideas introduced in [13,14] to create a CG model with channel base current as an input. Cooray [19] and Cooray *et al.* [20] extended the concept to include first return strokes with connecting leaders.

In CG models, one has the choice of selecting the channel base current, $I_b(t)$, the distribution of the charge deposited by the return stroke along the channel, $\rho(z)$, the return stroke speed, $v(z)$, and the magnitude and variation of the corona discharge time constant with height, $\tau(z)$ as input parameters. Any set of three of these four input parameters will provide a complete description of the temporal and spatial variation of the return stroke current. Most of the CG models use $v(z)$ and either the $\rho(z)$ or $\tau(z)$ in combination with $I_b(t)$ as input parameters. Recently, Cooray and Rakov [21] developed a model in which $\rho(z)$, $\tau(z)$, and $I_b(t)$ are selected as input parameters. The model could generate $v(z)$ as a model output.

3.3.2 Expression for the current at any height

We assume that the return stroke channel is straight and vertical. Our goal is to find an expression for the total current in the channel at a height z. Here we consider the case where the return stroke speed is uniform (denoted by v) but the equations to be presented below can be easily changed for non-uniform speed. Consider a channel element $d\xi$ located at height ξ which is located above the level z. Let us represent the corona current per unit length generated by the channel element by $i_c(\xi, t)$. The contribution of corona current from this channel element to the total current at height z is given by

$$dI(z,t) = i_c(\xi, t - \xi/v - (\xi - z)/v_c)d\xi \tag{3.5}$$

Thus, the total current at level z due to all the channel elements located above that height is

$$I(z,t) = \int_z^{z_{cg}} i_c(\xi, t - \xi/v - (\xi - z)/v_c)d\xi \tag{3.6}$$

In the above equation, z_{cg} is the maximum height where the corona currents could contribute to the current at level z. This height is given by

$$\frac{z_{cg}}{v} + \frac{(z_{cg} - z)}{v_c} = t \tag{3.7}$$

That is

$$z_{cg} = \left(t + \frac{z}{v_c}\right) / \left(\frac{1}{v} + \frac{1}{v_c}\right) \tag{3.8}$$

The current at the channel base, $I_b(t)$, can be obtained from the above equations by making $z = 0$. That is

$$I_b(t) = \int_0^{z_{cg}} i_c(\xi, t - \xi/v - \xi/v_c)d\xi \tag{3.9}$$

with

$$z_{cg} = t / \left(\frac{1}{v} + \frac{1}{v_c}\right) \tag{3.10}$$

These equations define the spatial and temporal variation of the current in a CG model which is needed in calculating electromagnetic fields from a CG type model.

3.4 Current dissipation models (CD models)

3.4.1 General description

As mentioned previously, if a current pulse is propagating without corona along a transmission line, it will travel along the line without any attenuation and modification of the current waveshape. This concept is used as a base in creating current propagation models. When the current amplitude is larger than the threshold current necessary for corona generation, each element of the transmission line acts as a corona current source. Half of the corona current generated by the sources travels downwards and the other half travels upwards. The upward moving corona currents interact with the front of the injected current pulse in such a way that the speed of the upward moving current pulse is reduced, and for a transmission line in air, to a value less than the speed of light [22]. In a recent publication Cooray [23] showed that the upward moving corona current concept can also be used to create return stroke models. He coined the term "Current Dissipation Models" for the same. The basic features of the current dissipation models are depicted in Figure 3.1. The main assumptions of the current dissipation models are the following: The return stroke is initiated by a current pulse injected into the leader channel from the grounded end. The arrival of the return stroke front at a given channel element will turn on a current source that will inject a corona current into the central core. It is important to stress here that by the statement *the arrival of the return stroke front at a given channel element* it is meant the onset of the return stroke current in that channel element (i.e. point B in Figure 3.1). Once in the core

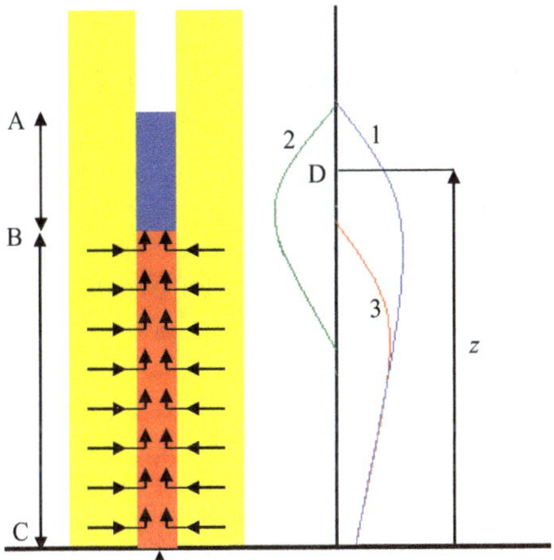

Figure 3.1 Pictorial description of the processes associated with a current dissipation model at a given time t. The injected current (waveform 1) and the sum of corona currents (waveform 2) travel upwards with speed v_c*. Point A is the front of these current waveforms. In the region* A–B*, these two currents cancel each other making the current above point* B *equal to zero. The cancellation is not complete below point* B *and therefore the net current below point* B *is finite (waveform 3). Thus point* B *is the front of the net current (i.e. return stroke front) moving upwards. Distance* AC *is equal to* $v_c t$ *and the distance* BC *is* vt *where* v *is the speed of propagation of the net current front (i.e. return stroke front). Note that the current waveforms are not drawn to scale. Adapted from [23]*

this corona current will travel upward along the channel. In the case of negative return strokes, the polarity of the corona current is such that it will deposit positive charge on the corona sheath and transport negative charge along the central core. Let us now incorporate mathematics into this physical scenario.

Assume that the return stroke is initiated by a current pulse injected into the leader channel at ground level. This current pulse propagates upward along the channel with speed v_c. When the return stroke front (i.e. the net current front) reaches a given channel element a corona source is turned on. This source will generate a corona current that will travel upward along the central core with the same speed as the current pulse injected at the channel base (i.e. v_c). Note that the polarity of the upward moving corona current is opposite to that of the upward moving current injected at the channel base. For example, in the case of negative return stroke, the current injected at the channel base carry positive charge upward whereas the corona current transports negative charge upward. According to this model, the total current at a given point of the channel consists of two parts – upward moving current pulse injected at the channel base and the total contribution of the upward moving corona currents. The upward moving corona current being of opposite polarity leads to the dissipation of the current pulse injected at the channel base.

3.4.2 Expression for the current at any height

Consider the diagram to the right in Figure 3.1. This depicts a situation at any given time t. At this time, the tip of the injected current is located at point A and the return stroke front is located at point B. As in the case of a current pulse propagating along a transmission line under corona, at points above current front the current is zero. Of course in the case of transmission line the current above the point B is clamped below the corona threshold. However, as pointed out by Cooray [23], since the corona in the return stroke process is caused by a neutralization process, the threshold current necessary for corona is close to zero. The next task for us is to derive an expression for the total current at any given level z along the channel. The description presented hear is much easier to visualize than the one presented in [23]. But, the mathematical treatment remains the same. Consider a channel element $d\xi$ locate at a height ξ from ground level. This channel element is located below the reference level z. The corona current per unit length generated by this element is denoted by $-i_c(\xi, t)$. Note that the corona current is defines as negative here because it transports negative charge towards the cloud whereas the injected current transports positive charge upwards. The contribution of this channel element to the corona current at level z is given by

$$dI_c(z,t) = -i_c(\xi, t - \xi/v - (z - \xi)/v_c)d\xi \tag{3.11}$$

Thus the total corona current at level z is given by

$$I_c(z,t) = -\int_0^z i_c(\xi, t - \xi/v - (z - \xi)/v_c)d\xi \tag{3.12}$$

Consider the diagram shown in Figure 3.2. Let us denote the injected current at level z at time t by $I_i(z,t)$. At the same time, the front of the return stroke is located

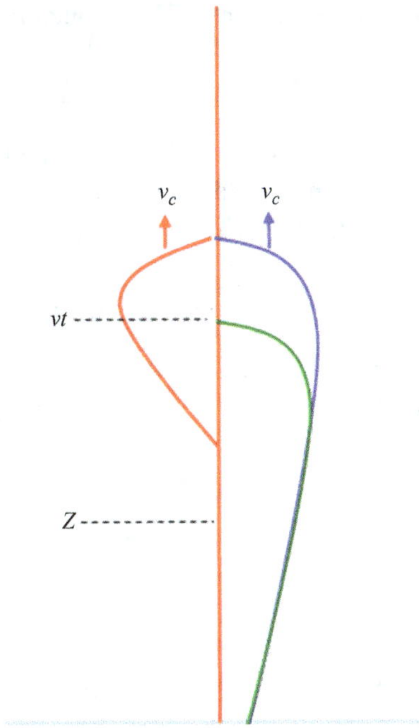

Figure 3.2 The physical scenario that is useful in deriving the total current at any height in the return stroke channel in the CD type models

at a height vt. Now, when the injected current at level z reaches the front of the return stroke, let the front of the return stroke be at a height z_{cd}. Then we can write

$$\frac{z_{cd} - vt}{v} = \frac{z_{cd} - z}{v_c} \tag{3.13}$$

That is

$$z_{cd} = \frac{t - z/v_c}{\left(\frac{1}{v} - \frac{1}{v_c}\right)} \tag{3.14}$$

Now the corona current at the return stroke front when the return stroke front is at a height z_{cd} is given by

$$I_c\left(z_{cd}, \frac{z_{cd}}{v}\right) = -\int_0^{z_{cd}} i_c(\xi, \frac{z_{cd}}{v} - \xi/v - (z_{cd} - \xi)/v_c)d\xi \tag{3.15}$$

Since at the return stroke front the injected current amplitude is completely neutralized by the corona current at the return stroke front, we can write

$$I_i(z,t) = \int_0^{z_{cd}} i_c(\xi, \frac{z_{cd}}{v} - \xi/v - (z_{cd} - \xi)/v_c)d\xi \tag{3.16}$$

After substituting to z_{cd} inside the integral, we obtain

$$I_i(z,t) = \int_0^{z_{cd}} i_c(\xi, t - z/v_c - \xi/v + \xi/v_c)d\xi \tag{3.17}$$

Now we are ready to write down the expression for the complete current at level z. Observe that the total current at this level at time t consists of the sum of the injected current and the corona current. Thus

$$I(z,t) = I_i(z,t) - \int_0^{z} i_c(\xi, t - \xi/v - (z-\xi)/v_c)d\xi \tag{3.18}$$

Note that the second term on the right hand side is the corona current at level z at time t. This can be written as

$$I(z,t) = I_i(z,t) - \int_0^{z_{cd}} i_c(\xi, t - \xi/v - (z-\xi)/v_c)d\xi$$
$$+ \int_z^{z_{cd}} i_c(\xi, t - \xi/v - (z-\xi)/v_c)d\xi \tag{3.19}$$

Now appealing to (3.17) the first two terms on the right hand side cancels off each other leaving with us the final expression for the current at level z at time t, namely

$$I(z,t) = \int_z^{z_{cd}} i_c(\xi, t - \xi/v - (z-\xi)/v_c)d\xi \tag{3.20}$$

Observe from (3.18), the current at the channel base is given by the injected current i.e.

$$I(z,t) = I_i(0,t) \tag{3.21}$$

3.5 Comparison of CG and CD

In the CG type model, the total current at any level along the channel is given by

$$I(z,t) = \int_z^{z_{cg}} i_c(\xi, t - \xi/v - (\xi - z)/v_c)d\xi \tag{3.22}$$

With

$$z_{cg} = \left(t + \frac{z}{v_c}\right) / \left(\frac{1}{v} + \frac{1}{v_c}\right) \tag{3.23}$$

In the CD type model, the total current at any level along the channel is given by

$$I(z,t) = \int_z^{z_{cd}} i_c(\xi, t - \xi/v + (\xi - z)/v_c)d\xi \tag{3.24}$$

With

$$z_{cd} = \left(t + \frac{z}{v_c}\right) / \left(\frac{1}{v} - \frac{1}{v_c}\right) \tag{3.25}$$

Note that the only change that one has to make to change the current of a CG model to a CD model is to change the speed of propagation of the corona currents and the injected current from v_c to $-v_c$. The same change in a CG model will give rise to the corresponding current in a CD model. In this respect, CD type models can be regarded as a time reversal version of the CG type model. As we have seen in Chapter 2, an identical change will convert the field equation of CG models to CD or vice versa. Observe that in all CG and CD type models it is assumed that v_c is equal to the speed of light.

3.5.1 Generalization of any model to current generation type

Cooray [24] showed that any return stroke model can be converted to a current generation model by introducing an effective corona current. Here we will illustrate the mathematical analysis that led to that conclusion.

Consider a channel element of length dz at height z and let $I(z,t)$ represent the temporal variation of the total return stroke current at that height. In the case of CG models, this current is generated by the action of corona current sources located above this height. Assume for the moment that the channel element does not generate any corona current. In this case, the channel element will behave as a passive element that will just transport the current that is being fed from the top. Thus, one can write

$$I(z + dz, t) = I(z, t + dz/v_c) \tag{3.26}$$

That is, the current injected at the top of the element will appear without any change at the bottom of the channel element after a time dz/v_c which is the time taken by the current to travel from the top of the channel element to the bottom.

Now let us consider the real situation in which the channel element dz will also generate a corona current. As the current injected at the top passes through the channel element the corona sources will add their contribution resulting in a larger current appearing at the bottom than the amount injected at the top. The difference in these two quantities will give the corona current injected by the channel element. Thus the average corona current generated by the element dz is given by

$$I_c(z, t)dz = I(z, t + dz/v_c) - I(z + dz, t) \tag{3.27}$$

Using Taylor's expansion, the above equation can be rewritten as

$$I_c(z, t)dz = I(z, t) - I(z + dz, t) + \frac{dz}{v_c}\frac{\partial I(z, t)}{\partial t} \tag{3.28}$$

Dividing both sides by dz and taking the limit $dz \to 0$, the corona current per unit length, $I_c(z,t)$, injected into the return stroke channel at height z is given by

$$I_c(z,t) = -\frac{\partial I(z,t)}{\partial z} + \frac{1}{v_c}\frac{\partial I(z,t)}{\partial t} \tag{3.29}$$

This equation can be utilized to transfer any return stroke model to a current generation model with an equivalent corona current. It is important to stress here that even though the distribution of the return stroke current as a function of height remains the same during this conversion, there is a radical change in the corona current. Now, in reality the corona current at a given height is dictated by the physics of the charge neutralization process in the corona sheath at that height. A change in the corona current therefore requires a change in the physics of the neutralization process. This can easily be illustrated using the Transmission Line Model [3]. In the current propagation scenario of the TL model, the upward propagating current will not give rise to any corona and therefore the corona current is zero. On the other hand, if the same model is converted to a current generation model then the equivalent corona current associated with the converted model (obtained from 3.29) becomes bipolar [24]. The physics of the neutralization process pertinent to this equivalent corona current is the following: As the rising part of the upward moving current passes through a given channel element the corona sheath located around that channel element will be neutralized by injection of positive charge into it. During the decaying part of the upward moving current, all the deposited positive charge will be removed bringing the corona sheath back to its original state. Thus the physics of corona dynamics in the two scenarios is completely different even though the longitudinal distribution of current and the charge along the channel at any given time is the same in the two formulations. This shows that conversion of a model from one type to another will change the underlying physics even though both descriptions are identical from a point of view of the total current as a function of height. Thus one has to apply caution in deriving the physics of corona neutralization process using these models because the information extracted concerning it will be model dependent.

3.6 Generalization of any model to a current dissipation type model

An analysis similar to the one presented in Section 3.5 was conducted by Cooray [23] for current dissipation type models. That analysis is presented below.

Consider a channel element of length dz at height z and let $I(z,t)$ represent the temporal variation of the total return stroke current at that height. Assume for the moment that the channel element does not generate any corona current. In this case the channel element will behave as a passive element that will just transport the current that is being fed from the bottom. In this case one can write

$$I(z+dz,t) = I(z,t-dz/v_c) \tag{3.30}$$

That is, the current injected at the bottom of the channel element will appear without any change at the top of the channel element after a time dz/v_c which is the time taken by the current to travel from the bottom of the channel element to the top.

Now let us consider the real situation in which the channel element dz will also generate a corona current. As the current injected at the bottom passes through the channel element the corona sources will add their contribution, and since the polarity of the corona current is opposite to that of the injected current, resulting in a smaller current appearing at the top than the amount of current injected at the bottom. The difference in these two quantities will give the corona current injected by the channel element. Thus the average corona current generated by the element dz is given by

$$I_c(z,t)dz = I(z, t - dz/v_c) - I(z + dz, t) \tag{3.31}$$

Using Taylor's expansion, the above equation can be rewritten as

$$I_c(z,t)dz = I(z,t) - I(z + dz, t) - \frac{dz}{v_c}\frac{\partial I(z,t)}{\partial t} \tag{3.32}$$

Dividing both sides by dz and taking the limit $dz \rightarrow 0$, the corona current per unit length, $I_c(z,t)$, injected into the return stroke channel at height z is given by

$$I_c(z,t) = -\frac{\partial I(z,t)}{\partial z} - \frac{1}{v_c}\frac{\partial I(z,t)}{\partial t} \tag{3.33}$$

Note that this equation is completely symmetrical to the one derived for the current generation model (i.e. (3.25)) except that second term has a negative sign. This equation can be utilized to transfer any return stroke model to a current dissipation model with an equivalent corona current. The discussion given at the end of Section 3.5 is also applicable here.

As we have pointed out previously, the input parameters of the CG or CD models can be any three of the following: channel base current, return stroke speed, parameter that defines the temporal variation of the corona current and the charge deposited by the return stroke along the leader channel. Observe that the complete definition of the corona current requires a parameter that defines its temporal variation and the charge associated with the corona current. Since the total charge associated with corona current per unit length at any height is the same as the charge deposited per unit length by the return stroke at that height, once the temporal variation of the corona current and the charge deposited by the corona current is given, the complete knowledge of the spatial and temporal variation of the corona current is defined. It is standard practice to define the temporal variation of the corona current as an exponential function i.e. its temporal variation is defined by the decay time constant of the exponential variation. However, in general, this need not be the case. Moreover, in CG and CD models, the speed of propagation of the corona current (in CG) or the speed of propagation of the corona current and the injected current (in CD) are usually assumed to be the speed of light in free space, however, in general one is free to select any other value for this speed. Now, once three of the parameters of the CG or CD models are specified the fourth parameter can be numerically estimated from the equations pertinent to the current at any level for these models. How, this is done is described in Refs. [16,17,25].

3.7 Current dissipation models and the modified transmission line models

In the case of CP type models the deposition of charge by the return stroke current along the return stroke channel can be described using a radial corona current that flows into the corona sheath from the central core. In these models, corona current does not propagate along the return stroke channel as in the CG or CD type models. However, consider the situations depicted in Figure 3.3. The left panel illustrates what happens when the return stroke current reaches a channel element as described by CP type models. As the current passes through the channel element, corona currents flow radially depositing positive charge in the corona sheath. Due to this removal of charge, the amplitude of the return stroke will be decreased by an amount depending on the magnitude of the radial corona current. Now, the same scenario can be describe as depicted in the right panel. In this scenario, the injected current passes through the channel element without any attenuation. But, it will trigger a negative corona current having an identical shape to the radial corona current in the left panel and it will propagate upwards with the same speed as the injected current. The net result is that the current coming out of the element will be reduced by exactly the same amount as depicted in the left panel. The two scenarios give identical results while the left panel describing the CP type model whereas the right panel describing a CD type model. This illustrates pictorially the fact that the basic concepts behind the CP and the CD models are similar.

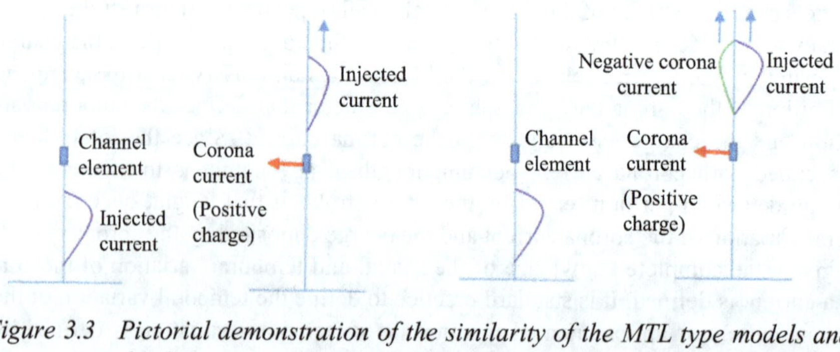

Figure 3.3 Pictorial demonstration of the similarity of the MTL type models and the CD type models. The left-hand pannel illustrate what happens when the injected current passes through a channel element in the MTL type models and the right-hand pannel depicts the events taking place when the same process is described by the CD type model

One difference between the CD and the CP description as depicted in the right-hand panel of Figure 3.3 is the following. In the CD, the corona currents and the injected current propagate upwards with speed v_c while the return stroke front (the net current front) propagates with another speed denoted by v. Whereas in the CP model, all the current components propagate up with the same return stroke speed v. Thus, if one selects the input parameters in such a way that it makes the speed v_c equal to v, then the CD type models will reduce to a very general CP type model. Here by "very general," we mean a CP model where both the amplitude and the waveshape of the return stroke current varies along the channel. The spatial and temporal variation of the current is of course decided by the input parameters. For such a very general CP model, the corona current per unit length at any given height z is given by (3.33) with $v_c = v$. That is

$$I_c(z,t) = -\frac{\partial I(z,t)}{\partial z} - \frac{1}{v}\frac{\partial I(z,t)}{\partial t} \tag{3.34}$$

Now, we consider a special case of these CP models. Assume that the corona current is selected in such a way that

$$I_c(z,t) = -I_i(0, t - z/v)\frac{dA(z)}{dz} \tag{3.35}$$

where $A(z)$ is some function of z and $I_b(t)$ is the channel base current. The negative sign is used in the equation to ascertain that a negative value of $dA(z)/dz$ leads to a deposition of positive charge in the corona sheath. According to this equation the corona current at a given height is proportional to the injected current at that height. Substituting this expression in (3.34) one finds that

$$-I_i(0, t - z/v)\frac{dA(z)}{dz} = -\frac{\partial I(z,t)}{\partial z} - \frac{1}{v}\frac{\partial I(z,t)}{\partial t} \tag{3.36}$$

One can easily show by substitution that the solution of this equation is given by

$$I(z,t) = A(z)I_i(t - z/v) \tag{3.37}$$

According to the above equation, the current at any given level maintains the same shape as the injected current (or the channel base current) while its amplitude changes with height according to the function $A(z)$. Indeed, (3.37) describes a MTL model. This analysis shows that when the speed of propagation of the injected current and the corona currents is assigned the same speed as the return stroke, the CD models reduces to CP models. In addition, if the corona current is assumed to be a product of the channel base current and a function of height, the CP models will reduce to standard MTL models used in the literature. In this special case (3.33) reduces to (3.34) which was derived previously by Mazlowski and Rakov [26], because the return stroke speed v becomes equal tov_c. Thus, (3.34) is a special case of (3.33) and the latter reduces to the former in the case of MTL models. The above also demonstrates that all the current propagation models available in the literature are special cases of current dissipation models.

3.8 Unification of engineering return stroke models

Consider the injection of a current into the transmission line. If the current amplitude is less than the critical amplitude necessary for corona emission, the current pulse will propagate with speed of light and, in the case of an ideal transmission line, without attenuation. If the current amplitude is larger than the critical current necessary for corona emission, each element of the transmission line acts as a corona source injecting a corona current into the return stroke channel. Half of the corona current propagates with speed of light into the forward direction (i.e. the direction of propagation of the injected current) and the other half propagates backwards with the same speed. The polarity of the upward moving corona current is opposite to that of the injected current. In general, this corona current interacts with the injected current, which is also moving with the same speed, in such a way that the net current propagates in the forward direction with a reduced speed. The cumulative sum of the downward moving corona current appears as a current pulse at the base of the channel. In summary, this scenario gives rise to three current waveforms in the return stroke channel, namely, upward moving injected current, downward moving corona current, and upward moving corona current. At the base of the transmission line, the total current consists of two parts, namely, injected current and the cumulative current generated by the downward moving corona current. Both these current waveforms have the same polarity. A model based on this complete scenario was introduced by Cooray and Diendorfer [25] and the model was referred to as the Current-model or C-model.

3.9 Concluding remarks

In this chapter, we have presented the basic principles underlying engineering return stroke models together with the information necessary to use available return stroke models to evaluate the spatial and temporal variation of the return stroke current and to use that information to calculate the electromagnetic fields generated by return strokes.

It is important to note here that any new return stroke model that is introduced into the scientific literature should be able to present a new way of studying the return stroke process. On the other hand, the model parameters should be considered as information that should or could be changed when more experimental data becomes available concerning the return stroke process. Unfortunately, some scientists give more emphasis to the model parameters and by doing so losses the important message that a model builder is trying to convey to the scientific establishment. This wrong way of looking at the models also leads to the creation of “new models” by changing one or two parameters of an existing model.

References

[1] Gomes, C. and Cooray, V., Concepts of lightning return stroke models, *IEEE Trans. Electromagn. Compat.*, 42(1), 82–96, 2000.

[2] Rakov, V. A and Uman, M. A., Review and evaluation of lightning return stroke models including some aspects of their application, *IEEE Trans. Electromagn. Compat.*, 40, 403–426, 1998.

[3] Uman, M. A. and McLain, D. K., Magnetic field of lightning return stroke, *J. Geophys. Res.*, 74, 6899–6910, 1969.

[4] Dennis, A. S., and E. T. Pierce, The return stroke of lightning flash to earth as a source of atmospherics, *Radio Sci.*, 68, 777–794, 1964.

[5] Nucci, C. A., Mazzetti, C., Rachidi, F., and Ianoz, M., On lightning return stroke models for LEMP calculations, Paper presented at 19th International Conference on Lightning Protection, Graz, Austria, 1988.

[6] Rakov, V. A. and Dulzon, A. A., A modified transmission line model for lightning return stroke field calculation, in *Proceedings of the 9th International Symposium on EMC*, Zurich, Switzerland, 44H1, 1991, pp. 229–235.

[7] Cooray, V., and Orville, R. E., The effects of variation of current amplitude, current risetime and return stroke velocity along the return stroke channel on the electromagnetic fields generated by return strokes, *J. Geophys. Res.*, 95 (D11), 18617–18630, 1990.

[8] Wagner, C. F., A new approach to the calculation of the lightning performance of transmission lines, *AIEE Trans.*, 75, 1233–1256, 1956.

[9] Lin, Y. T., Uman, M. A., and Standler, R. B., Lightning return stroke models, *J. Geophys. Res.*, 85, 1571–1583, 1980.

[10] Master, M, Lin, Y. T., Uman, M. A., and Standler, R. B., Calculations of lightning return stroke electric and magnetic fields above ground, *J. Geophys. Res.*, 86, 12127–12132, 1981.

[11] Heidler, F., Traveling current source model for LEMP calculation, in *Proceedings of the 6th International Symposium on EMC*, Zurich, Switzerland, 29F2, 1985, pp. 157–162.

[12] Hubert, P. New model of lightning return stroke – Confrontation with triggered lightning observations, in *Proceedings of the 10th International Aerospace and Ground Conference on Lightning and Static Electricity*, Paris, 1985, pp. 211–215.

[13] Cooray, V., A return stroke model, in *Proceedings of the International Conference on Lightning and Static Electricity*, University of Bath, September, 1989.

[14] Cooray, V., A model for the subsequent return strokes, *J. Electrostatics*, 30, 343–354, 1993.

[15] Deindorfer, G. and Uman, M. A., An improved return stroke model with specified channel base current, *J. Geophys. Res.*, 95, 13,621–13,644, 1990.

[16] Thottappillil, R., Mclain, D. K., Uman, M. A., and Diendorfer, G., Extension of Diendorfer-Uman lightning return stroke model to the case of a variable

upward return stroke speed and a variable downward discharge current speed, *J. Geophys. Res.*, 96, 17143–17150, 1991.

[17] Thottappillil, R. and Uman, M. A., Lightning return stroke model with height-variable discharge time constant, *J. Geophys. Res.*, 99, 22,773–22,780, 1994.

[18] Cooray, V., Predicting the spatial and temporal variation of the current, the speed and electromagnetic fields of subsequent return strokes, *IEEE Trans. Electromag. Compat.*, 40, 427–435, 1998.

[19] Cooray, V., A model for negative first return strokes in negative lightning flashes, *Phys. Scripta*, 55, 119–128, 1997.

[20] Cooray, V., Montano, R., and Rakov, V., A model to represent first return strokes with connecting leaders, *J. Electrostatics*, 40, 97–109, 2004.

[21] Cooray, V. and Rakov, V., A current generation type return stroke model that predicts the return stroke velocity, *J. Lightning Res.*, 1, 32–39, 2007.

[22] Cooray, V. and Theethayi, N., Pulse propagation along transmission lines in the presence of corona and their implication to lightning return strokes, *IEEE Trans. Antennas Propag.*, 56(7), 1948–1959, 2008, doi: 10.1109/TAP.2008.924678.

[23] Cooray, V., A novel procedure to represent lightning strokes – current dissipation return stroke models, *Trans. IEEE (EMC)*, 51, 748–755, 2009.

[24] Cooray, V. On the concepts used in return stroke models applied in engineering practice, *Trans. IEEE (EMC)*, 45, 101–108, 2003.

[25] Cooray, V. and Diendorfer, G., Merging of current generation and current dissipation lightning return stroke models, *Electric Power Syst. Res.*, 153, 10–18, 2017.

[26] Maslowski, G. and Rakov, V. A., Equivalency of lightning return stroke models employing lumped and distributed current sources, *Trans. IEEE (EMC)*, 49, 123–132, 2007.

Chapter 4

Electromagnetic models of lightning return strokes

Yoshihiro Baba[1] *and Vladimir A. Rakov*[2]

In this chapter, electromagnetic (full-wave) models of the lightning return stroke, which have been used in lightning electromagnetic field and surge simulations, are reviewed and evaluated. In this class of models, using a numerical technique such as the method of moments or the finite-difference time-domain method, Maxwell's equations are solved to yield the distribution of current along a vertical wire that represents the lightning return-stroke channel. Lightning models are needed for specifying the source in studying lightning interaction with various systems and with the environment. Here, it is shown that a current wave necessarily suffers distortion as it propagates upward along a vertical non-zero-thickness wire above perfectly conducting ground excited at its bottom by a lumped source, even if the wire has no ohmic losses, which is a distinctive feature of this class of models. Electromagnetic models proposed to date are classified into six types depending on lightning channel representation. Channel-current distributions and resultant electromagnetic fields calculated for these channel representations are presented. Further, methods of excitation, representative numerical procedures for solving Maxwell's equations, and applications of lightning return-stroke electromagnetic models are reviewed.

4.1 Introduction

Lightning return-stroke models are needed in studying lightning effects on various objects and systems, and in characterizing the lightning electromagnetic environment. Clearly, conclusions drawn from these studies are influenced by the choice and validity of lightning return-stroke model employed. Rakov and Uman [1], based on governing equations, have categorized return-stroke models into four classes: gas dynamic models, electromagnetic models, distributed-circuit models, and "engineering" models.

Engineering return-stroke models are equations relating the longitudinal current along the lightning channel at any height and any time to the current at the

[1]Department of Electrical Engineering, Doshisha University, Japan
[2]Department of Electrical and Computer Engineering, University of Florida, USA

channel origin (the origin is usually situated at ground level, but can be at the top of a tall grounded strike object (e.g., [2]). The return-stroke wavefront speed in these models can be set arbitrarily, since it is one of the input parameters. Engineering return-stroke models have been reviewed by Nucci *et al.* [3], Thottappillil and Uman [4], Thottappillil *et al.* [5], Rakov and Uman [1], and Gomes and Cooray [6].

Distributed-circuit models of the lightning return stroke usually consider the lightning channel as an *R–L–C* transmission line (e.g., [7,8]), where *R*, *L*, and *C* are series resistance, series inductance, and shunt capacitance, all per unit length, respectively. In an *R–L–C* transmission line model, voltage and current are the solutions of the telegrapher's equations. Note that the telegrapher's equations can be derived from Maxwell's equations assuming that the electromagnetic waves guided by the transmission line have a transverse electromagnetic (TEM) field structure. Strictly speaking, the latter assumption is not valid for a vertical conductor above ground. Indeed, any current wave suffers attenuation as it propagates upward along a vertical conductor, except for the special (unrealistic) case of a zero-radius conductor [9] and the resultant electromagnetic field structure is non-TEM (e.g., [10–12]). Clearly, an incorrect assumption on the electromagnetic field structure (e.g., TEM when it is actually non-TEM) in the vicinity of lightning channel will result in an incorrect current distribution along the channel (e.g., [11]). Distributed-circuit models have been reviewed by Rakov and Uman [1]. There has been a renewed interest in developing distributed-circuit models (e.g., [13,14]).

Electromagnetic return-stroke models are based on full-wave solution of Maxwell's equations [15,16]. These are relatively new and most rigorous (no TEM assumption) models suitable for specifying the source in studying lightning interaction with various systems and with the environment. In this class of models, Maxwell's equations are solved to yield the distribution of current along the lightning channel using numerical techniques, such as the method of moments (MoM) [17–19] and the finite-difference time-domain (FDTD) method [20]. The resultant distribution of lightning channel current can be used to compute electric and magnetic fields radiated by the channel. In order to reduce the speed of the current wave propagating along the channel-representing vertical wire to a value lower than the speed of light in air, *c*, a wire is loaded by additional distributed series inductance (e.g., [21]), surrounded by a dielectric medium (other than air) that occupies the entire half space above ground (e.g., [22]), coated by a dielectric material (e.g., [23]), or coated by a fictitious material having high relative permittivity and high relative permeability (e.g., [24]). Two parallel wires having additional distributed shunt capacitance have been also suggested [25]. In contrast with distributed-circuit and engineering models, electromagnetic return-stroke models allow a self-consistent full-wave solution for both lightning-current distribution and resultant electromagnetic fields. One of the advantages of the use of electromagnetic models, although it may be computationally expensive, is that one does not need to employ any model of field-to-conductor coupling in analyzing lightning-induced effects on electrical circuits (e.g., [26,27]). Electromagnetic models are generally capable of reproducing most salient features of observed electric and magnetic fields at distances ranging from tens of meters to hundreds of kilometers (e.g., [28,29]).

Note that the so-called hybrid electromagnetic/circuit (HEM) model (e.g., [30]) has been also applied to representing lightning return strokes. It employs electric scalar and magnetic vector potentials for taking into account electromagnetic coupling but is formulated in terms of circuit quantities, voltages and currents. Since the HEM model, on the one hand, yields a non-TEM close electromagnetic field structure (as do electromagnetic models) and, on the other hand, apparently considers electric and magnetic fields as decoupled (as in distributed-circuit models), it occupies an intermediate place between electromagnetic and distributed-circuit models. Applications of HEM model to lightning return-stroke studies and the interaction of lightning with grounded objects are described by Visacro and Silveira [30] and by Visacro and Silveira [31], respectively. Baba and Rakov [15] have shown that the current distribution along a vertical resistive wire, representing a lightning channel, predicted by the HEM model, is consistent with that obtained using electromagnetic models. Also note that the partial-element equivalent-circuit (PEEC) method [32], which is based on a similar concept and has essentially the same features as the HEM model, has recently been applied to lightning electromagnetic field and surge simulations (e.g., [33,34]).

In this chapter, electromagnetic models of the lightning return stroke, proposed to date, are reviewed. This chapter is organized as follows. In Section 4.2, it is shown that a current wave necessarily suffers attenuation (dispersion to be exact) as it propagates upward along a vertical non-zero-thickness wire above perfectly conducting ground excited at its bottom by a lumped source, even if the wire has no ohmic losses. This is a distinctive feature of electromagnetic return-stroke models. In Section 4.3, lightning return-stroke electromagnetic models are classified into six types depending on lightning channel representation used to find the distribution of current along the channel. In Section 4.4, distributions of current along a vertical channel and electromagnetic fields calculated for different channel representations are presented. In Section 4.5, methods of excitation used in electromagnetic return-stroke models are described. In Section 4.6, representative numerical procedures for solving Maxwell's equations used in electromagnetic models of the lightning return stroke are compared. In Section 4.7, applications of lightning return-stroke electromagnetic models are reviewed.

4.2 General approach to finding the current distribution along a vertical perfectly conducting wire above ground

All electromagnetic return-stroke models involve a representation of the lightning channel as a non-zero-thickness vertical wire. In this section, using Chen's analytical equation [35] we show that a current wave necessarily suffers attenuation as it propagates along a vertical wire of uniform non-zero thickness that is located above perfectly conducting ground and excited at its bottom by a lumped source, even if the wire has no ohmic losses. This effect, well known in the radio science community, but not in the lightning research community (e.g., [36]), is usually attributed to radiation losses. Here, it is shown that current attenuation (dispersion) is necessary to satisfy the boundary condition on the tangential component of electric field on the surface of vertical wire.

4.2.1 Current distribution along a vertical perfectly conducting wire above ground

Chen [35] has derived an approximate analytical equation for the transient current $I(z',t)$ along an infinitely long perfectly conducting cylinder in air excited in the middle by a zero-length voltage source generating step voltage V. This equation is reproduced below:

$$I(z',t) = \frac{2V}{\eta}\tan^{-1}\left(\frac{\pi}{2\ln(\sqrt{c^2t^2 - z'^2}/a)}\right), \tag{4.1}$$

where η is free space impedance (120π Ω), ln is the natural logarithm, and a is the radius of the cylinder. If one applies (4.1) to a vertical cylinder on flat perfectly conducting ground excited at its bottom by a zero-length step-voltage source, one has only to multiply the magnitude of resultant current by 2 in order to account for the image source. Chen's analytical equation (4.1) can be used in testing the accuracy of numerical techniques employed in electromagnetic models of the lightning return stroke.

Figure 4.1 shows current waveforms at different heights along a vertical perfectly conducting wire of radius 0.23 m in air above ground excited at its bottom by a zero-length source that produces a ramp-front wave having a magnitude of 5 MV and a risetime of 1 μs. Note that the response to this ramp-front voltage wave was obtained using numerical convolution since (4.1) yields the solution for a step voltage excitation.

It is clear from Figure 4.1 that a current wave suffers attenuation as it propagates along the vertical perfectly conducting wire above ground. It will be shown in Section 4.6.3 that current waveforms calculated using the MoM in the time and frequency domains and the FDTD method agree well with those calculated using Chen's equation.

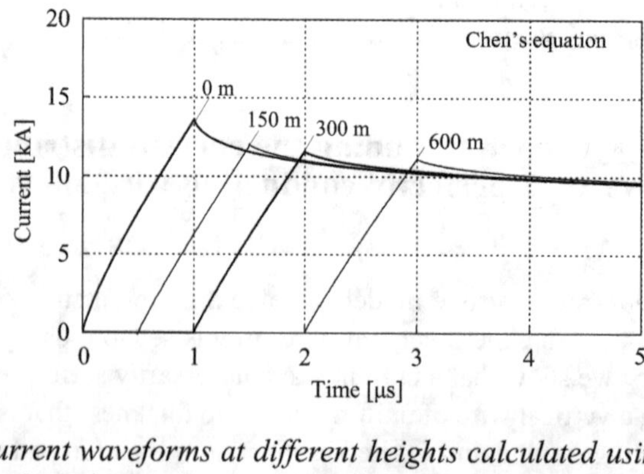

Figure 4.1 Current waveforms at different heights calculated using Chen's analytical equation [see (4.1)] for a vertical perfectly conducting cylinder of radius 0.23 m in air above perfectly conducting ground excited at its bottom by a zero-length voltage source. The source produces a ramp wave having a magnitude of 5 MV and a risetime of 1 μs. Adapted from Baba and Rakov [15]

4.2.2 Mechanism of attenuation of current wave in the absence of ohmic losses

According to analytical equation (4.1), any current wave suffers attenuation as it propagates upward along a vertical perfectly conducting wire above flat perfectly conducting ground excited at its bottom by a lumped source (the same result follows from numerical solution of Maxwell's equations; e.g., [10–12]), except for the ideal (unrealistic) case of a zero-thickness wire excited by a zero-length source [9] or a conical conductor excited at its apex [37]. In this section, we discuss the mechanism of current attenuation in the absence of ohmic losses.

Baba and Rakov [12] have visualized the mechanism of attenuation of the current wave propagating along a vertical nonzero-thickness perfectly conducting wire as illustrated in Figure 4.2. A reference (no interaction with the wire, no attenuation) positive current pulse I_{inc} propagating upward generates an incident spherical TEM wave [9], with vertical electric field component on the surface of the wire being directed downward. Cancellation of this field, as required by the boundary condition on the tangential component of electric field on the surface of a perfectly conducting

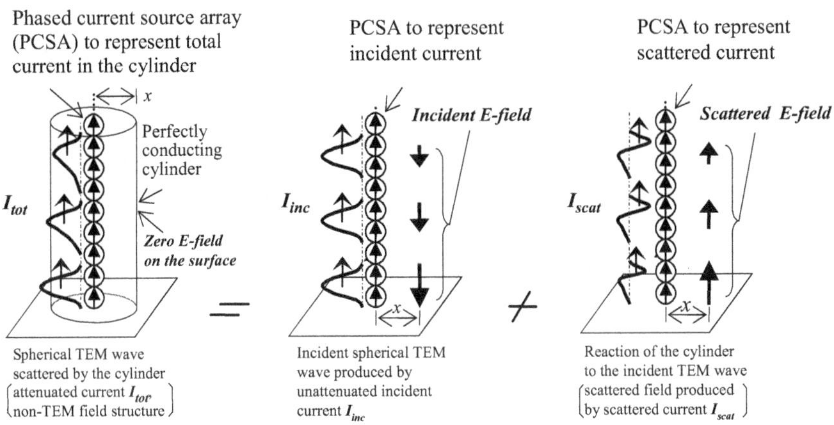

Figure 4.2 Conceptual picture to explain the mechanism of current attenuation along a vertical non-zero-thickness perfectly conducting wire above perfectly conducting ground. All currents are assumed to flow on the axis. An attenuated "total" current pulse I_{tot} is separated into an "incident" unattenuated current pulse I_{inc} and an induced or "scattered" current pulse I_{scat}. I_{inc} generates an incident downward vertical electric field at a horizontal distance x from the axis (on the lateral surface of the cylinder). I_{scat} produces a scattered upward vertical electric field that cancels the incident downward vertical electric field on the surface of the cylinder, and modifies the incident current I_{inc}. The resultant current pulse, $I_{tot} = I_{inc} + I_{scat}$, appears attenuated and its tail is lengthened as this pulse propagates along the wire. Adapted from Baba and Rakov [12]

wire, gives rise to an induced or "scattered" current I_{scat}. This scattered current I_{scat} modifies I_{inc}, so that the resultant total current pulse I_{tot} appears attenuated as it propagates along the vertical wire. The attenuation of the total current pulse is accompanied by the lengthening of its tail, such that the total charge transfer is independent of height. The electromagnetic field structure associated with an attenuated current distribution along a vertical wire is non-TEM. Baba and Rakov [12] have shown that the current attenuation becomes more pronounced as (1) the thickness of vertical wire increases, (2) the source height decreases, (3) the frequency increases, and (4) the height above the excitation point decreases.

In summary, current attenuation (or more generally, dispersion) is necessary to satisfy the boundary condition on the tangential component of electric field on the surface of vertical wire. The resultant field structure is non-TEM, particularly in the vicinity of the excitation point.

4.3 Representation of the lightning return-stroke channel

There are six types of electromagnetic representation of lightning return-stroke channel used in lightning electromagnetic field and surge simulations, which are listed below:

1. a perfectly conducting/resistive wire in air above ground;
2. a wire loaded by additional distributed series inductance in air above ground;
3. a wire surrounded by a dielectric medium (other than air) that occupies the entire half space above ground (the artificial dielectric medium is used only for finding current distribution along the lightning channel and is then removed for calculating electromagnetic fields in air);
4. a wire coated by a dielectric material in air above ground;
5. a wire coated by a fictitious material having high relative permittivity and high relative permeability in air above ground; and
6. two parallel wires having additional distributed shunt capacitance in air (this fictitious configuration is used only for finding current distribution and is then applied to a vertical wire in air above ground for calculating electromagnetic fields).

All the representations, except for type 1, are used to reduce the speed of the current wave propagating along the channel-representing wire to a value lower than the speed of light in air. Table 4.1 gives a list of papers on electromagnetic models of the lightning return stroke that are grouped into six categories depending on the channel representation.

In the following, we will review the return-stroke speed and channel characteristic impedance corresponding to each of the six types of channel representation. The return-stroke speed, along with the current peak, largely determines the radiation field initial peak (e.g., [64]), while the characteristic impedance of lightning channel influences the magnitude of lightning current and/or the current reflection coefficient

Table 4.1 List of papers on electromagnetic models of the lightning return stroke grouped into four categories depending on the lightning channel representation

Representation	Papers	Channel radius, mm	ε_r	R, Ω/m	L, μH/m	C, pF/m	Phase velocity
Perfectly conducting or resistive wire in air above ground	Podgorski and Landt [38]	Unknown	1	0.7	0	–	c
	Kordi *et al.* [10]	50	1	0	0	–	c
	Mozumi *et al.* [39]	100	1	0	0	–	c
	Baba and Ishii [40]	50	1	0, 0.1	0	–	c
	Kordi *et al.* [41]	100	1	0.07	0	–	c
	Baba and Rakov [11]	230	1	0	0	–	c
	Baba and Rakov [12]	2,000[a]	1	0	0	–	c
	Pokharel *et al.* [42]	100	1	0.1	0	–	c
	Baba and Rakov [43]	68	1	0	0	–	c
	Baba and Rakov [29]	680	1	0	0	–	c
	Matsuura *et al.* [44]	29	1	0	0	–	c
	Ishii *et al.* [45]	100	1	1	0	–	c
	Takami *et al.* [46]	230	1	0	0	–	c
	Du *et al.* [47]	5	1	0	0	–	c
	Diaz *et al.* [48]	200	1	1	0	–	c
	Tatematsu [49]	5	1	0	0	–	c
Wire loaded by additional distributed series inductance in air above ground	Kato *et al.* [21]	10	1	0	0.1	–	0.33c
	Baba and Ishii [50]	300	1	1	3	–	0.56c
	Baba and Ishii [40]	50	1	1	6	–	0.43c
	Pokharel *et al.* [26]	10	1	0.5	6	–	0.43c
	Pokharel *et al.* [42]	100	1	1	9	–	0.37c
	Noda *et al.* [51]	230	1	0	10	–	0.33c
	Pokharel and Ishii [52]	Unknown	1	0.6	6	–	0.5c
	Bonyadi-Ram *et al.* [53]	20	1	0.45	Nonuniform	–	0.53c to 0.3c
	Miyazaki and Ishii [54]	100	1	Unknown	Unknown	–	Unknown
	Miyazaki and Ishii [55]	50	1	1	6	–	0.5c

(Continues)

Table 4.1 (*Continued*)

Representation	Papers	Channel radius, mm	ε_r	R, Ω/m	L, μH/m	C, pF/m	Phase velocity
	Baba and Rakov [29]	680	1	0.5	2.5	–	0.5*c*
	Aniserowicz and Maksimowicz [56]	50	1	1	4.5	–	0.43*c*
	Janani *et al.* [27]	20	1	0.5	8	–	Unknown
	Khosravi-Farsani *et al.* [57]	690	1	1	3	–	0.5*c*
	Saito *et al.* [58]	10	1	0.3	6	–	0.5*c*
	Khosravi *et al.* [59]	230	1	1	4.3	–	0.56*c*
	Tatematsu and Ueda [60]	58	1	0	7.6	–	0.36*c*
	Yamanaka *et al.* [61]	135	1	0.7	13	–	0.5*c*
Wire surrounded by a dielectric medium of $\varepsilon_r > 1$ that occupies the entire half space above ground	Moini *et al.* [62]	unknown	4	0	0	–	0.5*c*
	Moini *et al* [22]	unknown	5.3	0.07	0	–	0.43*c*
	Shoory *et al.* [28]	50	5.3	0.1	0	–	0.43*c*
	Geranmayeh *et al.* [63]	5	5.3	0	0	–	0.43*c*
	Baba and Rakov [29]	680	4	0	0	–	0.5*c*
Wire coated by a dielectric material of $\varepsilon_r > 1$ in air above ground	Kato *et al.* [23] 4-m-radius coating	10 ($\varepsilon_r = 200$)	1	0	0	–	0.7*c*
	Baba and Rakov [29] 10-m-radius coating	680 ($\varepsilon_r = 400$)	1	0	0	–	0.7*c*
	Miyazaki and Ishii [24]	10 (ε_r, μ_r: Unknown)	1	0	0	–	Unknown
	Baba and Rakov [29] 10-m-radius coating	680 ($\varepsilon_r = 5$, $\mu_r = 5$)	1	0.25	0	–	0.5*c*
Two parallel wires having additional distributed shunt capacitance in air	Bonyadi-Ram *et al.* [25]	20	1	0.2	0	50	0.43*c*

R, *L*, and *C* are the additional resistance, inductance, and capacitance (each per unit length), respectively, of the equivalent lightning channel

Underlining is used to identify conference papers

[a]2 m × 2 m rectangular cross-section

at the top of strike object when a lumped voltage source is employed. It is desirable that the following two features are reproduced by models:

- typical values of return-stroke wavefront speed are in the range from $c/3$ to $c/2$ [65], as observed using optical techniques, where c is the speed of light;
- the equivalent impedance of the lightning return-stroke channel is expected to be in the range from 0.6 to 2.5 kΩ [66], as estimated from measurements of lightning current at different points along the 530-m-high Ostankino Tower in Moscow.

Most of the values of the radius of the lightning channel in Table 4.1 are larger than expected, about 30 mm (e.g., [67]), but this geometry was necessary to achieve agreement of the characteristic impedance of the simulated channel with the expected equivalent channel impedance values (0.6–2.5 kΩ).

Note that the resistance per unit length of a lightning return-stroke channel (behind the return-stroke front) is estimated to be about 0.035 Ω/m and about 3.5 Ω/m ahead of the return-stroke front [67]. Values of distributed resistance (for the case of resistive channel) in Table 4.1 are between these two expected values.

There is one more representation of the lightning return-stroke channel, which has been employed in lightning electromagnetic field and surge simulations using numerical techniques such as the MoM and the FDTD method. It is the vertical phased-current-source array [11] in air above ground. This channel representation, which also includes excitation, can be employed for simulation of "engineering" lightning return-stroke models such as the transmission-line (TL) model [68], the traveling-current-source (TCS) model [69], or the modified TL model with linear current decay with height (MTLL model) [64]. Each current source of the phased-current-source array is activated successively by the arrival of lightning return-stroke wavefront that progresses upward at specified speed. Although the impedance of this channel model is infinitely large, appropriate reflection coefficients at the impedance discontinuity points can be implemented when a tall strike object and/or upward-connecting leader are involved (e.g., [70]).

4.3.1 Type 1: a perfectly conducting/resistive wire in air above ground

Podgorski and Landt [38], using the modified thin-wire time-domain (TWTD) code [17], have represented a lightning strike to the 553-m-high CN Tower in Toronto by a resistive (0.7 Ω/m) vertical wire with a nonlinear resistor (10 kΩ prior to the attachment and 3 Ω after the attachment) connected between the bottom of the resistive wire and the top of the CN Tower (or the top of uncharged wire simulating the upward connecting leader from the tower).

The speed of the current wave propagating along such a wire is nearly equal to the speed of light, which is 2–3 times higher than typical measured values of return stroke wavefront speed: $c/3$ to $c/2$ (e.g., [65]). This discrepancy is the main deficiency of the type-1 model. This should result in overestimation of remote electric and magnetic fields, since their magnitudes are expected be proportional to the current wave propagation speed (e.g., [64,68]).

The characteristic impedance of the channel-representing vertical wire of radius 3 cm is estimated to be around 0.6 kΩ at a height of 500 m above ground (it varies with height above ground). This is right at the lower bound of its expected range of variation (0.6–2.5 kΩ). Note that, as stated in Section 4.2, a current wave suffers attenuation and distortion as it propagates along a vertical wire even if the wire has no ohmic losses. Further attenuation can be achieved by loading the wire with distributed series resistance.

This type of representation has been used by Kordi *et al.* [10], Mozumi *et al.* [39], Baba and Ishii [40], Kordi *et al.* [41], Baba and Rakov [11,12,29,43], Pokharel *et al.* [42], Matsuura *et al.* [44], Ishii *et al.* [45], Takami *et al.* [46], Du *et al.* [47], Diaz *et al.* [48], and Tatematsu [49].

4.3.2 *Type 2: a wire loaded by additional distributed series inductance in air above ground*

In this section, the TEM-wave-based *R-L-C* uniform transmission line theory is reviewed. Then, based on this theory, representation of the lightning return-stroke channel using a vertical wire loaded by additional distributed series inductance is discussed. Note that applying the *R–L–C* transmission-line theory to describing a vertical wire above ground is an approximation, since inductance L and capacitance C, both per unit length, vary with height along the vertical wire, and the resultant electromagnetic field structure is non-TEM.

The propagation constant γ_0 of the *R–L–C* uniform transmission line, the phase velocity v_{p0} of a wave propagating along this line, and the characteristic impedance Z_{c0} of the line are given by (e.g., [67,71]),

$$\gamma_0 = \sqrt{j\omega C_0 \ (R_0 + j\omega L_0)}, \tag{4.2}$$

$$v_{p0} = \frac{\omega}{\mathrm{Im}(\gamma_0)} = \frac{1}{\sqrt{L_0 C_0}} \left[\frac{2}{\sqrt{1 + (R_0/\omega L_0)^2} \ + \ 1} \right]^{1/2} \tag{4.3}$$

$$Z_{c0} = \sqrt{\frac{R_0 + j\omega L_0}{j\omega C_0}} \tag{4.4}$$

where $\mathrm{Im}(\gamma_0)$ stands for the imaginary part of γ_0, ω is the angular frequency ($2\pi f$), R_0 is the series resistance per unit length, L_0 is the intrinsic series inductance per unit length, and C_0 is the intrinsic shunt capacitance per unit length. If ωL_0 is much larger than R_0 at a frequency of interest, (4.3) and (4.4) reduce to

$$v_{p0} \simeq 1/\sqrt{L_0 C_0} \tag{4.5}$$

$$Z_{c0} \simeq \sqrt{L_0 / C_0} \tag{4.6}$$

As an example, the assumption that (4.5) and (4.6) are based on is satisfied at frequencies f = 1 MHz or higher for L_0 = 2.1 μH/m (evaluated for a 3-cm-radius horizontal wire at a height of 500 m above ground [67] and $R_0 = 1\ \Omega$/m, where ωL_0 (=13 Ω/m) >> R_0 (=1 Ω/m). If the transmission line is surrounded by air, v_{p0} given by (4.5) is equal to c.

If the transmission line has additional distributed series inductance L, the phase velocity v_{pi} and the characteristic impedance Z_{ci} for such a line are

$$v_{pi} \simeq \frac{1}{\sqrt{(L_0 + L)\ C_0}} = \sqrt{\frac{L_0}{L_0 + L}}\ v_{p0} = \sqrt{\frac{L_0}{L_0 + L}}\ c \tag{4.7}$$

$$Z_{ci} \simeq \sqrt{\frac{L_0 + L}{C_0}} = \sqrt{\frac{L_0 + L}{L_0}}\ Z_{c0} = \frac{c}{v_{pi}}\ Z_{c0} \tag{4.8}$$

Equations (4.7) and (4.8) show that if $L = 3L_0$, v_{pi} becomes $0.5c$ and Z_{ci} becomes $2Z_{c0}$. In this representation, Z_{ci} increases linearly with decreasing v_{pi}. Note that additional inductance has no physical meaning and is invoked only to reduce the speed of the current wave propagating along the wire to a value lower than the speed of light. Electromagnetic waves radiated from the vertical inductance-loaded wire into air propagate at the speed of light. The use of this representation allows one to calculate both the distribution of current along the channel-representing wire and the radiated electromagnetic waves in a single, self-consistent procedure.

If the natural inductance of a vertical wire is assumed to be $L_0 = 2.1$ μH/m (evaluated as for a 3-cm-radius horizontal wire at a height of 500 m above ground by [67]), the additional inductance needed to obtain typical values of $v_{pi} = 0.5c$ and $0.33c$ is estimated from (4.7) to be L = 6 and 17 μH/m, respectively. These inductance values (6 and 17 μH/m) are not much different from those employed to date, which range from 2.5 [29] to 10 μH/m [51], except for that employed by Kato *et al.* [21], who used 0.1-μH/m additional inductance. The speeds of the current wave propagating along the wire are about $0.5c$ and $0.33c$ for the wire loaded by $L = 2.5$ and 10 μH/m, respectively. In summary, in order to simulate a typical speed of return-stroke wavefront, values of additional distributed inductance should be in the range roughly from 1 to 20 μH/m.

From (4.8), Z_{ci} is 1.2 kΩ for $v_{pi} = 0.5c$, and 1.8 kΩ for $v_{pi} = 0.33c$, respectively, if the characteristic impedance of a vertical wire of radius 3 cm is 0.6 kΩ (estimated at a height of 500 m above ground). The characteristic impedance of the inductance-loaded wire (Z_{ci} = 1.2 to 1.8 kΩ) is within the range of expected values (from 0.6 to 2.5 kΩ [66]).

Baba and Ishii [40,50] added distributed series resistance of 1 Ω/m to the inductance-loaded wire in order to stabilize non-physical oscillations caused by the employed numerical procedure, which also caused current attenuation along the wire. This same approach was used by Aniserowicz [72], Miyazaki and Ishii [54,55], Pokharel *et al.* [26,42], Pokharel and Ishii [52], Aniserowicz and Maksimowicz [56], Janani *et al.* [27], Khosravi-Farsani *et al.* [57], Saito *et al.* [58], Khosravi *et al.* [59], Tatematsu and Ueda [60], and Yamanaka *et al.* [61]. Note that Bonyadi-Ram *et al.* [53] have incorporated additional distributed series inductance that increases with increasing height in order to simulate the optically observed reduction in return-stroke speed with increasing height (e.g., [73]).

4.3.3 Type 3: a wire embedded in a dielectric (other than air) above ground

In this section, (4.5) and (4.6), which are based on the *R–L–C* uniform transmission line approximation, are used to examine phase velocity and characteristic impedance of a vertical wire above ground surrounded by a dielectric (other than air) that has relative permittivity ε_r and occupies the entire half space above ground. From (4.5) and (4.6), the phase velocity v_{pd} and the characteristic impedance Z_{cd} for this representation are

$$v_{pd} \simeq \frac{1}{\sqrt{L_0 \ \varepsilon_r C_0}} = \frac{v_{p0}}{\sqrt{\varepsilon_r}} = \frac{c}{\sqrt{\varepsilon_r}} \tag{4.9}$$

$$Z_{cd} \simeq \sqrt{\frac{L_0}{\varepsilon_r C_0}} = \frac{Z_{c0}}{\sqrt{\varepsilon_r}} = \frac{v_{pd}}{c} Z_{c0} \tag{4.10}$$

Equations (4.9) and (4.10) show that Z_{cd} decreases linearly with decreasing v_{pd}. When ε_r ranges from 4 to 9, v_{pd} ranges from 0.5c to 0.33c, which corresponds to typical measured speeds of the lightning return-stroke wavefront (e.g., [65]). The corresponding characteristic impedance Z_{cd} is 0.3 kΩ for v_{pd} = 0.5c, and 0.2 kΩ for v_{pd} = 0.33c, if Z_{c0} = 0.6 kΩ. This characteristic impedance (Z_{cd} = 0.2 to 0.3 kΩ) is smaller than values of the expected equivalent impedance of the lightning return stroke channel (0.6–2.5 kΩ) [66]. This does not cause significant differences in resultant current distributions in analyzing a branchless subsequent lightning stroke terminating on flat ground, in which upward connecting leaders are usually neglected and the return-stroke current wave propagates upward from the ground surface. However, in analyzing lightning strikes to a grounded metallic object [74], one needs to insert several-hundred-ohm lumped resistance between the lightning channel and the strike object in order to obtain a realistic impedance of the lightning return-stroke channel seen by waves entering the channel from the strike object. Note that the artificial dielectric medium is used only for finding current distribution along the lightning channel, which is then removed for calculating electromagnetic fields in air.

Aniserowicz [72] has found a useful relation between a resistive wire loaded by additional distributed series inductance and a resistive wire (characterized by R_0, L_0, and C_0) embedded in a dielectric. From (4.2), the propagation constant for a resistive transmission line embedded in a dielectric of relative permittivity ε_r is given by

$$\gamma_d = \sqrt{j\omega \ \varepsilon_r \ C_0 \ (R_0 + j\omega L_0)} \tag{4.11}$$

which can be written as

$$\gamma_i = \sqrt{j\omega \ C_0 \ (\varepsilon_r R_0 + j\omega \varepsilon_r L_0)} \tag{4.12}$$

Equations (4.11) and (4.12) show that the effect of distributed resistance $\varepsilon_r R_0$ (= 5.3 R_0) of a wire in air loaded by additional distributed series inductance of L = (ε_r−1)L_0 (= 4.3 L_0) on the propagation constant is the same as that of R_0 of a wire embedded in a dielectric of relative permittivity ε_r (= 5.3). For example, the effect of R_0 = 0.07 Ω/m of a wire embedded in a dielectric of ε_r = 5.3 (e.g., [22]) on the propagation constant is the same as that of $\varepsilon_r R_0$ = 0.37 Ω/m of a wire in air loaded by $L = 4.3L_0$.

It follows from (4.3) that the phase velocity v_{pi} for a wire above ground surrounded by air having a distributed series resistance of $\varepsilon_r R_0$ and an additional distributed series inductance of $L = (\varepsilon_r-1)L_0$ is the same as the phase velocity v_{pd} for a wire having a distributed series resistance of R_0 and being embedded in a dielectric of ε_r. The characteristic impedances of these two representations are given below:

$$Z_{ci} = \sqrt{\frac{\varepsilon_r R_0 + j\omega\varepsilon_r L_0}{j\omega C_0}} = \sqrt{\varepsilon_r} Z_{c0} \tag{4.13}$$

$$Z_{cd} = \sqrt{\frac{R_0 + j\omega L_0}{j\omega\varepsilon_r C_0}} = \frac{Z_{c0}}{\sqrt{\varepsilon_r}} = \frac{Z_{ci}}{\varepsilon_r} \tag{4.14}$$

Equations (4.13) and (4.14) show that the effect of distributed resistance $\varepsilon_r R_0$ of a wire in air loaded by an additional distributed series inductance of $L = (\varepsilon_r-1)L_0$ on the characteristic impedance, relative to that of total inductance $\varepsilon_r L_0$, is the same as that of R_0 of a wire embedded in a dielectric of relative permittivity ε_r, relative to that of natural inductance L_0.

This type of representation has been used by Moini *et al.* [62], Shoory *et al.* [28], Geranmayeh *et al.* [63], and Baba and Rakov [29].

4.3.4 *Type 4: a wire coated by a dielectric material in air above ground*

Kato *et al.* [23] have represented the lightning channel by a vertical perfectly conducting wire, which is placed along the axis of a 4-m-radius dielectric cylinder of relative permittivity equal to 200. This dielectric cylinder was surrounded by air. The speed of the current wave propagating along the wire is about 0.7*c*. Such a representation allows one to calculate both the distribution of current along the wire and the remote electromagnetic fields in a single, self-consistent procedure, while that of a vertical wire surrounded by an artificial dielectric medium occupying the entire half space (type 3 described above) requires two steps to achieve the same objective. However, electromagnetic fields produced by this configuration (dielectric-coated wire in air) are influenced by the presence of coating, which will be shown in Section 4.4.2. Note that a conductor with dielectric coating is also known as the Goubau waveguide [75].

4.3.5 *Type 5: a wire coated by a fictitious material having high relative permittivity and high relative permeability in air above ground*

Miyazaki and Ishii [24] have represented the lightning channel by a vertical wire, which is placed along the axis of a cylinder or parallelepiped-shaped block having high relative permittivity and high relative permeability (transverse dimensions of the block, and the values of relative permittivity and permeability are not given). This structure is surrounded by air. The speed of the current wave propagating along the wire was about 0.5c. Similar to the type-4 model, this representation allows one to calculate both the distribution of current along the wire and the remote electromagnetic fields in a single, self-consistent procedure. For the same speed of current wave, the characteristic impedance value for this channel representation is higher than that for the type-4 model, since both relative permittivity and permeability are set at higher values in the type-5 model.

4.3.6 Type 6: two wires having additional distributed shunt capacitance in air

In this section, we use (4.5) and (4.6) to examine phase velocity and characteristic impedance of a transmission line having additional distributed shunt capacitance. From (4.5) and (4.6), the phase velocity v_{pc} and the characteristic impedance Z_{cc} for this case are given by

$$v_{pc} \simeq \frac{1}{\sqrt{L_0(C_0 + C)}} = \sqrt{\frac{C_0}{C_0 + C}}\, v_{p0} = \sqrt{\frac{C_0}{C_0 + C}}\, c \tag{4.15}$$

$$Z_{cc} \simeq \sqrt{\frac{L_0}{C_0 + C}} = \sqrt{\frac{C_0}{C_0 + C}}\, Z_{c0} = \frac{v_{pc}}{c}\, Z_{c0} \tag{4.16}$$

Bonyadi-Ram *et al.* [25] have evaluated the distribution of current along a lightning channel approximating this channel and its image by two 7-km-long parallel wires with additional shunt capacitance between them and excited at their one end by a lumped voltage source. Each wire has a radius of 2 cm, and the separation between the wires is 30 m. The resultant parallel-wire transmission line also has a distributed series resistance of 0.2 Ω/m. The additional shunt capacitance is C = 50 pF/m, which allows one to reduce the speed of the current wave propagating along the parallel wires to v = 0.43c. The current distribution, obtained for the two capacitively loaded parallel wires is used to calculate electric and magnetic fields 0.5, 5, and 100 km from a vertical lightning channel above ground.

Similar to the type-3 model described above, the type-6 model employs a fictitious configuration for finding a reasonable distribution of current along the lightning channel, and then this current distribution is applied to the actual configuration (vertical wire in air above ground). The type-6 model is not further considered in this section.

4.4 Comparison of model-predicted current distributions and electromagnetic fields for different channel representations

In this section, current distributions along a vertical channel above perfectly conducting ground excited at its bottom by a lumped current source, which are predicted by electromagnetic models of types 1–5 described above, are presented. Then, vertical electric and azimuthal magnetic field waveforms, calculated for these models, are presented and compared with typical measured electric and magnetic field waveforms due to natural and triggered lightning return strokes at different distances.

4.4.1 Comparison of distributions of current for different channel representations

The five representations of lightning channel excited at its bottom above flat perfectly conducting ground, to be analyzed using the FDTD method [20] in the 2D-cylindrical coordinate system, are illustrated in Figure 4.3. Figure 4.3(a) shows a vertical perfectly conducting wire in air (type 1). Figure 4.3(b) shows a vertical wire in air loaded by additional distributed series inductance $L = 2.5$ μH/m and distributed series resistance $R = 0.5$ Ω/m (type 2). Note that the 0.5-Ω/m resistance is employed in order to stabilize high-frequency oscillations that appear when additional series distributed inductance is included. Figure 4.3(c) shows a vertical perfectly conducting wire embedded in dielectric of $\varepsilon_r = 4$, which occupies the entire half space (type 3). Figure 4.3(d) shows a vertical perfectly conducting wire embedded in a 10-m-radius dielectric cylinder of $\varepsilon_r = 400$ surrounded by air (type 4). Figure 4.3(e) shows a vertical wire loaded by $R = 0.25$ Ω/m and embedded in a 10-m-radius cylinder of $\varepsilon_r = 5$ and $\mu_r = 5$ surrounded by air (type 5). The length of the channel-representing wire is set to 30 km. This unrealistically long wire is employed in order to avoid effects of any reflections from the upper end of the channel in field waveforms during at least the first 100 μs (=30 km/v for $v = c$). The length of vertical channel section is expected to be 5–8 km, with contributions from higher sections being relatively small (particularly at close distances). At 50 km, however, the computed fields after 50 μs or so may be influenced by the unrealistic vertical channel section above 8 km and, hence, should be viewed with caution. The assumed higher resistance at larger heights (see Figures 4.7 and 4.8c) serves to alleviate this problem.

The lumped current source located at the bottom of simulated channel has a length of 10 m and produces a current waveform having a peak of 11 kA, a 10-to-90% risetime (RT) of 1 μs, and a time to half-peak value of 30 μs (see the waveform labeled "$z' = 0$" in Figure 4.4). This channel-base current waveform is the same as the waveform, proposed by Nucci *et al.* [3] and thought to be typical for subsequent lightning return strokes, except for its rising portion ($RT = 1$ μs here vs. 0.15 μs in Nucci *et al.* [3]). For the FDTD calculations, the vertical conducting wire is represented by a zero-radius wire placed at the left-side boundary ($r = 0$) in the working space of 65 km × 31 km, which is divided into 5 m × 10 m rectangular cells. When cells having a lateral side length of 5 m are used, the vertical (z-directed) zero-radius perfectly conducting wire placed at $x = 0$ has an equivalent radius of 0.675 m = 0.135 × 5 m [76]. Liao's second-order absorbing boundaries [77] are set at the bottom, top, and right-side boundaries in order to minimize reflections there. The time increment is set to 10 ns.

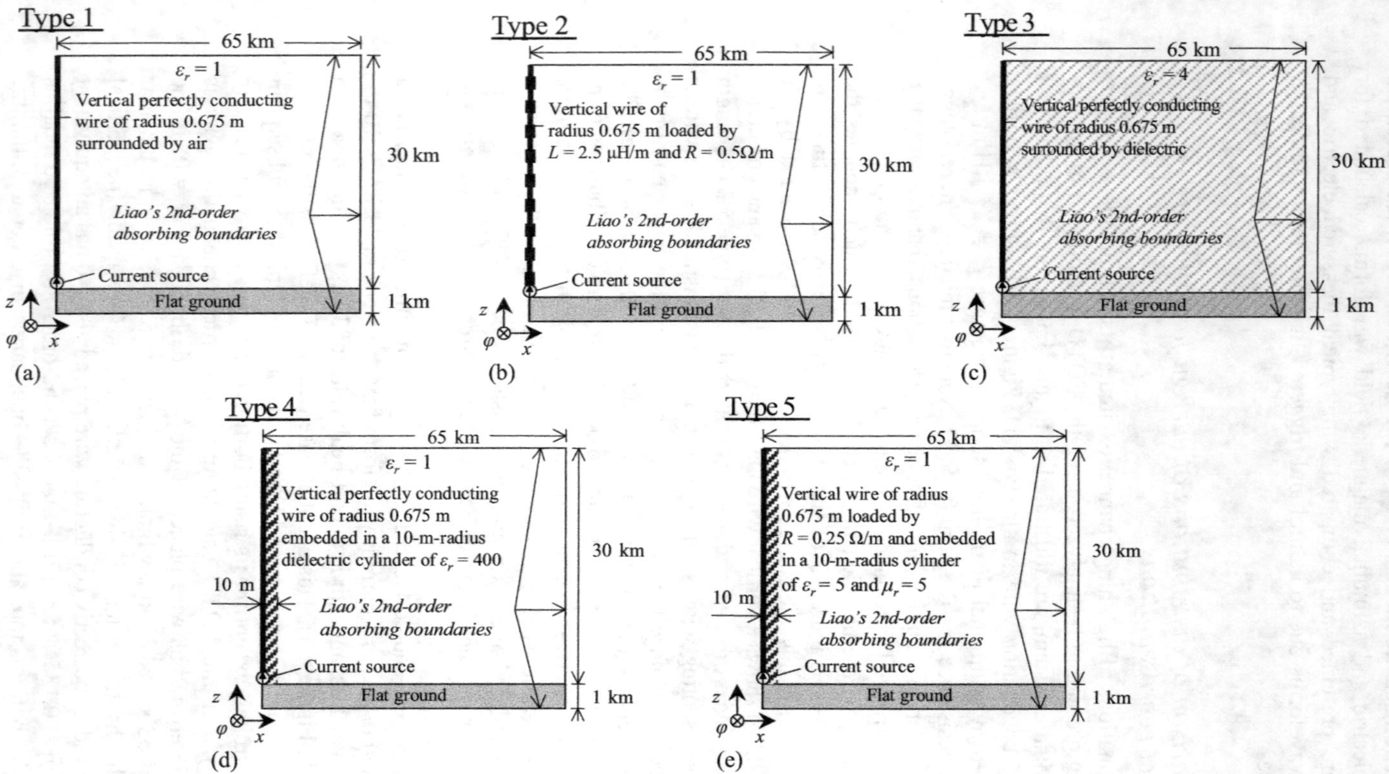

Figure 4.3 Five representations of lightning return-stroke channel excited at its bottom by a 10-m current source above flat perfectly conducting ground, to be analyzed using the 2D-cylindrical FDTD method: (a) a vertical perfectly conducting wire surrounded by air (type 1), (b) a vertical wire loaded by additional distributed series inductance $L = 2.5$ μH/m and distributed series resistance $R = 0.5$ Ω/m in air (type 2), (c) a vertical perfectly conducting wire embedded in dielectric of $\varepsilon_r = 4$, which occupies the entire half space (type 3), (d) a vertical perfectly conducting wire embedded in a 10-m-radius dielectric cylinder of $\varepsilon_r = 400$ surrounded by air (type 4), and (e) a vertical wire loaded by $R = 0.25$ Ω/m and embedded in a 10-m-radius cylinder of $\varepsilon_r = 5$ and $\mu_r = 5$ surrounded by air (type 5). For representation (a) $v = c$, for representations (b), (c), and (e) $v = 0.5c$, and for representation (d) $v = 0.7c$. Adapted from Baba and Rakov [29]

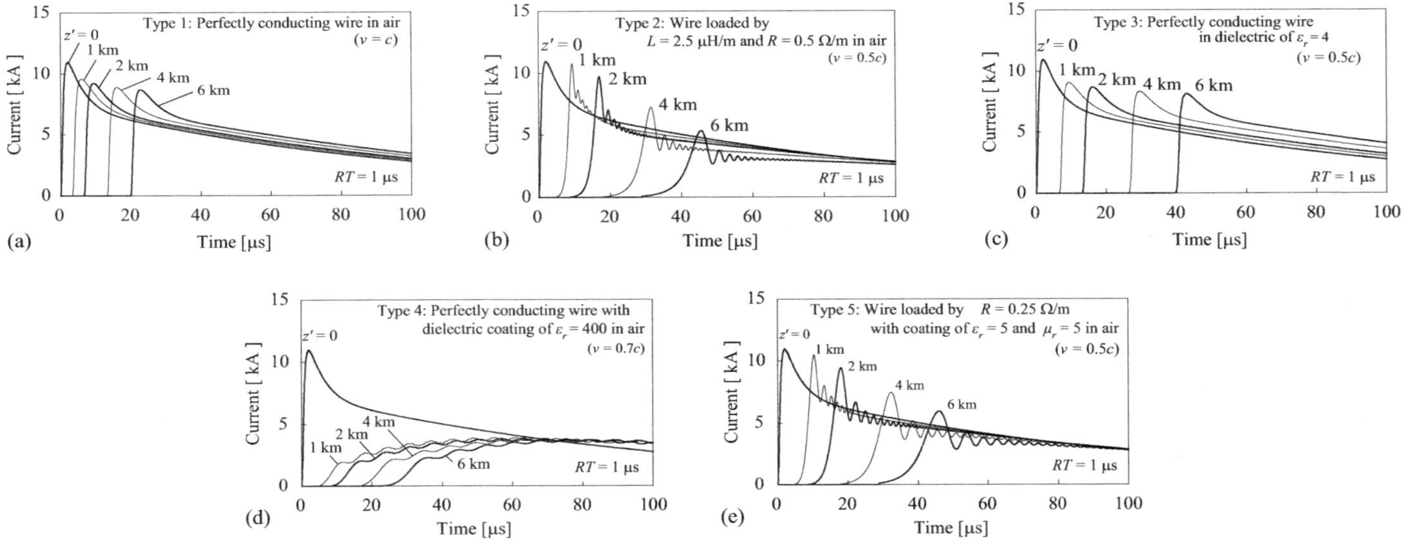

Figure 4.4 Current waveforms at different heights calculated using the FDTD method for five representations of the lightning return-stroke channel shown in Figure 4.3: (a) type 1, (b) type 2, (c) type 3, (d) type 4, and (e) type 5. Adapted from Baba and Rakov [29]

Figure 4.4 shows FDTD-calculated distributions of current along the lightning return-stroke channel for its different representations shown in Figure 4.3. For representation (a) $v = c$, for representations (b), (c), and (e) $v = 0.5c$, and for representation (d) $v = 0.7c$. Note that these speeds were calculated based on times needed for current waves to propagate from $z' = 0$ to 2 km along the vertical channel, which were determined by tracking an intersection point between a straight line passing through 10% and 90% points on the rising part of the current waveform and the time axis. In Figure 4.4(b), (c), and (e), parameters of channel representation were adjusted to achieve the same value of $v = 0.5c$.

It appears from Figure 4.4(d) that in order to reduce the speed of current wave propagating along a vertical wire having a dielectric coating (type 4), which is surrounded by air, to a value similar to the typical measured return-stroke speed [65], the relative permittivity of the dielectric coating would need to be set to a very high value (much higher than in the case of dielectric half space). The current-wave propagation speed decreases with increasing the thickness of the dielectric coating and its relative permittivity, but the dependence is weak [15]. The initial peak of the longitudinal current decays significantly owing to the presence of dielectric coating having very high relative permittivity. It follows from comparison of Figures 4.4(d) and (e) that increasing both relative permittivity and relative permeability of the coating (the effect of $R = 0.25\ \Omega/\text{m}$ is minor) is more efficient in reducing the current-wave propagation speed (also yields more realistic current profile) than increasing the relative permittivity only.

4.4.2 *Comparison of model-predicted electric and magnetic fields with measurements*

In this section, vertical electric and azimuthal magnetic field waveforms, calculated using the FDTD method for different channel representations, are compared with typical measured electric and magnetic field waveforms due to first and subsequent natural lightning return strokes at distances $d = 5$ and 50 km, and triggered lightning strokes at $d = 50$ m.

The following five features, labeled in Figure 4.5, have been identified in electric and magnetic field waveforms measured at distances ranging from 1 to 200 km from first and subsequent natural lightning return strokes [78] and at tens to hundreds of meters from triggered lightning strokes [79]:

- characteristic flattening of vertical electric field at tens to hundreds of meters within 15 μs or so of the beginning of return stroke;
- sharp initial peak in both electric and magnetic field waveforms at a few kilometers and beyond;
- slow ramp following the initial peak in electric field waveforms measured within a few tens of kilometers;
- hump following the initial peak in magnetic field waveforms measured within several tens of kilometers; and
- zero-crossing within tens of microseconds in both electric and magnetic field waveforms measured at 50 km and beyond.

These features have been used as a benchmark in testing the validity of various lightning return-stroke models (e.g., [1]).

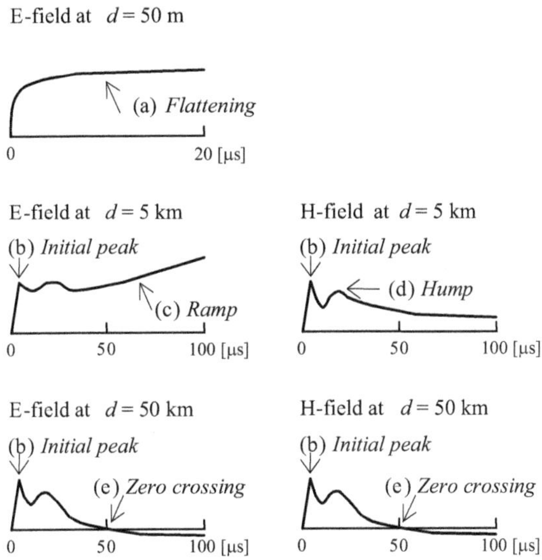

Figure 4.5 Typical features of vertical electric and azimuthal magnetic field waveforms measured at different distances from lightning return strokes [1]

Figures 4.6(a), (b), and (c) show vertical electric field waveforms at $d = 50$ m, 5 km, and 50 km, calculated using the FDTD method for different representations of the lightning return-stroke channel shown in Figure 4.3 (types 1, 2, 3, 4, and 5). Figure 4.6(d) shows FDTD-calculated waveforms of azimuthal magnetic field at $d = 5$ km.

Feature (a) is reproduced by type-2 and 5 models. Feature (b) is reproduced by all model types considered, except for type 4. Feature (c) is reproduced by type-2 and 5 models. Features (d) and (e) are not reproduced by any model considered. Overall, it follows that fields predicted by type-2 and 5 models better match experimental data than those predicted by any other model considered here.

It follows from the electromagnetic field calculations that no electromagnetic model with the input parameters considered can reproduce either feature (d), hump of the magnetic field, or feature (e), zero-crossing of the remote fields. Baba *et al.* [80] have shown, using their engineering return-stroke model, that significant current attenuation within a few tens of meters of the return-stroke channel base is needed to reproduce feature (d) and appreciable attenuation along the upper part of the channel is needed to reproduce feature (e). Note that Thottappillil *et al.* [81] have shown that the TCS model [69] and the DU model [82] with a somewhat different channel-base current waveform whose time to half peak value is 20 μs reproduce features (b)–(e). Further, Cooray *et al.* [83] have found that a horizontal section of the channel inside the cloud may be responsible for the observed zero-crossing in distant fields.

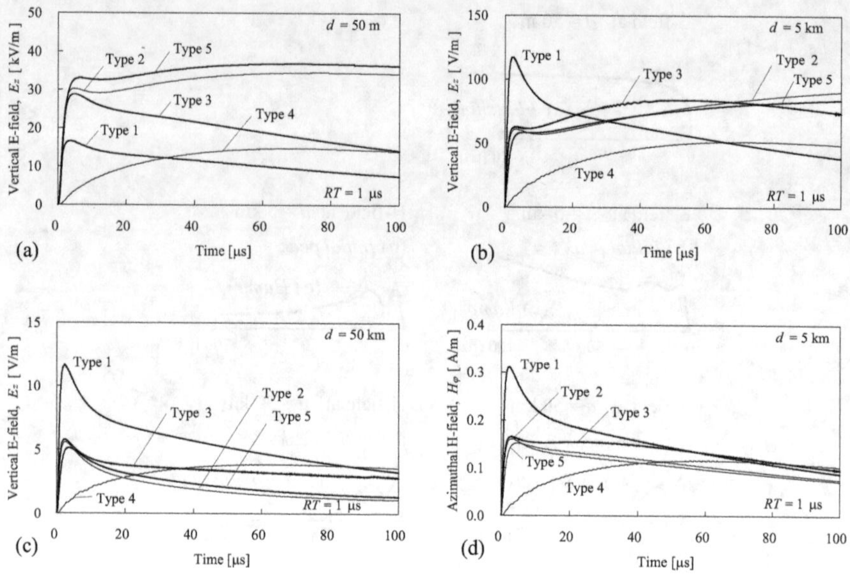

Figure 4.6 Vertical electric field waveforms at (a) d = 50 m, (b) d = 5 km, and (c) d = 50 km, and (d) azimuthal magnetic field waveforms at d = 5 km, calculated using the FDTD method for different representations of the lightning return-stroke channel shown in Figure 4.3 (types 1–5). Adapted from Baba and Rakov [29]

Vertical electric and azimuthal magnetic field waveforms at d = 50 m, 5 km, and 50 km from the lightning channel, calculated using the most promising type-2 and 5 models that are modified here to include nonuniformly distributed series resistance, are considered next. The modified type-2 model has nonuniformly distributed series resistance: R = 2 Ω/m for z' = 0 to 0.5 km, 0.65 Ω/m for z' = 0.5 to 4 km, 1 Ω/m for z' = 4 to 7.5 km, and 10 Ω/m for z' > 7.5 km. The modified type-5 model has nonuniformly distributed series resistance: R = 1 Ω/m for z' = 0 to 0.5 km, 0.3 Ω/m for z' = 0.5 to 4 km, 0.5 Ω/m for z' = 4 to 7.5 km, and 5 Ω/m for z' > 7.5 km. The specified resistance profiles with relatively high resistance within the bottom 0.5 km appear to be consistent with observed light profiles along both natural subsequent return-stroke channels [84] and rocket-triggered lightning channels [85], showing that the light intensity decays significantly with height near ground. The reason for assuming the quite high channel resistance above 7.5 km is to diminish the current above 7.5 km. Note that it is not expected for the vertical lightning channel to extend above 7.5 km.

Figures 4.7(a) and (b) show FDTD-calculated distributions of current along the lightning channel for type-2 and 5 models. Comparisons of Figures 4.7(a) and 4.4(b), and of Figures 4.7(b) and 4.4(e) show that, when distributed resistance is not uniform, a current wave attenuates more significantly below z' = 1 km and above z' = 4 km. The current propagation speed is only slightly influenced by the

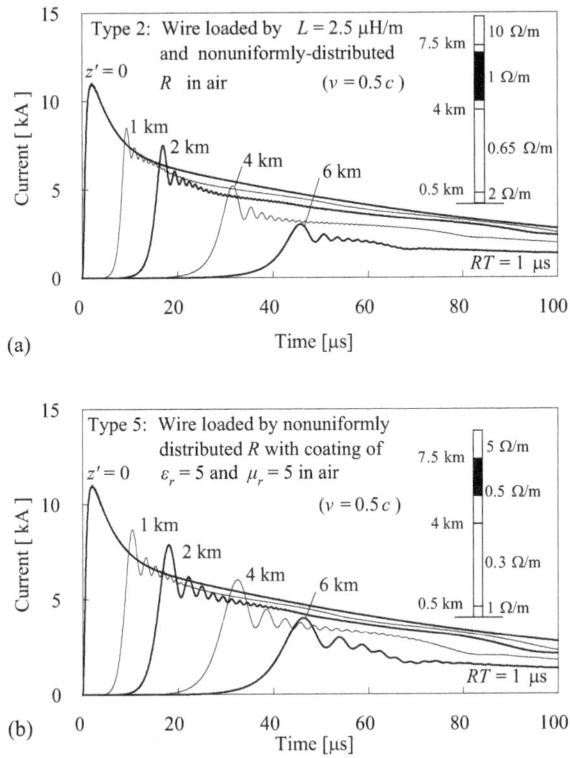

Figure 4.7 Current waveforms at different heights calculated using the FDTD method for (a) type-2 model and (b) type-5 model shown in Figure 4.3 (b) and (e) but with nonuniformly-distributed series resistance. For type-2 model, uniformly distributed additional series inductance is L = 2.5 μH/m, and nonuniformly distributed series resistance is R = 2 Ω/m for z' = 0 to 0.5 km, 0.65 Ω/m for z' = 0.5 to 4 km, 1 Ω/m for z' = 4 to 7.5 km, and 10 Ω/m for $z' >$ 7.5 km. For type-5 model, the relative permittivity and permeability of 10-m-radius cylinder containing the vertical wire are $\varepsilon_r = 5$ and $\mu_r = 5$, and nonuniformly distributed series resistance is R = 1 Ω/m for z' = 0 to 0.5 km, 0.3 Ω/m for z' = 0.5 to 4 km, 0.5 Ω/m for z' = 4 to 7.5 km, and 5 Ω/m for $z' >$ 7.5 km. Adapted from Baba and Rakov [29]

higher distributed resistance or the nonuniformly distributed resistance, and remains at about $v = 0.5c$.

Figures 4.8(a), (b), and (c) show FDTD-calculated vertical electric field waveforms at d = 50 m, 5 km, and 50 km for type-2 and 5 models for the case of nonuniform R. Figure 4.8(d) shows FDTD-calculated waveforms of azimuthal magnetic field at d = 5 km. It is clear from Figure 4.8 that all five features are well reproduced by both type-2 and 5 models with nonuniformly-distributed series resistance.

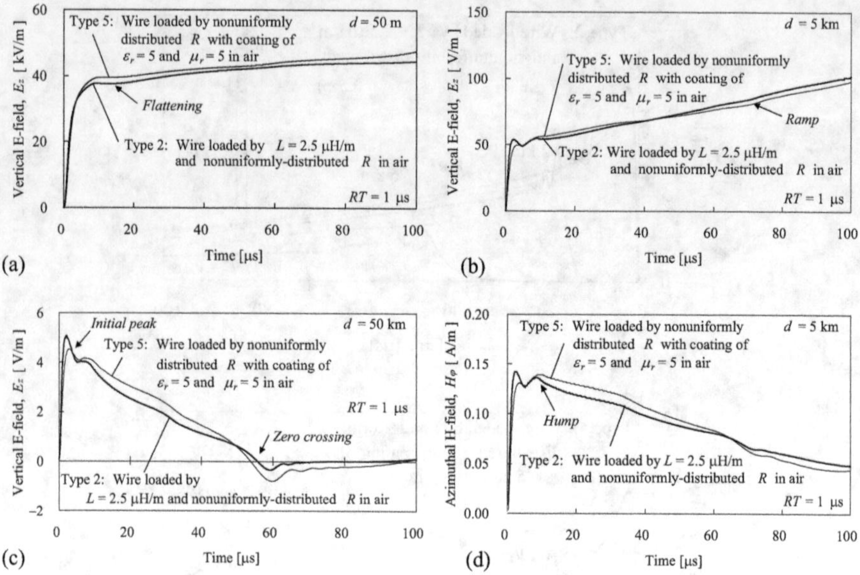

Figure 4.8 Vertical electric field waveforms at (a) d = 50 m, (b) d = 5 km, and (c) d = 50 km, and (d) azimuthal magnetic field waveforms at d = 5 km, calculated using the FDTD method for type-2 and type-5 models shown in Figure 7.3(b) and (e) but with nonuniformly-distributed series resistance. Adapted from Baba and Rakov [29]

4.5 Excitations used in electromagnetic models of the lightning return stroke

In this section, methods of excitation used to date in electromagnetic return-stroke models are described, and distributions of current along a vertical perfectly conducting wire above perfectly conducting ground corresponding to different excitation methods are compared. Methods of excitation used in electromagnetic models are listed below.

1. closing a charged vertical wire at its bottom end with a specified impedance (or circuit),
2. a lumped voltage source (delta-gap electric-field source), and
3. a lumped current source.

Table 4.2 gives a list of journal papers on electromagnetic models of the lightning return stroke that are grouped into three categories depending on the method of excitation.

4.5.1 Closing a charged vertical conducting wire at its bottom end with a specified circuit

Podgorski and Landt [38] have represented a leader/return-stroke sequence by a pre-charged vertical resistive wire representing the lightning channel connected,

Table 4.2 List of journal papers on electromagnetic models of the lightning return stroke grouped into three categories depending on the method of excitation employed

Excitation	**Papers**
Closing charged channel with a specified impedance	Podgorski and Landt [38]
Lumped voltage source (Delta-gap electric-field source)	Moini *et al.* [22,62]
	Baba and Ishii [40,50]
	Kordi *et al.* [10,41]
	Mozumi *et al.* [39]
	Pokharel *et al.* [26,42]
	Pokharel and Ishii [52]
	Miyazaki and Ishii [54,55]
	Baba and Rakov [43]
	Aniserowicz and Maksimowicz [56]
	Khosravi-Farsani *et al.* [57]
	Saito *et al.* [58]
	Tatematsu and Ueda [60]
	Yamanaka *et al.* [61]
	Tatematsu [49]
	Baba and Rakov [11,12,29]
	Shoory *et al.* [28]
Lumped current source	Geranmayeh *et al.* [63]
	Noda *et al.* [51]
	Bonyadi-Ram *et al.* [53]
	Matsuura *et al.* [44]
	Janani *et al.* [27]
	Ishii *et al.* [45]
	Takami *et al.* [46][a]
	Khosravi *et al.* [59]
	Du *et al.* [47]
	Diaz *et al.* [48]

[a]400-Ω lumped resistor is connected in parallel with the current source.

via a nonlinear resistor, to the top of a vertical perfectly conducting wire representing the 553-m-high CN Tower or to the tip of the upward connecting leader emanated from the top of the tower. In their model, closing a charged vertical wire in a grounded circuit constitutes the return-stroke excitation of lightning channel.

4.5.2 Lumped voltage source

A lumped voltage source (delta-gap electric-field source) is located at ground surface (e.g., [62]) or at the top of a grounded strike object (e.g., [58]). This type of source generates a specified electric field, which is independent of magnetic field surrounding the source or current flowing through it. Since such a lumped voltage source has zero internal impedance, its presence in series with the lightning channel and a strike object does not disturb any transient processes in them. If necessary, one could insert a lumped resistor in series with the lumped voltage source to adjust the impedance seen by waves entering the channel from the strike object to a value consistent with the expected equivalent impedance of the lightning channel.

This type of excitation has also been used by Moini *et al.* [22], Baba and Ishii [40,50], Kordi *et al.* [10,41], Mozumi *et al.* [39], Pokharel *et al.* [26,42], Pokharel and Ishii [52], Miyazaki and Ishii [54,55], Baba and Rakov [43], Aniserowicz and Maksimowicz [56], Khosravi-Farsani *et al.* [57], Tatematsu and Ueda [60], Yamanaka *et al.* [61], and Tatematsu [49].

4.5.3 Lumped current source

A lumped current source is located at ground surface (e.g., [28]) or at the top of a grounded strike object (e.g., [51]). If reflected waves returning to the current source are negligible, the use of a lumped current source inserted at the attachment point does not cause any problem. This is the case for a branchless subsequent lightning stroke terminating on flat ground, in which upward connecting leaders are usually neglected and the return-stroke current wave propagates upward from the ground surface. The primary reason for the use of a lumped current source at the channel base is a desire to use directly the channel-base current, known from measurements for both natural and triggered lightning, as an input parameter of the model. When one employs a lumped ideal current source at the attachment point in analyzing lightning strikes to a tall grounded object, the lightning channel, owing to the infinitely large impedance of the ideal current source, is electrically isolated from the strike object, so that current waves reflected from ground cannot be directly transmitted to the lightning channel. Since this is physically unreasonable, a series ideal current source is not suitable for modeling of lightning strikes to tall grounded objects.

This type of excitation has also been used by Baba and Rakov [11,12,29], Geranmayeh *et al.* [63], Bonyadi-Ram *et al.* [53], Matsuura *et al.* [44], Janani *et al.* [27], Ishii *et al.* [45], Takami *et al.* [46], Khosravi *et al.* [59], Du *et al.* [47], and Diaz *et al.* [48].

4.5.4 Comparison of current distributions along a vertical perfectly conducting wire excited by different sources

In this section, distributions of current along a vertical perfectly conducting wire in air energized by different methods of excitation described above are compared. Figure 4.9 shows three representations of lightning return stroke attached to flat perfectly conducting ground by (a) closing a pre-charged vertical perfectly conducting wire of radius 0.23 m in air with a nonlinear resistor (left panel may be viewed as representing leader process and right panel the return-stroke process), (b) a vertical perfectly conducting wire of radius 0.23 m in air excited at its bottom by a 10-m-long lumped voltage source, and (c) a vertical perfectly conducting wire of radius 0.23 m in air excited at its bottom by a 10-m-long lumped current source. The lumped voltage source of the model shown in Figure 4.9(a) generates a ramp-front wave having a magnitude of 10 MV/m (100 MV along the 10-m-long source) and a risetime of 1 μs, while that of the model shown in Figure 4.9(b) generates a ramp-front wave having a magnitude of 500 kV/m (5 MV along the 10-m-long

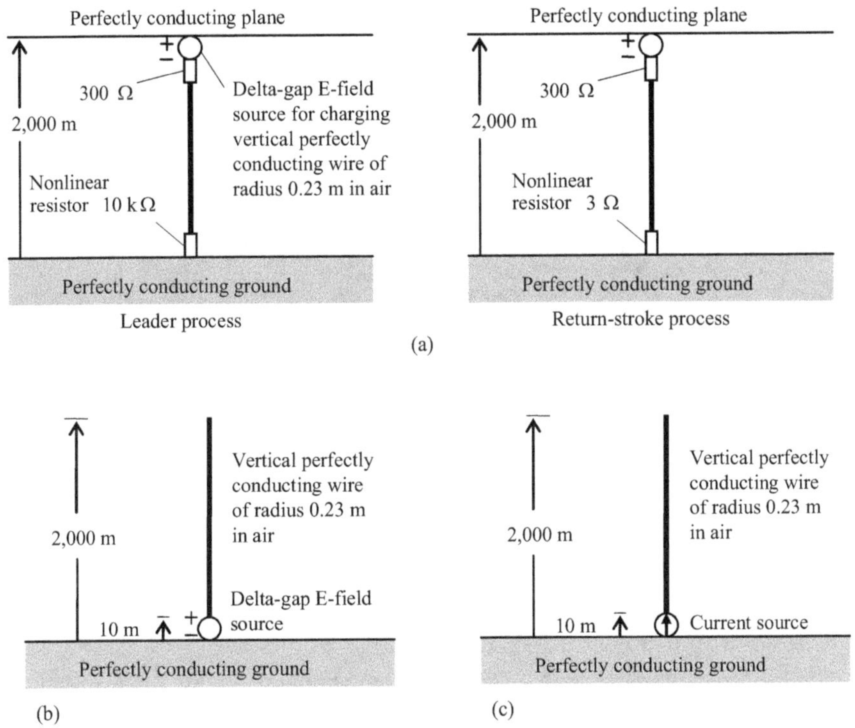

Figure 4.9 Three representations of the lightning return stroke channel above flat perfectly conducting ground by (a) a vertical perfectly conducting wire of radius 0.23 m in air that is essentially open-circuited at its bottom end when being charged by a lumped voltage source at its top (left panel) and then connected to flat ground via a 3-Ω resistor during its discharging (right panel), (b) a vertical perfectly conducting wire of radius 0.23 m in air excited at its bottom by a 10-m-long lumped voltage source, and (c) a vertical perfectly conducting wire of radius 0.23 m in air excited at its bottom by a 10-m-long lumped current source. Adapted from Baba and Rakov [15]

source) and a risetime of 1 μs. The current waveform injected by the lumped current source in Figure 4.9(c) is set to be the same as the resultant current waveform injected by the lumped voltage source in Figure 4.9(b).

Figure 4.10 shows distributions of current along the 0.23-m-radius vertical perfectly conducting wire in air excited by different sources (see Figure 4.9) that are calculated using the FDTD method. It is clear from Figures 4.10(b) and (c) that the distributions of current along the vertical wire excited at its bottom by the lumped voltage source and by the lumped current source are identical. Therefore, resultant electric and magnetic fields generated around the vertical wire are also identical (the use of either voltage or current source makes no difference in electric and magnetic

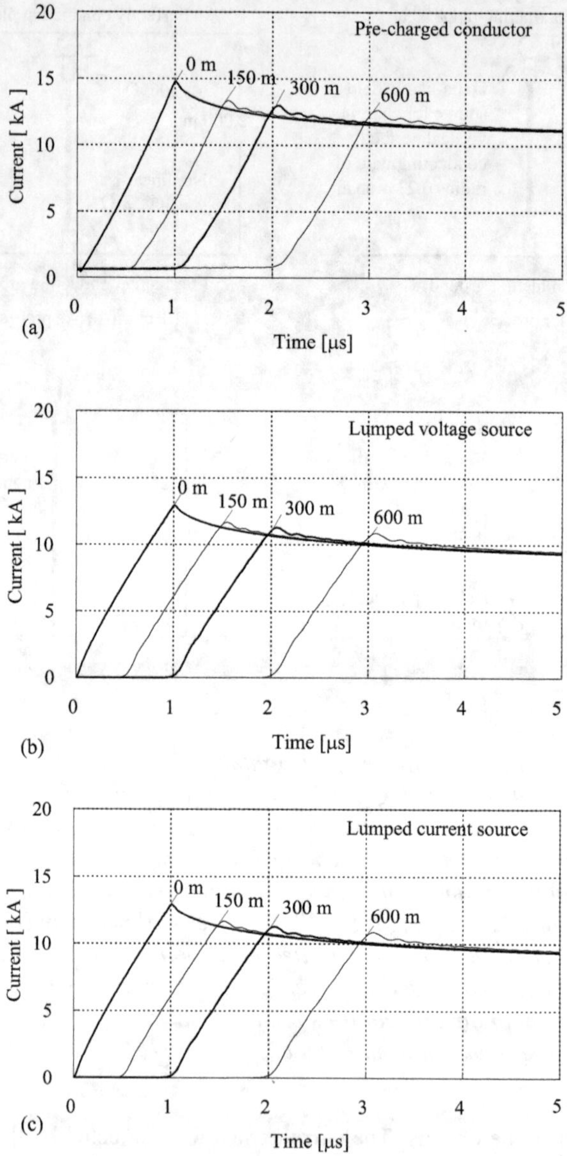

Figure 4.10 Current waveforms at different heights calculated using the FDTD method for three methods of excitation of the lightning return stroke channel shown in Figures 4.9(a), (b), and (c). Adapted from Baba and Rakov [15]

fields in the case of lightning strike to flat ground, if there are no downward reflections in the channel). It is clear from Figures 4.10(a) and (b) that the distribution of current along the charged vertical wire closed with the nonlinear resistor is similar to that along the same vertical wire excited at its bottom by the lumped voltage source.

4.6 Numerical procedures used in electromagnetic models of the lightning return stroke

In this section, numerical procedures used in electromagnetic models of the lightning return stroke are briefly explained. They include (in chronological order of their usage in electromagnetic models):

1. the MoM in the time domain,
2. the MoM in the frequency domain, and
3. the FDTD method.

Table 4.3 includes a list of journal papers on electromagnetic models of the lightning return stroke grouped depending on the numerical procedure used.

Although the finite element method (FEM) (e.g., [71]) has also been employed in simulations of lightning-induced currents in shield conductors of buried cables

Table 4.3 List of journal papers on electromagnetic models of the lightning return stroke grouped into three categories depending on the numerical procedure used

Numerical technique	**Papers**
MoM in the time domain	Podgorski and Landt [38]
	Moini *et al.* [22,62]
	Kordi *et al.* [10], [41]
	Mozumi *et al.* [39]
	Pokharel and Ishii [52]
	Bonyadi-Ram *et al.* [53]
	Janani *et al.* [27]
MoM in the frequency domain	Baba and Ishii [40,50]
	Pokharel *et al.* [26], [42]
	Shoory *et al.* [28]
	Geranmayeh *et al.* [63]
	Pokharel and Ishii [52]
	Miyazaki and Ishii [54,55]
	Aniserowicz and Maksimowicz [56]
	Saito *et al.* [58]
FDTD method	Baba and Rakov [11,12,29,43]
	Noda *et al.* [51]
	Matsuura *et al.* [44]
	Ishii *et al.* [45]
	Khosravi-Farsani *et al.* [57]
	Takami *et al.* [46]
	Khosravi *et al.* [59]
	Tatematsu and Ueda [60]
	Du *et al.* [47]
	Diaz *et al.* [48]
	Yamanaka *et al.* [61]
	Tatematsu [49]

and lightning-induced voltages on overhead wires [86–89], no electromagnetic model of lightning return stroke was employed. A phased-current-source array [11] was used instead. Similarly, in simulations of lightning-induced voltages on an overhead wire and lightning electromagnetic pulses over lossy ground [90,91] based on the transmission-line-matrix or transmission-line-model (TLM) method [92], the lightning return stroke is represented by the phased-current-source array. Further, the HEM model [30, 31] has been used in lightning electromagnetic field and surge simulations, in which the current propagation speed along the lightning channel-representing conductor is reduced by increasing the radius of the conductor for calculating the capacitance (but not the inductance). This is more or less similar to Type 3 or Type 4 of channel representation discussed in Section 4.3. In more recent studies based on the HEM model (e.g. [93–97]), the current propagation speed along the lightning channel is reduced by increasing relative permittivity and permeability in calculating the coupling between segments of lightning channel (Silveira, personal communication 2017). This is more or less similar to Type 5 of channel representation discussed in Section 4.3. Although electromagnetic models of the lightning return stroke can be used in simulations based on the FEM, the TLM method, the HEM model or the PEEC method, the corresponding computational procedures are not discussed here, but they can be found in Baba and Rakov [98].

4.6.1 *Methods of moments (MoMs) in the time and frequency domains*

4.6.1.1 MoM in the time domain

The MoM in the time domain [17, 19] is widely used in analyzing responses of thin-wire metallic structures to external time-varying electromagnetic fields. The entire conducting structure representing the lightning channel is modeled by a combination of cylindrical wire segments whose radii are much smaller than the wavelengths of interest. It is assumed that current I and charge q are confined to the wire axis (thin-wire approximation). Then the so-called electric-field integral equation for a

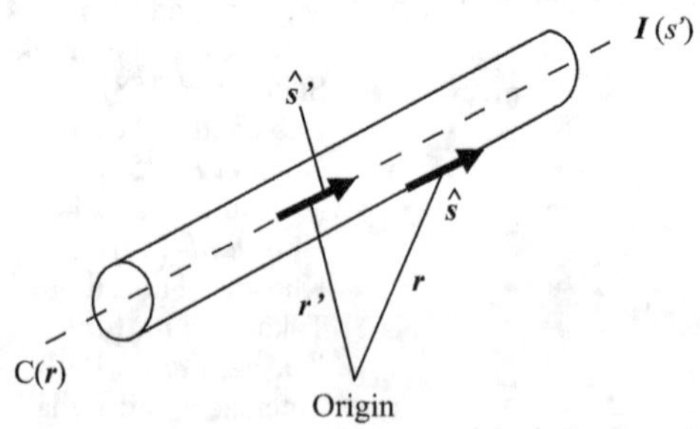

Figure 4.11 Thin-wire segment for MoM-based calculations. Current is confined to the wire axis, and the tangential component of electric field on the surface of the wire is set to zero

perfectly conducting thin wire in air (see Figure 4.11), which is based on Maxwell's equations and expresses the boundary condition on the tangential electric field on the surface of the wire (this field component must be equal to zero), is given by

$$\widehat{s} \cdot \boldsymbol{E}_{inc}(\boldsymbol{r}, t) = \frac{\mu_0}{4\pi} \int_C \left[\frac{\widehat{s} \cdot \widehat{s}'}{R} \frac{\partial I(s', t')}{\partial t'} + c \frac{\widehat{s} \cdot \boldsymbol{R}}{R^2} \frac{\partial I(s', t')}{\partial s'} - c^2 \frac{\widehat{s} \cdot \boldsymbol{R}}{R^3} q(s', t') \right] ds' \tag{4.17}$$

where

$$q(s', t') = -\int_{-\infty}^{t'} \frac{\partial I(s', \tau)}{\partial s'} d\tau,$$

C is an integration path along the wire axis, $\boldsymbol{E}_{inc}$ denotes the incident electric field that induces current I, $\boldsymbol{R} = \boldsymbol{r} - \boldsymbol{r}'$, $\boldsymbol{r}$ and t denote the observation point location (a point on the wire surface) and time, respectively, $\boldsymbol{r}'$ and t' denote the source point location (a point on the wire axis) and time, respectively, s and s' denote the distance along the wire surface at $\boldsymbol{r}$ and that along the wire axis at $\boldsymbol{r}'$, $\widehat{s}$ and $\widehat{s}'$ denote unit vectors tangential to path $\boldsymbol{C}$ in (4.17) at $\boldsymbol{r}$ and $\boldsymbol{r}'$, μ_0 is the permeability of vacuum, and c is the speed of light. The time-dependent current distribution along the wire structure (lightning channel), excited by a lumped source, can be obtained by numerically solving (4.17).

The thin-wire time-domain (TWTD) code [17] (available from the Lawrence Livermore National Laboratory) is based on the MoM in the time domain. One of the advantages of the use of the time-domain MoM is that it can incorporate highly nonlinear effects such as the lightning attachment process (e.g., [38]), although it does not allow lossy ground and wires buried in lossy ground to be incorporated.

This method has also been used by Moini *et al.* [22, 62], Kordi *et al.* [10,41], Mozumi *et al.* [39], Pokharel and Ishii [52], Bonyadi-Ram *et al.* [53], and Janani *et al.* [27].

4.6.1.2 MoM in the frequency domain

The MoM in the frequency domain [18] is widely used in analyzing the electromagnetic scattering by antennas and other metallic structures. In order to obtain the time-varying responses, Fourier and inverse Fourier transforms are employed. The electric-field integral equation derived for a perfectly conducting thin wire in air (see Figure 4.11) in the frequency domain is given by

$$-\widehat{s} \cdot \boldsymbol{E}_{inc}(\boldsymbol{r}) = \frac{j\eta}{4\pi k} \int_C I(s') \left(k^2 \widehat{s} \cdot \widehat{s}' - \frac{\partial^2}{\partial s \, \partial s'} \right) g(\boldsymbol{r}, \boldsymbol{r}') \, ds' \tag{4.18}$$

where

$$g(\boldsymbol{r}, \boldsymbol{r}') = \exp\left(\frac{-jk|\boldsymbol{r} - \boldsymbol{r}'|}{|\boldsymbol{r} - \boldsymbol{r}'|} \right), \quad k = \omega\sqrt{\mu_0 \varepsilon_0}, \quad \eta = \sqrt{\frac{\mu_0}{\varepsilon_0}}$$

ω is the angular frequency, μ_0 is the permeability of vacuum, and ε_0 is the permittivity of vacuum. Other quantities in (4.18) are the same as those in (4.17). Current distribution along the lightning channel can be obtained by numerically solving (4.18).

This method allows lossy ground and wires in lossy ground (for example, grounding of a tall strike object) to be incorporated into the model [99]. The commercially available numerical electromagnetic codes (e.g., NEC-2 [100], and NEC-4 [101]) are based on the MoM in the frequency domain.

This method has been used by Baba and Ishii [40, 50], Pokharel *et al.* [26,42], Shoory *et al.* [28], Geranmayeh *et al.* [63], Pokharel and Ishii [52], Miyazaki and Ishii [54,55], Aniserowicz and Maksimowicz [56], and Saito *et al.* [58].

4.6.2 Finite-difference time-domain (FDTD) method

The FDTD method [20] employs a simple way to discretize Maxwell's equations in differential form. In the Cartesian coordinate system, it requires discretization of the entire space of interest into small cubic or rectangular-parallelepiped cells. Cells for specifying or computing electric field (electric field cells) and magnetic field cells are placed relative to each other as shown in Figure 4.12. Electric and magnetic fields of the cells are calculated using the discretized Maxwell's equations given below:

$$
\begin{aligned}
E_z^{\,n}\left(i,j,k+\frac{1}{2}\right) &= \frac{1-\sigma(i,j,k+1/2)\Delta t/[2\varepsilon(i,j,k+1/2)]}{1+\sigma(i,j,k+1/2)\Delta t/[2\varepsilon(i,j,k+1/2)]} \\
&\quad \times E_z^{\,n-1}\left(i,j,k+\frac{1}{2}\right) \\
&\quad + \frac{\Delta t/\varepsilon(i,j,k+1/2)}{1+\sigma(i,j,k+1/2)\Delta t/[2\varepsilon(i,j,k+1/2)]}\frac{1}{\Delta x\Delta y} \\
&\quad \times \begin{bmatrix} H_y^{\,n-\frac{1}{2}}\left(i+\frac{1}{2},j,k+\frac{1}{2}\right)\Delta y - H_y^{\,n-\frac{1}{2}}\left(i-\frac{1}{2},j,k+\frac{1}{2}\right)\Delta y \\ -H_x^{\,n-\frac{1}{2}}\left(i,j+\frac{1}{2},k+\frac{1}{2}\right)\Delta x + H_x^{\,n-\frac{1}{2}}\left(i,j-\frac{1}{2},k+\frac{1}{2}\right)\Delta x \end{bmatrix}
\end{aligned}
\tag{4.19}
$$

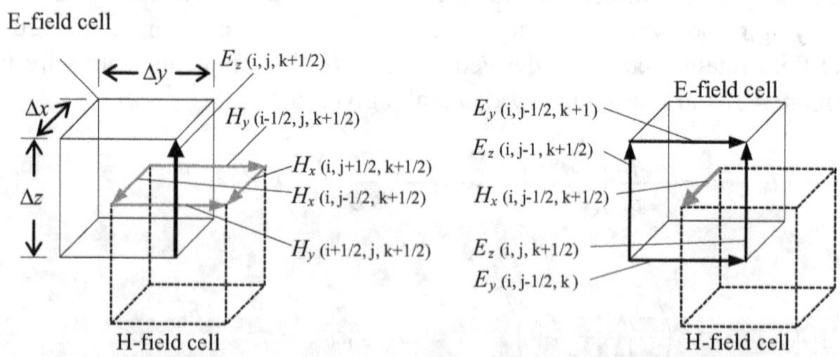

Figure 4.12 Placement of electric-field and magnetic-field cells for solving discretized Maxwell's equations using the FDTD method

$$
\begin{aligned}
H_x^{n+\frac{1}{2}}\left(i,j-\frac{1}{2},k+\frac{1}{2}\right) &= H_x^{n-\frac{1}{2}}\left(i,j-\frac{1}{2},k+\frac{1}{2}\right) \\
&+ \frac{\Delta t}{\mu(i,j-1/2,k+1/2)}\frac{1}{\Delta y \Delta z} \\
&\times \begin{bmatrix} -E_z^n\left(i,j,k+\frac{1}{2}\right)\Delta z + E_z^n\left(i,j-1,k+\frac{1}{2}\right)\Delta z \\ +E_y^n\left(i,j-\frac{1}{2},k+1\right)\Delta y - E_y^n\left(i,j-\frac{1}{2},k\right)\Delta y \end{bmatrix}
\end{aligned}
\tag{4.20}
$$

Equation (4.19), which is based on Ampere's law, is an equation updating the z component of electric field, $E_z(i, j, k+1/2)$, at point $x = i\Delta x$, $y = j\Delta y$, and $z = (k+1/2)\Delta z$, and at time $t = n\Delta t$. Equation (4.20), which is based on Faraday's law, is an equation updating the x component of magnetic field, $H_x(i, j-1/2, k+1/2)$, at point $x = i\Delta x$, $y = (j-1/2)\Delta y$, and $z = (k+1/2)\Delta z$, and at time $t = (n+1/2)\Delta t$. Equations updating x and y components of electric field, and y and z components of magnetic field can be written in a similar manner. Note that $\sigma(i, j, k+1/2)$ and $\varepsilon(i, j, k+1/2)$ are the conductivity and permittivity at point $x = i\Delta x$, $y = j\Delta y$, and $z = (k+1/2)\Delta z$, respectively, $\mu(i, j-1/2, k+1/2)$ is the permeability at point $x = i\Delta x$, $y = (j-1/2)\Delta y$, and $z = (k+1/2)\Delta z$. By updating electric and magnetic fields at every point using (4.19) and (4.20), transient fields throughout the computational domain are obtained. Since the material constants of each cell can be specified individually, a complex inhomogeneous medium can be analyzed easily.

In order to analyze fields in unbounded space, an absorbing boundary condition (e.g., [77]) has to be set on each plane which limits the space to be analyzed, so as to minimize reflections there. The FDTD method allows one to incorporate wires buried in lossy ground, such as strike-object grounding electrodes [51], and nonlinear effects [102]. Note that more details about the FDTD method and its application to lightning electromagnetic field and surge simulations are found in works of Baba and Rakov [98,103].

This method was also used by Baba and Rakov [11,12,29,43], Matsuura *et al.* [44], Ishii *et al.* [45], Khosravi-Farsani *et al.* [57], Takami *et al.* [46], Khosravi *et al.* [59], Tatematsu and Ueda [60], Du *et al.* [47], Diaz *et al.* [48], Yamanaka *et al.* [61], and Tatematsu [49].

4.6.3 Comparison of current distributions along a vertical perfectly conducting wire calculated using different numerical procedures with those predicted by Chen's analytical equation

In this section, distributions of current along a channel-representing vertical wire, calculated using MoMs in the time and frequency domains and the FDTD method, are compared with that based on Chen's analytical equation (4.1) that is shown in Figure 4.1. Figure 4.13 shows configuration to be used for comparison of different

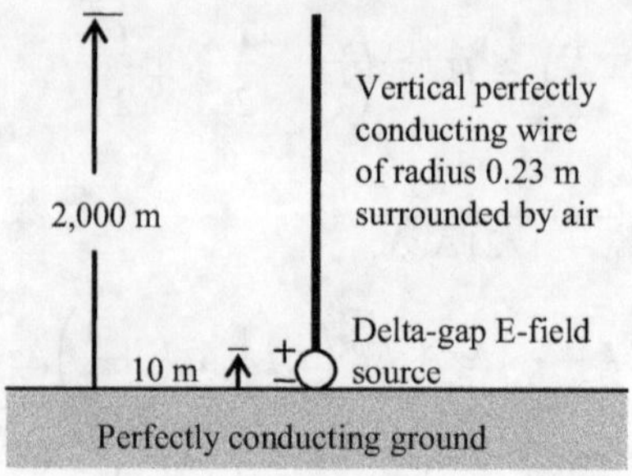

Figure 4.13 A vertical perfectly conducting wire of radius 0.23 m in air above perfectly conducting ground excited at its bottom by a 10-m-long lumped voltage source. The source produces a ramp-front wave having a magnitude of 500 kV/m (5 MV along the 10-m-long source) and a risetime of 1 μs. This configuration was used for comparison of different numerical procedures employed in electromagnetic models. Adapted from Baba and Rakov [15]

numerical procedures: a vertical perfectly conducting wire of radius 0.23 m in air located above perfectly conducting ground and excited at its bottom by a 10-m-long lumped voltage source. The lumped voltage source produces a ramp-front wave having a magnitude of 500 kV/m (5 MV along the 10-m-long source) and a risetime of 1 μs. Figure 4.14(a), (b), and (c) shows current waveforms at different heights calculated using the TWTD code (MoM in the time domain), the NEC-2 code (MoM in the frequency domain), and the FDTD method, respectively. As expected, waveforms calculated using the three different approaches agree well. Further, they all agree reasonably well with those calculated using Chen's analytical equation (see Figure 4.1).

For the TWTD calculations, the 2,000-m-long vertical conducting wire was divided into 10-m-long segments and the response was calculated up to 5 μs with a 33.3-ns increment. For the NEC-2 calculations, the vertical conducting wire was divided into 10-m-long segments, and the responses were calculated from 9.77 kHz to 10 MHz with a 9.77-kHz increment (corresponded to a time range from 0 to 102 μs with a 50-ns increment). In order to suppress non-physical oscillations caused by the NEC-2 numerical procedure, the vertical wire above 1,500 m from the bottom was loaded by distributed series resistance. This resistive loading did not influence the current at the bottom ($z' = 0$) for the first 10 μs and that at a height of 600 m for the first 8 μs. For the FDTD calculations, the vertical wire of 0.23 m radius was replaced by a zero-radius perfectly conducting wire placed in the working volume of 60 m × 60 m × 2,300 m, which was divided into 1 m × 1 m × 10 m cells. When cells having a cross-sectional area of 1 m × 1 m are used, the vertical (z-directed) zero-radius perfectly conducting wire in air has an equivalent radius of 0.23 m [104]. Perfectly matched layers (PML) [105] (absorbing boundaries) were set at the top and sides of the working volume in order to minimize reflections there. The time increment was set to 2 ns.

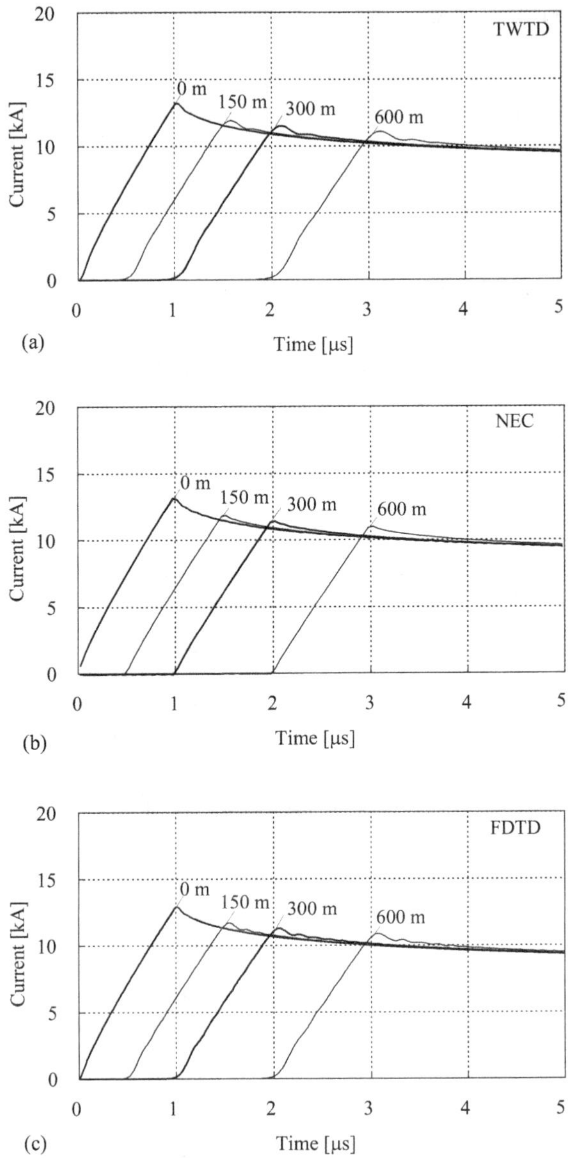

Figure 4.14 Current waveforms at different heights calculated using (a) the TWTD code (based on the MoM in the time domain), (b) the NEC-2 code (based on the MoM in the frequency domain), and (c) the FDTD method for the configuration shown in Figure 4.13. Adapted from Baba and Rakov [15]

4.7 Applications of electromagnetic models of the lightning return stroke

In this section, applications of electromagnetic models of the lightning return stroke to studying lightning and its effects are reviewed [16, 98, 106]. These include

1. strikes to flat ground;
2. strikes to free-standing tall objects;
3. strikes to overhead power transmission lines;
4. strikes to overhead power distribution lines; and
5. strikes to wire-mesh-like structures.

Table 4.4 gives a list of journal papers for each of these four configurations.

Table 4.4 List of journal papers on applications of electromagnetic models of the lightning return stroke

Configuration	Papers
Strike to flat ground	Moini *et al.* [22,62]
	Kordi *et al.* [10]
	Baba and Ishii [40]
	Pokharel *et al.* [26][a]
	Baba and Rakov [11,12]
	Shoory *et al.* [28][a]
	Geranmayeh *et al.* [63]
	Pokharel and Ishii [52]
	Bonyadi-Ram *et al.* [53]
	Baba and Rakov [29][a]
	Janani *et al.* [27]
	Khosravi-Farsani *et al.* [57][a]
	Podgorski and Landt [38]
	Baba and Ishii [50]
	Kordi *et al.* [41]
	Pokharel *et al.* [42][a]
	Baba and Rakov [43][a]
	Saito *et al.* [58][a]
	Khosravi *et al.* [59][a]
	Du *et al.* [47]
	Yamanaka *et al.* [61][a]
Strike to power transmission line	Mozumi *et al.* [39]
	Noda *et al.* [51][a]
	Pokharel and Ishii [52]
	Miyazaki and Ishii [55][a]
	Takami *et al.* [46]
	Tatematsu and Ueda [60][a]
	Tatematsu [49][a]
Strike to power distribution line	Matsuura *et al.* [44]

[a]Lossy ground is considered in finding current distribution along the lightning channel.

4.7.1 Strikes to flat ground

Moini *et al.* [22], Kordi *et al.* [10], Baba and Ishii [40], Shoory *et al.* [28], Geranmayeh *et al.* [63], Bonyadi-Ram *et al.* [53], and Baba and Rakov [29] have calculated waveforms of vertical electric and azimuthal magnetic fields due to lightning return strokes at different distances from the lightning channel attached to flat ground, and compared them with typical measured waveforms of electric and magnetic fields [78,79]. In these calculations, a typical subsequent-stroke waveform of channel-base current [3] and a typical propagation speed of return-stroke wavefront were used. Note that Bonyadi-Ram *et al.* [53] incorporated additional distributed series inductance, which increased with increasing height along the channel. Typical features of vertical electric and azimuthal magnetic field waveforms measured at different distances from lightning return strokes are described in Section 4.4.2 and illustrated in Figure 4.5. Khosravi-Farsani *et al.* [57], using the FDTD method, have calculated the horizontal electric field due to lightning strikes to flat ground of conductivity ranging from 0.1 mS/m to 10 mS/m to examine the validity of the approximate formula for horizontal electric field, the so-called Cooray-Rubinstein formula [107]. The lightning channel was represented by a vertical conductor having additional distributed series inductance of 3 μH/m and distributed series resistance of 1 Ω/m.

Moini *et al.* [62], Pokharel *et al.* [26], Pokharel and Ishii [52], and Janani *et al.* [27] have calculated transient induced voltages on overhead wires due to lightning strikes to flat ground. These studies are reviewed below.

Moini *et al.* [62] have calculated transient voltages on overhead perfectly-conducting wires of different geometries such as parallel and non-parallel wires above flat perfectly conducting ground using the MoM in the time domain. In order to find the distribution of current along the lightning channel, they represented it by a vertical perfectly-conducting wire, which was excited at its bottom by a lumped voltage source. The wire was surrounded by a dielectric medium with a relative permittivity of 4 that occupied the entire half space above ground. The speed of the current wave propagating along the wire was about 0.5*c*. Using the resultant distribution of current along this channel-representing vertical wire and replacing the artificial dielectric medium by air, they calculated transient voltages induced on the overhead wires. The authors conclude that scattering-theory approach is more appropriate in calculating induced effects on nonuniform wires or complex-shape wires than that based on field-to-conductor electromagnetic coupling models (e.g., [108]) based on transmission line theory (telegrapher's equations with source terms).

Pokharel *et al.* [26] have calculated transient voltages on a 25-m-long horizontal overhead perfectly-conducting wire above flat ground having conductivity 0.06 S/m shown in Figure 4.15, using the Numerical Electromagnetic Code (NEC-2) [100] that is based on the MoM in the frequency domain. They represented the lightning channel by a 28-m-long vertical 0.5-Ω/m resistive wire having additional distributed series inductance of 6 μH/m. The wire was excited at its bottom by a lumped voltage source in series with 750-Ω lumped resistance. The speed of the

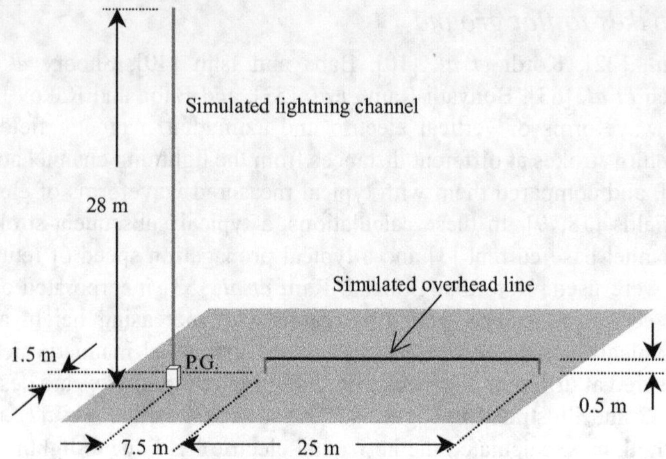

Figure 4.15 *Lightning interaction with a 25-m-long horizontal perfectly conducting wire above flat ground having conductivity 0.06 S/m represented using the NEC-2 code. Lightning channel is represented by a vertical wire loaded by distributed series resistance of 0.5 Ω/m and additional distributed series inductance of 6 μH/m, with the current-wave propagation speed being about 0.43c. One end of the horizontal wire is located at distances x = 7.5 m and y = 1.5 m from the lightning channel, as in Ishii* et al. *'s (1999) small-scale experiment. Both ends of the horizontal wire are terminated in 430-Ω resistance in parallel with 20-pF capacitance. Adapted from Pokharel* et al. *[26]*

current wave propagating along the wire was about 0.43*c*. Induced voltages were computed within the first 300 ns, so that they are not influenced by reflections from the open end of the 28-m-long vertical wire. Figure 4.16(a) and (b) shows calculated induced-voltage waveforms at the close and remote ends of the horizontal wire, respectively, and those measured by Ishii *et al.* [109]. Calculated waveforms agree well with corresponding measured waveforms. This work showed for the first time that voltages induced on an overhead wire above lossy ground could be calculated reasonably accurately using the NEC-2 code.

Pokharel and Ishii [52] have calculated transient voltages on a 500-m-long horizontal overhead perfectly-conducting wire above flat perfectly-conducting ground, using the thin-wire time-domain (TWTD) code [17] based on the MoM in the time domain. A nonlinear element simulating a surge arrestor was connected between the wire at its center point and ground. The lightning return-stroke channel was represented by a vertical 0.6-Ω/m resistive wire having additional distributed series inductance of 6 μH/m that was excited at its bottom by a lumped voltage source. The speed of the current wave propagating along the wire was about 0.48*c*. The use of TWTD code allows one to incorporate nonlinear elements, but makes it impossible to consider the frequency-dependent effects of lossy ground.

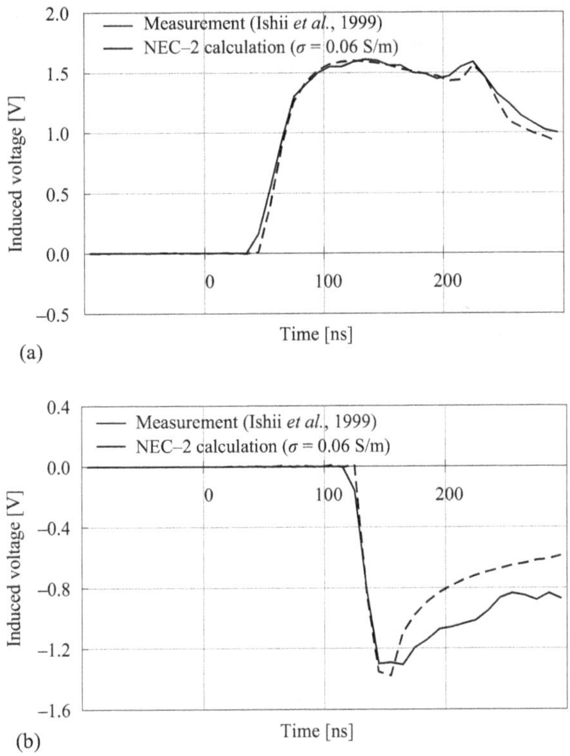

Figure 4.16 Waveforms of voltage induced at (a) close and (b) remote ends of the 25-m-long horizontal wire above flat ground measured by Ishii et al. *[109] and those calculated, using the NEC-2 code, for ground conductivity 0.06 S/m by Pokharel* et al. *[26]. Adapted from Pokharel* et al. *[26]*

Janani *et al.* [27], using the MoM in the time domain, have calculated transient voltages on vertically and horizontally configured three-phase overhead power lines above flat perfectly-conducting ground. The lines were equipped with a shield wire and surge arresters. The lightning return-stroke channel was represented by a vertical 0.5-Ω/m resistive wire having additional distributed series inductance of 8 μH/m that was excited at its bottom by a lumped current source.

4.7.2 Strikes to free-standing tall object

Podgorski and Landt [38], using the modified TWTD code [17] that is based on the MoM in the time domain, have represented a lightning strike to the 553-m-high CN Tower by a resistive (0.7 Ω/m) vertical wire with a nonlinear resistance (10 kΩ prior to the attachment and 3 Ω after the attachment) connected between the bottom of the wire and the top of the CN Tower (or the top of uncharged wire simulating the upward connecting leader from the tower). The CN Tower was represented by a perfectly conducting wire. The calculated waveform of current near the top of the tower was found to be similar to the corresponding measured waveform.

Kato *et al.* [21] have calculated waveforms of lightning current and associated electric and magnetic fields 200 m from the strike point, assuming perfectly conducting ground, due to lightning strikes to the 553-m-high CN Tower and to the 168-m-high Peissenberg tower using the MoM in the time domain, and compared them with corresponding measured waveforms. Baba and Ishii [50] have calculated electric and magnetic fields 2 km and 630 m from the strike point, assuming perfectly conducting ground, due to lightning strikes to the CN Tower and the 200-m-high Fukui chimney using the NEC-2 code, and compared those with corresponding measured waveforms. Figure 4.17 shows NEC-2-calculated waveforms for the Fukui-chimney case along with the corresponding waveforms measured by Goshima *et al.* [110]. NEC-2-calculated waveforms agree well with the corresponding measured waveforms. Miyazaki and Ishii [24] have carried out similar calculations using the FDTD method and shown that FDTD-calculated waveforms agree well with corresponding measured waveforms. Kordi *et al.* [41] have calculated waveforms of lightning current and associated electric and magnetic fields 2 km away from the strike point, assuming perfectly conducting ground, due to a lightning strike to the CN Tower using the MoM in the time domain, and compared them with corresponding measured waveforms. In these works, except for Kordi *et al.*'s one, the lightning channel was represented by a vertical wire having additional distributed series inductance, and the lightning channel and the tall strike object were excited by a delta-gap electric-field source inserted between them. In Kordi *et al.*'s work [41], the lightning channel was represented by a resistive wire in air and excited by a delta-gap electric-field source.

Pokharel *et al.* [42] have calculated, using the NEC-2 code, induced voltages on an overhead wire due to a lightning strike to the 200-m-high Fukui chimney and compared those with corresponding measured voltage waveforms [111]. They represented the lightning channel by a vertical wire having distributed series resistance of 1 Ω/m and additional distributed series inductance 9 μH/m. The lightning channel and the chimney were excited by a delta-gap electric-field source in series with a lumped 100-Ω resistor inserted between them. The propagation speed of current wave along the channel was about 0.37*c*. Figure 4.18 shows the plan view of the overhead wire and the chimney. Figure 4.19 shows waveforms of current measured by Michishita *et al.* [111] at the top of the chimney and voltages induced on the overhead wire near its terminations. Figure 4.20 shows voltage waveforms calculated for the same configuration by Pokharel *et al.* [42] assuming the ground conductivity to be 0.02 S/m, for which the best agreement with measured waveforms was found. Induced voltages corresponding to perfectly conducting ground are also shown for comparison.

Kato *et al.* [21] studied the influence of the inclination of lightning channel attached to the Peissenberg tower on lightning electromagnetic fields. Pokharel *et al.* [42] investigated the influence of the inclination of lightning channel attached to the Fukui chimney on lightning-induced voltages.

Baba and Rakov [43] have calculated, using the FDTD method in the 2D-cylindrical coordinate system, the close vertical and horizontal electric fields and

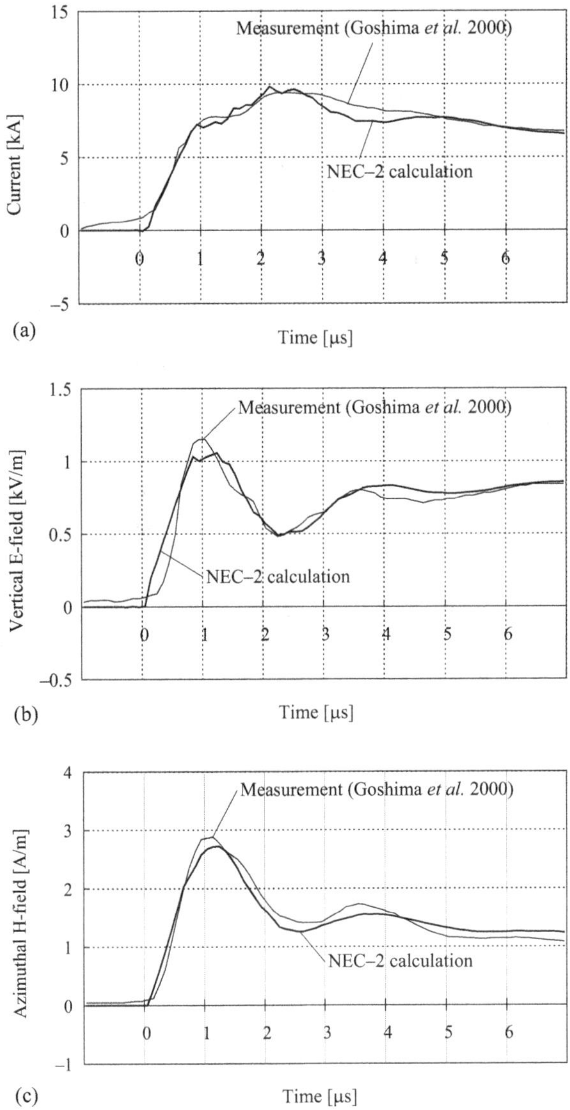

Figure 4.17 *Waveforms of (a) current at the top of the 200-m-high Fukui chimney, (b) vertical electric field, and (c) azimuthal magnetic field 630 m from the chimney, calculated by Baba and Ishii [50] using the NEC-2 code and assuming perfectly conducting ground, and those measured by Goshima* et al. *[110]. The lightning channel is represented by a vertical conductor having distributed series resistance of 1 Ω/m and additional distributed series inductance of 3 μH/m, with the current-wave propagation speed being about 0.5c. The 200-m-high chimney is represented by a vertical perfectly conducting wire. The lightning channel and the chimney are excited by a delta-gap electric-field source in series with a 400-Ω lumped resistor. Adapted from Baba and Ishii [50]*

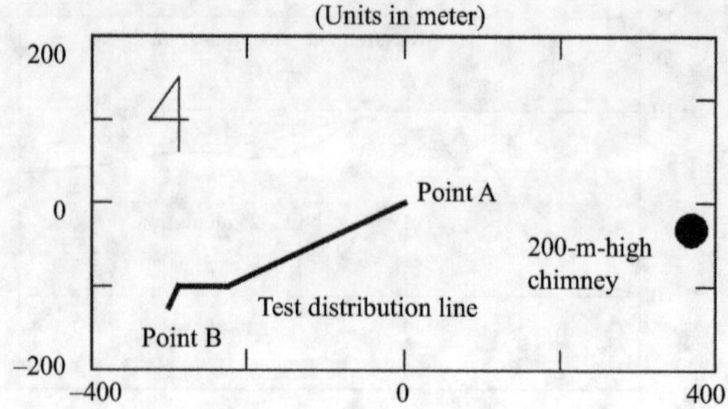

Figure 4.18 Plan view of a single overhead wire and a nearby 200-m-high chimney. Voltages on the overhead wire induced by a lightning strike to the chimney were calculated using the NEC-2 code by Pokharel et al. *[42]. The lightning channel is represented by a vertical conductor having distributed series resistance of 1 Ω/m and additional distributed series inductance of 9 μH/m, with the current-wave propagation speed being about 0.37c. The 200-m-high chimney is represented by a vertical perfectly conducting wire. The lightning channel and the chimney are excited by a delta-gap electric-field source in series with a lumped 100-Ω resistor inserted between them. Adapted from Pokharel* et al. *[42]*

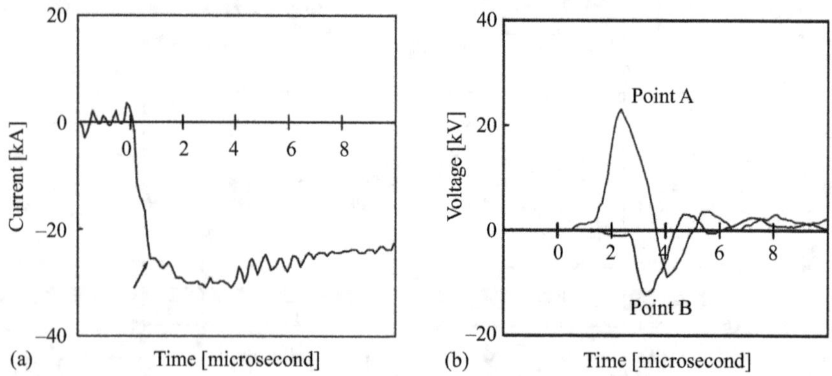

Figure 4.19 Waveforms of (a) current at the top of the 200-m-high chimney and (b) voltages induced on the overhead wire near its terminations, measured by Michishita et al. *(2003). Adapted from Pokharel* et al. *[42]*

the azimuthal magnetic fields generated on the lossy ground surface by lightning strikes to 160-m- and 553-m-high towers that represent the Peissenberg tower in Germany and the CN Tower in Canada, respectively. The lightning channel was represented by a vertical perfectly conducting wire of radius 68 mm in air. The fields were calculated at distances from the bottom of the tower ranging from 5 to

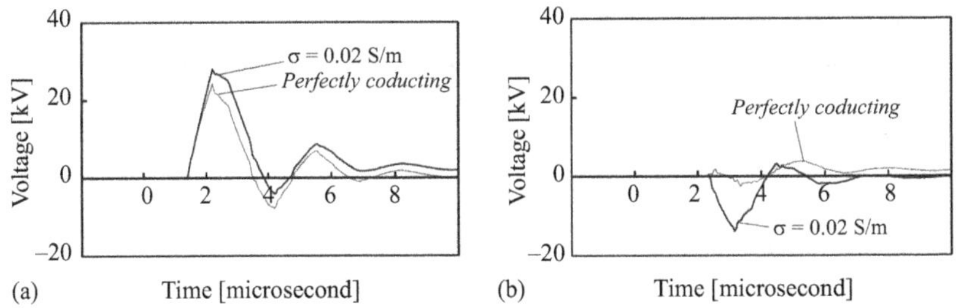

Figure 4.20 Waveforms of voltage induced on the overhead wire near the terminations (a) closer to the chimney (point A in Figure 4.18), and (b) farther from it (point B in Figure 4.18), calculated by Pokharel et al. *[42] assuming perfectly conducting ground and ground conductivity 0.02 S/m. Adapted from Pokharel* et al. *[42]*

100 m for the 160-m-high tower and from 10 to 300 m for the 553-m-high tower. In all cases considered, waveforms of horizontal electric field and azimuthal magnetic field are not much influenced by the presence of strike object, while waveforms of vertical electric field are. Waveforms of vertical electric field are essentially unipolar (as they are in the absence of strike object) when the ground conductivity is 10 mS/m (the equivalent transient grounding impedance is several ohm) or greater. For the 160-m-high tower and for the ground conductivity being equal to 1 and 0.1 mS/m, waveforms of vertical electric field become bipolar (exhibit polarity change) at distances within 10 m and within 50 m, respectively. The source of opposite polarity vertical electric field is the potential rise at the object base.

Saito *et al.* [58] have calculated, using the NEC-4 code, lightning currents in the Tokyo Skytree, the height of which is 634 m, and vertical electric fields at distances of 27, 57, and 101 km from the tower. They also compared calculated waveforms with the corresponding measured waveforms. The ground conductivity values of 0.003 S/m, 0.02 S/m, and infinity were considered. A non-vertical 3D lightning channel, reconstructed on the basis of optical observations, was employed, which was represented by a wire having distributed series resistance of 0.3 Ω/m and additional distributed series inductance 6 μH/m. The propagation speed of current wave along the channel was about 0.5*c*. The lightning channel and the tower were excited by a lumped voltage source. The computed waveforms of electric field with 0.003 or 0.02 S/m agree well with the corresponding measured ones.

Khosravi *et al.* [59] have calculated, using the FDTD method in the 2D-cylindrical coordinate system, lightning currents in a 200-m high tower on a 400-m high hill, as well as the horizontal electric field and azimuthal magnetic field at a distance of 1 km from the tower. The ground conductivity was assumed to be 1 mS/m. The lightning channel was represented by a vertical wire having distributed series resistance of 1 Ω/m and additional distributed series inductance of 4.25 μH/m. The propagation speed of current wave along the channel was about 0.56*c*. The lightning channel and the tower were excited by a lumped current source. Magnitudes of horizontal electric field and azimuthal magnetic field were enhanced by the presence of both the tower and the hill.

Du *et al.* [47] have calculated, using the FDTD method, currents and voltages on electrical wiring in a tall building struck by lightning. The building structural material was ignored, a 90-m long, single down-conductor of its lightning protective system was represented by a simple vertical conductor, and electrical wiring was represented by parallel perfect conductors whose terminals were open or connected via lumped elements (or surge protective devices). Lightning channel was represented by a vertical perfectly conducting wire, and thus, the speed of the current wave propagating along the wire was essentially equal to the speed of light. The lightning channel and the down-conductor were excited by a lumped current source. The ground was assumed to be perfectly conducting. They have shown that induced currents and voltages are higher on the top and bottom floors of the building.

Yamanaka *et al.* [61] have calculated, using the FDTD method in the 2D-cylindrical coordinate system, lightning currents in the Tokyo Skytree, and vertical electric fields at distances of 27, 57, and 101 km from the tower. The ground conductivity was set to 0.02 S/m. The lightning channel was represented by a vertical straight wire having distributed series resistance of 0.7 Ω/m and additional distributed series inductance of 13 μH/m. The propagation speed of current wave along the channel was about 0.5*c*. The lightning channel and the tower were excited by a lumped voltage source. The computed waveforms of electric field with ground conductivity value of 0.02 S/m generally agree with corresponding measured ones. Note that they have also carried out simulations with a vertical phase-current-source array with a height-dependent return-stroke speed, which equivalently represents the Tokyo Skytree and a tortuous lightning channel attached to its top.

4.7.3 Strikes to overhead power transmission lines

Mozumi *et al.* [39] have calculated, using the TWTD code, voltages across insulators of a 500-kV double-circuit power transmission line tower with two overhead ground wires, for the case when the tower top is struck by lightning and thereby back-flashover occurs across the insulator of one phase. Ground was assumed to be perfectly conducting. In order to analyze back-flashover using the TWTD code, they modified it to incorporate a flashover model [112]. For the TWTD calculations, the lightning channel was represented by a vertical perfectly conducting wire of radius 0.1 m in air. The lightning channel and the tower were excited by a lumped voltage source in series with a 5-kΩ lumped resistor inserted between them.

Noda *et al.* [51] have calculated, using the FDTD method, voltages across insulators of a 500-kV double-circuit power transmission line tower, in the case that the tower top is struck by lightning. In their calculations, lightning channel was represented by a 0.23-m-radius [104] vertical perfectly conducting wire having additional distributed series inductance of 10 μH/m, and the speed of the current wave propagating along the wire was 0.33*c*. The lightning channel and the tower were excited by a lumped current source inserted between them. Ground was assumed to have conductivity of 10 mS/m.

Pokharel and Ishii [52] have calculated, using the TWTD and NEC-2 codes, the transient voltage across an insulator of a simplified 60-m-high transmission-line tower, assuming perfectly conducting ground. The lightning channel was represented by a vertical perfectly conducting wire. The speed of the current wave propagating along the wire was essentially equal to the speed of light. The lightning channel and the tower were excited by a lumped voltage source in series with a 5-kΩ resistor inserted between the lightning channel and the tower. The TWTD-calculated voltage waveform agrees well with the NEC-2-calculated waveform.

Miyazaki and Ishii [55] have calculated, using the NEC-4 code, vertical electric and azimuthal magnetic fields on the ground surface due to lightning strikes to free-standing towers of height 50, 100, 120, 150, or 200 m and to the midpoint of a 3-km-long transmission line with seven 120-m-high towers. The ground conductivity was set to 3 mS/m, 10 mS/m, and infinity. Fields were calculated at distances ranging from 100 m to 500 km from the lightning channel. The lightning channel was represented by a vertical 0.1-m-radius wire having additional distributed series inductance of 6 μH/m and distributed series resistance of 1 Ω/m. The speed of the current wave propagating along the wire was 0.5*c*. The lightning channel and the tower or the transmission line were excited by a lumped voltage source inserted between them. They showed that the magnitude of vertical electric field in the vicinity of the isolated tower or the transmission line was reduced, compared to the case of strike to flat ground, while at far distances it was enhanced.

Takami *et al.* [46] have studied, using the FDTD method, lightning surges invading a 500-kV air-insulated substation. The surges resulted from lightning strikes to the top of double-circuit transmission-line tower on flat, perfectly conducting ground. Non-horizontal conductors and power frequency operating voltages were considered in the simulations. The lightning return-stroke channel was represented by a vertical, 0.23-m-radius perfectly conducting wire. The lightning channel and the transmission line were excited by a lumped current source inserted between the lightning channel and the top of the tower. They compared FDTD-calculated voltages in the substation with their counterparts calculated using the Electro-Magnetic Transients Program (EMTP) (e.g., [113]) and found that the peaks of FDTD-calculated voltages were lower than those of EMTP-calculated voltages.

Tatematsu and Ueda [60] have analyzed, using the FDTD method, lightning surges invading a 77-kV air-insulated substation as a result of lightning strikes to the top of double-circuit transmission-line tower on flat ground of conductivity of 14.3 mS/m. Models of surge arrester and flashover were included. Power-frequency operating voltages were also considered. The lightning channel was represented by a vertical, 58-mm-radius wire loaded by additional distributed series inductance of 7.6 μH/m. Their FDTD-computed waveforms of voltage at the entrance of the substation agree well with those observed at a real substation.

Tatematsu [49] has calculated, using the FDTD method, currents flowing into the top of a 1/20-small-scale 77-kV transmission-line tower and in the shield wires and voltages at the cross-arms. The ground was assumed to be perfectly conducting. The lightning channel was represented by a vertical straight perfectly conducting wire, and thus, the propagation speed of current wave along the channel was essentially equal to the speed of light. The lightning channel and the tower were excited by a lumped voltage source in series with a 560-Ω resistor. The computed waveforms of current and voltage agree well with corresponding measured ones. Further, Tatematsu [49] has carried out simulations of surges in a full-scale 77-kV transmission-line tower with a cross-linked polyethylene (XLPE) insulated power cable and metallic cleats over ground with different values of conductivity, 0.1, 1, 10, and 100 mS/m.

4.7.4 Strikes to overhead power distribution lines

Matsuura *et al.* [44] have analyzed, using the FDTD method, transient voltages due to direct lightning strikes to the overhead ground wire of a three-phase distribution line over flat ground of conductivity of 10 mS/m. The lightning channel attached to the ground wire is represented by a vertical perfectly conducting wire, excited by a lumped current source inserted between the simulated lightning channel and the ground wire.

4.7.5 Strikes to wire-mesh-like structures

Miyazaki and Ishii [54] have also calculated, using the NEC-4 code, time derivatives of magnetic field inside a 30-m-high building struck by lightning, assuming perfectly conducting ground. They represented the vertical lightning channel attached to the building top by a vertical 0.1-m-radius wire having distributed series resistance and additional distributed series inductance and the building by a perfectly-conducting-wire grid. They inserted a lumped voltage source between the lightning channel and building. They showed that time derivatives of magnetic field in upper parts of the building were largest but could be reduced by installing a finer conducting mesh on the building roof.

Aniserowicz and Maksimowicz [56], using several computer codes such as NEC-2 and NEC-4 based on the MoM in the frequency domain, have calculated electric and magnetic fields, as well as currents in a loop conductor in a building struck by lightning. The building was modeled by a perfectly conducting grid located on a flat, perfectly conducting ground. The lightning channel was represented by a vertical, 50-mm-radius wire having additional distributed series inductance of 4.5 μH/m and distributed series resistance of 1 Ω/m. The lightning channel and the building were excited by a lumped voltage source inserted between them.

Ishii *et al.* [45] have calculated, using the FDTD method, currents and voltages on electrical wiring in a 25- or 30-m high building whose lightning rod is struck by lightning. Lightning channel was represented by a vertical conductor having a series distributed resistance of 1 Ω/m, so that the speed of the current wave propagating along the wire was equal or close to the speed of light. The lightning channel and the building were excited by a lumped current source. The ground was assumed to be perfectly conducting. They have shown that induced currents and voltages are higher on the top and bottom floors of the building.

Diaz *et al.* [48] have calculated, using the FDTD method, voltages and currents in a buried coaxial cable entering a building struck by lightning. Lightning channel was represented by a vertical conducting wire having a series distributed resistance of 1 Ω/m, so that the current wave propagation speed along the wire was equal or close to the speed of light. The lightning channel and the building were excited by a lumped current source. The ground conductivity was set to 10 mS/m. They have shown that FDTD-computed results agree well with the corresponding ones computed with the method of moments.

4.8 Summary

In this chapter, electromagnetic (full-wave) models of the lightning return stroke have been reviewed and evaluated. This relatively new class of models can be used for specifying the source in studying lightning interaction with various systems and with the environment.

It has been shown that a current wave necessarily suffers distortion as it propagates upward along a vertical non-zero-thickness wire above perfectly conducting ground excited at its bottom by a lumped source, even if the wire has no ohmic losses, which is a distinctive feature of this class of models. Mechanism of this attenuation is related to the boundary condition for the tangential component of electric field on the surface of the wire.

Electromagnetic models proposed to date have been classified into six types depending on lightning channel representation: (1) a perfectly conducting/resistive wire in air above ground; (2) a wire loaded by additional distributed series inductance in air above ground; (3) a wire surrounded by a dielectric medium (other than air) that occupies the entire half space above ground (the artificial dielectric medium is used only for finding current distribution along the lightning channel and is then removed for calculating electromagnetic fields in air); (4) a wire coated by a dielectric material in air above ground; (5) a wire coated by a fictitious material having high relative permittivity and high relative permeability in air above ground; and (6) two parallel wires having additional distributed shunt capacitance in air (this fictitious configuration is used only for finding the current distribution which is then applied to a vertical wire in air above ground for calculating electromagnetic fields). Type-2 and 5 models reproduce the maximum number (three out of five) of characteristic features of electric and magnetic field waveforms observed at distances ranging from 1 to 200 km from natural lightning, and at distances ranging from tens to hundreds of meters from rocket-triggered lightning. Modifications of type-2 and 5 models, in which distributed channel resistance is not uniform (relatively high within the bottom 0.5 km and above 4 km), can reproduce all of the five features. It is also desirable that models are capable of reproducing the following two features of the lightning return stroke: typical values of optically measured return-stroke wavefront speed in the range from $0.33c$ to $0.5c$, and the expected equivalent impedance of the lightning return-stroke channel in the range from 0.6 to 2.5 kΩ.

Different methods of excitation used to date in lightning return-stroke electromagnetic models have been compared: (1) closing a charged vertical wire at its bottom with a specified impedance, (2) a lumped voltage source (or a delta-gap electric-field source), and (3) a lumped current source. Distributions of current along the vertical perfectly conducting wire in air excited at its bottom by the lumped voltage source and by the lumped current source are identical, as expected. Also, the distribution of current along the charged vertical perfectly conducting wire closed with the nonlinear resistor is similar to that along the same wire excited at its bottom by the lumped voltage source.

Distributions of current along the vertical perfectly conducting wire in air excited at its bottom by a lumped voltage source, calculated using different numerical procedures, MoMs in the time and frequency domains, and the FDTD method, have been compared. As expected, current distributions calculated using these three procedures agree well. They also agree reasonably well with those calculated using Chen's analytical equation.

Applications of lightning return-stroke electromagnetic models to analyzing (1) lightning strikes to flat ground; (2) strikes to free-standing tall objects; (3) strikes to overhead power transmission lines; (4) strikes to overhead power distribution lines; and (5) strikes to wire-mesh-like structures have been reviewed.

References

[1] Rakov, V. A., and M. A. Uman (1998), Review and evaluation of lightning return stroke models including some aspects of their application, *IEEE Trans. Electromagn. Compat.*, EMC-40 (4), 403–426.

[2] Rachidi, F., V. A. Rakov, C. A. Nucci, and J. L. Bermudez (2002), Effect of vertically extended strike object on the distribution of current along the lightning channel, *J. Geophys. Res.*, 107 (D23), 16-1–16-6. 10.1029/2002JD002119.

[3] Nucci, C. A., G. Diendorfer, M. A. Uman, F. Rachidi, M. Ianoz, and C. Mazzetti (1990), Lightning return stroke current models with specified channel-base current: a review and comparison, *J. Geophys. Res.*, 95 (D12), 20395–20408.

[4] Thottappillil, R., and M. A. Uman (1993), Comparison of lightning return-stroke models, *J. Geophys. Res.*, 98, 22903–22914.

[5] Thottappillil, R., V. A. Rakov, and M. A. Uman (1997), Distribution of charge along the lightning channel: relation to remote electric and magnetic fields and to return-stroke models, *J. Geophys. Res.*, 102, 6987–7006.

[6] Gomes, C. and V. Cooray (2000), Concepts of lightning return stroke models, *IEEE Trans. Electromagn. Compat.*, EMC-42 (1), 82–96.

[7] Mattos M. A. da F., and C. Christopoulos (1988), A nonlinear transmission line model of the lightning return stroke, *IEEE Trans. Electromagn. Compat.*, EMC-30 (3), 401–406.

[8] Baum C. E. and L. Baker (1990) *Analytic return-stroke transmission-line model, in Lightning Electromagnetics*, New York: Hemisphere, pp. 17–40.

[9] Thottappillil, R., J. Schoene, and M. A. Uman (2001), Return stroke transmission line model for stroke speed near and equal that of light, *Geophys. Res. Lett.*, 28 (18), 3593–3596.

[10] Kordi, B., R. Moini, and, V. A. Rakov (2002), Comment on 'Return stroke transmission line model for stroke speed near and equal that of light' by R. Thottappillil, J. Schoene, and M. A. Uman, *Geophys. Res. Lett.*, 29 (10), 10.1029/2001GL014602.

[11] Baba, Y. and V. A. Rakov (2003), On the transmission line model for lightning return stroke representation, *Geophys. Res. Lett.*, 30 (24), 2294, 10.1029/2003GL018407.

[12] Baba, Y. and V. A. Rakov (2005), On the mechanism of attenuation of current waves propagating along a vertical perfectly conducting wire above ground: application to lightning, *IEEE Trans. Electromagn. Compat.*, EMC-47 (3), 521–532.

[13] Theethayi, N. and V. Cooray (2005), On the representation of the lightning return stroke as a current pulse propagating along a transmission line, *IEEE Trans. Power Delivery*, PWRD-20 (2), 823–837.

[14] De Conti, A., F. H. Silveira, and S. Visacro (2010), Influence of a nonlinear channel resistance on lightning-induced voltages on overhead lines, *IEEE Trans. Electromagn. Compat.*, EMC 52 (3), 676–683.

[15] Baba, Y. and V. A. Rakov (2007), Electromagnetic models of the lightning return stroke, *J. Geophys. Res.*, 112 (D04102), 18 pages, doi:10.1029/2006JD007222.

[16] Baba, Y. and V. A. Rakov (2008), Applications of electromagnetic models of the lightning return stroke, *IEEE Trans. Power Delivery*, PWRD-23 (2), 800–811.

[17] Van Baricum, M. and E. K. Miller (1972), TWTD – A Computer Program for Time-Domain Analysis of Thin-Wire Structures, UCRL-51-277, Lawrence Livermore Laboratory, California.

[18] Harrington, R. F. (1968), *Field Computation by Moment Methods*, Macmillan Company, New York, NY.

[19] Miller, E. K., A. J. Poggio, and G. J. Burke (1973), An integro-differential equation technique for the time-domain analysis of thin wire structures, *J. Computational Phys.*, 12, 24–48.

[20] Yee, K. S. (1966), Numerical solution of initial boundary value problems involving Maxwell's equations in isotropic media, *IEEE Trans. Antennas Propagat.*, AP-14 (3), 302–307.

[21] Kato, S., T. Narita, T. Yamada, and E. Zaima (1999), Simulation of electromagnetic field in lightning to tall tower, Paper Presented at 11th International Symposium on High Voltage Engineering, 467, London, UK.

[22] Moini, R., B. Kordi, G. Z. Rafi, and V. A. Rakov (2000), A new lightning return stroke model based on antenna theory, *J. Geophys. Res.*, 105 (D24), 29693–29702.

[23] Kato, S., T. Takinami, T. Hirai, and S. Okabe (2001), A study of lightning channel model in numerical electromagnetic field computation, Paper Presented at 2001 IEEJ National Convention, 7-140, Nagoya, Japan (in Japanese).

[24] Miyazaki, S. and M. Ishii (2004), Reproduction of electromagnetic fields associated with lightning return stroke to a high structure using FDTD method, Paper Presented at 2004 IEEJ National Convention, no. 7-065, 98, Kanagawa, Japan (in Japanese).

[25] Bonyadi-Ram, S., R. Moini, S. H. H. Sadeghi, and V. A. Rakov (2005), *Incorporation of distributed capacitive loads in the antenna theory model of lightning return stroke, Paper Presented at 16th International Zurich Symposium and Technical Exhibition on Electromagnetic Compatibility*, Swiss Federal Institute of Technology, Zurich, Switzerland, pp. 213–218.

[26] Pokharel, R. K., M. Ishii, and Y. Baba (2003), Numerical electromagnetic analysis of lightning-induced voltage over ground of finite conductivity, *IEEE Trans. Electromagn. Compat.*, EMC-45 (4), 651–666.

[27] Janani, J., R. Moini, and S. H. H. Sadeghi (2012), Evaluation of lightning-induced voltage on overhead lines with nonlinear load using the scattering theory, *IEEE Trans. Electromagn. Compat.*, EMC-27 (1), 317–324.

[28] Shoory, A., R. Moini, S. H. H. Sadeghi, and V. A. Rakov (2005), Analysis of lightning-radiated electromagnetic fields in the vicinity of lossy ground, *IEEE Trans. Electromagn. Compat.*, EMC-47 (1), 131–145.

[29] Baba, Y. and V. A. Rakov (2009), Electric and magnetic fields predicted by different electromagnetic models of the lightning return stroke versus measured fields, *IEEE Trans. Electromagn. Compat*, EMC-51 (3), 479–483.

[30] Visacro, S. and F. H. Silveira (2004), Evaluation of current distribution along the lightning discharge channel by a hybrid electromagnetic model, *J. Electrostatics*, 60 (2), 111–120.

[31] Visacro, S. and F. H. Silveira (2005), Lightning current waves at short instrumented towers: the influence of sensor position, *Geophys. Res. Lett.*, 32 (L18804), doi:10.1029/2005GL023255.

[32] Ruehli, A. (1974), Equivalent circuit models for three-dimensional multi-conductor systems, *IEEE Trans. Microwave Theory and Techniques*, MTT-22 (3), 216–221.

[33] Wang, S., J. He, B. Zhang, R. Zeng, and Z. Yu (2010), A time-domain multiport model of thin-wire system for lightning transient simulation, *IEEE Trans. Electromagn. Compat.*, EMC-52 (1), 128–135.

[34] Yutthagowith, P., A. Ametani, F. Rachidi, N. Nagaoka, and Y. Baba (2013), Application of a partial element equivalent circuit method to lightning surge analyses, *Electric Power Systems Research*, 94, 30–37.

[35] Chen, K. C. (1983), Transient response of an infinite cylindrical antenna, *IEEE Trans. Antennas Propagat.*, AP-31 (1), 170–172.

[36] Bermudez, J. D., M. Rubinstein, F. Rachidi, F. Heidler, and M. Paolone (2003), Determination of reflection coefficients at the top and bottom of elevated strike objects struck by lightning, *J. Geophys. Res.*, 108(D14), 10.1029/2002JD002973.

[37] Baba, Y. and V. A. Rakov (2005), On the interpretation of ground reflections observed in small-scale experiments simulating lightning strikes to towers, *IEEE Trans. Electromagn. Compat., EMC*-47 (3), 533–542.

[38] Podgorski, A. S. and J. A. Landt (1987), Three dimensional time domain modelling of lightning, *IEEE Trans. Power Delivery*, PWRD-2 (3), 931–938.

[39] Mozumi, T., Y. Baba, M. Ishii, N. Nagaoka, and A. Ametani (2003), Numerical electromagnetic field analysis of archorn voltages during a back-flashover on a 500 kV twin-circuit line, *IEEE Trans. Power Delivery*, PWRD-18 (1), 207–213.

[40] Baba, Y. and M. Ishii (2003), Characteristics of electromagnetic return-stroke models, *IEEE Trans. Electromagn. Compat., EMC*-45 (1), 129–135.

[41] Kordi, B., R. Moini, W. Janischewskyj, A. M. Hussein, V. O. Shostak, and V. A. Rakov (2003), Application of the antenna theory model to a tall tower struck by lightning, *J. Geophys. Res.*, 108 (D17), 4548, 10.1029/2003JD003398.

[42] Pokharel, R. K., Y. Baba, and M. Ishii (2004), Numerical electromagnetic field analysis of transient induced voltages associated with lightning to a tall structure, *J. Electrostatics*, 60 (2), 141–147.

[43] Baba, Y. and V. A. Rakov (2008), Influence of strike object grounding on close lightning electric fields, *J. Geophys. Res.*, 113 (D12109), 17 pages, doi:10.1029/2008JD009811.

[44] Matsuura, S., A. Tatematsu, T. Noda, and S. Yokoyama (2009), A simulation study of lightning surge characteristics of a distribution line using the FDTD method, *IEEJ Trans. Power and Energy*, 129 (10), 1225–1232.

[45] Ishii, M., K. Miyabe, and A. Tatematsu (2012), Induced voltages and currents on electrical wirings in building directly hit by lightning, *Electric Power Sys-tems Research*, 85, 2–6.

[46] Takami, J., T. Tsuboi, K. Yamamoto, S. Okabe, and Y. Baba (2015), FDTD simulation considering an AC operating voltage for air-insulation substation in terms of lightning protective level, *IEEE Trans. Dielectr. Insul.*, DEI-22 (2), 806–814.

[47] Du, Y., B. Li, and M. Chen (2016), Surges induced in building electrical systems during a lightning strike, *Electric Power Syst. Res.*, 139, 68–74.

[48] Diaz, L., C. Miry, P. Baraton, C. Guiffaut, and A. Reineix (2017), Lightning transient voltages in cables of a large industrial site using a FDTD thin wire model, *Electric Power Syst. Res.*, 153, 94–103.

[49] Tatematsu, A. (2020), Lightning surge analysis of a transmission line tower with an XLPE power cable and metallic cleats using the FDTD method, *IEEE Trans., Electromagn. Compat.*, EMC-62 (5), 1796–1806.

[50] Baba, Y. and M. Ishii (2001), Numerical electromagnetic field analysis of lightning current in tall structures, *IEEE Trans. Power Delivery, PWRD*-16 (2), 324–328.

[51] Noda, T., A. Tetematsu, and S. Yokoyama (2007), Improvements of an FDTD-based surge simulation code and its application to the lightning overvoltage calculation of a transmission tower, *Electric Power Syst. Res.*, 77 1495–1500.

[52] Pokharel, R. K. and M. Ishii (2007), Applications of time-domain numerical electromagnetic code to lightning surge analysis, Pokharel, *IEEE Trans. Electromagn. Compat.*, EMC-49 (3), 623–631.

[53] Bonyadi-Ram, S., R. Moini, S. H. H. Sadeghi, and V. A. Rakov (2008), On representation of lightning return stroke as a lossy monopole antenna with inductive loading, *IEEE Trans. Electromagn. Compat.*, EMC-50 (1), 118–127.

[54] Miyazaki, S. and M. Ishii (2008), Role of steel frames of buildings for mitigation of lightning-induced magnetic fields, *IEEE Trans. Electromagn. Compat.*, EMC-50 (2), 333–339.

[55] Miyazaki, S. and M. Ishii (2008), Influence of independent towers and transmission lines on lightning return stroke current and associated fields, *IEEE Trans. Electromagn. Compat.*, EMC-50 (2), 358–368.

[56] Aniserowicz, K. and T. Maksimowicz (2011), Comparison of lightning-induced current simulations in the time and frequency domains using different computer codes, *IEEE Trans. Electromagn. Compat., EMC*-53 (2), 446–461.

[57] Khosravi-Farsani, M., R. Moini, S. H. H. Sadeghi, and F. Rachidi (2013), On the validity of approximate formulas for the evaluation of the lightning electromagnetic fields in the presence of a lossy ground, *IEEE Trans. Electromagn. Compat.*, EMC-55 (2), 362–370.

[58] Saito, M., T. Miki, T. Shindo, *et al.* (2015), Reproduction of electromagnetic field waveforms of subsequent return strokes hitting Tokyo Skytree over lossy ground, *IEEJ Trans. Power and Energy*, 135 (7), 472–478.

[59] Khosravi, R., S. H. H. Sadeghi, and R. Moini (2016), Electromagnetic field due to lightning strike to a tall tower sitting on a mountainous terrain, *IEEE Trans. Electromagn. Compat.*, EMC-58 (4), 1090–1099.

[60] Tatematsu, A. and T. Ueda (2016), FDTD-based lightning surge simulation of an HV air-insulated substation with back-flashover phenomena, *IEEE Trans. Electromagn. Compat.*, EMC-58 (5), 1549–1560.

[61] Yamanaka, A., N. Nagaoka, Y. Baba, M. Saito, and T. Miki (2019), Lightning strike to a tall-grounded object: Part 2—LEMP calculation with a channel approximation, *IEEE Trans., Electromagn. Compat.*, EMC-61 (3), 736–744.

[62] Moini, R., B. Kordi, and M. Abedi (1998), Evaluation of LEMP effects on complex wire structures located above a perfectly conducting ground using electric field integral equation in time domain, *IEEE Trans. Electromagn. Compat.*, EMC-40 (2), 154–162.

[63] Geranmayeh, A., R. Moini, and S. H. H. Sadeghi (2006), Numerical simulation of electromagnetic fields radiated by lightning return stroke channels: a wavelet-based approach, *IEEE Trans., Electromagn. Compat.*, EMC-48 (1), 225–233.

[64] Rakov, V. A. and A. A. Dulzon (1987), Calculated electromagnetic fields of lightning return stroke, *Tekh. Electrodinam.*, 1, 87–89 (in Russian).

[65] Rakov, V. A. (2007), Lightning return stroke speed, *J. Lightning Res.*, 1, 80–89.

[66] Gorin, B. N. and A. V. Shkilev (1984), Measurements of lightning currents at the Ostankino tower, *Electrichestrvo*, 8, 64–65 (in Russian).

[67] Rakov, V. A. (1998), Some inferences on the propagation mechanisms of dart leaders and return strokes, *J. Geophys. Res.*, 103 (D2), 1879–1887.

[68] Uman, M. A., D. K. McLain, and E. P. Krider (1975), The electromagnetic radiation from a finite antenna, *Amer. J. Phys.*, 43, 33–38.

[69] Heidler, F. (1985), Traveling current source model for LEMP calculation, Paper Presented at 6th International Symposium on EMC, Zurich, Switzerland, pp. 157–162.

[70] Baba, Y. and V. A. Rakov (2005), On the use of lumped sources in lightning return stroke models, *J. Geophys. Res.*, 110 (D3), D03101, doi:10.1029/2004JD005202.

[71] Sadiku, M. N. O. (1994), *Elements of Electromagnetics*, Oxford University Press, UK.

[72] Aniserowicz, K. (2004), A new algorithm for antenna theory modeling of a lightning return stroke, Paper Presented at 27th International Conference on Lightning Protection, Soc. de l'Electr., de l'Electron., et des Technol. de l'Inf. et de la Commun., Avignon, France.

[73] Idone, V. P. and R. E. Orville (1982), Lightning return stroke velocities in the Thunder-storm Research International Program (TRIP), *J. Geophys. Res.*, 87, (C7), 4903–4915.

[74] Rakov, V. A. (2001), Transient response of a tall object to lightning, *IEEE Trans. Electromagn. Compat.*, EMC-43 (4), 654–661.

[75] Goubau, G. (1950), Surface waves and their application to transmission lines, *J. Applied Phys.*, 21, 1119–1128.

[76] Taniguchi, Y. Y. Baba, N. Nagaoka, and A. Ametani (2008), Representation of an arbitrary-radius wire for FDTD calculations in the 2D cylindrical coordinate system, *IEEE Trans. Electromagn. Compat.*, EMC-50 (4), 1114–1118.

[77] Liao, Z. P., H. L. Wong, B. -P. Yang, and Y. -F. Yuan (1984), A transmitting boundary for transient wave analysis, *Sci. Sin.*, A27 (10), 1063–1076.

[78] Lin, Y. T., M. A. Uman, J. A. Tiller, *et al.* (1979), Characterization of lightning return stroke electric and magnetic fields from simultaneous two-station measurements, *J. Geophys. Res.*, 84 (C10), 6307–6314.

[79] Crawford, D. E., V. A. Rakov, M. A. Uman, *et al.* (2001), The close lightning electromagnetic environment: dart-leader electric field change versus distance, *J. Geophys. Res.*, 106 (D14), 14909–14917.

[80] Baba, Y., S. Miyazaki, and M. Ishii (2004), Reproduction of lightning electromagnetic field waveforms by engineering model of return stroke, *IEEE Trans. Electromagn. Compat., EMC*-46 (1), 130–133.

[81] Thottappillil, R., M. A. Uman, and G. Diendorfer (1991), Influence of channel base current and varying return stroke speed on the calculated

fields of three important return stroke models, Paper Presented at International Conference on Lightning and Static Electricity, Cocoa Beach, FL, USA.

[82] Diendorfer, G. and M. A. Uman (1990), An improved return stroke model with specified channel-base current, *J. Geophys. Res.*, 95 (D9), 13621–13644.

[83] Cooray, V., V. A. Rakov, F. Rachidi, R. Montano, and C. A. Nucci (2008), On the relationship between the signature of close electric field and the equivalent corona current in lightning return stroke models, *IEEE Trans. Electromagn. Compat.*, EMC-50 (4), 921–927.

[84] Jordan, D. M., V. A. Rakov, W. H. Beasley, and M. A. Uman (1997), Luminosity characteristics of dart leaders and return strokes in natural lightning, *J. Geophys. Res.*, 102 (D18), 22,025–22,032.

[85] Olsen III, R. C., D. M. Jordan, V. A. Rakov, M. A. Uman, and N. Grimes (2004), Observed one-dimensional return stroke propagation speeds in the bottom 170 m of a rocket-triggered lightning channel, *Geophys. Res. Lett.*, 31, L16107, doi:10.1029/2004GL02187.

[86] Paknahad, J., K. Sheshyekani, and F. Rachidi (2014), Lightning electromagnetic fields and their induced currents on buried cables. Part I: the effect of an ocean–land mixed propagation path, *IEEE Trans. Electromagn. Compat.*, EMC-56 (5), 1137–1145.

[87] Paknahad, J., K. Sheshyekani, and F. Rachidi (2014), Lightning electromagnetic fields and their induced currents on buried cables. Part II: the effect of a horizontally stratified ground, *IEEE Trans. Electromagn. Compat.*, EMC-56 (5), 1146–1154.

[88] Paknahad, J., K. Sheshyekani, F. Rachidi, M. Paolone, and A. Minouni (2014c), Evaluation of lightning-induced currents on cables buried in a lossy dispersive ground, *IEEE Trans. Electromagn. Compat.*, EMC-56 (6), 1522–1529.

[89] Sheshyekani, K. and J Paknahad (2015), The effect of an ocean-land mixed propagation path on the lightning electromagnetic fields and their induced voltages on overhead lines, *IEEE Trans. Power Delivery*, PWRD-30 (1), 229–236.

[90] Yuda, S., S. Sawaki, Y. Baba, N. Nagaoka, and A. Ametani (2013), Application of the TLM method to transient simulations of a conductor system with a lossy ground: grounding electrodes and an overhead wire, *IEEE Trans. Electromagn. Compat.*, EMC-55 (1), 175–182.

[91] Tanaka, Y., Y. Baba, N. Nagaoka, and A. Ametani (2014), Computation of lightning electromagnetic pulses with the TLM method in the 2-D cylindrical coordinate system, *IEEE Trans. Electromagn. Compat.*, EMC-56 (4), 949–955.

[92] Johns, P. B. and R. B. Beurle (1971), Numerical solutions of 2-dimensional scattering problems using a transmission-line matrix, *Proc. IEE*, 118 (9), 1203–1208.

[93] Silveira, F. H. and S. Visacro (2008), The influence of attachment height on lightning-induced voltages, *IEEE Trans. Electromagn. Compat.*, EMC-50 (3), 743–747.

[94] Silveira, F. H. and S. Visacro (2009), On the lightning-induced voltage amplitude: first versus subsequent negative strokes, *IEEE Trans. Electromagn. Compat.*, EMC-51 (3), 741–747.

[95] Silveira, F. H., S. Visacro, J. Herrera, and H. Torres (2009), Evaluation of lightning-induced voltages over a lossy ground by the hybrid electromagnetic model, *IEEE Trans. Electromagn. Compat.*, EMC-51 (1), 156–160.

[96] Silveira, F. H., A. De Conti, and S. Visacro (2011), Voltages induced in single-phase overhead lines by first and subsequent negative lightning strokes: influence of the periodically grounded neutral conductor and the ground resistivity, *IEEE Trans. Electromagn. Compat.*, EMC-53 (2), 414–420.

[97] Silveira, F. H., S. Visacro, R. Alipio, and A. De Conti (2014), Lightning-induced voltages over lossy ground: the effect of frequency dependence of electrical parameters of soil, *IEEE Trans. Electromagn. Compat.*, EMC-56 (5), 1129–1136.

[98] Baba, Y. and V. A. Rakov (2016), *Electromagnetic Computation Methods for Lightning Surge Protection Studies*, John Wiley & Sons, Singapore, 315 pages.

[99] Burke, G. J. and E. K. Miller (1984), Modeling antennas near to and penetrating a lossy interface, *IEEE Trans. Antennas Propagat.*, AP-32 (10), 1040–1049.

[100] Burke, G. J. and A. J. Poggio (1980), Numerical Electromagnetic Code (NEC) – Method of Moments, Technical Document 116, Naval Ocean Systems Center, San Diego, CA.

[101] Burke, G. J. (1992), *Numerical Electromagnetic Code (NEC-4) – Method of Moments, UCRL-MA-109338*, Lawrence Livermore National Laboratory, California.

[102] Tatematsu, A. and T. Noda (2014), Three-dimensional FDTD calculation of lightning-induced voltages on a multiphase distribution line with the lightning arresters and an overhead shielding wire, *IEEE Trans. Electromagn. Compat.*, EMC-56 (1), 159–167.

[103] Baba, Y. and V. A. Rakov (2020), *Lightning-Induced Effects in Electrical and Telecommunication Systems*, IET, 243 pages.

[104] Noda, T. and S. Yokoyama (2002), Thin wire representation in finite difference time domain surge simulation, *IEEE Trans. Power Delivery*, PWRD-17 (3), 840–847.

[105] Berenger, J. P. (1994), A perfectly matched layer for the absorption of electromagnetic waves, *J. Comput. Phys.*, 114, 185–200.

[106] Baba, Y. and V. A. Rakov (2014), Applications of the FDTD method to lightning electromagnetic pulse and surge simulations, *IEEE Trans. Electromagn. Compat.*, EMC-56 (6),1506–1521.

[107] Rubinstein, M. (1996), An approximate formula for the calculation of the horizontal electric field from lightning at close, intermediate, and long range, *IEEE Trans. Electromagn. Compat.*, EMC-38 (3), 531–535.

[108] Agrawal, A. K., H. J. Price, and S. H. Gurbaxani (1980), Transient response of multiconductor transmission lines excited by a nonuniform electromagnetic field, *IEEE Trans. Electromagn. Compat., EMC*-22 (2), 119–129.

[109] Ishii, M., K. Michishita, and Y. Hongo (1999), Experimental study of lightning-induced voltage on an overhead wire over lossy ground, *IEEE Trans. Electromagn. Compat.*, EMC-41 (1), 39–45.

[110] Goshima, H., H. Motoyama, A. Asakawa, A. Wada, T. Shindo, and S. Yokoyama (2000), Characteristics of electromagnetic fields due to lightning stroke current to a high stack in winter lightning, *IEEJ Trans. PE*, 120 (1), 44–49.

[111] Michishita, K., M. Ishii, A. Asakawa, S. Yokoyama, and K. Kami (2003), Voltage induced on a test distribution line by negative winter lightning strokes to a tall structure, *IEEE Trans. Electromagn. Compat.*, EMC-45 (1), 135–140.

[112] Motoyama, H. (1996), Development of a new flashover model for lightning surge analysis, *IEEE Trans. Power Delivery*, PWRD-11 (2), 972–979.

[113] Dommel, H. (1969), Digital computer solution of electromagnetic transients in single- and multiphase networks, *IEEE Trans. Power Apparatus Syst.*, PAS-88 (4), 388–399.

Chapter 5

Antenna models of lightning return-stroke: an integral approach based on the method of moments

Rouzbeh Moini[1], Seyed Hossein Hesamedin Sadeghi[2], Simon Fortin[1], Moein Nazari[1] and Farid P. Dawalibi[1]

5.1 Introduction

Lightning is a transient, high-current electric discharge with a height in the kilometer range. Cloud-to-Ground (CG) lightning is less common than other kinds of lightning (i.e., cloud-to-cloud and intracloud lightning), but they are more important for protection studies of electric and electronic apparatus used in power systems, information technology systems, etc. The complete discharge, known as a flash, has a time duration of about 0.5 sec and is made up of various components, including a stepped leader, return strokes, and dart leaders. The part of a flash which has been of great interest for protection purposes is the return-stroke phase.

The expression "lightning return-stroke model" is generally used to describe a specification of the time- and height-dependent current in the return-stroke channel so as to enable the calculation of the resultant remote electromagnetic fields [1]. A suitable model should be characterized by a minimum number of adjustable parameters and be consistent with the measured characteristics of the return-stroke namely:

- Current at the base of the channel
- Variation of light intensity with height
- Upward propagation speed of the luminosity front
- Electromagnetic fields at various distances from the channel

It is worth mentioning that other physical parameters (such as an existing space charge in the vicinity of the lightning channel) could also be included in a realistic model for describing lightning phenomena. However, considering each of these

[1]SafEngServices & Technologies Ltd., Canada
[2]Department of Electrical Engineering, Amirkabir University of Technology, Iran

parameters will add to the complexity of the model, making it impractical for most applications.

Many models have been proposed to describe the behavior of lightning and predict its effects. All of these models can be placed in four categories [1], namely: physical models, distributed-circuit models, engineering models, and electromagnetic models.

Physical models or gas dynamic models solve the gas dynamic equations [2,3]. They do not consider the longitudinal evolution of the lightning channel. They also usually ignore the electromagnetic skin effect (found to be negligible) [4], the corona sheath which presumably contains the bulk of the leader charge, and any heating of the air surrounding the current-carrying channel by preceding lightning processes [5]. An attempt to include the previous heating in a gas dynamic model is also made in [6].

Distributed circuit models assume a transmission line representation of the lightning Return-stroke channel (RSC) considering only the TEM wave propagation along the channel. They solve the so-called Telegrapher's expressions to obtain a current distribution along the channel [7]. It is known that the representation of a vertical conductor such as a non-uniform transmission line is able to lead to a spatial and temporal current distribution essentially close to those obtained by the full-wave solutions [8]. Recently, proposed distributed-circuit models [9,10] have shown to predict electromagnetic fields in good agreement with experimental data. In [10], formulas are proposed for calculating corona sheath and core losses for a lightning return-stroke model that is based on a distributed circuit approach. Also recently proposed transmission line equations in air in the presence of corona (considered as a series of corona current sources distributed along the line) have been derived demonstrating that the measured return-stroke speed is less than the speed of light [11].

Engineering models [1] specify a closed-form relation between the current distribution along the channel and current at the channel-base. The most commonly employed engineering models can be classified into two categories. The first category uses the concept of the distributed-source. The Diendorfer–Uman (DU) model and the traveling current source (TCS) model are placed in this category. The second category contains transmission line type models, namely: the transmission line (TL) model, the modified transmission line model with linear current decay with height (MTLL), and the modified transmission line model with exponential current decay with height (MTLE).

A great deal of attention has been recently devoted to the electromagnetic models [12–14]. In these models, the lightning return-stroke channel is usually considered as a monopole wire antenna above a perfectly conducting ground. Maxwell's equations are first solved numerically to obtain a current distribution along the channel. Then, the electromagnetic fields due to the lightning channel are readily computed. Electromagnetic models provide several degrees of freedom to simulate all known behaviors of the lightning RSC [15]. These models are relatively new and considered to be the most rigorous. They solve Maxwell's equations for current distributions along the lightning channel. Due to the lack of an analytical solution, these are usually done with the aid of numerical techniques. The first category of electromagnetic models solve the time-domain Maxwell's equations using the finite difference time-domain technique [16] assuming a distributed array of current sources representing the lightning RSC [17]. The model is used to describe some wave propagation effects prescribed by the transmission line theory.

The second category of electromagnetic models, known as antenna theory (AT) models, uses the method of moments (MoM) solution of the electric field integral equation (EFIE) form of the electromagnetic problem in either the time-domain [18] or the frequency-domain [19]. In these models, the lightning RSC is modeled by a thin wire antenna [13,20]. The antenna can be excited through a voltage source [13] or a current source [21]. The idea behind the AT models was initially proposed by Podgorski and Landt [22] in 1985. They worked on a frequency domain code for modeling a lightning strike to the CN tower. In 1987 they used a Thin-Wire Time Domain (TWTD) code [23] to analyze lightning strikes to tall structures [24]. In 1997 Moini *et al.* [13] considered a monopole antenna above a perfectly conducting ground to represent a lightning return-stroke channel. They adjusted the propagation speed and the value of the distributed resistance along the channel to make EM fields similar to measured fields [25], especially at close ranges. In 2001, Baba and Ishii [20] applied a set of distributed inductive and resistive loads on the wire representing the lightning channel. These distributed inductive loads reduced the propagation speed of the current wave along the channel consistent with optical observation. The channel was analyzed in the frequency domain and with the help of Numerical Electromagnetics Code (NEC-2), requiring a huge number of calculations to be performed for a wide range of frequency sampling points.

Modified time-domain AT models have also been developed in recent years. Bonyadi-Ram *et al.* [26,27] considered a monopole antenna above a perfectly conducting ground to model the lightning return-stroke channel. They controlled the propagation speed of the current wave along the monopole antenna by applying distributed inductances to it [27]. Therefore, the current distribution predicted by their model exhibited features (such as current dispersion) that were more consistent with optical observations of lightning compared to the predictions of the original AT model. In fact, in this model, values of inductance and resistance applied along the channel were chosen by trial and error, and the velocity of the current wave propagation along the channel was calculated as $v = 1.3 \times 10^8$ m/s. The dispersion of this current was consistent with the optical observations of Jordan and Uman [28] and Wang *et al.* [29].

Using a similar approach, another treatment based on capacitive loading has also been proposed in [30], where the distributed capacitance is applied to a two-wire parallel transmission line to obtain the desired characteristics of the current distribution along the channel. The resultant current waveform is then applied to a dipole antenna to calculate the radiated electromagnetic fields. Introducing inductive or capacitive energy storing elements further provides the dispersion of the current wave as it propagates upward along the channel. Moosavi *et al.* [31] have modified the time-domain ATIL-F electromagnetic model to include nonlinear resistive loading in the presence of fixed inductive loading of the channel. Resistive elements are considered as a nonlinear distributed load whose resistance is a function of both current and time. This is done in a consistent manner from an available physical model with respect to the varying value of the channel radius from its base to the cloud [31]. Considering the channel load resistance as a nonlinear element, they attempt to improve the modeling of losses along the channel, using the Braginskii model [2,32] to provide a nonlinear resistive load along the channel by the ATIL-F model.

An improved antenna theory model (AT) approach in the frequency domain has been proposed for the electromagnetic analysis of the lightning return stroke channel and nearby structures [83]. With this approach, a long wire antenna is used to model the lightning channel. Unlike conventional AT models, which employ a voltage or current source localized at the channel base, the new model considers a distributed current source along the antenna. The implementation of appropriate phase shifts between consecutive current sources in the frequency domain accounts for the current wave propagation speed along the channel. Contrary to conventional AT models, the implemented methodology permits the establishment of an accurate propagation speed along the channel without the necessity of adjusting the electrical parameters of the elements representing the channel. Finally, a solution of the electric field integral equation (EFIE) by the method of moments (MoM) in the frequency domain is adopted for modeling the lightning channel and nearby structures. The presence of a multilayer soil is accounted for by using Sommerfeld integrals, enabling an efficient analysis of a lightning channel in the presence of a complex electromagnetic environment.

The chapter relating to this model focuses on the problem of an integral approach for the AT modeling of a lightning return-stroke, using the method of moments. First, we give a general formulation of the problem where the governing frequency and time domain electromagnetic field equations are derived. The method of moments used for numerical solutions of these equations is then described. Special treatments required for channel excitation, channel path and lossy ground will be addressed next. Finally, we evaluate the performance of the modeling technique by presenting results for a number of case studies.

5.2 General formulation

The schematic of a lightning RSC located above a half-space ground is shown in Figure 5.1(a). In this figure, the RSC is modeled by a long conductive thin wire excited either by a voltage or current source at its lower end at the ground surface, or by a distributed current source along the wire (Figure 5.1(b)). Using the zero-tangential electric field boundary condition along the surface of a straight thin wire leads to the electric field integral equation (EFIE) which is solved for the distribution of current along the channel.

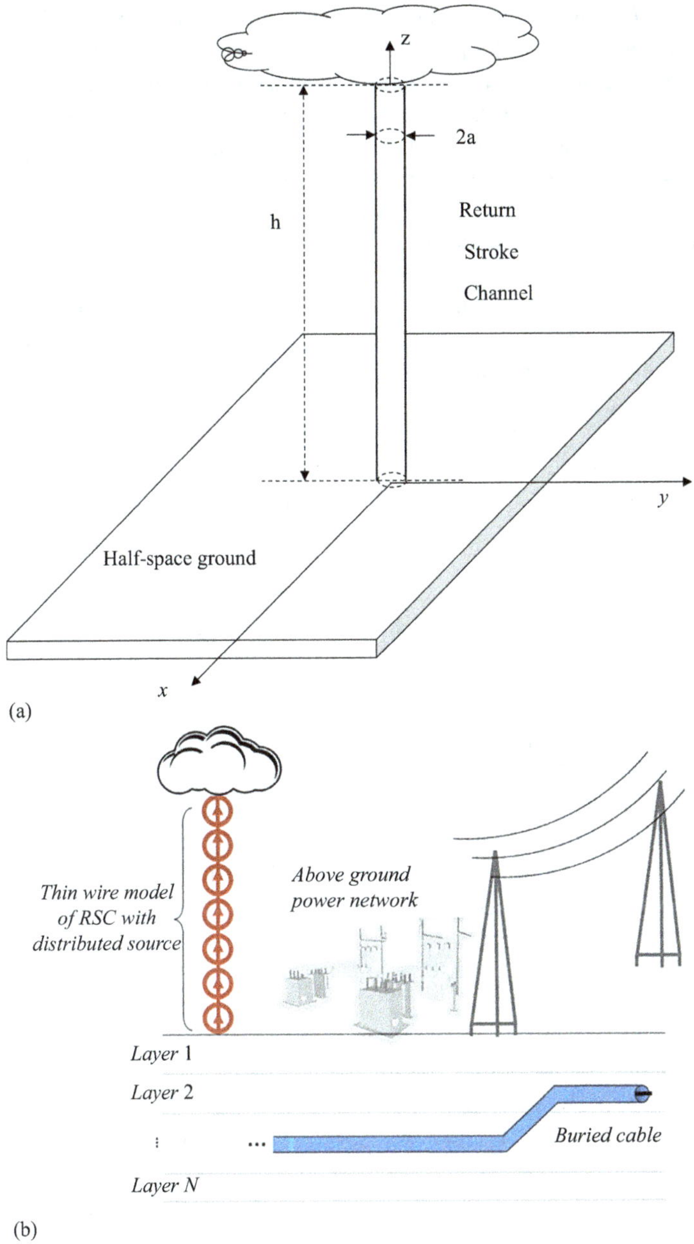

Figure 5.1 Schematic of the problem. (a) Lightning channel above half-space ground. (b) Lightning channel and its nearby structure above a multilayer soil. Adapted from Moini et al. *[83]*

The methods used for the analysis of the above-mentioned problems can be classified as either time-domain or frequency-domain methods. Time-domain methods generate accurate results and are well suited to the analysis of transients and nonlinear problems [33]. However, they cannot be easily used to deal with problems involving a lossy ground. Since the lightning phenomenon provides a narrow pulse, the frequency domain analysis of the problem should be performed for many frequencies, which is a time-consuming process [34]. Additionally, the treatment of non-linearity in the frequency domain is not straightforward. As with the other frequency-domain analysis methods of transient phenomena, it is also necessary to appropriately adjust the sampling rate and the total analysis time to avoid undesirable aliasing or circular convolution. Alternately, frequency-domain methods counterparts can use a lossy ground assumption [21] and take multilayer soils into account [83].

5.2.1 Time-domain formulation

The analysis begins with the time-domain Maxwell's equations for a linear, homogenous, and time-invariant medium:

$$\begin{aligned} \nabla \times \mathbf{e}(\mathbf{r},t) &= -\mu \frac{\partial\, \mathbf{h}(\mathbf{r},t)}{\partial\, t}, \\ \nabla \times \mathbf{h}(\mathbf{r},t) &= \mathbf{j}(\mathbf{r},t) + \varepsilon \frac{\partial \mathbf{e}(\mathbf{r},t)}{\partial t}, \\ \nabla \cdot \mathbf{e}(\mathbf{r},t) &= \frac{\rho(\mathbf{r},t)}{\varepsilon}, \\ \nabla \cdot \mathbf{h}(\mathbf{r},t) &= 0, \end{aligned} \tag{5.1}$$

where ε and μ are the electric permittivity and the magnetic permeability of a medium, **e** and **h** are the electric and magnetic field intensities, and **j** and ρ are the volume current and charge densities, respectively. **r** is the position vector of a point in space and t denotes time of observation. The relation between **j** and ρ is specified by the current continuity equation:

$$\nabla \cdot \mathbf{j}(\mathbf{r},t) = -\frac{\partial \rho(\mathbf{r},t)}{\partial t} \tag{5.2}$$

Combining Maxwell's equations yields the following expressions (wave equations):

$$\begin{aligned} \nabla^2 \mathbf{e}(\mathbf{r},t) - \frac{1}{v^2}\frac{\partial^2 \mathbf{e}(\mathbf{r},t)}{\partial\, t^2} &= \mu \frac{\partial \mathbf{j}(\mathbf{r},t)}{\partial t} + \frac{1}{\varepsilon}\nabla \rho(\mathbf{r},t), \\ \nabla^2 \mathbf{h}\,(\mathbf{r},t) - \frac{1}{v^2}\frac{\partial^2 \mathbf{h}(\mathbf{r},t)}{\partial\, t^2} &= -\nabla \times \mathbf{j}\,(\mathbf{r},t), \end{aligned} \tag{5.3}$$

where $v = \frac{1}{\sqrt{\varepsilon\mu}}$

If we consider a perfectly conducting antenna in a homogenous medium, the volume current and charge densities, **j** and ρ, can be replaced by the surface current and charge densities, $\mathbf{j}_s$ and ρ_s:

$$\mathbf{j}(\mathbf{r},t) = \mathbf{j}_s(\mathbf{r}_0,t)\,\delta_s(\mathbf{r}-\mathbf{r}_0), \tag{5.4}$$

$$\rho(\mathbf{r},t) = \rho_s(\mathbf{r}_0,t)\,\delta_s(\mathbf{r}-\mathbf{r}_0),$$

$$\delta_s(\mathbf{r}-\mathbf{r}_0) = \begin{cases} \infty & \textit{if } r = r_0 \\ 0 & \text{otherwise.} \end{cases}$$

where $\mathbf{r}_0$ is the position vector of a point on the surface of the antenna, and δ_s is the two-dimensional impulse function.

The Green's solution (g) of a nonhomogeneous wave equation

$$\nabla^2 g - \frac{1}{v^2}\frac{\partial^2 g}{\partial t^2} = -\delta(\mathbf{r}-\mathbf{r}_0)\,\delta(t-t_0), \tag{5.5}$$

$$\delta(\mathbf{r}-\mathbf{r}_0) = \delta(x-x_0)\,\delta(y-y_0)\,\delta(z-z_0),$$

is given by Stratton [35]:

$$g(\mathbf{r},\mathbf{r}_0,t-t_0) = \frac{1}{4\pi}\frac{\delta\left(t-t_0-\frac{R}{v}\right)}{R} \tag{5.6}$$

In this equation, $R = |\mathbf{r}-\mathbf{r}_0|$; $\mathbf{r}_0$ is the spatial location, and t_0 is the initial temporal point of the source (current or charge density on the surface of the conductor).

Applying the above solution to (5.3), we obtain the radiated fields produced by the surface current density $\mathbf{j}_s$ [36]:

$$\mathbf{e}(\mathbf{r},t) = -\iint_{S_0}\left[\mu\frac{\partial \mathbf{j}_s(\mathbf{r}_0,t)}{\partial t} \underset{(t)}{*} g(\mathbf{r},\mathbf{r}_0,t) + \frac{1}{\varepsilon}\int_0^t \nabla\cdot\mathbf{j}_s(\mathbf{r}_0,\tau)\,d\tau \underset{(t)}{*} \nabla_0 g(\mathbf{r},\mathbf{r}_0,t)\right] ds_0, \tag{5.7}$$

$$\mathbf{h}(\mathbf{r},t) = \iint_{S_0}\mathbf{j}_s(\mathbf{r}_0,t) \underset{(t)}{\hat{*}} \nabla_0 g(\mathbf{r},\mathbf{r}_0,t)\,ds_0,$$

where the subscript (t) refers to convolution operator variable (time), and the subscript zero is related to the differentiation variable ($\mathbf{r}_0$). The symbol $\underset{(t)}{\hat{*}}$ denotes the cross product with respect to space and convolution in time, and integration is performed over the outer surface of the antenna (S_0).

The response of a monopole antenna (see Figure 5.2) above a perfectly conducting ground to an electromagnetic wave produced by an external source can be found considering the scattering of electromagnetic fields by a metallic object. The boundary condition on the surface of a perfect conductor can be expressed as

$$\mathbf{n}\times(\mathbf{e}^a + \mathbf{e}) = 0, \tag{5.8}$$

where $\mathbf{e}^a$ is the applied field, and **n** is the normal vector to the surface of conducting object. For a non-ideal conductor, the right-hand side of (5.8) is not equal to zero

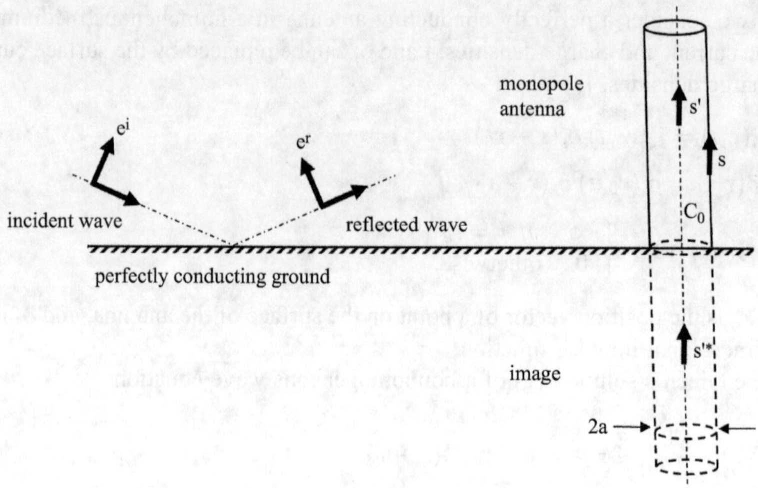

Figure 5.2 Geometry of vertical monopole antenna in receiving mode located above a perfectly conducting ground

and depends on resistance per unit length. Combining (5.7) and (5.8), we have

$$\mathbf{n} \times \mathbf{e}^a(\mathbf{r}, t) = \mathbf{n} \times \mathrm{Vp}\iint_{S_0}\left[\mu \frac{\partial \mathbf{j}_s(\mathbf{r}_0, t)}{\partial t} \underset{(t)}{*} g(\mathbf{r}, \mathbf{r}_0, t)\right.$$
$$\left. + \frac{1}{\varepsilon}\int_0^t \nabla \cdot \mathbf{j}_s(\mathbf{r}_0, \tau)\, d\tau \underset{(t)}{*} \nabla_0 g(\mathbf{r}, \mathbf{r}_0, t)\right] ds_0, \tag{5.9}$$

where Vp is the principal value of the integral operator [13].

In the case of homogenous, linear, and time-invariant media, inserting (5.6) into (5.9) will yield the following equation:

$$\mathbf{n} \times \mathbf{e}^a(\mathbf{r}, t) = \frac{\mathbf{n}}{4\pi} \times \mathrm{Vp}\iint_{S_0}\left[\frac{\mu}{R}\frac{\partial \mathbf{j}_s(\mathbf{r}_0, t)}{\partial t} + \frac{\mathbf{R}}{\varepsilon v R^2}\nabla \cdot \mathbf{j}_s(\mathbf{r}_0, t - R/v)\right.$$
$$\left. + \frac{\mathbf{R}}{\varepsilon R^3}\int_0^t \nabla \cdot \mathbf{j}_s(\mathbf{r}_0, \tau)\, d\tau\right] ds_0, \tag{5.10}$$

where $\mathbf{R} = \mathbf{r} - \mathbf{r}_0$.

We will use the thin-wire approximation [13] according to which the current i (s,t) on a wire structure of radius (a) satisfies the equations:

$$\begin{aligned} i(s,t) &= 2\pi a j(s,t), \\ \mathbf{j}(s,t) &= \tfrac{i(s,t)}{2\pi a}\mathbf{s}, \end{aligned} \tag{5.11}$$

where s is the location of a point on the wire structure and $\mathbf{s}$ is the corresponding tangential unit vector (see Figure 5.2).

Because of the presence of the ground the total excitation (or applied) field produced by the source above ground is:

$$\mathbf{e}^{\mathrm{a}} = \mathbf{e}^{\mathrm{i}} + \mathbf{e}^{\mathrm{r}}, \tag{5.12}$$

where $\mathbf{e}^{\mathrm{i}}$ is the incident field and $\mathbf{e}^{\mathrm{r}}$ is the reflected field from the ground. The excitation field $\mathbf{e}^{\mathrm{a}}$ induces current i(**s**,t) at each point of the antenna. According to (5.10) and the thin-wire approximation, (5.11), one can determine the scattered field **e**, which can be described in the following form:

$$\mathbf{e}(\mathbf{s}, t) = \mathbf{L}[i(\mathbf{s}, t)], \tag{5.13}$$

where **L** is an integro-differential operator [37]. Note that $\mathbf{e}^{\mathrm{a}}$ is independent of the presence of the antenna, and **e** can be viewed as a reaction of the antenna to $\mathbf{e}^{\mathrm{a}}$. For the receiving antenna mode (illustrated in Figure 5.2), $\mathbf{e}^{\mathrm{a}}$ is produced by an external source and exists everywhere in space. For the transmitting antenna mode, $\mathbf{e}^{\mathrm{a}}$ is produced by a lumped source connected between the lower end of the antenna and the ground and is zero everywhere except for the position of the source. Similar to the receiving mode, $\mathbf{e}^{\mathrm{a}}$ is totally independent of the presence of antenna. The source, whose voltage is related to $\mathbf{e}^{\mathrm{a}}$ as described later in this section, launches a current wave along the antenna. The electric field produced by this current wave is the scattered field **e** which, similar to the receiving mode, can be viewed as a reaction of the antenna to $\mathbf{e}^{\mathrm{a}}$.

The continuity of the tangential component of the total electric field at any point on the antenna surface requires that:

$$\mathbf{s} \cdot (e^{\mathrm{a}}(\mathbf{s}, t) \; + \; \mathbf{L}[i(\mathbf{s}, t)]) = 0, \tag{5.14}$$

which is the same as (5.8). As stated above for (5.8), for a non-ideal conductor the right-hand side of (5.14) is not equal to zero. Using the definition of the **L** operator and the thin-wire approximation [37,38], we can write:

$$\begin{aligned} \mathbf{s} \cdot \mathbf{e}^{\mathrm{a}}(\mathbf{s}, t) = \frac{\mu_0}{4\pi} \int_{C_0} \Bigg[& \frac{\mathbf{s} \cdot \mathbf{s}'}{R} \frac{\partial i(\mathbf{s}', t')}{\partial t'} + v \frac{\mathbf{s} \cdot \mathbf{R}}{R^2} \frac{\partial i(\mathbf{s}', t')}{\partial s'} + v^2 \frac{\mathbf{s} \cdot \mathbf{R}}{R^3} \int_0^{t'} \left(\frac{\partial i(\mathbf{s}', \tau)}{\partial s'} d\tau \right) \\ & - \frac{\mathbf{s} \cdot \mathbf{s}'^{*}}{R^{*}} \frac{\partial i(\mathbf{s}', t'*)}{\partial t'*} - v \frac{\mathbf{s} \cdot \mathbf{R}^{*}}{R^{*2}} \frac{\partial i(\mathbf{s}', t'*)}{\partial s'} - v^2 \frac{\mathbf{s} \cdot \mathbf{R}^{*}}{R^{*3}} \int_0^{t'*} \frac{\partial i(\mathbf{s}', \tau)}{\partial s'} d\tau \Bigg] ds', \end{aligned} \tag{5.15}$$

where

$$R = (|\mathbf{s} - \mathbf{s}'|^2 + a^2)^{1/2},$$

$$R^{*} = (|\mathbf{s} - \mathbf{s}'^{*}|^2 + a^2)^{1/2},$$

$$t' = t - \frac{R}{v}, t'* = t - \frac{R^{*}}{v},$$

where C_0 is the path along which the current is flowing, a is the radius of the antenna (Figure 5.1), and v is the wave propagation speed for the case of a non-resistive channel. Further, s and s' are the observation and source points on the antenna, respectively, s'* is the image of the source point s', and **s, s'**, and **s'*** are the corresponding unit tangential vectors. The last three terms in the right-hand side of (5.15) represent the effect of a perfect ground. The left-hand side of (5.15) represents the applied (excitation) electric field, tangential to the surface of the antenna. For the case of a transmitting antenna, the applied field $\mathbf{e}^a(\mathbf{s}, t)$ is produced by a voltage source (not shown in Figure 5.1, which illustrates the receiving antenna mode). The relation between $e^a(\mathbf{s}, t)$ and source voltage $v(\mathbf{s}, t)$ according to Herault *et al.* [37] is:

$$\mathbf{e}^a(\mathbf{s}, t) = \begin{cases} -\nabla v(\mathbf{s}', t) & \text{if } s = s' = 0 \\ 0 & \text{elsewhere.} \end{cases} \tag{5.16}$$

The numerical solution of (5.15), the electric field integral equation (EFIE) in the time domain, by the method of moments (MOM) [18], yields the time–space distribution of current along the antenna.

5.2.2 *Frequency-domain formulation*

In the previous section, we presented Maxwell's equations for general time-varying electromagnetic fields. To derive a frequency domain formulation, the time variations of electromagnetic waves can be represented by $e^{j\omega t}$ where $j = \sqrt{-1}$ and ω is the angular frequency. The starting point is the frequency-domain Maxwell's equations for a linear, homogenous, and time-invariant medium, i.e.:

$$\begin{aligned} \nabla \times \mathbf{E}(\mathbf{r}) &= -j\omega\mu\mathbf{H}(\mathbf{r}), \\ \nabla \times \mathbf{H}(\mathbf{r}) &= \mathbf{J}(\mathbf{r}) + j\omega\varepsilon\mathrm{E}(\mathrm{r}), \\ \nabla \cdot \mathbf{E}(\mathbf{r}) &= \frac{\mathrm{P}(\mathbf{r})}{\varepsilon}, \\ \nabla \cdot \mathbf{H}(\mathbf{r}) &= 0, \end{aligned} \tag{5.17}$$

where **E** and **H** are the complex electric and magnetic field intensities, and **J** and P are the complex volume current and charge densities, respectively. **r** is the position vector of a point in space. The relation between **J** and P is specified by the current continuity equation

$$\nabla \cdot \mathbf{J}(\mathbf{r}) = -j\omega\mathrm{P}. \tag{5.18}$$

The electric field induced by a volume current density is as follows:

$$\mathbf{E}(\mathbf{r}) = \nabla(\nabla \cdot \Pi) - \gamma^2\Pi, \tag{5.19}$$

where the Hertz vector potential is defined as:

$$\mathbf{\Pi}(\mathbf{r}) = \frac{\gamma^2}{4\pi j\omega\mu}\int_V \mathbf{J}(\mathbf{r}')\,\mathbf{g}(\mathbf{r}, \mathbf{r}')dv', \tag{5.20}$$

where $\mathbf{r}'$ is a position vector on the source. Thus:

$$\mathbf{E}(\mathbf{r}) = \frac{\gamma^2}{4\pi j\omega\mu}\int_V \mathbf{J}\cdot\overline{\overline{\mathbf{G}}}(\mathbf{r},\mathbf{r}')dv', \tag{5.21}$$

where the Dyadic Green's function $\overline{\overline{\mathbf{G}}}$ is defined as follows [39]:

$$\begin{aligned} \overline{\overline{\mathbf{G}}}(\mathbf{r},\mathbf{r}') &= \left(\nabla\nabla\cdot - \gamma^2\overline{\overline{\mathbf{I}}}\right)\mathbf{g}(\mathbf{r},\mathbf{r}'), \\ g(\mathbf{r},\mathbf{r}') &= \frac{e^{-\gamma|\mathbf{r}-\mathbf{r}'|}}{|\mathbf{r}-\mathbf{r}'|}, \\ \gamma &= \sqrt{j\omega\mu(\sigma + j\omega\varepsilon)}, \end{aligned} \tag{5.22}$$

and γ is the medium complex wave number. In (5.22) the identity dyadic tensor $\overline{\overline{\mathbf{I}}}$ in three dimensions is defined as:

$$\overline{\overline{\mathbf{I}}} = \hat{x}\hat{x} + \hat{y}\hat{y} + \hat{z}\hat{z} \tag{5.23}$$

If we consider a perfectly conducting antenna in a homogenous medium, the volume current $\mathbf{J}$ can be replaced by the surface current $\mathbf{Js}$, simplifying (5.21) as:

$$\mathbf{E}(\mathbf{r}) = \frac{\gamma^2}{4\pi j\omega\mu}\iint_{S_0} \mathbf{J_s}(\mathbf{r}')\cdot\overline{\overline{\mathbf{G}}}(\mathbf{r},\mathbf{r}')ds_0. \tag{5.24}$$

The boundary condition on the surface of a perfect conductor can be expressed as:

$$\mathbf{n}\times\left[\mathbf{E^a}(\mathbf{r}) + \mathbf{E}(\mathbf{r})\right] = 0, \tag{5.25}$$

where $\mathbf{E^a}$ is the applied field, and $\mathbf{n}$ is the normal vector to the surface of the conducting object. Combining (5.24) and (5.25), we have the electric field integral equation in the frequency domain:

$$\begin{aligned} \frac{\gamma^2}{4\pi j\omega\mu}\mathbf{n}\times\iint_{S_0} \mathbf{J_s}(\mathbf{r}')\cdot\left(\nabla\nabla\cdot - \gamma^2\overline{\overline{\mathbf{I}}}\right)\mathbf{g}(\mathbf{r},\mathbf{r}')ds_0 \\ = -\mathbf{n}\times\mathbf{E}^i(\mathbf{r}). \end{aligned} \tag{5.26}$$

The vector equation (5.26) is an electric field equation for surfaces and is converted to a scalar equation by the thin-wire approximation. The wire antenna representing a lightning channel satisfies the thin-wire approximation [40] since its radius is much smaller than the wavelength, and its length is much greater than the radius. This implies that: (1) current on the wire flows only in the longitudinal direction, and (2) circumferential or radial variations in the axial current are negligible. According to the thin-wire approximation, the current I(s) on a wire structure of radius (a) satisfies the following equations [40],

$$I(s) = 2\pi a J_s(s),$$

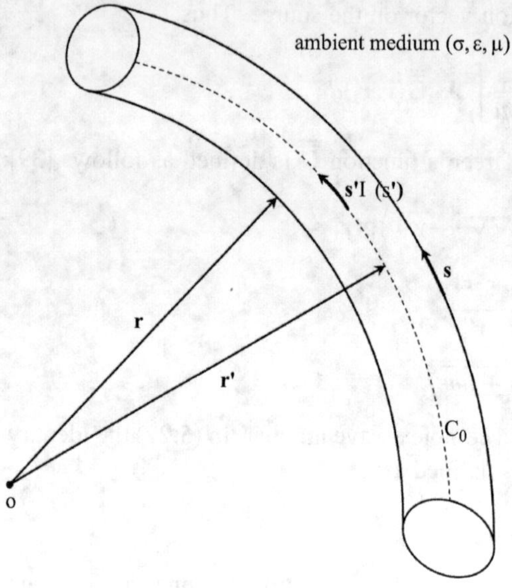

Figure 5.3 Thin-wire approximation

$$\mathbf{J}_0(s) = \frac{I(s)}{2\pi a}\mathbf{s}, \tag{5.27}$$

where s is the location of a point on the wire structure and **s** is the corresponding tangential unit vector (see Figure 5.3).

Now, consider a perfectly conducting wire placed in a medium with conductivity σ, permittivity ε, and permeability μ, shown in Figure 5.3. As mentioned in the previous section, EFIE expresses the fact that the total tangential electric field on the wire surface vanishes:

$$\mathbf{s} \cdot (\mathbf{E}^{\mathbf{a}} + \mathbf{E}) = 0, \tag{5.28}$$

where the terms in parentheses are the applied $\mathbf{E}^{\mathbf{a}}$ and scattered **E** electric fields.

The electric field **E** is related to the vector potential **A** and scalar potential $\mathbf{\Phi}$ by:

$$\mathbf{E} = -\nabla\mathbf{\Phi} - j\omega\mathbf{A}, \tag{5.29}$$

where:

$$\mathbf{A}(\mathbf{r}) = \frac{\mu}{4\pi}\int_{C_0} \mathbf{s}' I(s') g(\mathbf{r}, \mathbf{r}') ds',$$

$$\Phi(\mathbf{r}) = \frac{j\omega}{4\pi(\sigma + j\omega\varepsilon)}\int_{C_0} \rho_l(s') g(\mathbf{r}, \mathbf{r}') ds', \tag{5.30}$$

and ρ_l is the linear charge density and C_0 is the path along which the current is flowing. s' is the source point on the antenna and $\mathbf{s}'$ is the corresponding unit tangential vector. The relation between $I(s')$ and $\rho_l(s')$ is specified by the current continuity equation:

$$\nabla \cdot \mathbf{s}' I(s') + j\omega\rho_l(s') = 0. \tag{5.31}$$

Combining (28)–(31) yields the expression for the electric field due to the current of a thin wire in a homogeneous medium:

$$\mathbf{E}(\mathbf{r}) = \frac{\gamma^2}{4\pi j\omega\mu}(\nabla\nabla\cdot - \gamma^2)\int_{C_0} \mathbf{s}' I(s') g(\mathbf{r}, \mathbf{r}') ds'. \tag{5.32}$$

The term being integrated in (5.32) is always continuous because the minimum distance between the source and observation points cannot be less than the radius of the thin wire. In addition, integration is done for the source while differentiation is done for the observation point. Therefore, it is permitted to change the order of integration and differentiation in (5.32):

$$\mathbf{E}(\mathbf{r}) = \frac{\gamma^2}{4\pi j\omega\mu}\int_{C_0} (\nabla\nabla\cdot - \gamma^2)\mathbf{s}' I(s') g(\mathbf{r}, \mathbf{r}') ds'. \tag{5.33}$$

By applying the vector identity:

$$\begin{aligned} \nabla\nabla\cdot(\beta\mathbf{C}) = {} & (\nabla\beta)\nabla\cdot\mathbf{C} \\ & + \beta\nabla\nabla\cdot\mathbf{C} + \nabla\beta\times\nabla\times\mathbf{C} \\ & + (\mathbf{C}\cdot\nabla)\nabla\beta + (\nabla\beta\cdot\nabla)\mathbf{C}, \end{aligned} \tag{5.34}$$

and with the use of some mathematical operations, (5.33) is reduced to:

$$\mathbf{E}(\mathbf{r}) = \frac{\gamma^2}{4\pi j\omega\mu}\int_{C_0}\left\{ \begin{array}{l} [\mathbf{s}' I(s').\nabla]\nabla g(\mathbf{r}, \mathbf{r}') \\ -\gamma^2 \mathbf{s}' I(s') g(\mathbf{r}, \mathbf{r}') \end{array} \right\} ds'. \tag{5.35}$$

Combining (5.28) and (5.35), we have

$$-\mathbf{s}\cdot\mathbf{E}^a = \frac{\gamma^2}{4\pi j\omega\mu}\int_{C_0} I(s') G(\mathbf{r}, \mathbf{r}') ds', \tag{5.36}$$

where $G(\mathbf{r}, \mathbf{r}')$ is the kernel of integral equation and is related to the Green's function by:

$$G(\mathbf{r}, \mathbf{r}') = \left[(\mathbf{s}'\cdot\nabla)(\nabla\cdot\mathbf{s}) - \gamma^2\mathbf{s}\cdot\mathbf{s}'\right] g(\mathbf{r}, \mathbf{r}'). \tag{5.37}$$

The numerical solution of (5.37), the electric field integral equation (EFIE) in the frequency domain, by the method of moments (MOM) [18], yields the space distribution of current along the antenna at a given frequency.

5.3 Numerical treatment

Wire antennas have been investigated by researchers for more than a century [41,42], continue to be used for modeling electromagnetic problems including: antenna design, scattering, and electromagnetic compatibility. In problems involving antennas, parameters such as current distribution, input impedance, radiation efficiency, gain and radiated field are of great interest, whereas in the modeling of scattering problems, a different set of parameters such as radar cross section, scattering matrix, and scattered field are to be determined. In the case of large structures, high-frequency approximation techniques have been recommended due to limitations in available memory and processing speeds. Examples of such techniques are: geometrical optics, physical optics, and geometrical theory of diffraction [43].

Electromagnetic compatibility problems involving lightning return-stroke channels, lightning protection systems and grounding systems can be analyzed by wire antenna modeling. The main objective in wire antenna modeling is to obtain the current distribution on each wire. In the preceding sections, we have shown that the current on the wire is the kernel of the electric field integral equations obtained from Maxwell's equations in the time or frequency domain. Since there is no closed-form solution for such integral equations, a numerical method is to be employed. Due to the nature of the wire antennas, a numerical method based on the Method of Moments (MoM) is found to be appropriate. For transient studies, the time domain MoM is preferred [44,45] whereas the frequency domain technique is more appropriate for narrow band frequency analyses [46].

In the following sections, we first outline the steps required to implement the method of moments for solving integral equations. We will then present necessary formulations for treating time-domain and frequency-domain problems. Finally, we describe how a lossy ground is treated.

5.3.1 Method of moments

The use of MoM for solving electromagnetic problems was first introduced by Richmond in 1965 [47] and later extended by Harrington in 1967 [48]. This method, also called Method of Weighted Residuals, is a general solution for a linear-operator equation, i.e.:

$$\mathrm{L(f) = g} \tag{5.38}$$

where L is an integro-differential operator, f is an unknown function (response), and g is a known function (excitation).

Solving (5.38) by the MoM generally involves the following steps:

- Extract an appropriate integral equation.
- Transform the integro-differential equation into a matrix equation using proper basis and test functions.

- Acquire the so-called impedance matrix elements.
- Solve the impedance matrix equation and obtain the unknown quantities.

To this end, the unknown function f is initially expanded by a set of basis functions

f_n, $n = 1, \ldots, N$,

$$f = \sum_{n=1}^{N} \alpha_n f_n. \tag{5.39}$$

Given that the operator L is linear, the replacement in this series into (5.37) leads to:

$$\sum_{n=1}^{N} \alpha_n L(f_n) = g. \tag{5.40}$$

The objective is to obtain the unknown coefficients α_n, which lead to determine f in the linear-operator equation. Equation (5.40) has N unknowns. To ensure a unique solution, one needs to establish N equations. A set of equations for the coefficients α_n are then obtained by taking the inner product of (5.38) with a set of test functions $w_m, m = 1, \ldots, N$,

$$\sum_{n=1}^{N} \alpha_n \langle w_m, L(f_n) \rangle = \langle w_m, g \rangle, \tag{5.41}$$

or,

$$\begin{pmatrix} \langle w_1, L(f_1) \rangle & \langle w_1, L(f_2) \rangle & \cdots & \langle w_1, L(f_N) \rangle \\ \langle w_2, L(f_1) \rangle & \langle w_2, L(f_2) \rangle & \cdots & \langle w_2, L(f_N) \rangle \\ \vdots & \vdots & \ddots & \vdots \\ \langle w_N, L(f_1) \rangle & \cdots & \cdots & \langle w_N, L(f_N) \rangle \end{pmatrix} \begin{pmatrix} \alpha_1 \\ \alpha_2 \\ \vdots \\ \alpha_N \end{pmatrix} = \begin{pmatrix} \langle w_1, g \rangle \\ \langle w_2, g \rangle \\ \vdots \\ \langle w_N, g \rangle \end{pmatrix}, \tag{5.42}$$

which can be summarized as follows:

$$[z_{mn}][\alpha_n] = [v_m]. \tag{5.43}$$

Here, $[z_{mn}]$, $[\alpha_n]$, and $[v_m]$ are referred to as the impedance matrix, coefficient matrix, and excitation matrix, respectively. Notice that the choice of the basis and test functions is very important for an accurate and efficient solution.

5.3.2 Time-domain formulation

In this section, we describe how the MoM [34–45] is used to treat time-domain lightning problems by solving (5.15). In this regard, we first divide the thin wire into N elementary segments of length Δ_i while the time span is divided into N_T equal steps of length Δ_t. Then, a set of rectangular basis functions is defined for expressing the known current in each segment, i.e.:

$$i(s_0,t_0) = \sum_{i=1}^{N}\sum_{j=1}^{N_T} I_{ij}(s_i'',t_j'')U(s_i'')V(t_j''),$$

where $s_i'' = s_0 - s_i, t_i'' = t_0 - t_i$ and (5.44)

$$U(s_i'') = \begin{cases} 1 & |s_i''| \le \Delta_i/2 \\ 0 & \text{otherwise} \end{cases} \tag{5.45}$$

$$V(t_i'') = \begin{cases} 1 & |t_i''| \le \Delta_i/2 \\ 0 & \text{otherwise} \end{cases}$$

A second-order polynomial representation is used to evaluate $I_{ij}\left(s_i'',t_j''\right)$ and the interpolation is chosen to be Lagrangian:

$$I_{ij}\left(s_i'',t_j''\right) = \sum_{l=-1}^{+1}\sum_{m=v}^{v+2} B_{ij}{}^{(l,m)}\mathrm{I}_{i+1,j+1}, \tag{5.46}$$

with:

$$B_{ij}^{(l,m)} = \prod_{p=-1}^{+1}\prod_{q=v}^{v+2}\frac{(s_\circ - s_{i+p})(t_\circ - t_{j+q})}{(s_{i+1} - s_{i+p})(t_{j+m} - t_{j+q})}, \tag{5.47}$$

$$\begin{cases} v = -1; & \Delta R = \dfrac{R}{c(t_j - t_{j-1})} > 0.5 \\ v = -2; & \Delta R < 0.5 \end{cases}$$

Notice that $I_{i+1,j+m}$ is the current value at the center of the (i+1)th space segment and the (j+m)th time step.

The next step consists of choosing the test functions in order to obtain a system of linear equations [45]. Here, we use the point matching method, based on Dirac distributions, i.e.:

$$\begin{aligned} &\delta(t - t_u) \text{ in the space} \\ &\delta(t - t_v) \text{ in the time} \end{aligned} \tag{5.48}$$

Applying the test functions to (5.15), we obtain the following system of linear equations:

$$\begin{bmatrix} z_{11} & \cdots & z_{1N} \\ \vdots & \ddots & \vdots \\ z_{N1} & \cdots & z_{NN} \end{bmatrix}\begin{bmatrix} I_{1,j} \\ \vdots \\ I_{N,j} \end{bmatrix} = \begin{bmatrix} V_{1,j} \\ \vdots \\ V_{N,j} \end{bmatrix} \times \frac{1}{\Delta_i}, \tag{5.49}$$

or equivalently

$$\mathbf{ZI}_j = \mathbf{V}_j \times \frac{1}{\Delta_i}, \tag{4.50}$$

where **Z** is called per-unit-length impedance matrix, Δ_i is the wire structure segment length, $\mathbf{I}_j$ is the vector of current distribution at each wire segment i (= 1, 2, ..., N) at time step j (= 1, 2, ..., N_T), and $\mathbf{V}_j$ is the vector of segment voltages at the corresponding time step.

For convenience, we use the electric fields of each segment instead of voltage:

$$\mathbf{E}_j^{total} = \mathbf{E}_j^a + \mathbf{E}_j = \mathbf{ZI}_j\ , \tag{5.51}$$

As stated earlier, **Z** is a spatial–temporal matrix related to the structure that is generated considering the electromagnetic interaction between segments. Matrix **Z** also expresses the relation between the current and the total fields for the structure, which is time-invariant only if the shape of the structure remains unchanged during computations.

In the case where the segments are loaded with additional elements, (5.50) should be updated as follows:

$$\mathbf{E}_j^{total} = \mathbf{ZI}_j + \mathbf{E}_j^{Load} = (\mathbf{ZI}_j + \mathbf{Z}_j^{Load}\mathbf{I}_j) = (\mathbf{Z} + \mathbf{Z}_j^{Load})\mathbf{I}_j, \tag{5.52}$$

where $\mathbf{Z}_j^{Load}$ describes the effect of the segment loads in time step j. For resistive loading, the values of segment resistive loads are added to diagonal elements of the **Z** matrix. Regarding the voltage–current relation of an inductor in series with a resistor, an additional tangential electric field due to inductive load at segment i at time-step j can be discretized using the classical backward finite difference approximation as follows [27]:

$$E_{i,j}{}^R = \frac{V_{i,j}^R}{\Delta i} = \frac{1}{\Delta i} R_i I_{i,j} = \frac{1}{\Delta i}(R_i^F \ \ + R_{i,j-1}^N) I_{i,j}, \tag{5.53}$$

$$E_{i,j}{}^L = \frac{V_{i,j}^L}{\Delta i} = \frac{1}{\Delta i} L_i \frac{dI_{i,j}}{dt} = \frac{L_i}{\Delta i}\frac{3}{2\Delta t} I_{i,j} + \frac{L_i}{\Delta i}\left\{\frac{1}{2\Delta t} I_{i,j-2} - \frac{2}{\Delta t} I_{i,j-1}\right\}, \tag{5.54}$$

where R_i^F is a fixed resistance, $R_{i,j-1}^N$ is a nonlinear resistance being a function of current, $I_{i,j-1}$, and time, $(j-1)\Delta t$, and L_i is a fixed inductance all calculated in segment i, respectively, Δt is the duration of time interval j, Δi is the segment length, and $I_{i,j}$ denotes the current in segment i at time-step j.

We can now separate $\mathbf{E}_j^{Load}$ into two parts, namely, $E_{i,j}^{Load_1}$ for the unknown part as [27]:

$$E_{i,j}^{Load_1} = \frac{V_{i,j}^{L_1}}{\Delta i} = \left\{\frac{R_i^F}{\Delta i} + \frac{R_{i,j-1}^N}{\Delta i} + \frac{L_i}{\Delta i}\frac{3}{2\Delta t}\right\} I_{i,j} = Z_{ii}^{Load_1} I_{i,j}, \tag{5.53}$$

and $E_{i,j}^{Load_2}$ for the known part (available at time step j) as:

$$E_{i,j}^{Load_2} = \frac{V_{i,j}^{L_2}}{\Delta i} = \frac{1}{\Delta i}\left\{\frac{L_i}{2\Delta t}I_{i,j-2} - \frac{2L_i}{\Delta t}I_{i,j-1}\right\} = E_{i,j}^{add_{Load}}. \quad (5.54)$$

Substituting (5.53) and (5.54) in (5.52) leads to the matrix form of the discretized EFIE, i.e.:

$$\mathbf{E}_j^{scattered} + \mathbf{E}_j^{incident} - \mathbf{E}_j^{add_{Load}} = (\mathbf{Z} + \mathbf{Z}_j^{Load_1})\mathbf{I}_j. \quad (5.55)$$

When the thin-wire is not loaded or when it is loaded with static elements such as resistors, inductors, or capacitors rather than with dynamic elements, the impedance matrix is not a time-dependent matrix, hence one needs to invert it once in a whole time-domain solution and then use the inverted matrix at each time step to obtain the corresponding solution. As was stated earlier, the effect of these types of loads only results in modifying the diagonal or semi-diagonal elements of the impedance matrix. However, for the case of a dynamic nonlinear load, the impedance matrix is time-dependent, and thus must be inverted at each time step, introducing an inevitable computational cost [49].

5.3.3 Frequency-domain formulation for uniform soil

To solve the governing EFIE (5.36) for longitudinal current distribution on the wire, I(s), it is expanded in a finite series of N overlapped sinusoidal currents and an additional sinusoidal monopole at the injection point, as depicted in Figure 5.4:

$$I(s) = \sum_{k=0}^{N} I_k F_k(s), \quad (5.56)$$

where I_k are the unknowns to be determined, s is the distance along the wire, and $F_k(s)$ are the normalized sinusoidal currents over the length of one dipole or

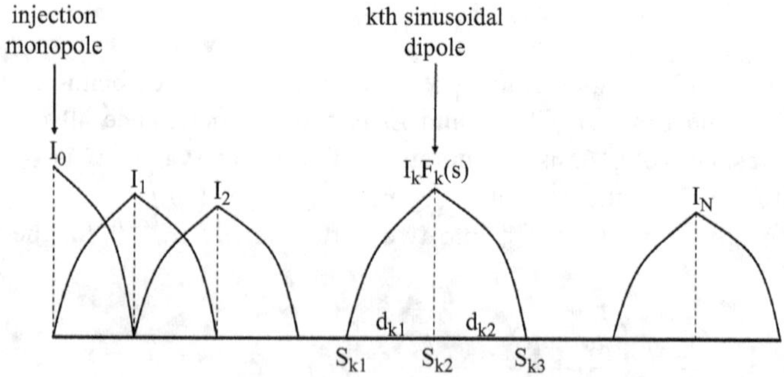

Figure 5.4 Sinusoidal current expansion

injection monopole, expressed as follows [21]:

$$F_k(s) = \frac{P_{k1}(s)\sinh\,\gamma(s - s_{k1})}{\sinh\,\gamma d_{k1}} + \frac{P_{k2}(s)\sinh\,\gamma(s_{k3} - s)}{\sinh\,\gamma d_{k2}},$$
$$P_{k1}(s) = \begin{cases} 1, & s_{k1} \le s \le s_{k2} \\ 0, & \text{elsewhere} \end{cases}, \quad P_{k2}(s) = \begin{cases} 1, & s_{k2} \le s \le s_{k3} \\ 0, & \text{elsewhere} \end{cases} \tag{5.57}$$

where for the kth dipole, s_{k2} is the mid (terminal) point, s_{k1}, s_{k3} are the end points ($k = 1,2\ldots N$), and d_{k1}, d_{k2} are the dipole-arm lengths. For the injection monopole ($k = 0$), only the second term of (5.54) is used.

Applying (5.56)–(5.57) to (5.36) and using the Galerkin method [21], one can obtain a system of N linear algebraic equations of the form **ZI** = **V**, where **I** is the current vector to be solved for, **V** is the excitation vector, and **Z** is the impedance matrix. Evaluating the elements of **Z** and **V** requires numerical double integrations. To increase computational efficiency and accuracy, it is desirable, when possible, to replace numerical integrations with appropriate analytic expressions. Such an alternative exists only for the sinusoidal current expansion [21] whose near-field expressions are available in rigorous form. Rigorous field expressions of a sinusoidal monopole in a cylindrical coordinate system (see Figure 5.5) are given

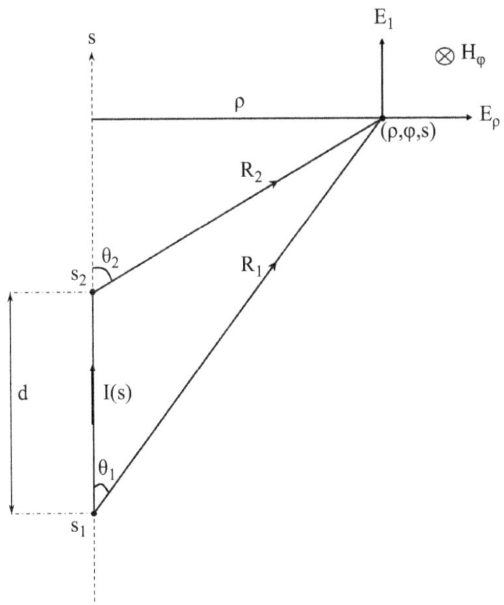

Figure 5.5 The geometry of sinusoidal monopole and its electromagnetic fields

below [21]:

$$E_\rho = \frac{\gamma}{\rho(\sigma + j\omega\varepsilon)} \frac{1}{4\pi \sinh \gamma d} \begin{bmatrix} (I_1 e^{-\gamma R_1} - I_2 e^{-\gamma R_2}) \sinh \gamma d \\ +(I_1 \cosh \gamma d - I_2) e^{-\gamma R_1} \cos \theta_1 \\ +(I_2 \cosh \gamma d - I_1) e^{-\gamma R_2} \cos \theta_2 \end{bmatrix}, \tag{5.58}$$

$$E_l = \frac{\gamma}{(\sigma + j\omega\varepsilon)} \frac{1}{4\pi \sinh \gamma d} \begin{bmatrix} (I_1 - I_2 \cosh \gamma d) \dfrac{e^{-\gamma R_2}}{R_2} \\ +(I_2 - I_1 \cosh \gamma d) \dfrac{e^{-\gamma R_1}}{R_1} \end{bmatrix}, \tag{5.59}$$

$$H_\varphi = \frac{1}{4\pi\rho \sinh \gamma d} \begin{bmatrix} (I_1 \sinh \gamma d \cos \theta_1 + I_1 \cosh \gamma d - I_2) e^{-\gamma R_1} \\ -(I_2 \sinh \gamma d \cos \theta_2 - I_2 \cosh \gamma d + I_1) e^{-\gamma R_2} \end{bmatrix}, \tag{5.60}$$

where I_1 and I_2 are the values of the monopole current, I(s), at s_1 and s_2, respectively:

$$I(s) = \frac{I_1 \sinh \gamma(s_2 - s) + I_2 \sinh \gamma(s - s_1)}{\sinh \gamma d}. \tag{5.60}$$

Other quantities are depicted in Figure 5.5. Using these field components, one can replace the inner integrals for **Z** and **V** by their analytic expressions.

To show the advantages of the current-source excitation, we first consider the voltage source case. In this case [13,21,27], the source voltage, $V(f)$, is obtained as follows:

$$V(f) = Z_{in}(f) \cdot I(f), \tag{5.61}$$

where $Z_{in}(f)$ is the input impedance of the wire antenna, and $I(f)$ is the Fourier transform of the specified channel-base current. Once the excitation voltage is obtained, it can be used in the following (N + 1) × (N + 1) linear matrix equation to obtain the current distribution:

$$\begin{pmatrix} z_{00} & z_{01} & \cdots & z_{0N} \\ z_{10} & z_{11} & \cdots & z_{1N} \\ \vdots & \vdots & \ddots & \vdots \\ z_{N0} & z_{N1} & \cdots & z_{NN} \end{pmatrix} \begin{pmatrix} I_0 \\ I_1 \\ \vdots \\ I_N \end{pmatrix} = \begin{pmatrix} V(f) \\ 0 \\ \vdots \\ 0 \end{pmatrix}. \tag{5.62}$$

The input impedance calculation increases the computation time and may introduce an additional error in the solution. In this study, we use a current source to excite the wire antenna, so that (5.62) reduces to:

$$\begin{pmatrix} z_{00} & z_{01} & \cdots & z_{0N} \\ z_{10} & z_{11} & \cdots & z_{1N} \\ \vdots & \vdots & \ddots & \vdots \\ z_{N0} & z_{N1} & \cdots & z_{NN} \end{pmatrix} \begin{pmatrix} I_0 \\ I_1 \\ \vdots \\ I_N \end{pmatrix} = \begin{pmatrix} 0 \\ 0 \\ \vdots \\ 0 \end{pmatrix}, \tag{5.63}$$

where I_0 is determined by the current source. In this case, there are $(N + 1)$ equations with N unknowns. Consequently, (5.63) can be transformed to the following $N \times N$ linear matrix equation:

$$\begin{pmatrix} z_{11} & z_{12} & \cdots & z_{1N} \\ z_{21} & z_{22} & \cdots & z_{2N} \\ \vdots & \vdots & \ddots & \vdots \\ z_{N1} & z_{N2} & \cdots & z_{NN} \end{pmatrix} \begin{pmatrix} I_1 \\ I_2 \\ \vdots \\ I_N \end{pmatrix} = \begin{pmatrix} -z_{10}I_0 \\ -z_{20}I_0 \\ \vdots \\ -z_{N0}I_0 \end{pmatrix} \text{ or } \mathbf{Z}\,\mathbf{I} = \mathbf{V}, \tag{5.64}$$

where z_{mn} $(m,\ n = 1,2 \ldots N)$ are the mutual impedances between the sinusoidal dipoles defined as follows:

$$z_{mn} = \int_m F_m(s) E_n^t(s) ds, \tag{5.65}$$

and z_{m0} $(m = 1,2 \ldots N)$ are the mutual impedances between the injection monopole and sinusoidal dipoles given below:

$$z_{m0} = \int_m F_m(s) E_0^t(s) ds, \tag{5.66}$$

where $E_n^t(s)$ $(n = 1,2 \ldots N)$, and $E_0^t(s)$ are, respectively, the electric fields tangential to the wire surface due to the nth dipole (basis function) and the injection monopole. Also, $F_m(s)$ is the mth test dipole and integrations are carried out over its length.

In the case that it is necessary to include distributed resistance, R_D, along the wire antenna, leading to the new impedance matrix given below [21]:

$$Z_{mn} = z_{mn} - \underbrace{R_D \int_m F_m(s) H_n^\varphi(s) ds}_{z'_{mn}}, \tag{5.67}$$

where $H_n^\varphi(s)$ is the azimuthal magnetic field due to the nth dipole, and z'_{mn} is the contribution from the distributed resistance. The introduction of distributed resistance requires modification of only diagonal and semi-diagonal elements in the impedance matrix [50]. Also, the limits of the integral in the expression for z'_{mn} extends over two wire segments in the domain of the test dipole $F_m(s)$.

Utilizing simplifying approximations, one can obtain [35]:

$$H_n^\varphi = \frac{F_n(s)}{2\pi a}. \tag{5.68}$$

Hence,

$$z'_{mn} = \frac{R_D}{2\pi a} \int_{m,n} F_n(s) F_m(s) ds, \tag{5.69}$$

where a is the radius of the wire antenna, and region (m,n) is the wire surface shared by dipoles m and n. The expression for z'_{mn} reduces to [21]:

$$z'_{mn} = \begin{cases} \dfrac{R_D}{4\pi\gamma a} \dfrac{\sinh(2\gamma d) - 2\gamma d}{\sinh^2(\gamma d)} & \text{if}(m = n) \\ \dfrac{R_D}{4\pi\gamma a} \dfrac{\gamma d \cosh(\gamma d) - \sinh(\gamma d)}{\sinh^2(\gamma d)} & \text{if}(m = n \pm 1) \\ 0 & \text{elsewhere} \end{cases} \tag{5.70}$$

5.3.4 Lossy half-space problem

So far, we have assumed that the ground is a perfect conductor. In this section, we describe how a lossy ground is treated in wire-antenna modeling. Figure 5.6 schematically shows a lightning channel above a lossy half-space. The thin-wire model of a lightning channel can be considered as the superposition of infinitesimal dipoles. Electromagnetic radiation of a dipole above a lossy half-space was first investigated by Sommerfeld [51]. The approach involves semi-infinite integrals of Bessel functions that are characterized by highly oscillatory and weakly damping behavior. For very close observation points, various approximations have been proposed. In this work, when evaluating elements of **Z** and **V** matrices in (5.64) for each sinusoidal dipole, its modified image is also included. Modified image theory was first introduced in [52]. As an illustration of this simple theory, consider a point

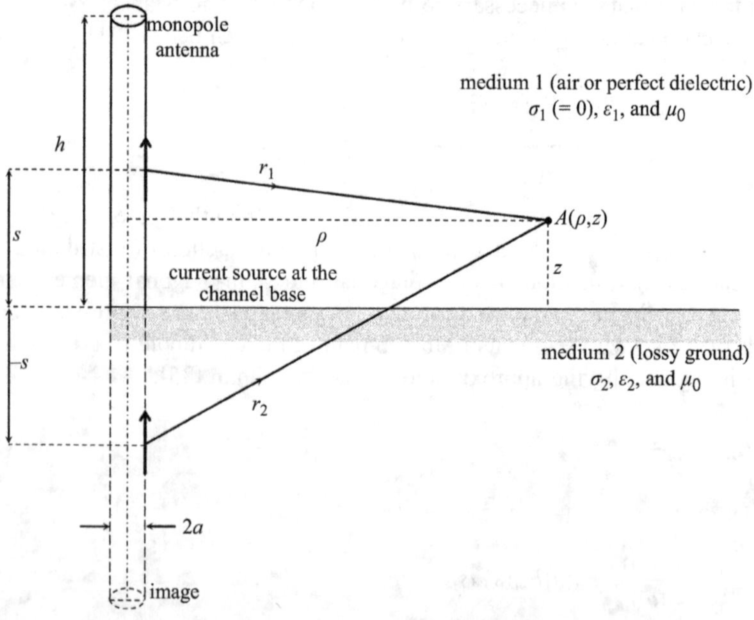

Figure 5.6 Lightning channel above a lossy half-space

current source placed in a dielectric medium above the interface of a lossy half-space (see Figure 5.6). The field in the upper half-space, medium #1, is the superposition of contributions from the original source and its mirror image multiplied by a correction factor α_{11}, and the field in the lower half-space, medium #2, is due to the original source multiplied by a correction factor α_{12}. Applying the boundary conditions at the interface (i.e., the continuity of the tangential electric field vector and scalar potential), one can obtain correction factors given below [52]:

$$\alpha_{11} = \frac{j\omega\varepsilon_1 - (\sigma_2 + j\omega\varepsilon_2)}{j\omega\varepsilon_1 + (\sigma_2 + j\omega\varepsilon_2)}\ ;\ \alpha_{12} = \frac{2(\sigma_2 + j\omega\varepsilon_2)}{j\omega\varepsilon_1 + (\sigma_2 + j\omega\varepsilon_2)} \tag{5.71}$$

Once **Z** and **V** are known, **I** can be determined from (5.64) utilizing a suitable solution of linear algebraic equations. The resultant current distribution can be used to calculate various field components in air. Since surface waves become significant as the observation point is moved farther from the source, the modified image theory is not applicable to distant field calculations. Thus, a different treatment is necessary. An accurate solution of this problem requires the time-consuming evaluation of Sommerfeld integrals. For the situation examined here (where the magnitude of the ground's complex wave number is much greater than that of free space), these integrals can be represented by closed-form expressions. In this regard, King [53] derived the complete expressions for the electromagnetic fields of a vertical electric dipole over an imperfectly conducting (lossy) half-space. Also, King and Sandler [54] verified the suitability of these expressions for a dipole above certain types of lower half-space. The complete electromagnetic field expressions for a vertical straight wire antenna of length L above a lossy half-space in air (i.e., $\varepsilon_1 = \varepsilon_0$ in Figure 5.6) are reproduced below:

$$H_{a\varphi}(\rho, z, j\omega) = \int_{C_0} \frac{I(s)}{2\pi} \left[\begin{array}{l} \dfrac{e^{-\gamma_1 r_1}}{2}\left(\dfrac{\rho}{r_1}\right)\left(\dfrac{1+\gamma_1 r_1}{r_1^2}\right) + \dfrac{e^{-\gamma_1 r_2}}{2}\left(\dfrac{\rho}{r_2}\right)\left(\dfrac{1+\gamma_1 r_2}{r_2^2}\right) \\ -e^{-\gamma_1 r_2}\left(\dfrac{\gamma_1^3}{\gamma_2}\right)\left(\dfrac{\pi}{j\gamma_1\gamma_2}\right)^{1/2} e^{-jP}F(P) \end{array} \right] ds, \tag{5.72}$$

$$\begin{aligned} E_{a\rho}(\rho, z, j\omega) = &\frac{j\omega\mu_0}{2\pi\gamma_1}\int_{C_0} I(s)\left\{\frac{e^{-\gamma_1 r_1}}{2}\left(\frac{\rho}{r_1}\right)\left(\frac{z-s}{r_1}\right)\left(\frac{3+3\gamma_1 r_1+\gamma_1^2 r_1^2}{\gamma_1 r_1^3}\right)\right. \\ &+ \frac{e^{-\gamma_1 r_2}}{2}\left(\frac{\rho}{r_2}\right)\left(\frac{z+s}{r_2}\right)\left(\frac{3+3\gamma_1 r_2+\gamma_1^2 r_2^2}{\gamma_1 r_2^3}\right) \\ &\left. + \frac{\gamma_1}{\gamma_2}e^{-\gamma_1 r_2}\left[-\left(\frac{\rho}{r_2}\right)\left(\frac{1+\gamma_1 r_2}{r_2^2}\right) + \frac{\gamma_1^3}{\gamma_2}\left(\frac{\pi}{j\gamma_1\gamma_2}\right)^{1/2} e^{-jP}F(P)\right]\right\} ds, \end{aligned} \tag{5.73}$$

$$E_{az}(\rho,z,j\omega) = \frac{j\omega\mu_0}{2\pi\gamma_1}\int_{C_0} I(s)\left\{\frac{e^{-\gamma_1 r_1}}{2\gamma_1 r_1^3}\left[(1+\gamma_1 r_1+\gamma_1^2 r_1^2) - \left(\frac{z-s}{r_1}\right)^2(3+3\gamma_1 r_1+\gamma_1^2 r_1^2)\right]\right.$$
$$+\frac{e^{-\gamma_1 r_2}}{2\gamma_1 r_2^3}\left[(1+\gamma_1 r_2+\gamma_1^2 r_2^2) - \left(\frac{z+s}{r_2}\right)^2(3+3\gamma_1 r_2+\gamma_1^2 r_2^2)\right]$$
$$\left. - e^{-\gamma_1 r_2}\left(\frac{\gamma_1^3}{\gamma_2}\right)\left(\frac{\pi}{j\gamma_1\gamma_2}\right)^{1/2}\left(\frac{\rho}{r_2}\right)e^{-jP}F(P)\right\}ds, \tag{5.74}$$

where subscript (a) denotes observation points in air, ω is the angular frequency, and (ρ,φ,z) is the cylindrical coordinate system. Also:

$$\gamma_1 = j\omega\sqrt{\mu_0\varepsilon_0}\ ;\ \gamma_2 = \sqrt{j\omega\mu_0(\sigma_2 + j\omega\varepsilon_2)} \tag{5.75}$$

ε_0 and μ_0 are free space permittivity and permeability, respectively,

$$r_1 = \sqrt{\rho^2 + (z-s)^2}\ ;\ r_2 = \sqrt{\rho^2 + (z+s)^2} \tag{5.76}$$

$$P = j\frac{\gamma_1^3 r_2}{2\gamma_2^2}\left[\frac{\gamma_1 r_2 + \gamma_2(z+s)}{\gamma_1\rho}\right]^2, \tag{5.77}$$

$$F(P) = \frac{1}{\sqrt{2\pi}}\int_P^\infty \frac{e^{jz}}{\sqrt{z}}dz = \frac{1}{2}(1+j) - C_2(P) - jS_2(P), \tag{5.78}$$

where $C_2(P)$ and $S_2(P)$ are, respectively, the Fresnel cosine and sine integrals of a complex argument expressed as [55]:

$$C_2(P) = \frac{1}{\sqrt{2\pi}}\int_0^P \frac{\cos z}{\sqrt{z}}dz\ ;\ S_2(P) = \frac{1}{\sqrt{2\pi}}\int_0^P \frac{\sin z}{\sqrt{z}}dz \tag{5.79}$$

Equations (5.72)–(5.74) are accurate everywhere in air or on the air-ground interface provided that the following single condition is satisfied [21]:

$$|\gamma_2|^2 >> |\gamma_1|^2 \text{ or } |\gamma_2| \geq 3|\gamma_1| \tag{5.80}$$

Each of the field expressions (5.72)–(5.74) contains three terms: a term in r_1, which is the direct wave from the source dipole, a similar term in r_2, which is the reflected wave from the image dipole, and a term containing Fresnel integrals, which is the surface or lateral wave. This surface wave is defined as the total field minus the perfect-ground approximation (geometrical-optics field), as opposed to the Zenneck surface wave (ZSW), which never exists as the sole contribution for the ground wave from a localized source [56]. Our surface wave might be considered as the Norton surface wave (NSW) [57,58]. One can readily observe that these field expressions reduce to those assuming the perfect-ground approximation only when the first two terms of each expression are considered.

Under the condition of $|P| \geq 4$ (e.g., at far observation points [21]), the Fresnel integrals have asymptotic expansions that allow one to simplify the field expressions. Specifically,

$$\frac{\gamma_1^3}{\gamma_2}\left(\frac{\pi}{j\gamma_1 r_2}\right)^{1/2} e^{-jP}F(P) \overset{|P|\geq 4}{\longrightarrow} \frac{\gamma_1^3}{\gamma_2}\left(\frac{\pi}{j\gamma_1 r_2}\right)^{1/2} \frac{1}{(2\pi P)^{1/2}}\left(\frac{1}{2P}+j\right) \tag{5.81}$$

$$\frac{\gamma_1^3}{\gamma_2}\left(\frac{\pi}{j\gamma_1 r_2}\right)^{1/2} e^{-jP}F(P) \overset{|P|\geq 4}{\longrightarrow} \frac{\gamma_1^3}{\gamma_2}\left(\frac{\pi}{j\gamma_1 r_2}\right)^{1/2} \frac{1}{(2\pi P)^{1/2}}\left(\frac{1}{2P}+j\right), \tag{5.82}$$

While most of the conventional methods [59,60] use different treatments for the two components of electric field, the simple closed-form electric field expressions (5.73)–(5.74) employed here are parts of the same solution. It can be shown that the Cooray–Rubinstein formula for the horizontal electric field component is a special case of (5.73) when the Fresnel term is neglected. Also, the magnetic field component given by (5.72) is obtained using the same methodology [21].

5.3.5 Frequency-domain formulation for stratified media

The lightning return-stroke channel is represented by a monopole antenna above a multilayer soil (Figure 5.1(b)). In order to extend the study to more complex structures including the lightning channel and nearby structures, an array of conducting wires and metallic surfaces is used to represent the entire structure.

All numerical computations have been performed using the commercial CDEGS software package [85]. The high-frequency module of this package is based on a numerical solution of the EFIE by MoM for a network of wires and surfaces [86,97,98]. It is suitable for the electromagnetic scattering analysis of complex objects in a stratified medium over a wide band of frequencies.

The EFIE states that the total tangential electric field on the surface of wire conductors and surface elements follows the relation:

$$\mathcal{L}\mathfrak{J}|_{\tan} - Z_s(\omega) \quad \mathfrak{J} = \mathbf{0} \tag{5.83}$$

where $\mathfrak{J}$ represents the induced electric current densities on the wire conductors, metallic surfaces, and the impressed current sources and $Z_s(\omega)$ denotes the internal impedance of those conductors. The operator $\mathcal{L}$ returns the electric field at the point $\mathbf{r}$ scattered by the current densities located at $\mathbf{r}'$ in stratified media and is defined as [87]:

$$\mathcal{L}\mathfrak{J}(\mathbf{r}') = \int_{\Omega'} \overline{\overline{G}}_{\Pi}(\mathbf{r},\mathbf{r}') \cdot \mathfrak{J}(\mathbf{r}')d\mathbf{r}' - \frac{1}{\gamma_m^2}\nabla\int_{\Omega'} \left[\nabla \cdot \overline{\overline{G}}_{\Pi}\right] \cdot \mathfrak{J}(\mathbf{r}')d\mathbf{r}' \tag{5.84}$$

where $\overline{\overline{G}}_{\Pi}$ is the dyadic Green's function for the Hertz vector. In this equation, Ω' represents the problem's domains which are wire conductors or surface plates in general. For a stratified medium parallel to the xy plane, $\overline{\overline{G}}_{\Pi}(\mathbf{r},\mathbf{r}')$ can be conventionally expressed as [87]:

$$\overline{\overline{G}}_{\Pi}\left(\mathbf{r},\mathbf{r}'\right) = (\widehat{xx}+\widehat{yy})\mathbf{G}_{xx}+\widehat{zx}\mathbf{G}_{zx}+\widehat{zy}\mathbf{G}_{zy}+\widehat{zz}\mathbf{G}_{zz} \tag{5.85}$$

where $\mathbf{G}_{xx}$, $\mathbf{G}_{zx}$, $\mathbf{G}_{zy}$, and $\mathbf{G}_{zz}$ are obtained from (5.91)–(5.93).

The current densities on the surface of wires and patches are expanded in terms of linear basis functions $\mathbf{j}_n^w$ and $\mathbf{j}_p^s$ multiplied by unknown coefficients and impressed currents:

$$\mathcal{J} = \sum_{n=1}^{N_w} I_n^w \mathbf{j}_n^w + \sum_{p=1}^{N_s} I_p^s \mathbf{j}_p^s + \sum_{m=1}^{N_i} I_m^i \mathbf{j}_m^i \tag{5.86}$$

where N_s and N_w stand for the total number of introduced one-dimensional and two-dimensional degrees of freedoms in each domain, while N_i is the total number of impressed current sources ($\mathbf{j}^i$) for a specific problem. After substituting (5.91), (5.92), (5.93), and (5.86) into (5.84), we use appropriate line test functions in order to transform the boundary value problem into a matrix equation $AI = V$ of size $N = N_w + N_s + N_i$. The structure of the resulting MoM matrix equation has nine blocks, as follows:

$$\begin{bmatrix} A^{WW} & A^{WS} & A^{WI} \\ A^{SW} & A^{SS} & A^{SI} \\ A^{IW} & A^{IS} & A^{II} \end{bmatrix} \begin{bmatrix} I^W \\ I^S \\ I^I \end{bmatrix} = 0 \tag{5.87}$$

where the unknown coefficients of the wire and surface currents are denoted by I^W and I^S, respectively (e.g. $I^W = \left[I_1^w, I_2^w, \ldots, I_{N_w}^w\right]$). Block A^{IJ} describes the interactions between a type of test function (wire, surface) and the discretized currents of the structure. The blocks of the third row depict the interactions of the remaining elements of the network with each current source. As impressed currents are used to represent sources or known quantities, they are not affected by these interactions. Thus, all elements of the blocks in the third row can be disregarded. Consequently, $N_w + N_s$ equations remain and (5.87) can be transformed into a square linear matrix equation:

$$\begin{bmatrix} A^{WW} & A^{WS} \\ A^{SW} & A^{SS} \end{bmatrix} \begin{bmatrix} I^W \\ I^S \end{bmatrix} = \begin{bmatrix} -A^{WI} I^I \\ -A^{SI} I^I \end{bmatrix} \tag{5.88}$$

Once the elements of the square matrix are computed, its inversion provides the induced currents (I^W, I^S) on the whole structure for a specific channel current distribution (I^I). A detailed description of the adopted numerical approach can be found in [86].

5.3.6 Green's functions for stratified media

The electric and magnetic fields in layer m of a horizontally stratified medium can be expressed in terms of the Hertz vector potential $\mathbf{\Pi}$ [88]:

$$\mathbf{E}^m = -\gamma_m^2 \mathbf{\Pi}^m - \nabla\nabla \cdot \mathbf{\Pi}^m \tag{5.89}$$

$$\mathbf{H}^m = \theta_m \nabla \times \mathbf{\Pi}^m \tag{5.90}$$

where $\gamma_m^2 = j\omega\mu_m\theta_m$, $\theta_m = \sigma_m + j\omega\varepsilon_m$. In addition, σ_m, μ_m, and ε_m are the conductivity, permeability, and permittivity of the layer where the observation point is located.

The Hertz vector satisfies the wave equation. Thus, for a unitary electric dipole ($I = 1$ Am) located at the point (x_s, y_s, z_s) in layer s, the general expressions for the components of the Hertz vector potential at the point (x_m, y_m, z_m) in layer m are given by [88,89]:

$$\Pi_x^m = \frac{X_s'}{4\pi\theta_m}\int_0^\infty \begin{bmatrix} \delta_{ms}\dfrac{\lambda}{\sqrt{\lambda^2+\gamma_m^2}}e^{-\sqrt{\lambda^2+\gamma_m^2}|z_s-z_m|} \\ F_s^+(\lambda)e^{+\alpha_m z_m} + F_s^-(\lambda)e^{-\alpha_m z_m} \end{bmatrix} J_0(\lambda\rho)d\lambda \tag{5.91}$$

$$\Pi_y^m = \frac{Y_s'}{4\pi\theta_m}\int_0^\infty \begin{bmatrix} \delta_{ms}\dfrac{\lambda}{\sqrt{\lambda^2+\gamma_m^2}}e^{-\sqrt{\lambda^2+\gamma_m^2}|z_s-z_m|} \\ F_s^+(\lambda)e^{+\alpha_m z_m} + F_s^-(\lambda)e^{-\alpha_m z_m} \end{bmatrix} J_0(\lambda\rho)d\lambda \tag{5.92}$$

$$\Pi_z^m = \frac{Z_s'}{4\pi\theta_m}\int_0^\infty \begin{bmatrix} \delta_{ms}\dfrac{\lambda}{\sqrt{\lambda^2+\gamma_m^2}}e^{-\sqrt{\lambda^2+\gamma_m^2}|z_s-z_m|} \\ H_s^+(\lambda)e^{+\alpha_m z_o} + H_s^-(\lambda)e^{-\alpha_m z_m} \end{bmatrix} J_0(\lambda\rho)d\lambda$$
$$+ \begin{bmatrix} \dfrac{X_s'(x_m - x_s)}{4\pi\theta_m\rho} \\ \dfrac{Y_s'(y_m - y_s)}{4\pi\theta_m\rho} \end{bmatrix} \int_0^\infty \left[P_s^+(\lambda)e^{+\alpha_m z_m} + P_s^-(\lambda)e^{-\alpha_m z_m}\right]\lambda J_1(\lambda\rho)d\lambda \tag{5.93}$$

where (X_s', Y_s', Z_s') is the tangent vector to the dipole at its center, $\rho = \sqrt{(x_m - x_s)^2 + (y_m - y_s)^2}$ is the radial distance between the observation point and the dipole, $\alpha_m = \sqrt{\lambda^2 + \gamma_m^2}$, and J_n is the Bessel function of order n. In addition, δ_{ms} stands for the Kronecker Delta function, where s is the index of the layer containing the dipole source and m is the index of the layer containing the observation point. The continuity of the tangential components of the electric and magnetic fields in terms of the Hertz vector at each interface provides equations for the coefficients F_s^+, F_s^-, H_s^+, H_s^-, P_s^+, and P_s^- which are solved analytically using a recursive approach [88].

5.4 Various AT models

The expression "lightning return-stroke model" is generally used to describe a specification of the time- and height-dependent current in the RSC to make possible the calculation of the resultant remote electromagnetic fields [1]. Most of the return-stroke models specify an analytical relation between the current at each point of the channel and the channel-base (ground level) current. Such an analytical relation also describes the propagation of the current wave along the channel.

In the AT model, the lightning RSC is represented by a vertical, monopole antenna above a conducting ground. The antenna is fed at its lower end by a source whose voltage is determined using the assumed antenna input current waveform. The distribution of current along the antenna is determined by applying the boundary condition for the tangential component of the electric field on the surface of the antenna. The solution is found numerically, using the MoM. The resultant distribution of current along the antenna will radiate into free space.

In this section, various time- and frequency-domain models of RSC, based on antenna theory, i.e., on a complete solution of Maxwell's equations, are presented and used to obtain the channel current and charge density profiles.

5.4.1 Time-domain AT model

As mentioned above, in the AT model, the antenna representing the RSC is fed at its lower end by a source that launches a current wave along the antenna. In fact, the source radiates electromagnetic fields into a nonconducting medium whose electrical permittivity is selected such that the wave propagates at a specified speed (lower than the speed of light). The artificial change of permittivity is used to account for the effect of corona on the wave propagation speed.

The distribution of current along the antenna is determined by applying the boundary condition for the tangential component of the electric field on the surface of the antenna. The solution is found numerically, using the MoM, with ohmic losses in the antenna taken into account by introducing the resistive loading R (series resistance per unit length). The evolution of the current and charge along the channel is described by Maxwell's equations. In order to compare "engineering" models with the AT model quantitatively, we need to use the same input current in all the models. This means that the voltage source for the monopole antenna should produce the same current as the channel-base current assumed in the "engineering" models. The voltage of the source is given by the following equation:

$$v(t) = \mathrm{F}^{-1}[Z(f) \cdot I(0,f)], \tag{5.94}$$

where $I(0,f)$ is the Fourier transform of the specified channel-base current, $Z(f)$ is the input impedance of the lossy monopole antenna, and $\mathbf{F}^{-1}$ denotes the inverse Fourier transform. The input impedance of the monopole antenna, which is a function of its length and distributed resistance, is calculated by applying the MoM to the EFIE. To find the input impedance of the monopole antenna, a modulated Gaussian waveform, shown in Figure 5.7, was used. Figure 5.8 depicts this waveform in the frequency domain. Using (5.16) and solving the EFIE (5.15) by the MoM, which takes into account resistive loading, we obtained the input current of the monopole antenna corresponding to the modulated Gaussian voltage waveform. The input impedance of the antenna is then found by dividing the input voltage by the input current in the frequency domain. Figure 5.9 shows the source voltage, found using the computed input impedance and specified current at the channel

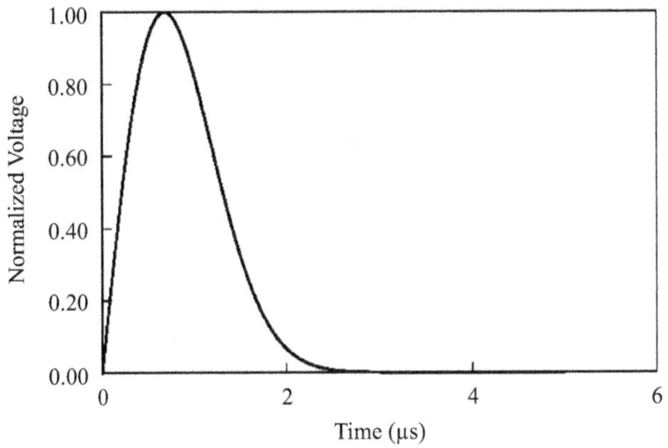

Figure 5.7 Normalized modulated Gaussian waveform in the time domain which has been used as a source voltage to calculate the input impedance of the monopole antenna. Adapted from Moini et al. *[13]*

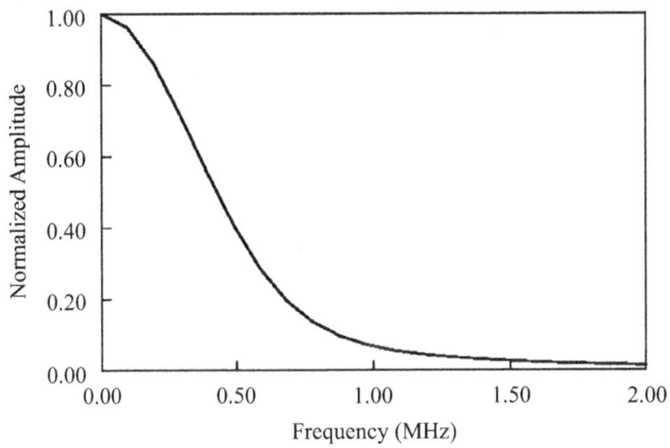

Figure 5.8 Normalized amplitude of the modulated Gaussian waveform in the frequency domain corresponding to the time domain waveform shown in Figure 5.7. Adapted from Moini et al. *[13]*

base, for different values of resistance R per unit length. The waveform of the source voltage is almost independent of R during the first tenths of microseconds. The value of $R = 0.07\ \Omega/m$ has been selected for the calculations presented in this work, since this value provides the best agreement between model-predicted and observed electric field waveforms at close (tens of meters) ranges. Interestingly, Rakov [61] estimated the resistance per unit channel length to be $0.035\ \Omega/m$ behind the return-stroke front.

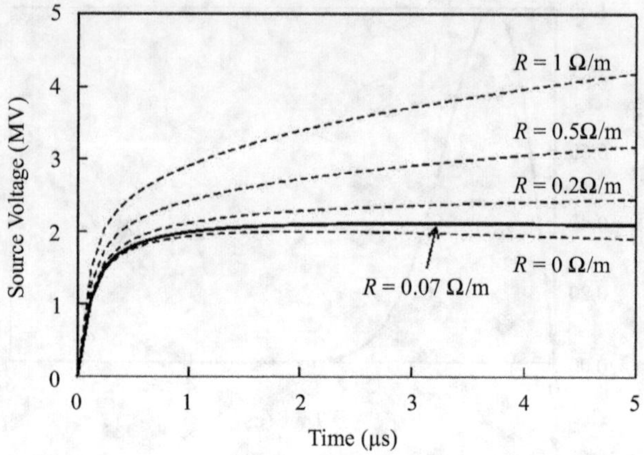

Figure 5.9 Calculated source voltage waveform corresponding the channel-base current for different values of the resistance per unit channel length. Adapted from Moini et al. *[13]*

Once the voltage $v(t)$ of the source is determined, the corresponding applied electric field $\boldsymbol{e^a}$, to be substituted in (5.15), is estimated as the ratio of this voltage and the length Δz of the excitation (source) segment of the antenna.

To reduce the propagation speed of the current wave in the AT model to a value consistent with observations, $v < 3 \times 10^8$ m/s, we use $\varepsilon > \varepsilon_0$ in calculating the current variation along the channel, and then use that current distribution to calculate the electromagnetic fields radiated by the antenna in free space ($\varepsilon = \varepsilon_0$). The arbitrary increasing of ε in determining the channel current distribution serves to account for the fact that channel charge is predominantly stored in the radial corona sheath whose radius is much larger than that of the channel core which carries the longitudinal channel current, resulting in $v < 3 \times 10^8$ m/s. This simulates an increase of shunt capacitance per unit antenna length due to corona. The use of $\varepsilon > \varepsilon_0$ additionally introduces the effect of radiation into the fictitious medium, but the resultant current distribution along the channel is unlikely to differ significantly from the case of no such effect (the transmission line current is expected to dwarf the antenna current). An alternative approach to modeling corona effect on propagation speed would be to introduce capacitive antenna loading. Ohmic losses in the antenna further reduce v, but for the selected value of resistance per unit length, this additional reduction in v is expected to be relatively small. In the AT model, there are only two parameters to be adjusted, the propagation speed v for the case of non-resistive antenna and the value of distributed resistance R. The value of resistance per unit length was selected (by trial and error) to provide an agreement between model-predicted and measured electric fields at close distances. It was assumed for the AT model that $R = 0.07$ Ω/m and that $v = 1.3 \times 10^8$ m/s, which corresponds to $\varepsilon_r = 5.3$. As stated above, the spatial and temporal distribution of current along the antenna was determined solving the EFIE, (5.15), using MoM.

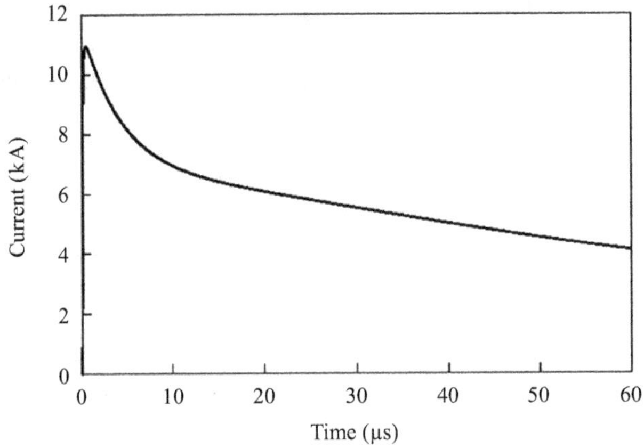

Figure 5.10 Channel-base current waveform used for the comparison of different models. The peak current is about 11 kA, and peak current rate of rise is about 105 kA/μs. Adapted from Nucci et al. *[62]*

In Section 5.5 (Numerical results), we compare the AT model with other models in terms of (1) spatial and temporal current distribution, (2) line charge density distribution, and (3) remote electromagnetic fields. In doing so, we assume the same lightning channel-base current waveform as that used by Nucci *et al.* [62], Rakov and Dulzon [63], and Thottappillil *et al.* [64]. This waveform is depicted in Figure 5.10.

5.4.2 Time-domain AT model with inductive loading

In the AT model, the assumption of $\varepsilon_r > 1$ was only used to find the current distribution along the channel, which was then allowed to radiate into free space with $\varepsilon_r = 1$ [27]. However, even the current distribution along the channel can be potentially influenced to some (presumably small) extent by the unrealistic assumption $\varepsilon_r > 1$.

In the AT model with inductive loading (ATIL), a modification is made in the original AT model [13] to include inductive loading of the channel in order to avoid the unrealistic assumption of higher permittivity of the surrounding medium. In this model, in order to control the return-stroke speed, inductive energy-storing elements are included in the antenna representation of the lightning channel, which is a monopole antenna with distributed resistance above a perfectly conducting ground [27]. In fact, the ATIL is a step forward to finding a method to reduce the speed of current waves propagating on a vertical conductor in air, such that the evolution of wave shape is consistent with optical observations of lightning.

The use of energy-storing elements in antenna and transmission line studies has been previously described in a number of works. Induction phenomenon in solving electrical circuits using the time-domain EFIE was studied by Bost *et al.* [65]. Guedira [66] applied local inductive and capacitive loads to the feeding point of a dipole antenna and determined resultant antenna currents. The additional distributed capacitance has been used to describe the effect of corona on transmission lines [67]. Applying shunt distributed capacitive loads, representing the radial corona sheath, to an antenna introduces some difficulties, as discussed by Bonyadiram *et al.* [30]. One difficulty is related to the fact that distributed shunt capacitors require a return conductor parallel to the antenna, which turns the monopole antenna to a transmission line. Since the phase velocity is a function of the product of the inductance and capacitance (each per unit length), distributed inductance can be used, instead of distributed shunt capacitance, to simulate the corona effect on the propagation speed [26].

An unloaded horizontal, perfectly conducting wire above a perfectly conducting ground behaves as a lossless uniform transmission line, and the propagation speed along the wire is constant and equal to the speed of light, $c = 1/\sqrt{L_0 C_0}$ where L_0 and C_0 are per unit length inductance and capacitance, respectively. Strictly speaking, this speed is not applicable to a vertical monopole antenna and its image because the per-unit-length capacitance and inductance of such an antenna vary with height. As a result, the equivalent transmission line is non-uniform and the propagation speed is slightly lower than the speed of light. Further, in the case of lightning, the speed is reduced (typically by a factor of two or three) relative to the speed of light due to the presence of a corona sheath and the transformation of the leader channel to the return-stroke channel. Also, as noted above, the optically observed speed decreases with increasing height (we do not consider here the non-monotonic variation of speed with height reported by Olsen *et al.* [68]). Ultimately, considering the lightning channel as a vertical wire above ground with its intrinsic capacitance and inductance per unit length does not allow one to reproduce observed lightning return-stroke speed profiles.

If we introduce an additional, height-variable distributed inductance along the antenna without any resistive loading, the resultant height-variable propagation speed along the simulated lightning channel will be given by:

$$v(z) = \frac{1}{\sqrt{L'(z) \cdot C_0(z)}} = \frac{1}{\sqrt{(L_0(z) + L^{add}(z)) \cdot C_0(z)}}, \tag{5.95}$$

where $L^{add}(z)$ is the additional height-dependent distributed inductance per unit length. As a result, for a specified height-variable speed, $v(z)$, $L^{add}(z)$ is given by:

$$L^{add}(z) = \frac{1}{v^2(z) \cdot C_0(z)} - L_0(z), \tag{5.96}$$

The capacitance, C_0, and inductance, L_0, per unit length for a cylindrical metallic wire of radius a can be estimated using the following equations given by Kodali *et al.* [69]:

$$C_0(z) = \frac{2\pi\varepsilon_0}{\ln(2z/a)} \quad (F/m), \tag{5.97}$$

$$L_0(z) = \frac{\mu_0}{2\pi} \ln(2z/a) \quad (H/m), \tag{5.98}$$

where z is the height above ground. The applicability of these equations, derived for a horizontal wire above ground, to a vertical conductor is discussed by Kodali *et al.* [69]. Considering typical speed profiles for lightning shows that the values of intrinsic capacitance and inductance of an ideal cylindrical antenna above a perfectly conducting ground given by (5.97) and (5.98) cannot yield the variation of speed as specified by (5.95), and hence an additional height-variable reactive element is needed to simulate such a speed profile. Note that any additional distributed inductance that is modeled along the channel does not correspond to additional inductance in a real-world system, but rather is invoked only to reduce the speed of the current wave to a value lower than the speed of light.

In the following, we will consider two types of inductive loading, namely, fixed (uniform) and height-varying (non-uniform). In the ATIL model with fixed inductive loading (ATIL-F), additional inductance per unit length was selected (by trial and error) so as to obtain a specified average speed. It was observed that fixed loading results in an almost uniform speed profile (see the Numerical results in Section 5.5). For example, $L^{add} = 8.0 \times 10^{-6}$ H/m results in $v = 1.3 \times 10^8$ m/s. Interestingly, this value of L^{add} is approximately equal to L^{add} that is computed from (5.96) using constant values C_0 and L_0 evaluated at $a = 0.02$ m and $z = 3{,}500$ m from (5.97) and (5.98), respectively. Note that the values of C_0 and L_0 at $z = 3{,}500$ m are applied to the entire channel. This is a commonly used simplification (e.g., [69]) based on the fact that the dependencies of C_0 and L_0 on z are weak (logarithmic).

In the ATIL model with variable inductive loading (ATIL-V), a speed profile was chosen, to achieve acceptable consistency with the published optical measurements. Optical observations are mainly limited to the visible part of the channel, usually extending from ground to a height of 1–3 km. The return-stroke speed typically decreases by 25% or more over the visible part of the channel with respect to the speed at the bottom of the channel [70]. Here, we use an exponentially decaying speed profile $v(z)$ that is described by the following equation:

$$v(z) = v_h - (v_h - v_0)e^{-\frac{z}{\lambda}}, \tag{5.99}$$

where λ is the decay height constant, v_0 is the propagation speed at the channel base (at ground level), and the final speed at the upper end of the channel (at $z = h$) asymptotically approaches v_{h}. Thottappillil and Uman [71] have used a similar relation in the modified Diendorfer–Uman model (MDU) in which $v_{\mathrm{h}} = 0$, that is at greater heights, the speed approaches zero. The use of a non-zero value of v_{h} gives us more control of the propagation speed profile, especially in the upper part of the channel. Optical observations have shown that the propagation speed near the bottom of the channel varies between c/2 and c/3 [72] and decreases with height. Here, we computed a speed profile along a 3.5 km lightning return-stroke channel using (5.99) with $\lambda = 450$ m, $v_0 = 1.6 \times 10^8$ m/s, and $v_{\mathrm{h}} = 0.9 \times 10^8$ m/s. This profile is shown in Figure 5.11. Note that at $z = 3.5$ km $v = 0.9 \times 10^8$ m/s, which is essentially equal to the assumed value of v_{h}. It should be noted that there is essentially no limitation on the length of the analyzed channel. Longer lightning

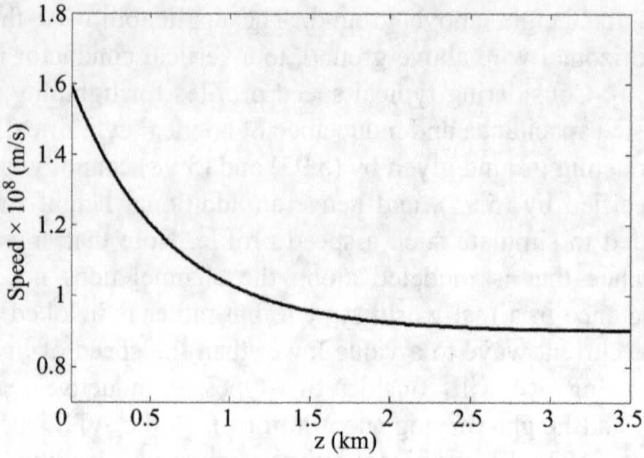

Figure 5.11 Propagation speed as a function of height for the ATIL-V model obtained using (5.88) with λ = 450 m, v_0 = 1.6 × 10^8 m/s, and v_h = 0.9 × 10^8 m/s. Adapted from Bonyadi-Ram et al. *[30]*

channels could be analyzed if the information on speed profile were available. In time domain methods, the length of an analyzed channel is limited by the duration of the analysis. We selected the analysis duration such that the traveling wave will never hit the top of the channel. This is a widely used approach in lightning research [1].

The corresponding additional distributed inductance as a function of height, calculated using (5.96), is depicted in Figure 5.12. The $L_0(z)$ and $L^{add}(z)$ profiles for the ATIL-F model with $v = 1.3 \times 10^8$ m/s are also shown in Figure 5.12 for comparison.

Let us consider a lightning channel with a height of 3.5 km and a radius of 0.02 m above a perfectly conducting ground. The current at the channel base is the same as that used in the previous section. The inductance per unit length for the ATIL-F model is set to $L^{add} = 8.0 \times 10^{-6}$ H/m. Distributed resistance of the channel for the ATIL-F model is 0.5 Ω/m, considerably larger than the 0.07 Ω/m in the AT model. For the AT model, ε_r is set to 5.3, while for both versions of the ATIL model, $\varepsilon_r = 1$.

For the ATIL-V model, one can use the profile of the distributed inductance (shown in Figure 5.12) while setting the distributed resistance to 0.45 Ω/m. The reason for using different values of distributed resistance in the AT (0.07 Ω/m), ATIL-F (0.5 Ω/m), and ATIL-V (0.45 Ω/m) models will be discussed in Section 5.5. A typical length of each channel segment for all three models is 10 m.

5.4.3 Time-domain AT model with nonlinear loading

The physical models, which are based on gas dynamic models of RSC, predict nonlinearities in the channel parameters, including per-unit-length inductance, capacitance and resistance [2,3]. To attain a more realistic model of RSC, attempts were made to include such nonlinearities in the AT modeling. This section

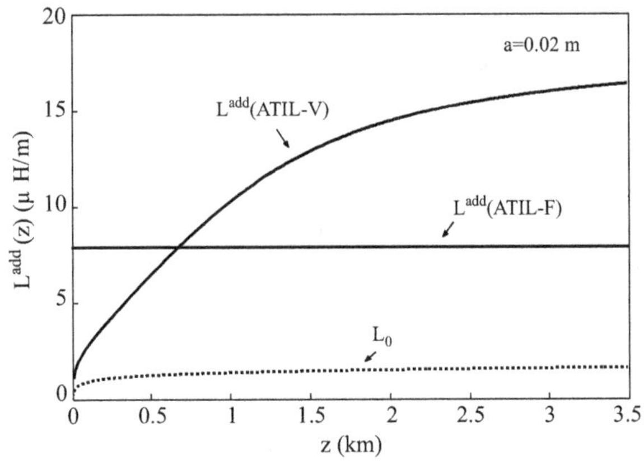

Figure 5.12 Additional distributed inductance per unit length, L^{add}, as a function of height for the ATIL-V (computed using (5.85)) and ATIL-F (determined by trial and error) models. The speed profile for the ATIL-V model is shown in Figure 5.11, and the speed for the ATIL-F is 1.3×10^8 m/s. Also shown is the intrinsic inductance of the monopole antenna computed using (5.87), L_0, as a function of height. Adapted from Bonyadi-Ram et al. *[30]*

describes the modifications made in the ATIL-F model to include nonlinear resistive loading in the presence of fixed inductive loading of the channel. Resistive elements are considered as a nonlinear distributed load whose resistance is a function of both current and time.

The nonlinear ATIL-F model adopts the method proposed by Braginskii which uses a distribution of nonlinear load for representing the nonlinear phenomenon occurring during a lightning strike to flat surfaces [2]. In this method, a strong-shock approximation is used to develop a spark channel model for describing the time variation for parameters such as radius, temperature, pressure, and resistance as a function of the input channel current. Initial conditions, which are meant to characterize the channel created by a lightning leader, include temperature (in the order of 10,000°K), channel radius (of the order of 1 mm), and either pressure equal to ambient (1 atm) or mass density equal to ambient (in the order of 10^{-3} g cm^{-3}), the latter two conditions representing, respectively, the older and the newly created channel sections.

The initial condition assuming ambient pressure probably best represents the upper part of the leader channel since that part has had sufficient time to expand and attain equilibrium with the surrounding atmosphere, while the initial condition assuming ambient density is more suitable for the recently created, bottom part of the leader channel. In the latter case, variations in the initial channel radius and initial temperature are claimed to have little influence on model predictions [2,73]. It is noted that the Braginskii's method assumes, as initial conditions, ambient density and pressure much higher than the surrounding ambient [74].

For a current $i(t)$ linearly increasing with time t, the channel radius, $a(i(t), t)$ can be obtained as follows [2]:

$$a(i(t), t) = 9.35[i(t)]^{\frac{1}{3}} t^{\frac{1}{2}}, \tag{5.100}$$

where $a(i(t), t)$ is in centimeters, $i(t)$ in amperes, and t in seconds.

In the derivation of (5.100), presumably applicable to the early stages of the discharge, the general expression of the arc channel radius is simplified [75], i.e.:

$$a \approx \sqrt{(4/\rho_0 \pi^2)^{1/3} \int_0^t (\sigma\, \xi)^{-1/3} i^{2/3} dt + a_i^2}, \tag{5.101}$$

where a_i is the initial radius in meters, σ is the channel conductivity in S/m, ρ_0 is the ambient atmospheric density in g cm^{-3}, and ξ is a factor describing the rate of radial expansion of the arc channel. This is done by assuming a linearly increasing current, which results in the expansion of the channel radius according to $a \sim t^{\text{constant}}$, $a_i = 0$, $\xi = 4.5$, $\rho_0 = 1.29 \times 10^{-3}\ g\ cm^{-3}$ and $\sigma = 2.22 \times 10^4\ (\Omega\ m)^{-1}$.

For a known value of $a(i(t), t)$, the resistance per unit channel length can be found as [2]:

$$R(i(t), t) = \frac{1}{\pi\, \sigma\, a^2(i(t), t)} \tag{5.102}$$

Equation (5.102) without any other constraints gives very large as well as very small values of resistance, thus we introduce upper R_{max} and lower R_{min} limits. These limits (corresponding to a_{min} and a_{max}, respectively) are physical limits and they bring the theoretical estimation closer to the experimental ones. Theoretical estimations give initial and final values of the channel radius as 1 and 20 mm, respectively [2,73]. Experimental estimates give a final channel radius from 10 to 35 mm [76,77]. To reach a model with realistic results consistent with the experimental data, we assume $\xi \cong 20$ and $\sigma = 5 \times 10^3\ S/m$. Using these values in (5.100) and (5.102) determines the values of a_{min} and a_{max}, and hence, the values of R_{max} and R_{min}, as 0.01 m and 0.18 m, and 0.6764 Ω/m and 0.040 Ω/m, respectively.

Here we need to apply a larger distributed resistance than that used in the original AT model. An additional distributed resistance is required for fixing the attenuation constant α at the same level as that of the AT model. Notice that inductive loading forces (α) to decrease [27].

The effect of resistive loading on the phase velocity of the propagation can be described using the transmission lines theory. The phase velocity, v_φ, for a transmission line is given by:

$$v_\varphi = \frac{\omega}{\beta}, \tag{5.103}$$

where ω is the angular frequency and β is the phase constant derived from the expression for the propagation constant:

$$\gamma = \alpha + j\beta = \sqrt{(R + j\omega L)(G + j\omega C)} \tag{5.104}$$

where $j = \sqrt{-1}$, and R, L, G, and C are, respectively, the per-unit-length resistance, inductance, conductance, and capacitance of the transmission line. By a proper definition of $p = \tan^{-1}\left(\frac{R}{\omega L}\right)$ and $q = \tan^{-1}\left(\frac{G}{\omega C}\right)$, (5.104) can be rewritten as:

$$\gamma = \sqrt{\omega L\left(\frac{R}{\omega L} + j\right) \cdot \omega C\left(\frac{G}{\omega C} + j\right)} = \omega\sqrt{LC}\sqrt{(\tan p + j) \cdot (\tan q + j)} \tag{5.105}$$

which can be further simplified to:

$$\gamma = \alpha + j\beta = \frac{\omega\sqrt{LC}}{\sqrt{\cos p \cos q}} \cdot \sin\frac{p+q}{2} + j\frac{\omega\sqrt{LC}}{\cos p \cos q} \cdot \cos\frac{p+q}{2} \tag{5.106}$$

Extracting β from (5.106) and inserting it in (5.103), one can determine v_φ as follows:

$$v_\varphi = \frac{\omega}{\beta} = \frac{1}{\sqrt{LC}} \cdot \frac{\cos p \cos q}{\cos \frac{p+q}{2}}. \tag{5.107}$$

Using the expressions discussed in [69] applied to a vertical conductor above the ground, for an RSC as a conductor with a length of 9 km and with a radius of 2 cm, the values of L and C are calculated as 2.74 μH/m and 4.056 pF/m, respectively. A typical value for the highest angular frequency of interest is 62,800 rad/s. Notice that this frequency coincides with the −3 dB frequency of a waveform spectrum defined by the sum of a Heidler function and a double exponential function, which is assumed to be the excitation current source at the channel base.

Substituting the selected values of L, C, and ω in (5.107), and assuming G = 0, it is deduced that, v_φ, becomes only a function of R for that frequency. To adjust the value of v_φ to a value lower than the speed of light, one can use a similar approach to that adopted in the ATIL-F model [27,69]. This is done by adding an additional distributed inductance, L_{add}, along the channel that is calculated by the classical transmission lines model. Using the new value of L together with other parameters, one can use (5.107) to plot the current wave propagation speed versus R for various locations along the RSC.

Figure 5.13 shows variations of v_φ versus R for both the channel base and tip. In the case of the channel base, L_{add} = 4.024 μH/m, L = 5.6 μH/m, and C = 6.94 pF/m, whereas for the channel tip L_{add} = 27.7 μH/m, L = 30.4 μH/m, and C = 4.05 pF/m. Knowing that the propagation speed of the current wave varies between c/2 and c/3 [31], one can use the results shown in Figure 5.13 to determine the approximate range of R. Table 5.1 summarizes calculated values of inductance

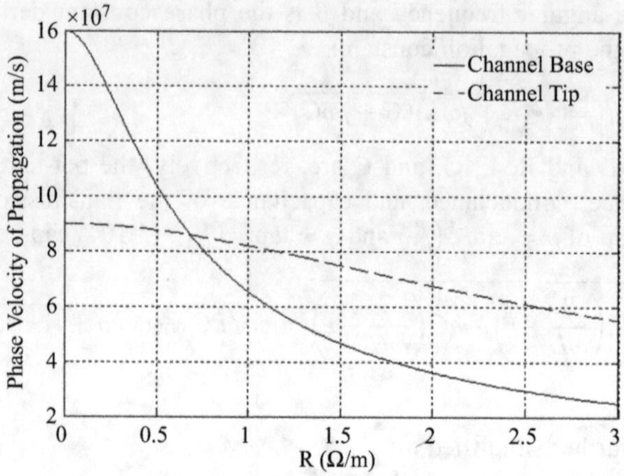

Figure 5.13 Phase velocity of wave propagation versus resistive loss at f = 10 kHz for a RSC modeled by a transmission line with length of 9 km and radius of 2 cm, both at the channel base (loaded by additional inductance 4.024 μH/m), and channel tip (loaded by an additional inductance 27.7 μH/m. Adapted from Seyed-Moosavi et al. *[31]*

Table 5.1 Calculated values of inductance and capacitance at the bottom and tip of the RSC

RSC parameter value	**RSC parameter unit**	**Formulation used**	**RSC bottom (channel base touching the ground)**	**RSC tip (channel top touching the clouds)**
z	m	z	25.7	8974.3
$C_0(z)$	pF/m	$C_0(z) = \dfrac{2\pi\varepsilon_0}{\ln(2z/a)} \quad (F/m)$	7.0816	4.0567
$L_0(z)$	μH/m	$L_0(z) = \dfrac{\mu_0}{2\pi}\ln(2z/a) \quad (H/m)$	1.569	2.739
$L^{add}(z)$	μH/m	$L^{add}(z) = \dfrac{1}{v^2(z)\cdot C_0(z)} - L_0(z)$	4.2252	27.694
$L(z)$	μH/m	$L(z) = L_0(z) + L^{add}(z)$	5.6	30.4
$v(z)$	m/s	[29]	1.6×10^8	0.9×10^8

and capacitance at the bottom and tip of the RSC incorporated with related formulations used.

Using Figure 5.13, a variation range in the resistive load value, R = 0.540–1.1764 Ω/m, is selected to ensure that the propagation speed of the current wave varies between c/2 and c/3. One can adopt a similar approach to that used in

[27] for the ATIL-F model, the value of R can be split into a fixed distributed resistive load of $R^F = 0.5$ Ω/m and a nonlinearly varying resistive load in the range of $R^N = 0.040–0.6764$ Ω/m, obtained from (5.102).

With reference to Figure 5.1, in all simulations presented in Section 5.5, we assume the value of a is changing with time and space, having an average value of $a_0 = 0.02$ m. We also assume that the electrical conductivity of the channel is $\sigma = 5 \times 10^3$ S/m and is unchanged everywhere along the channel. In addition, $L_{add} = 11.8$ μH/m, which is obtained by averaging out the values of L_{add} at the channel base and at one kilometer above the ground.

A comparison of this model with the ATIL-F model with fixed distributed resistance $R = 0.5$ Ω/m and distributed inductance $L = 11.8$ μH/m [27] will be presented.

In Section 5.5, the current distribution along the channel and the associated radiated electromagnetic fields, as well as the speed of propagation profile for the non-linear model, are discussed and compared with that predicted by the ATIL-F model.

5.4.4 *Frequency-domain AT model*

As mentioned earlier, the frequency-domain model of RSC is often preferred when dealing with problems involving a lossy ground. Consider the thin-wire antenna model of an RSC above a lossy ground, as illustrated in Figure 5.6. The antenna has length $h = 2{,}600$ m, radius a = 0.05 m, and distributed resistance $R_D = 0.1$ Ohm/m. It is driven at its bottom end by a current source whose waveform is the same as that used in [62]. This waveform (Figure 5.10) is characterized by a peak value of about 11 kA and a peak rate of current increase of about 105 kA/μs. In finding the distribution of current along the antenna, $\varepsilon_1 = 5.3\,\varepsilon_0$ is set for the ambient medium in order to obtain a current wave propagation velocity of 1.3×10^8 m/s [13]. Accordingly, a lossy ground effect is incorporated using this value of permittivity for the upper medium (medium 1) in (5.71).

Having determined the current distributions along the RSC, the respective current waveforms are used in field expressions [(5.72)–(5.74)] to calculate remote field components above a lossy ground assuming $\varepsilon_1 = \varepsilon_0$ (i.e., assuming that medium 1 is air or free space).

Frequency-domain calculations are usually carried out at 8,192 frequencies up to 10 MHz with frequency intervals of 2.44 kHz. This corresponds to a sampling interval of 0.05 μs over a time window of 409.6 μs. To ensure the convergence of the method of moments, the smallest segment length should not exceed one-fourth of the minimum wavelength. That means, the 2,600 m wire representing the lightning channel is to be divided into 800 segments, each 3.25 m in length.

In Section 5.5, current distributions and associated remote field components for the case of a perfect ground are computed using the proposed approach and compared with those obtained using the AT model described in [13]. Also, a comparison is made between the Cooray–Rubinstein formula (and the Cooray modified formula) for the horizontal electric field and the generalized expression for the horizontal electric field employed in this work. Finally, various field components over different lossy media are computed and discussed.

5.4.5 Frequency-domain AT model with distributed current source

A distributed current source along the monopole is considered for energizing the network [83,84]. Since the method presented in this work adopts a frequency-domain approach, a fast Fourier transform algorithm is used for the spectral representation of all involved quantities. To this end, a unit current is considered at the base of the channel. For each frequency ω, a phase shift between two adjacent sources separated by a distance defined as:

$$\Delta\Phi = \omega\frac{\Delta}{v} \tag{5.108}$$

provides the current wave-front speed v in the time domain. The value commonly used in the literature for v is 1.3×10^8 m/s. In the same way, the impressed (excitation) current at the location m is expressed by:

$$I_m^I = P_m(\omega, \mathrm{m})e^{-j\omega\frac{\mathrm{m}}{v}} \tag{5.109}$$

where P_m depicts the current amplitude of the distributed source at this location. In order to obtain a spatial–temporal distribution of current along the channel similar to the the engineering model referred to as modified transmission line with exponential decay (MTLE) engineering model [90], the current amplitude of the distributed source at the location m is given by:

$$P_m(\omega, \mathrm{m}) = e^{-\frac{\mathrm{m}\Delta}{\lambda}} \tag{5.110}$$

where λ is a decay constant which is assumed to be equal to 2 km. Furthermore, any type of channel current distribution similar to the engineering models can be obtained by an appropriate choice of I_m^I .

Finally, the application of the Inverse Fast Fourier Transform (IFFT) to the superposition of the values computed in the frequency domain, modulated by the amplitude and phase of the input current spectrum, provides the transient response of the system.

A comparison of the distributed current source model with conventional AT models in terms of current distribution and remote electromagnetic fields for a perfectly conducting soil is depicted in Section 5.5. The electromagnetic model of the lightning channel based on a thin-wire antenna with distributed current sources allows for the use of Sommerfeld integrals for the evaluation of electromagnetic fields due to lightning, in stratified media. Thus, the case of a multilayer soil and its comparison with a full-wave finite-element-based solution of Maxwell's equations is also presented. In order to extend the study to more complex systems including the lightning channel and nearby structures, an array of conducting wires and metallic surfaces is used to represent the entire system. An efficient and cost-effective MoM method capable of computing the variation of spatial 1D linear currents and 2D surface currents permits the evaluation of the electromagnetic interference effects of lightning in a complex environment, without the use of approximate formulas. The proposed method is capable of analyzing the electromagnetic interference effects of lightning for a wide range of applications. In this regard, several case studies, including the analysis of lightning-induced disturbances on overhead lines and underground cables, are described in Section 5.5.

5.5 Numerical results

In this section, we discuss the merits of various AT models in modeling a lightning return-stroke. In particular, we analyze the predicted current characteristics along the RSC as well as radiated electromagnetic fields at various distances from the channel. We also used the distributed source model of the lightning channel implemented in the MoM method to evaluate the inducing effects of lightning on complex electromagnetic structures.

5.5.1 Time-domain AT model

Current profiles

Current waveforms as a function of time at different heights along the channel for five models, the TL, MTLL, MTLE, DU, and AT models, are compared in Figures 5.14(a)–(e). For the TL model the current waveforms at different heights are the same, and for the MTLL and MTLE models, the current amplitude decreases with height, while the waveshape remains the same. For the DU model, both the attenuation and the dispersion of current waveform are observed. The current peak and current rise time, each as a function of height for the five models, are shown in Figure 5.15(a) and (b). For the TL model, neither the current peak nor the current rise time changes with height. The MTLL and MTLE models are characterized by a linear and an exponential decrease of the current peak with height, respectively, while the current rise time remains the same at all heights. For the AT model, the variation of current peak with height within the lower part (up to a height of approximately 4 km) of the channel is similar to that for the MTLL model. Both the DU and the AT models are characterized by an increase of current rise time with height, but for the DU model, the pronounced increase occurs only within the lower part (up to a height of approximately 1 km) of the channel. Note that for the AT model the current peak decreases with height due to ohmic losses in the antenna and radiation losses; that is, a decrease of current peak with increasing height would be observed even if the ohmic losses were neglected.

Line charge density profiles

The line charge density at any height z' on a straight vertical lightning channel at any time t is given by Thottappillil *et al.* [64]:

$$\rho_L(z',t) = \frac{i(z',z'/v)}{v} - \int_{z'/v}^{t} \frac{\partial i(z',\tau)}{\partial z'} d\tau \tag{5.111}$$

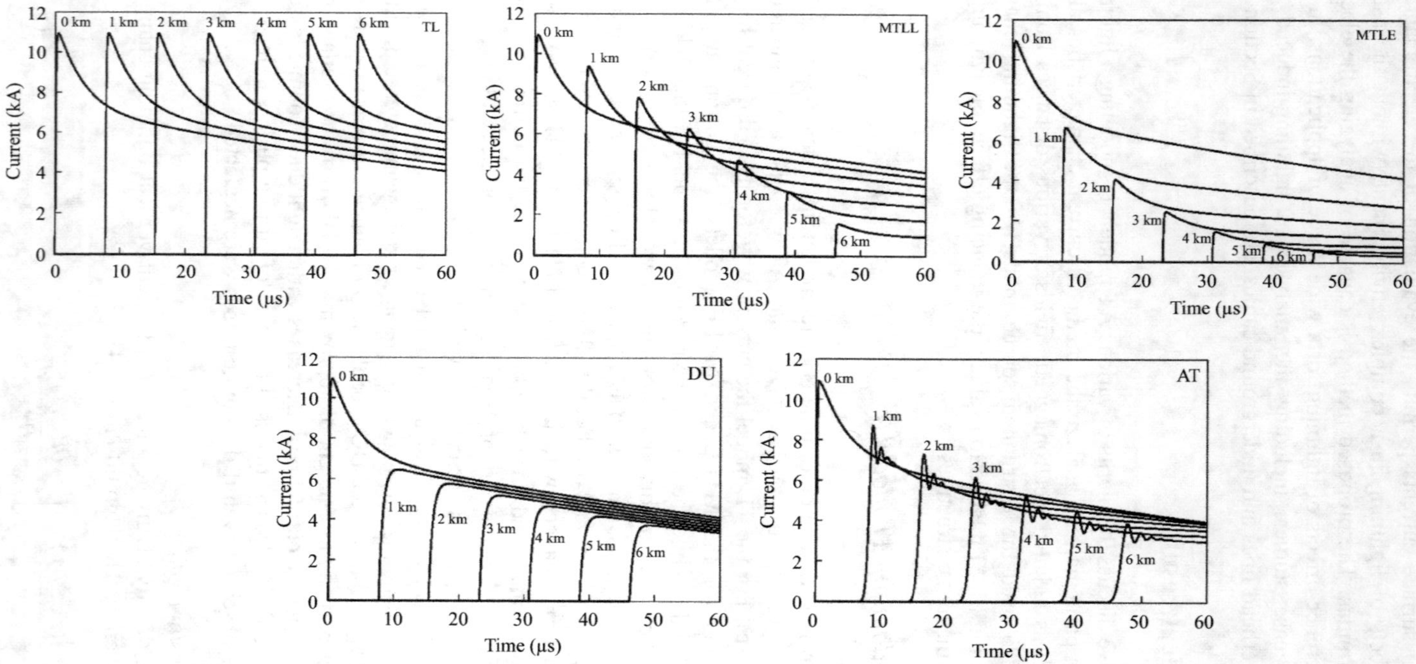

Figure 5.14 *Channel current as a function of time at different heights above ground as predicted by (a) the TL model, (b) the MTLL model, (c) the MTLE model, (d) the DU model, and (e) the AT model. For the AT model it is assumed that $\varepsilon_r = 5.3$ and $R = 0.07\ \Omega/m$. Adapted from Moini* et al. *[13]*

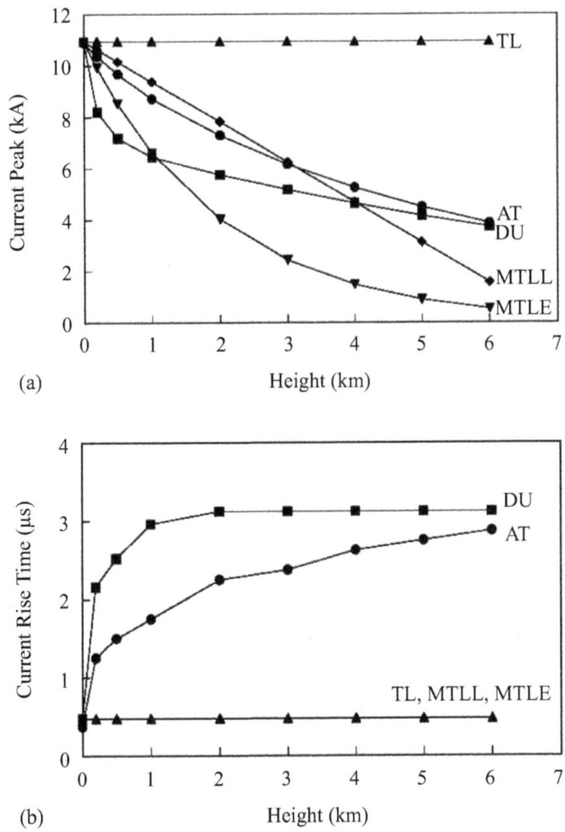

Figure 5.15 *(a) Current peak as a function of height predicted by the TL, MTLL, MTLE, DU, and AT models. (b) Current risetime to peak as a function of height predicted by the TL, MTLL, MTLE, DU, and AT models. Adapted from Moini* et al. *[13]*

The first term of (5.111) represents only the deposited charge density component, while the second term can contribute to both the transferred and the deposited charge density components, the two components being defined by Thottappillil *et al.* [64]. Applying the Leibnitz formula to (5.111), we obtain

$$\rho_L(z',t) = -\frac{d}{dz'}\int_{z'/v}^{t} i(z',\tau)d\tau \tag{5.112}$$

Equation (5.112) has been applied to five return-stroke models, and the resultant charge density profiles at $t = 60$ μs are shown in Figure 5.16. For the TL model, there is no deposited charge, and the total charge density is equal to the transferred charge density, which becomes equal to zero when the current ceases to flow in all channel sections of interest [64]. For the other four models, the total

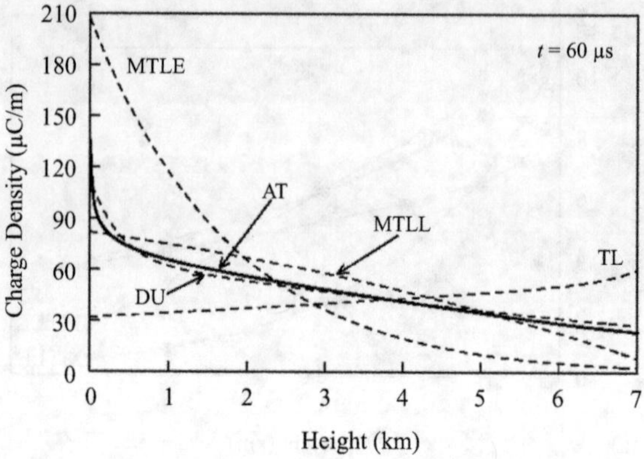

Figure 5.16 Line charge density distribution along the channel calculated for different models at t = 60 μs. Adapted from Moini et al. *[13]*

charge density at 60 μs is a combination of transferred and deposited charge density components. When the current ceases to flow everywhere in the channel, the transferred charge density component becomes zero and the total charge density becomes equal to the deposited charge density component. As seen in Figure 5.16, the MTLE model has a total charge density near ground 2–3 times greater than that predicted by the AT, MTLL, and DU models. This disparity translates into an appreciable difference in the return-stroke electric fields as predicted by the various models at close ranges, as shown in the simulation of remote electromagnetic field.

Electromagnetic fields

Figures 5.17(a)–(c) and 5.18(a)–(c) illustrate the calculated electric and magnetic fields at various distances from the channel, displayed on two different timescales. The fields were computed using traditional equations found, for example, in the work of Rakov and Uman [1]. The calculations were performed up to 60 μs, and for the AT model the channel segment length Δz was 15 m. The calculated fields can be compared with typical measurements presented in [25]. Except for the TL and MTLE models, the 500 m electric field waveforms predicted by all the models are generally consistent with experimental data. In particular, the electric fields predicted by the MTLL, DU, and AT models show little variation after approximately 10 μs, following the initial relatively rapid change, in keeping with observations. At 5 km, the electric field exhibits a ramp after the initial peak for all of the models, except for the TL model (see also [62]). In fact, at distances on the order of several kilometers, the TL model allows the reproduction of only the first few tens of microseconds of the characteristic electric field ramp observed in the experimental data to last for more than a hundred microseconds. Note that at distances exceeding several kilometers the initial rapid transition in the electric field is reasonably reproduced by all of the models, because (1) this feature is formed when the current wave is very close to the ground and (2) the same current waveform at ground level

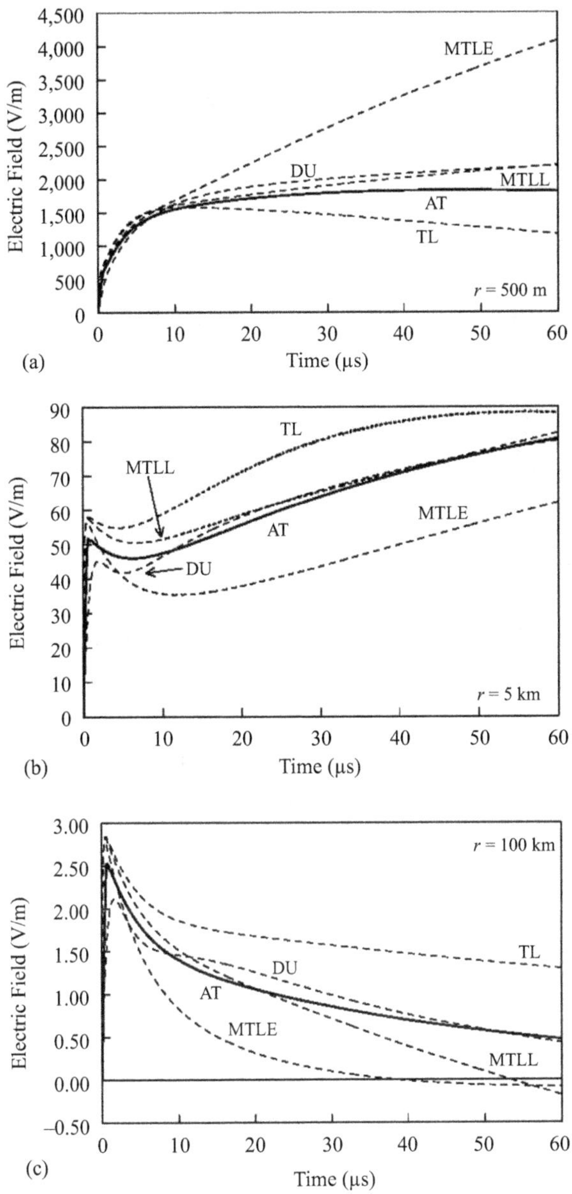

Figure 5.17 Vertical component of electric field calculated at different distances (r) from the channel: (a) r = 500 m, (b) r = 5 km, and (c) r = 100 km. Adapted from Moini et al. *[13]*

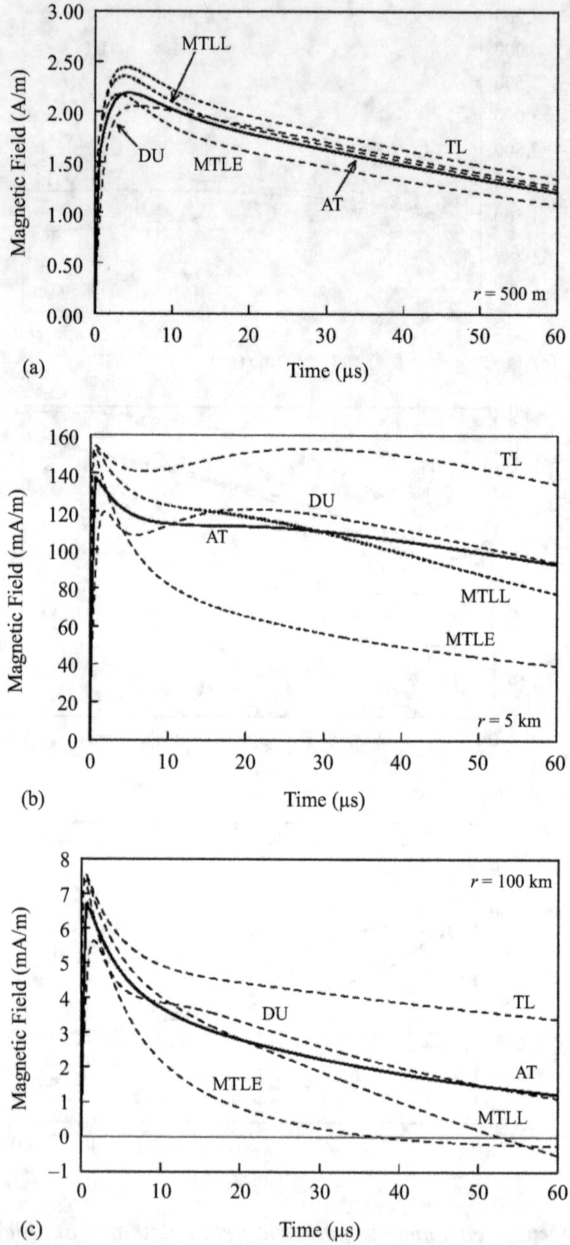

Figure 5.18 Horizontal component of magnetic field calculated at different distances (r) from the channel: (a) r = 500 m, (b) r = 5 km, and (c) r = 100 km. Adapted from Moini et al. *[13]*

is assumed for all models. In contrast with the other models, the DU model predicts a non-monotonic rise to the initial peak (a spike within the first hundreds of nanoseconds), as seen in Figures 5.17 and 5.18. At far distances from the channel base, e.g., $r = 100$ km, a similar field peak value is predicted by most of the models. On the other hand, the MTLL and MTLE models predict a zero crossing after a few tens of microseconds, a feature generally considered to be characteristic of distant fields [25], while the TL, DU, and AT models do not do so.

In summary, it appears that the TL model is not a realistic model for calculating lightning electric fields at times greater than several tens of microseconds at distances on the order of several kilometers (see also [62]) and after only a few microseconds at distances on the order of tens of meters from the channel [64]. The fields predicted by the MTLL model are consistent with observed fields at all ranges. The MTLE model is incapable of adequately reproducing the observed electric field waveforms at very close (tens to hundreds of meters) ranges [64]. The DU and AT models do not reproduce the zero crossing at far ranges. However, this latter feature depends on the assumed channel-base current waveform, in particular, on the rate of decrease of the current after the peak and on the channel geometry. Indeed, the field zero crossing occurs when the contribution from the leading positive and trailing negative portions of the spatial current derivative wave become equal in magnitude, the time of this event being a function of the channel inclination from the perspective of the observer. As an example, we present in Figure 5.19(a) and (b) electric field waveforms measured at distances of about 250 km in Florida, one of which showing and the other not showing zero crossing within 100 μs of the initial peak. Note that the two waveforms were apparently produced in the same thunderstorm within 3 min of each other.

5.5.2 Time-domain AT model with inductive loading

Current profiles

Figure 5.20(a)–(c) illustrates the current distribution along the RSC for the AT, ATIL-F, and ATIL-V models. There are appreciable differences between the AT and ATIL-F models in terms of the general shape of the current waveforms. The effect of speed, which is a function of height can also be observed in the current distribution for the ATIL-V model. Both versions of the ATIL model predict more a pronounced current dispersion than the AT model. In order to facilitate comparison with the AT-model results, we assumed the same attenuation rate for all models, which required larger distributed resistances in the ATIL models, compared to the AT model. This additional distributed resistance in the ATIL models can be described using the transmission line theory. The square of attenuation factor for a transmission line is given by the following formula:

$$\alpha^2 = \mathrm{Re}[(R + j\omega L)(G + j\omega C)] \tag{5.113}$$

where R, L, G, and C are, respectively, the per-unit-length resistance, inductance, conductance, and capacitance of the transmission line, ω is the angular frequency, $j = \sqrt{-1}$, and "Re" stands for the real part of a complex quantity. This equation

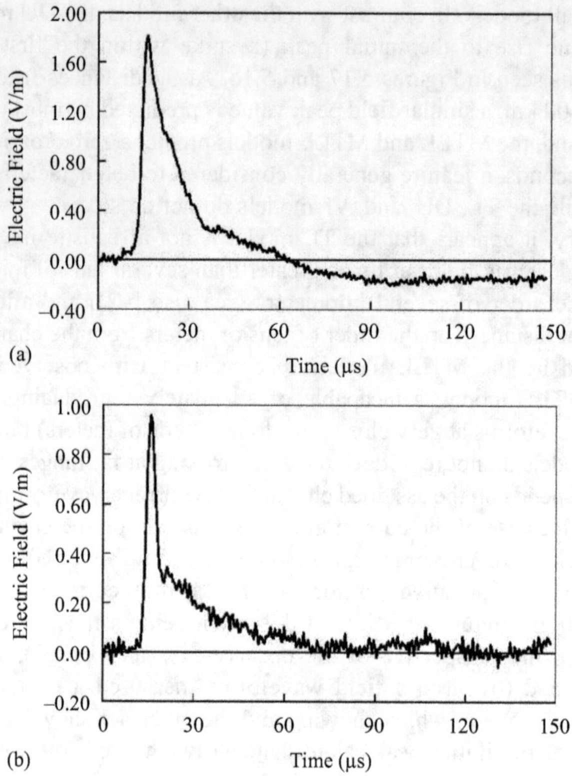

Figure 5.19 Measured electric field waveforms due to lightning discharges at distances of about 250 km in Florida. Note that one of the waveforms shows a zero crossing, while the other one does not. Adapted from Moini et al. *[13]*

can be rewritten in the following form:

$$\alpha^2 = \mathrm{Re}\left[(RG + jRC\omega + jLG\omega - LC\omega^2)\right] = RG - LC\omega^2, \tag{5.114}$$

or

$$\alpha = \sqrt{RG - LC\omega^2}. \tag{5.115}$$

Equation (5.115) shows that the attenuation factor decreases with increasing L and hence in order to have a constant α, R should be increased.

In the AT model, the current attenuation in the lower sections of the channel is more pronounced than for both versions of the ATIL model. In other words, the attenuation is reduced in the presence of inductive loads, although the distributed resistance for both versions of the ATIL model is greater than that for the AT model. As can be seen in Figure 5.12, the ATIL-V model employs smaller values of

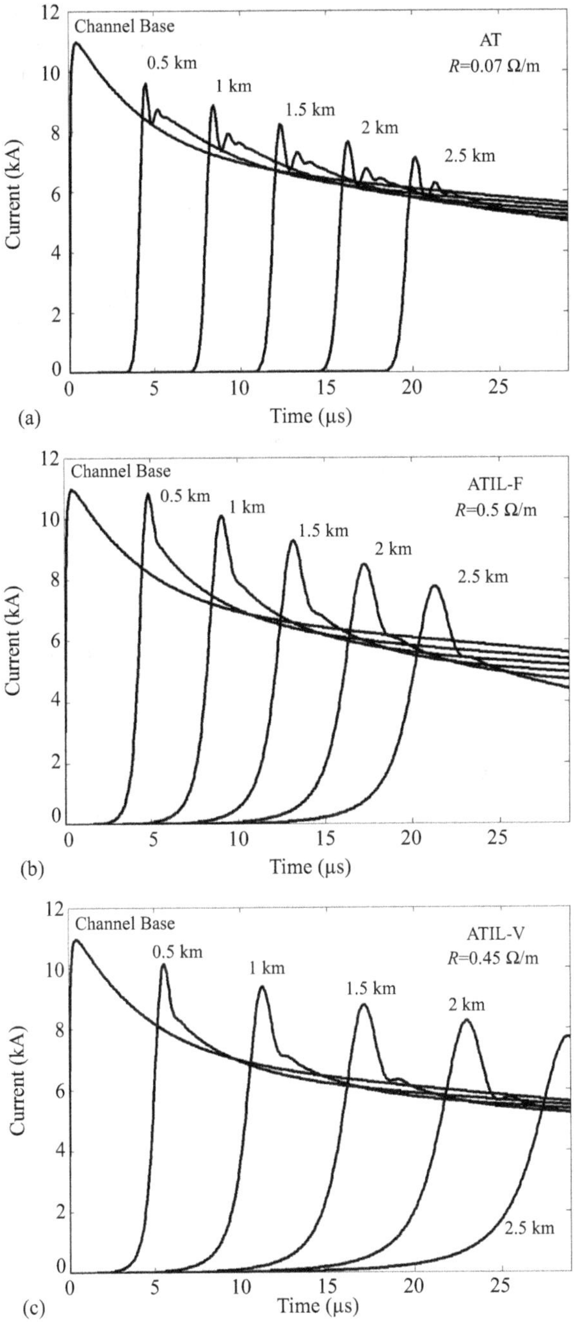

Figure 5.20 Current distributions along the channel for (a) AT, (b) ATIL-F, and (c) ATIL-V models. Shown are current versus time waveforms at the channel base and at heights of 500 m, 1 km, 1.5 km, 2 km, and 2.5 km above ground. Adapted from Bonyadi-Ram et al. *[30]*

inductance in the lower sections of the channel than the ATIL-F model does and, hence, more attenuation is observed for the ATIL-V model. Both versions of the ATIL model predict more or less similar attenuation in the higher sections of the channel due to similar amounts of inductive and resistive loading.

Current dispersion

It is clear that the ATIL-F and ATIL-V models (see Figure 5.20(b) and (c)) predict considerably larger current dispersion than the AT model does (see Figure 5.20(a)). This larger dispersion is in agreement with optical observations of Jordan and Uman [28] and Wang *et al.* [29]. The observed current dispersion in the inductively-loaded channel can be explained using the transmission line theory. The upward traveling current wave in the channel can be decomposed into two components [78].

- Antenna-mode current, which is governed by the scattering theory. The propagation speed of this component is equal to the electromagnetic wave propagation speed, υ, in the surrounding medium, which is the speed of light when $\varepsilon_r = 1$.
- Transmission line (TL)-mode current, which is governed by the transmission line theory and propagates at an adjusted speed $\upsilon(z)$.

The propagation speed of each of the two current components is a function of ε_r. The AT model yields a full wave solution for the current distribution along the metallic structure including both the TL-mode and antenna-mode currents [78]. The resultant current-wave propagation speed is a function of the relative permittivity of the medium. Adding distributed inductive loads as circuit elements to the discretized EFIE dramatically affects the propagation speed of the TL-mode current because the current passes through loaded segments. On the other hand, the antenna-mode component of the current is independent of the circuit elements because loaded segments do not play a dominant role in the electromagnetic coupling between the segments. Different propagation speeds for the two current components result in noticeable current dispersion along the channel, which increases with height.

Current wave propagation speed

The choice of tracking point will affect the model-predicted speed, since the shape of the current (or luminosity) waveform changes with height. Here, we analyze variations of the return-stroke speed along the channel predicted by the AT, ATIL-F, and ATIL-V models. Olsen *et al.* [68] obtained different speed values tracking different reference points, including 10%, 20%, 90%, and 100% of peak, as well as the maximum rate of rise of the return-stroke luminosity pulse. The point of initial deflection from zero level is usually masked by noise and, hence, is difficult to identify in optical measurements. We have chosen 10%, 20%, and 100% of peak, and the maximum time derivative of the current waveform (which corresponds to the time at which the current waveform reaches its maximum rate of rise, di/dt) as reference points. Speed υ_k in segment k is calculated at the center of the segment by dividing the vertical distance between adjacent viewed heights, $h_k–h_{k-1}$, by the tracked time interval, $t_k–t_{k-1}$. The resultant speed profiles for the AT, ATIL-F, and ATIL-V models are shown in Figure 5.21(a)–(c). In order to facilitate a direct comparison, the "theoretical" speed profiles (given by (5.99) for the ATIL-V model or a constant value equal to 1.3×10^8 m/s for the AT and ATIL-F models) are also shown in these figures. The current-peak and di/dt-peak speed profiles at heights exceeding several hundred meters above ground are similar to the theoretical profiles for the AT and the ATIL-V models. The estimated speed at lower sections of the channel is noticeably greater than the theoretical speed. This may be due to significant changes in the current wave shape in the lower sections of the channel (e.g., compare current wave shapes at the channel base and at a height of 500 m above ground in Figure 5.20(a)). Propagation speeds for the AT and ATIL-F, as expected, are almost constant over the entire channel length considered. The difference between speeds obtained from the four estimated profiles (current peak, peak of di/dt, 10%, and 20% of the peak) for the AT model is smaller than those predicted by the ATIL-F and ATIL-V models. This is because of the lower current dispersion predicted by the AT model. It can also be observed that tracking a lower percentage of peaks results in a higher propagation speed, which is in agreement with optical data reported in [68]. It is also observed that both versions of the ATIL model predict propagation speeds that appear to be nearly equal to the speed of light if the point of initial deflection from zero level is tracked, while in the AT model the propagation speed remains near 1.3×10^8 m/s. This difference is expected, since in both versions of the ATIL model, the antenna-mode current propagates at the speed of light, while in the AT model, it propagates at $\upsilon = 1.3 \times 10^8$ m/s.

Electromagnetic fields

Figures 5.22(a)–(c) and 5.23(a)–(c) illustrate the electric and magnetic fields, respectively, at three different distances, 500 m, 5 km, and 100 km from the lightning channel base calculated using the AT, ATIL-F, and ATIL-V models. At $r = 500$ m, after 10 μs, all three examined models predict similar magnetic field waveforms. The ATIL-F model predicts a higher initial peak of the magnetic field than the other two models, since the current in the lower parts of the channel with the ATIL-F is greater than with the other two models (see Figure 5.20). On the other hand, the time of the magnetic field peak with the AT model is greater than with the AILT-F and ATIL-V models. At $r = 500$ m, predicted electric fields prior to 1 μs for the ATIL-F and ATIL-V models are nearly identical, which suggests that the variation of the propagation speed cannot alter the electric field waveform at close distances within approximately the first microsecond. The AT model predicts the smallest electric field, while its rising slope after 5 μs is similar to that with the ATIL-F model. Due to the higher propagation speed in the lower part of the channel with the ATIL-F model, after 1 μs, this model predicts the steepest electric field slope as well as a greater final field value. Neither the AT model nor either version of the ATIL model shows the magnetic field hump at 50 μs which can be seen in typical measured waveforms of [25].

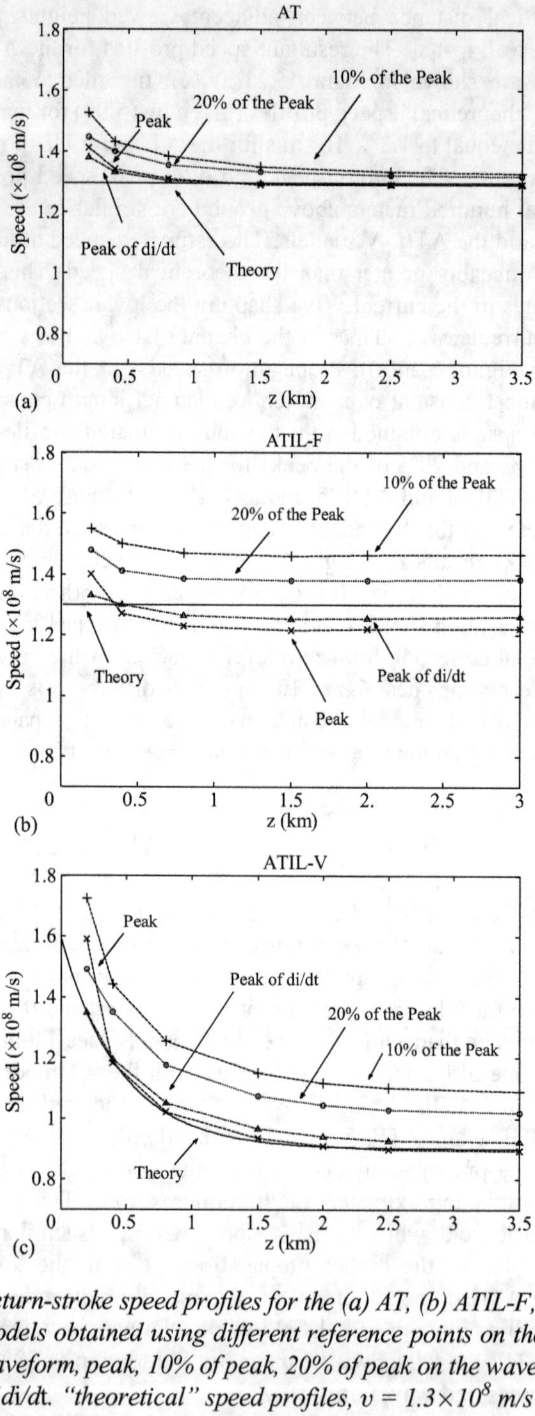

Figure 5.21 Return-stroke speed profiles for the (a) AT, (b) ATIL-F, and (c) ATIL-V models obtained using different reference points on the current waveform, peak, 10% of peak, 20% of peak on the wave front, and peak of di/dt. "theoretical" speed profiles, $v = 1.3 \times 10^8$ m/s = const, for the AT and ATIL-F models and the curve shown in Figure 5.11 for the ATIL-V, are also shown. Adapted from Bonyadi-Ram et al. *[30]*

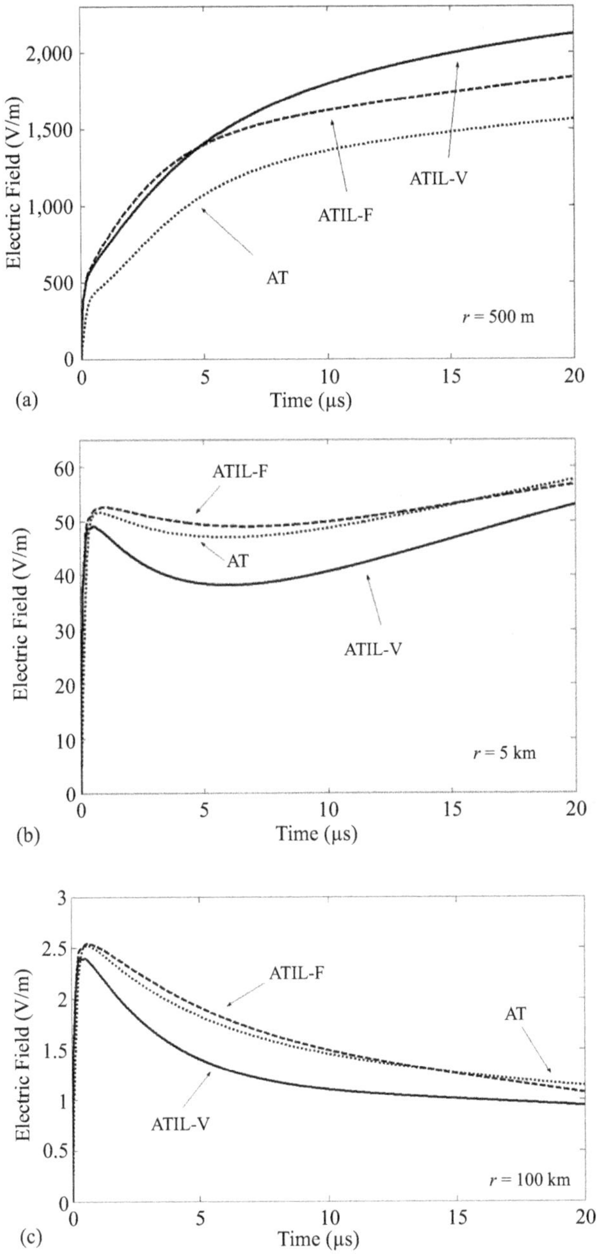

Figure 5.22 Vertical component of electric field for the AT, ATIL-F, and ATIL-V models calculated (a) r = 500 m, (b) r = 5 km, and (c) r = 100 km from the lightning channel. Adapted from Bonyadi-Ram et al. *[30]*

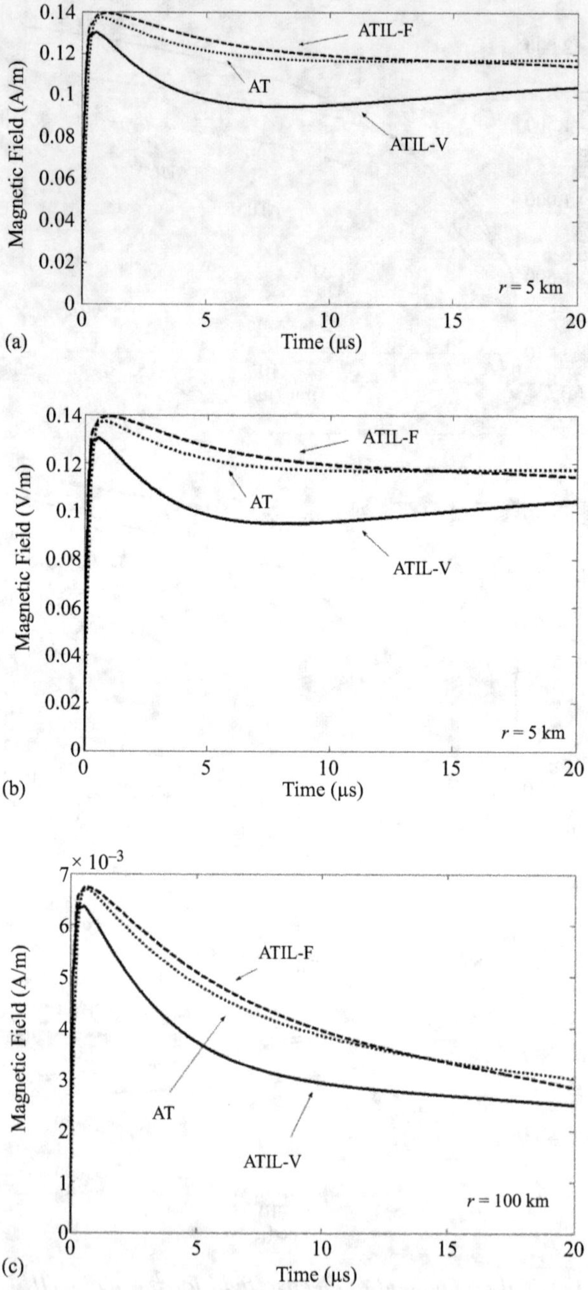

Figure 5.23 Horizontal component of magnetic field for the AT, ATIL-F, and ATIL-V models calculated (a) r = 500 m, (b) r = 5 km, and (c) r = 100 km from the lightning channel. Adapted from Bonyadi-Ram et al. *[30]*

The electric and magnetic fields at 5 km are for all the models considered. The ATIL-V model predicts a smaller overall electric and magnetic field value, a steeper falling slope after the initial peak, and a higher rising slope than the other two models, which is more consistent with the typical waveforms measured at 5 km [25]. This is because of a higher propagation speed in the lower sections of the lightning channel (causing a steeper falling slope) and a lower speed in higher parts of the channel (resulting in a higher rising slope) in the ATIL-V model.

At 100 km, the electric and magnetic field wave shapes are similar. At this distance, the ATIL-V predicts smaller fields than the other two models. This is due to a lower propagation speed in the upper part of the lightning channel which causes a greater delay in illuminating higher parts of the channel (compare the current at 2,500 m at 20 μs in Figure 5.20(a)–(c)). Within approximately the initial 13 μs, the ATIL-F and the AT models exhibit similar behavior in predicting distant electromagnetic fields, while the ATIL-V model predicts a lower peak and a steeper falling slope after the peak. Due to a lower propagation speed in the upper part of the channel, the final falling slope for the ATIL-V is less than that for the other two models considered. None of the three models considered here can predict zero crossing within about 50–60 μs, as typically seen in most waveforms measured at this distance.

5.5.3 Time-domain AT model with nonlinear loading

Current profiles

First, through the computational process for the time-domain current distribution along the channel, temporal–spatial curves for the RSC radius and the nonlinear per-unit length resistance utilized in the proposed model are shown in Figure 5.24(a) and (b), respectively. As seen in these figures, for a selected range of R = 0.040–0.6764 Ω/m, the value of the RSC radius, a, varies between 0.18 m and 0.01 m, respectively. Also from these figures, one can deduce that for a given time, the lower section of the channel generally holds a large cross-sectional area with small resistance whereas the higher section of the channel tends to be narrower with large values of resistance.

Next, the current distribution at the channel base and at distances of 2,250 m, 4,500 m, and 6,750 m from the channel base were obtained using the proposed model and the ATIL-F model. The temporal variation of the excitation current source at the channel base is assumed to be the sum of a Heidler function and a double exponential [62], i.e.:

$$i(t) = \frac{I_{o1}}{\eta} \cdot \frac{\left(\frac{t}{\tau_1}\right)^n}{1 + \left(\frac{t}{\tau_1}\right)^n} \cdot e^{-\frac{t}{\tau_2}} + I_{o2} \cdot \left(e^{-\frac{t}{\tau_3}} - e^{-\frac{t}{\tau_4}}\right), \tag{5.116}$$

where the related parameters are as considered as $I_{o1} = 9{,}900\ A$, $I_{o2} = 7{,}500\ A$, $\eta = 0.845$, $n = 4$, $\tau_1 = 0.072 \times 10^{-6}$ s, $\tau_2 = 5 \times 10^{-6}$ s, $\tau_3 = 100 \times 10^{-6}$ s and $\tau_4 = 6 \times 10^{-6}$ s.

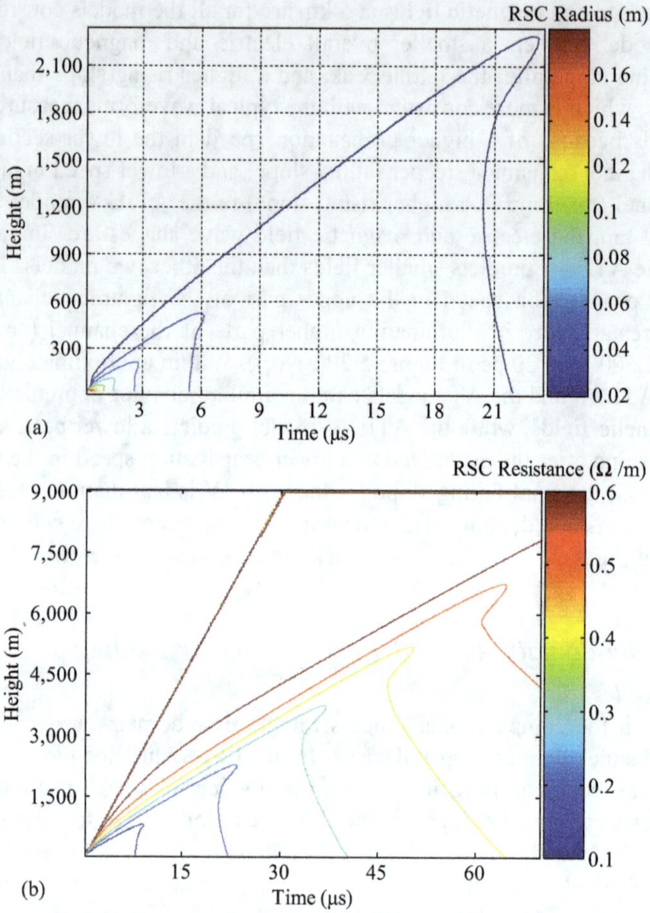

Figure 5.24 Temporal–spatial plot of return stroke channel (a) radius and (b) resistance per unit length. Adapted from Seyed-Moosavi et al. *[31]*

A comparison of the results shown in Figure 5.25 shows that both the current attenuation and the dispersion rates in the proposed model are higher than that predicted by the ATIL-F model.

Current peak and current rise time

Current peak values at different heights along the lightning channel and the corresponding rise time values are shown in Figures 5.26 and 5.27, respectively. The value of rise time is assumed to be 10–90% of the current peak time. These two parameters can be considered as major contributors to the zero crossing at far radiated electromagnetic fields and the hump in the intermediate magnetic field [13]. As is clearly observed in Figure 5.26, the current peak for the proposed method is often smaller than that predicted by the ATIL-F model, showing more

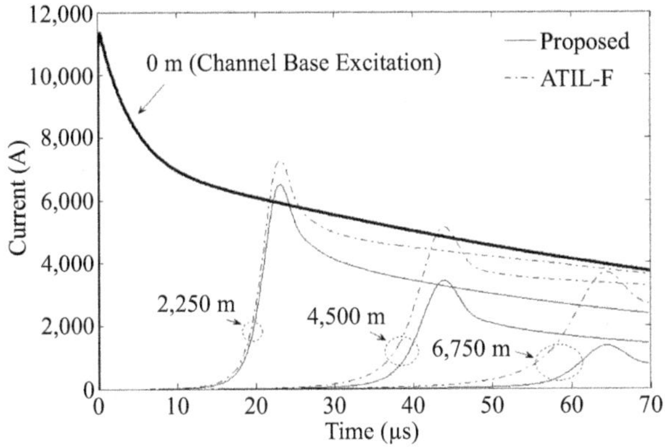

Figure 5.25 Current distribution along the channel for the ATIL model and the non-linear AT model, non-linear AT model (solid) and the ATIL-F model (dashed). Adapted from Seyed-Moosavi et al. *[31]*

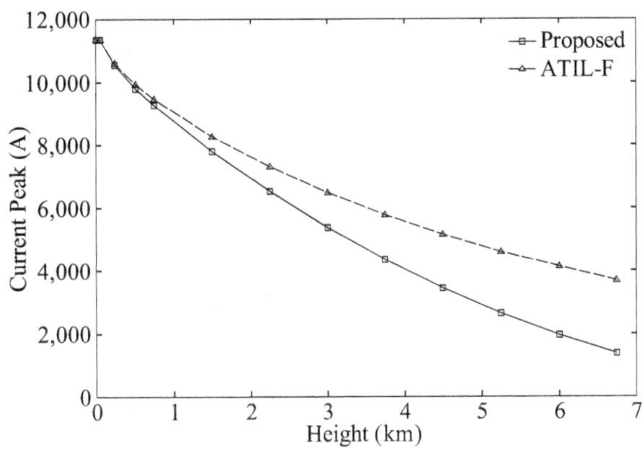

Figure 5.26 Peak for different heights along the lightning channel for the proposed method and the ATIL-F method. Adapted from Seyed-Moosavi et al. *[31]*

attenuation. Also, the rise time of the upward traveling current wave for the proposed method is often smaller than that of the ATIL-F model. In addition, both quantities exhibit nonlinear behaviors in the case of the proposed method whereas they are essentially linear in the case of the ATIL-F model.

Electromagnetic fields

Field waveforms are calculated for three different ranges of distances from the channel, namely, near distance (500 m), intermediate distance (5 km), and far distance (100 km). The electric field waveforms are shown in Figure 5.28(a)–(c), respectively. As can be seen in Figure 5.28(a), the electric field waveforms

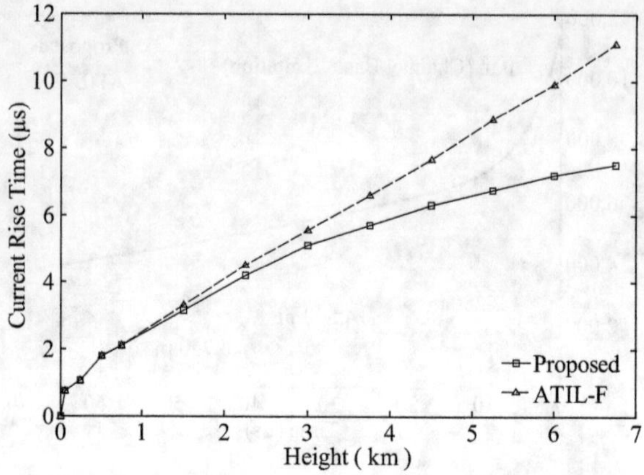

Figure 5.27 Current rise time for different heights along the lightning channel for the proposed method and the ATIL-F method. Adapted from Seyed-Moosavi et al. *[31]*

predicted by both models are initially (i.e., within the first 15 μs) consistent while departing from each other later on. At the intermediate distance (Figure 5.28(b)), the results are following each other and are essentially the same. The main feature of the proposed model is demonstrated in the case of the far distance (Figure 5.28 (c)) where, as opposed to the ATIL-F model, a zero crossing at $t = 50$ μs occurs in the predicted electric field waveform. Notice that the zero crossing is an important feature observed in all measurement results [25].

The magnetic field waveforms at near distance (500 m), intermediate distance (5 km), and far distance (100 km) for the two models are shown in Figure 5.29(a)–(c), respectively. Generally, similar observations to those described above for the case of electric field waveforms can be deduced from the results shown in Figure 5.29.

To study the effect of the electrical conductivity σ in resistive loads along the channel on the predicted field waveform, the magnitudes of the electric and magnetic fields at far distance (i.e., 100 km) versus time for various values of σ are calculated. The results are shown in Figure 5.30(a) and (b) where $\sigma = 5 \times 10^3$, 6×10^3, and 7×10^3 S/m. Also, a similar study is done for the predicted current peak and current rise time versus height and the results are shown in Figure 5.31(a) and (b). A comparison of the results shown in these figures shows that as the value of σ increases, both the electric and the magnetic field waveforms, as well as the current peak and the current rise time, tend to approach their ATIL-F counterparts. In particular, the zero crossings in both the electric and the magnetic field waveforms tend to disappear. This is thought to be due to the fact that a large value of σ results in a small value of the nonlinear part of channel resistance R^N, leading to a less-shared resistive nonlinearity out of the total resistive loading, i.e., $R = R^F + R^N$. With upper bound for the nonlinear resistance part of the channel loading, a behavioral study of the increase in conductivity is presented in Figure 5.32 for the time at 30.45 μs, while the

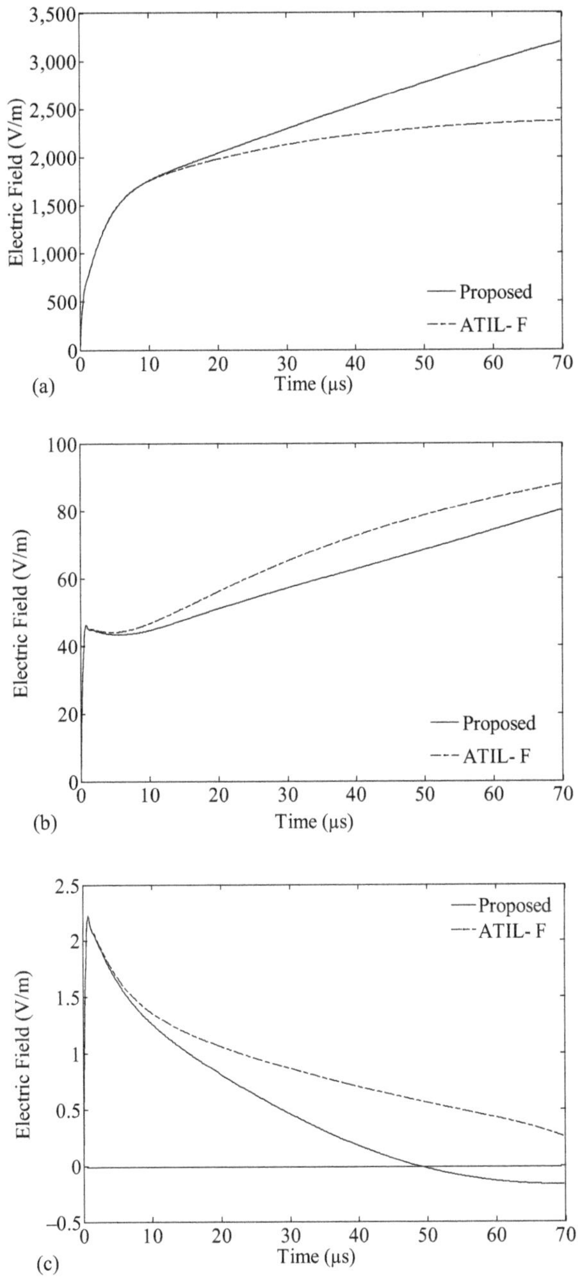

Figure 5.28 Electric field waveform at different distances from the channel base: (a) 500 m, (b) 5 km, and (c) 100 km. Adapted from Seyed-Moosavi et al. *[30]*

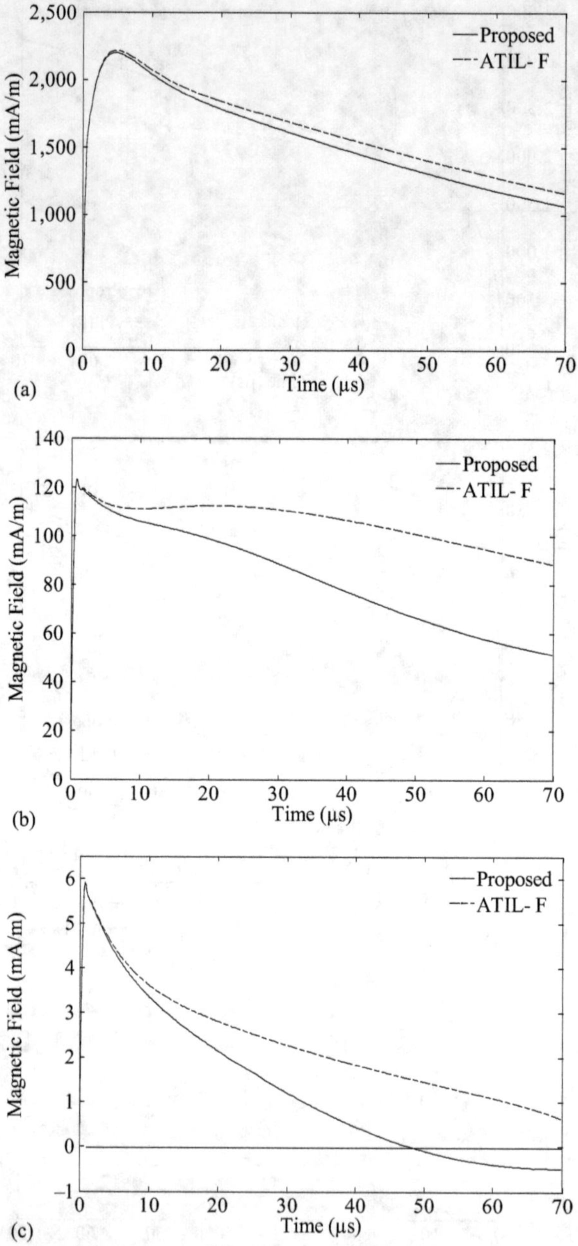

Figure 5.29 Magnetic field waveform at different distances from the channel base: (a) 500 m, (b) 5 km, and (c) 100 km. Adapted from Seyed-Moosavi et al. *[31]*

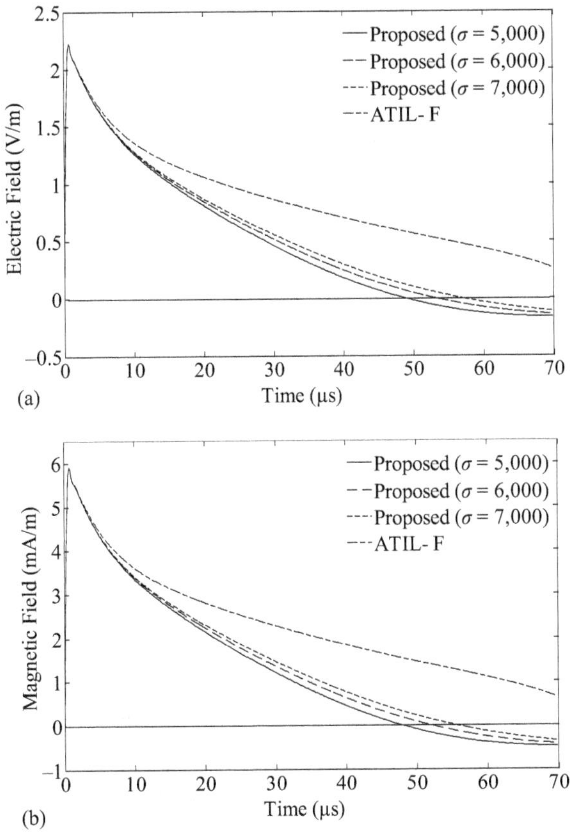

Figure 5.30 Electric field (a) and magnetic field (b) waveforms at 100 km far from the channel base for different conductivities of the channel (σ in S/m)). Adapted from Seyed-Moosavi et al. *[31]*

upward propagated initiated first return-stroke touches the clouds. The results shown in this figure demonstrate that as the value of conductivity increases (from $\sigma = 5 \times 10^3$ S/m to 6×10^3 S/m to 7×10^3 S/m), the resistance of the nonlinear part of the channel tends to decrease (respectively from R^N = 0.6764 Ω/m to 0.5703 Ω/m to 0.4946 Ω/m).

Current wave propagation speed

Current wave propagation speed variations along the RSC predicted by the proposed model are shown in Figure 5.33. Different speed values tracking different reference points including 10%, 20%, 90%, and 100% of the peak value and the maximum time derivative of the current waveform di/dt at which the current reaches its maximum rate of rise, are selected as reference points. Knowing that the value of the distributed inductive load in the proposed model is unchanged along

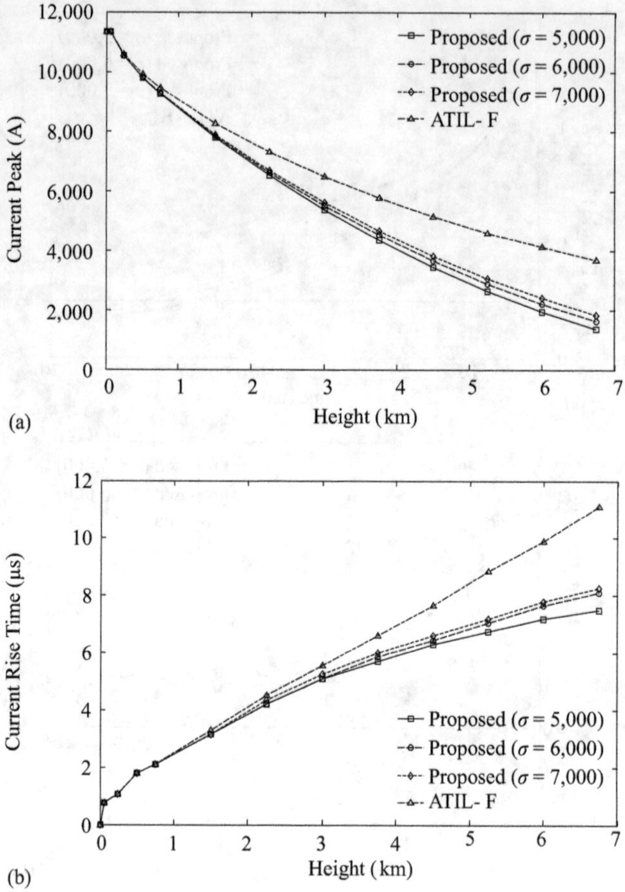

Figure 5.31 Current peak (a), current rise time (b) for different conductivities of the channel (σ in S/m)). Adapted from Seyed-Moosavi et al. *[30]*

the channel, one may expect a constant speed profile in the proposed model [27]. This is not, however, the case as demonstrated in Figure 5.13 where the dependency of phase velocity and resistance are clearly shown.

5.5.4 Frequency-domain AT model

Current profiles

The current source excitation of the antenna in the frequency-domain AT model is implemented by using a Dirac delta source [79] connected across a 3.25-m gap, whereas in the AT model a voltage source (delta-gap generator) is connected across the gap. There are other differences between the two excitation methods. For example, the channel-base current remains unchanged upon the arrival of a wave reflected from the wire's top end in the case of current excitation, although we

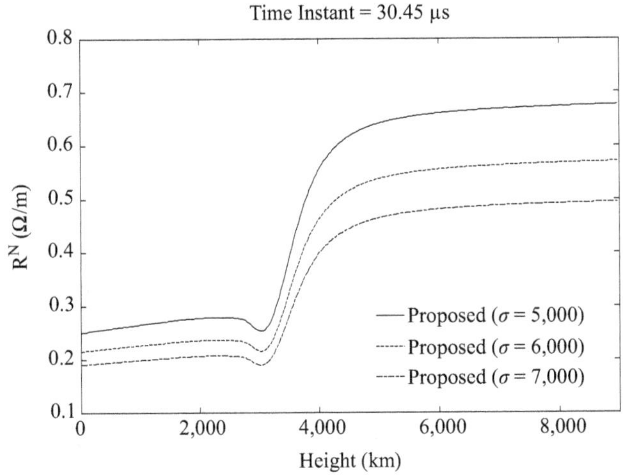

Figure 5.32 Nonlinear part of resistance along the channel for different conductivities of the channel (σ in S/m)). Adapted from Seyed-Moosavi et al. *[30]*

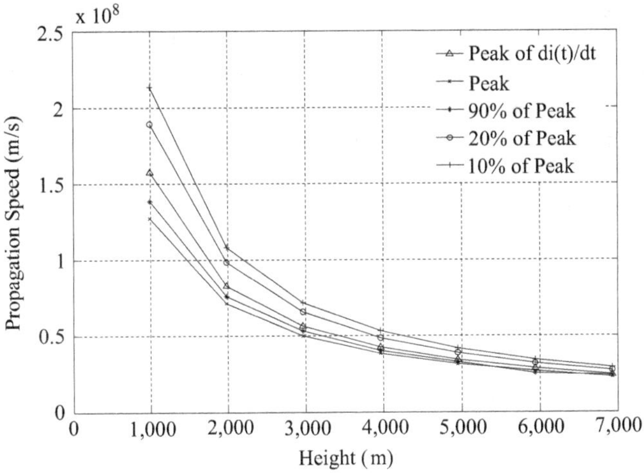

Figure 5.33 Return stroke speed profile along the channel for the proposed model. Adapted from Seyed-Moosavi et al. *[30]*

consider the times before the arrival of the first reflection from the top. This is not the case when voltage excitation is used. Also, the antenna theory model proposed here solves a modified version of the EFIE for space–time dependent currents with the excitation being the current source. Direct use of a channel-base current in the method of moments eliminates preliminary steps, including the calculation of the antenna's input impedance, and increases the method's reliability.

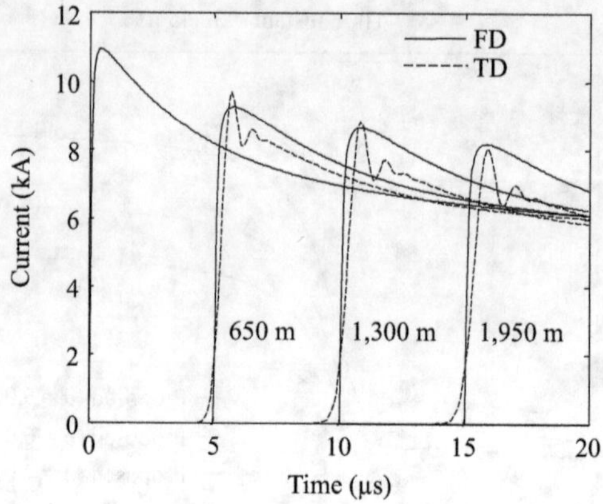

Figure 5.34 Current waveforms at three heights, 650 m, 1,300 m, and 1,950 m. Adapted from Shoory et al. *[21]*

To demonstrate the capabilities of the proposed model, it is compared with the AT model for the case of a perfectly conducting ground. Current distributions obtained using the two models are presented in Figure 5.34 as a function of time at heights of 650 m, 1,300 m, and 1,950 m from the channel base. It is worth mentioning that the plots for the AT model show oscillations around the current peak. These are due to numerical instabilities and could be eliminated by increasing the computation accuracy. As with the AT model in the time domain, the AT model in the frequency domain predicts attenuation and dispersion of the current pulse as it propagates along the lightning channel. Note that the observed variation of current with height follows from the solution of Maxwell's equations, in contrast with the commonly used engineering models in which imposed current distributions are used. A detailed comparison between the AT model and commonly used engineering models is found in Section 5.5.1.

Electromagnetic fields

Figure 5.35 depicts the distributions of electric and magnetic fields predicted by the frequency-domain and the original AT models at different distances from the channel-base on a perfectly conducting ground. As seen in this figure, except for small differences in the values of the field amplitude and rise time, the results predicted by the two models are similar. The differences between field amplitudes predicted by the two models may be related to the differences between the current waveforms which are mainly due to the different techniques implemented for the solution of the EFIE in the time and frequency domains. Also, differences between the rise times of the field waveforms are thought to be due to numerical instabilities in the time domain AT model.

When the electromagnetic field features tabulated in [1] are considered as a benchmark, both models reproduce all known attributes except for the hump following the initial peak in far magnetic fields and zero crossing in both the electric and magnetic fields at far ranges.

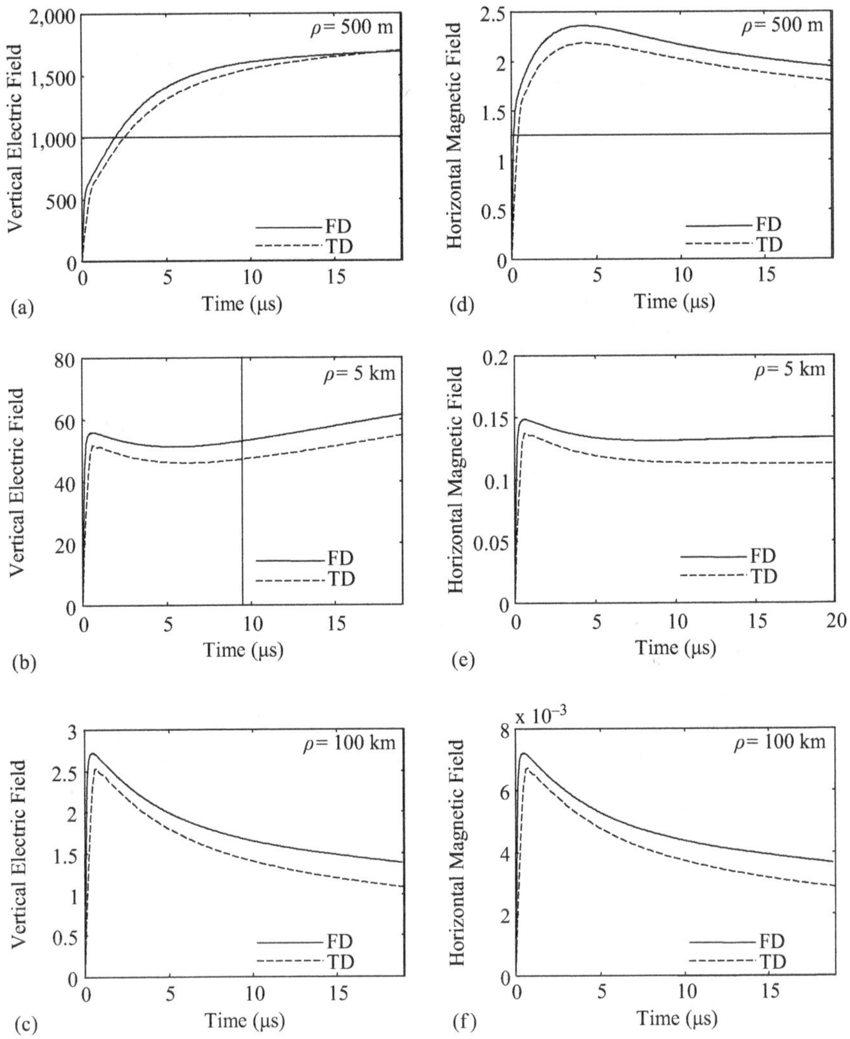

Figure 5.35 Vertical electric (a, b, and c) and horizontal magnetic (d, e, and f) fields at the air-ground interface 500 m, 5 km, and 100 km from the lightning channel-base for a perfect ground computed using the frequency-domain AT and the time-domain AT model. Adapted from Shoory et al. *[21]*

Electromagnetic fields above lossy half-space

(a) Horizontal electric field

In studies related to the coupling of lightning-radiated electromagnetic fields and overhead power lines, it is convenient to use the horizontal component of the electric field [80–82]. Up to now, the Cooray–Rubinstein formula [82] or the

Cooray-modified formula [59] have probably been the best simple approximations presented for the calculation of the horizontal electric field. For convenience, we rewrite the Cooray–Rubinstein formula as follows:

$$\begin{aligned} E_{ap}(\rho,z,j\omega) &= E_{app}(\rho,z,j\omega) - \sqrt{\frac{j\omega\mu_0}{\sigma_2 + j\omega\varepsilon_2}} H_{a\varphi p}(\rho,0,j\omega) \\ &= E_{app}(\rho,z,j\omega) + E_{ap}(\rho,0,j\omega), \end{aligned} \tag{5.117}$$

where ρ is the azimuthal distance from the channel-base, z is the height above ground, and the subscript p denotes the perfect-ground assumption (see Figure 5.6). Cooray [59] proposed a modification to the first term in (5.117) weakening the contribution of its radiation component:

$$\begin{aligned} E_{ap}(\rho,z,j\omega) = {} & 0.4E_{app,r}(\rho,z,j\omega) + E_{app,i}(\rho,z,j\omega) + E_{app,s}(\rho,z,j\omega) \\ & + E_{ap}(\rho,0,j\omega), \end{aligned} \tag{5.118}$$

where $E_{app,r}$, $E_{app,i}$, $E_{app,s}$ are, respectively, the radiation, induction, and static terms of $E_{app}(\rho,z,j\omega)$ in (5.117). In fact, in the expression for the horizontal electric field above a perfect ground, $E_{app,r}$ is the term with $1/r^3$ dependence, $E_{app,i}$ is the term with $1/r^4$ dependence, and $E_{app,s}$ is the term with $1/r^5$ dependence. Although this approach predicts a more accurate initial peak of the horizontal electric field at near ranges, it is not applicable to the case of a perfect ground [59].

Horizontal electric field waveforms at a height of 20 m above ground with conductivity $\sigma_2 = 0.04$ S/m and permittivity $\varepsilon_2 = 8\ \varepsilon_0$ for various distances from the channel base are shown (on a 5-μs time scale) in Figure 5.36. Fields are obtained using the method presented in previous sections, the Cooray–Rubinstein formula [82], and the Cooray-modified formula [59]. It is clearly seen from Figure 5.36 that at near ranges (e.g., within of the channel base 200 m), which are particularly critical for studies of lightning-induced overvoltages on power distribution lines, the fields predicted by these three methods are in close agreement. When the observation point is moved to intermediate and far ranges (e.g., within 100 km), differences between the approach of this work and the Cooray–Rubinstein or Cooray modified formulas become considerable. It is worth mentioning that at far ranges the Cooray–Rubinstein and the Cooray modified formulas yield results that are very close to those predicted by the wavetilt formula. It can be analytically shown that for a far observation point a few tens of meters above ground, the Cooray–Rubinstein (or Cooray-modified) formula approximately reduces to the wavetilt expression [56,82]. At these ranges, the limiting case of the Cooray–Rubinstein and wavetilt formulas is indeed exact if one deals with a pure Zenneck surface wave. However, it is never the case that a pure Zenneck surface wave is the only contribution from a localized source [56]. It follows from Figure 5.36 that the surface wave component, which in our approach is not necessarily the Zenneck surface wave, is not adequately accounted for in the Cooray–Rubinstein, Cooray-modified, and wavetilt formulas. While the Fresnel term is negligible at near

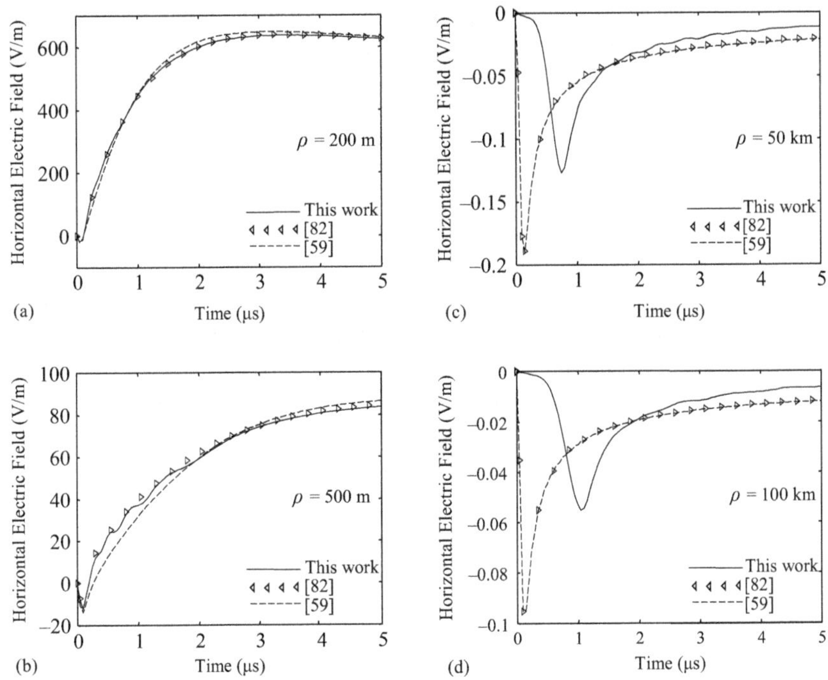

Figure 5.36 Horizontal electric field at 20 m height above ground with conductivity σ_2 = 0.04 S/m and permittivity $\varepsilon_2 = 8\,\varepsilon_0$ computed using the approach of this work, the Cooray-Rubinstein formula, and the Cooray modified formula (a) 200 m, (b) 500 m, (c) 50 km, and (d) 100 km from the lightning channel-base displayed on a 5 μs time scale. Adapted from Shoory et al. *[21]*

ranges, it becomes significant at intermediate and far ranges. For convenience, we rewrite the surface wave component of the horizontal electric field, the terms in the brackets in (5.73), as follows:

$$E_{\rho s}(\rho, z, j\omega) = -\frac{j\omega\mu_0}{2\pi\gamma_1}\int_0^h I(s)\left[\frac{\gamma_1}{\gamma_2}e^{-\gamma_1 r_2}\left(\frac{\rho}{r_2}\right)\left(\frac{1+\gamma_1 r_2}{r_2^2}\right)\right]ds$$
$$+\frac{j\omega\mu_0}{2\pi\gamma_1}\int_0^h I(s)\left[e^{-\gamma_1 r_2}\frac{\gamma_1^4}{\gamma_2^2}\left(\frac{\pi}{j\gamma_1\gamma_2}\right)^{1/2}e^{-jP}F(P)\right]ds, \tag{5.119}$$

where subscripts denotes the surface wave. Only the first term of (5.119) is included in the Cooray–Rubinstein formula. As Figure 5.36 indicates, the difference between the results of this work and those predicted by the Cooray–Rubinstein and Cooray-modified formulas at distant points is more pronounced at early times and gradually diminishes at later times. Further, when the value of the ground's conductivity and permittivity is large, our approach, the Cooray–Rubinstein formula, and the Cooray modified formula (and even the wavetilt formula for remote observation points a few tens of meters above ground) all predict similar fields, since the Fresnel term in this case is very small.

(b) Surface waves

We now consider three practical cases of the lower half-space, seawater, relatively high conductivity ground, and relatively low conductivity ground. The electrical parameters of these cases are given in Table 5.2 and are similar to those given in [54]. Figure 5.37 illustrates contributions of the surface waves to the vertical electric and horizontal magnetic fields. In order to better resolve all significant features of the waveforms occurring at early times, only the first 2 μs are shown. Calculations were carried out at the air-ground interface at two distances, 50 km and 100 km from the channel-base. This figure reveals four interesting features of surface waves:

- Surface waves exhibit flattening after the rapid variation (sharp initial peak) during the first several hundred nanoseconds.
- The amplitude of the surface waves tends to decrease as the observation point is moved farther away from the source. This is best seen from the curves of both the electric and the magnetic fields for a relatively low conductivity ground.
- For a given observation point, surface waves become more pronounced as the conductivity of the lower half-space decreases.
- For a given observation point, surface waves exhibit a smoother rise as the conductivity of the lower half-space decreases.

Figure 5.37 also shows that the initial rise-time of each field component increases as the observation point is moved farther away from the channel base. It follows that the neglecting of surface waves, as is common in the calculation of lightning-radiated electromagnetic fields, is only justified when the lower half-space has relatively high conductivity.

(c) Total electric and magnetic fields

The conductivity of the lower half-space could strongly influence the characteristics of lightning-radiated fields. Field components are affected by propagation effects as they travel along the interface. The lossy interface preferentially

Table 5.2 Electrical parameters of three realistic cases of the lower half-space

Medium	**σ_2, S/m**	**$\varepsilon_2/\varepsilon_0$**
Seawater	4.0	80
Relatively high conductivity earth	0.4	12
Relatively low conductivity earth	0.04	8

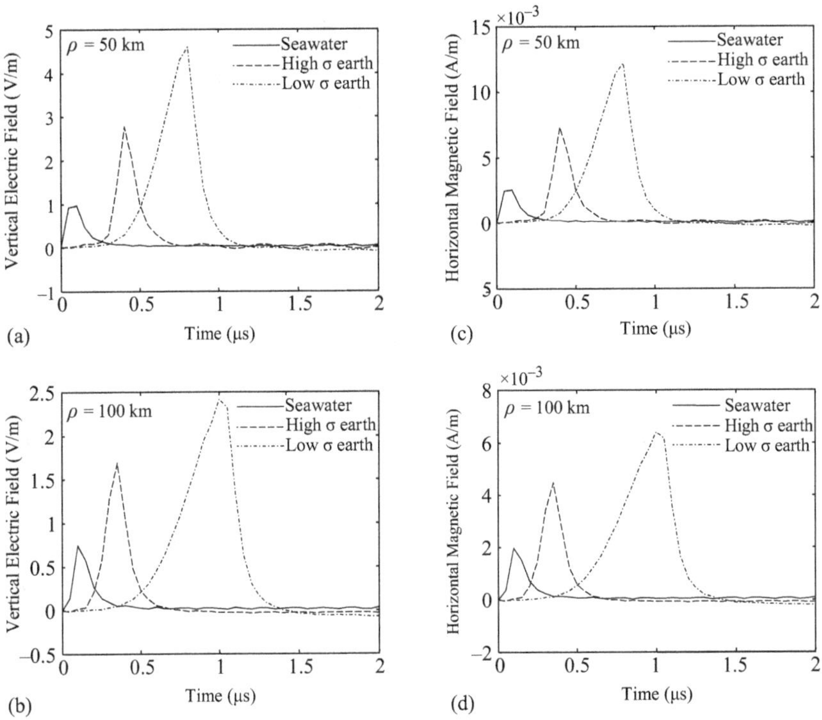

Figure 5.37 Surface wave components of the vertical electric (a and b) and horizontal magnetic (c and d) fields computed at the air-ground interface for three types of the lower half-space 50 km and 100 km from the lightning channel-base displayed on a 2 μs time scale. Adapted from Shoory et al. *[21]*

attenuates high frequency components of radiated electric and magnetic fields, which results in an increase in the rise time and a decrease in the magnitude. These are shown in Figure 5.38. Calculations are performed for three types of lower half-space: seawater, relatively high conductivity ground, and relatively low conductivity ground at distances of 200 m, 50 km, and 100 km from the lightning channel base (displayed on a 5-μs time scale). Figure 5.39 depicts horizontal electric fields for the same situation as in Figure 5.38. As seen from these figures, as the lower half-space becomes less conductive, distortions in the radiated vertical electric, horizontal electric, and horizontal magnetic fields become more significant, and horizontal electric fields become larger. The results of this study indicate that in the analysis of lightning-radiated electromagnetic fields over practical lower half-spaces, the perfect-ground assumption yields results that may be considerably different from those predicted by the more realistic approach described here, particularly at larger distances from the lightning channel.

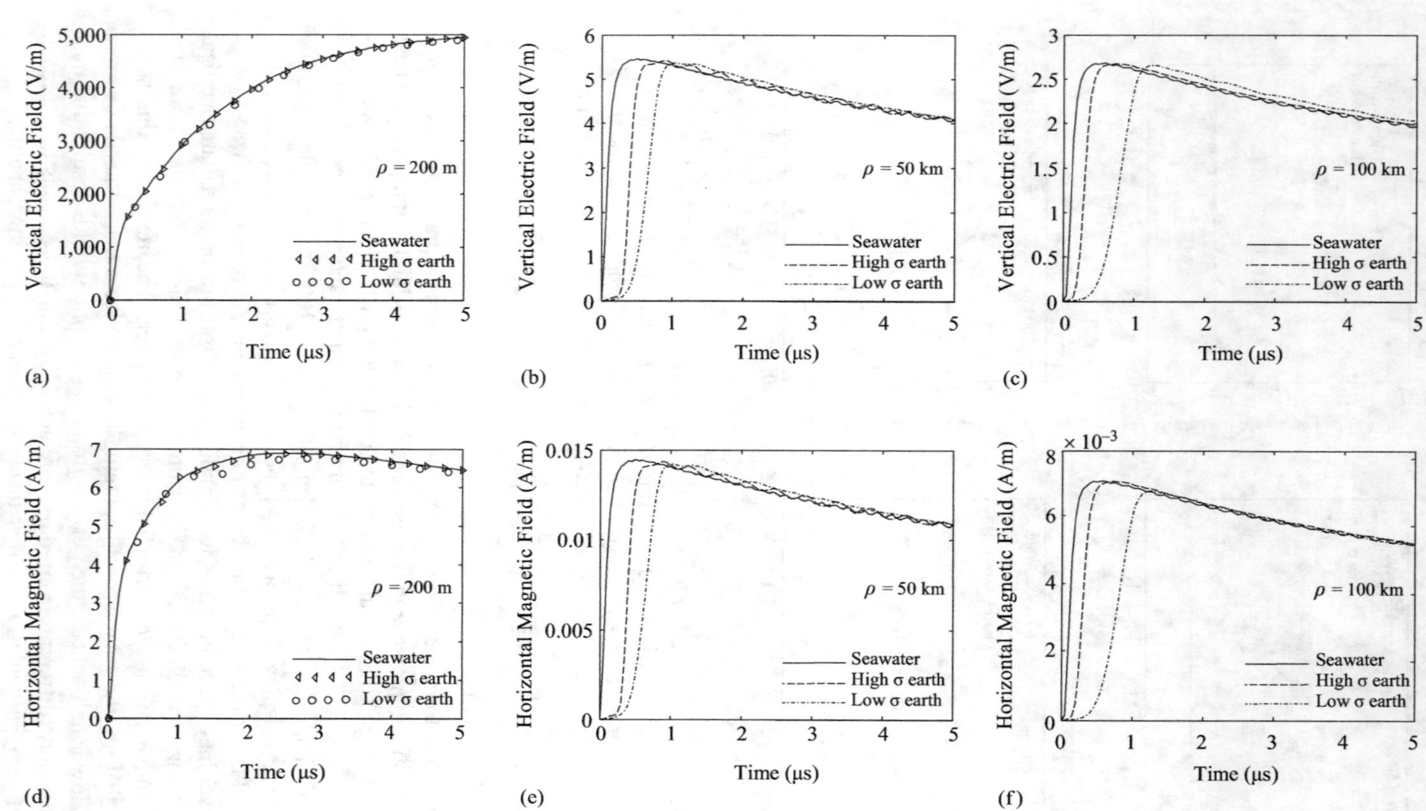

Figure 5.38 Vertical electric (a, b, and c) and horizontal magnetic (d, e, and f) fields at the air-ground interface for three types of the lower half-space 200 m, 50 km, and 100 km from the lightning channel-base displayed on a 5 μs time scale. Adapted from Shoory et al. *[21]*

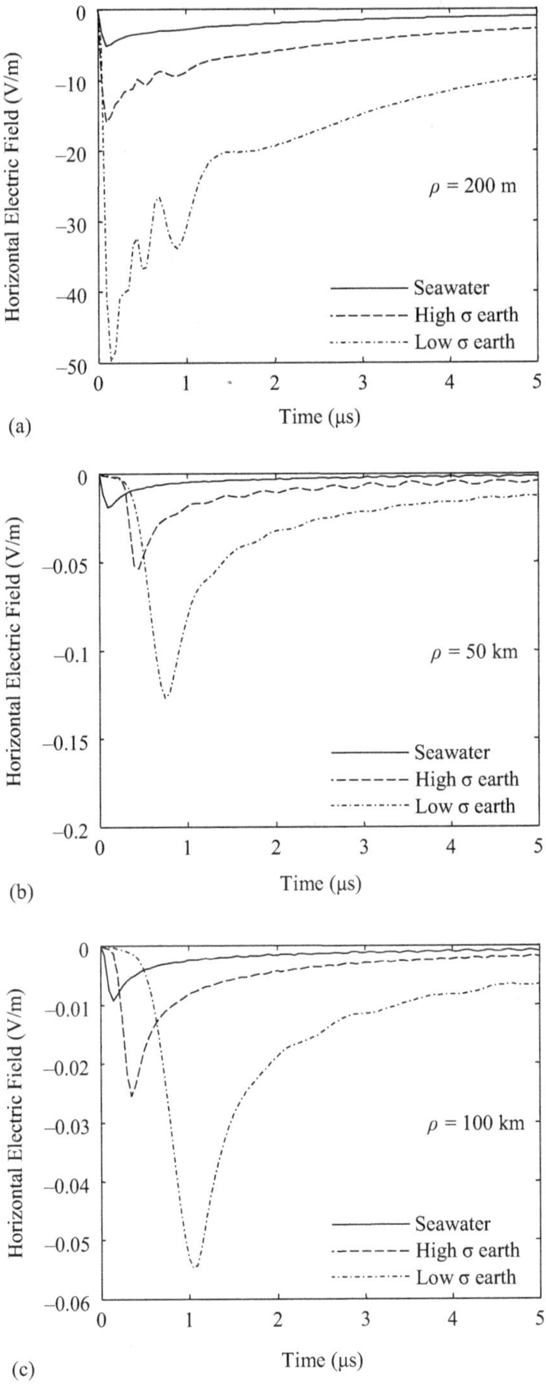

Figure 5.39 Horizontal electric fields at the air-ground interface for three types of the lower half-space (a) 200 m, (b) 50 km, and (c) 100 km from the lightning channel-base displayed on a 5 μs time scale. Adapted from Shoory et al. *[21]*

We now discuss the polarity of the horizontal electric field as a function of distance. The results of Figure 5.39 reveal that for the three types of lower half-space, the horizontal electric field at the air–ground interface exhibits the same (negative) polarity for the entire range of distances (200 m–100 km) considered. This is in contrast with the results of Figure 5.36 where the horizontal electric field waveforms are shown for observation points at a height of 20 m above ground. In this latter case, the horizontal electric field exhibits positive polarity for close distances and negative polarity for far distances. Such a dissimilarity is due to the difference in height of the observation points above ground. In the Cooray–Rubinstein expression for the horizontal electric field (5.117), or in the expression proposed in this work, the contribution from the first term (i.e., the horizontal electric field corresponding to the perfect-ground assumption) exhibits positive polarity [82], while the remainder of the expression shows negative polarity. The former dominates at near observation points few tens of meters above ground, resulting in the positive polarity of the total horizontal electric field, while the contribution from the latter becomes dominant as the observation point moves toward the air-ground interface or farther away from the channel base. Consequently, when the observation point is at the air–ground interface or at a large distance, the total horizontal electric field has negative polarity.

It is worth noting that the oscillations seen in Figures 5.38(d) and 5.39(a) during the initial 2 μs of field waveforms for a relatively low conductivity ground 200 m from the channel base are related to the oscillatory behavior of Fresnel integrals of complex argument.

5.5.5 *Frequency-domain AT model with distributed current source*

Input current wave-form

The sum of two Heidler functions [91] is commonly used in the literature to express the input current and it has been adopted for the lightning analysis in this study. Accordingly, the transient variation of the energization is depicted as follows:

$$i(t) = \frac{A}{\eta_1}\frac{(t/\tau_{11})^{n_1}}{1+(t/\tau_{11})^{n_1}}e^{\frac{-t}{\tau_{12}}} + \frac{B}{\eta_2}\frac{(t/\tau_{21})^{n_2}}{1+(t/\tau_{21})^{n_2}}e^{\frac{-t}{\tau_{22}}} \tag{5.120}$$

$$\eta_1 = e^{-\left(\frac{\tau_{11}}{\tau_{12}}\right)\left(n_1\frac{\tau_{12}}{\tau_{11}}\right)^{\frac{1}{n_1}}},\ \eta_2 = e^{-\left(\frac{\tau_{21}}{\tau_{22}}\right)\left(n_2\frac{\tau_{22}}{\tau_{21}}\right)^{\frac{1}{n_2}}} \tag{5.121}$$

The numerical values of two channel-base current waveforms corresponding, respectively, to typical first and subsequent return strokes, are given in Table 5.3. These values are based on observations made by Berger *et al.* [92] and are commonly used in lightning literature.

The first return-stroke channel-base current is characterized by a peak value of 30 kA and a maximum rise time of 12 kA/μs, whereas the subsequent return stroke current has a peak value of 12 kA and a maximum rise time of 40 kA/μs. Therefore, the following analysis will be restricted to a subsequent return-stroke current because of its higher frequency content compared to the current of a first return stroke.

Table 5.3 Parameters of Heidler's functions used to reproduce the channel-base current wave-shape

Parameter	A (kA)	τ_{11} (μs)	τ_{12} (μs)	n_1	B (kA)	τ_{21} (μs)	τ_{22} (μs)	n_2
First stroke	28	1.8	95	2	–	–	–	–
Subsequent stroke	10.7	0.25	2.5	2	6.5	2	230	2

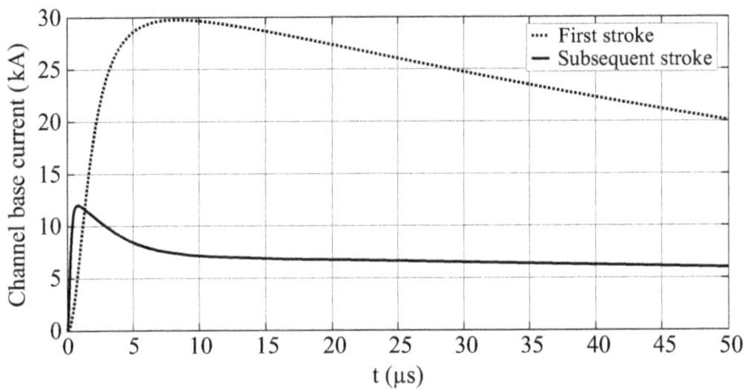

Figure 5.40 Input current wave-form. Adapted from Moini et al. *[83]*

Figure 5.40 depicts the characteristics of the channel-base current for the first and subsequent strokes. The total time of the analysis is set to 1,300 μs. This time window is subdivided into 16,384 time samples, giving a time step of 0.079 μs for the transient analysis. Thus, the maximum frequency of analysis is 6.3 MHz which corresponds to a minimum wavelength of 47.6 m. To ensure the convergence of the MoM, the smallest segment length should not exceed one-sixth of the minimum wavelength. Accordingly, the 2,400 m monopole representing the channel is divided into 480 segments giving a segment length of 5 m. The frequency window is subdivided into 8,192 frequency samples, giving a frequency step of 770 Hz for the frequency analysis. At each frequency, the electromagnetic fields are computed by solving the EFIE. Finally, the time-domain characteristics of the network are obtained by applying an inverse Fourier transform operation to the frequency-domain results.

As the top of the channel is left open, a nonphysical reflection from the top of the channel is expected. Accordingly, all time-domain signals will be contaminated by this reflection. For a channel with length H, if the speed of propagation of the lightning current along the channel is defined by v, the reflection occurs after $t = \mathrm{H/v}$ s. Thus all obtained results are valid (not contaminated by reflections) before t. If a longer observation time is required, a longer channel must be taken into account resulting in a longer computing time.

Comparison with conventional AT models for perfectly conducting soil
In this section, we compare the lightning currents along the channel and the remote electromagnetic fields predicted by the proposed model with those predicted by conventional AT models. All simulations are performed on a workstation of eight-core 3.50 GHz Intel Xeon E5-1620 processor with 64-GB memory. The thin-wire antenna model of a lightning channel above a perfectly conducting ground is illustrated in Figure 5.1(b). For all models, the antenna has length $L = 2{,}400$ m and radius $a = 0.05$ m.

The thin wire representing the return stroke channel (RSC) is divided into 480 segments. For each segment a current source excitation is considered. To define a spatial-temporal distribution of current along the channel similar to the engineering model referred to as MTLE, the amplitude and phase of each source are obtained from (5.109) and (5.110) by choosing $v = 1.3 \times 10^8$ m/s and $\lambda = 2$ km.

In the conventional AT models, the thin-wire monopole antenna representing the return stroke-channel (RSC) is fed at its lower end by a source that injects a current wave along the monopole. The current source waveform of a typical subsequent stroke used for the comparison of models is depicted in Figure 5.40. The spatial–temporal distribution of current along the RSC is then determined by solving Maxwell's equations for the thin-wire model located above a perfectly conducting soil for the chosen excitation. In order to define a model consistent with the observed characteristics of the return stroke, a few adjustable parameters have to be defined.

In the initially adopted AT model [21], to adjust the propagation velocity of the current wave to $v = 1.3 \times 10^8$ m/s (a value commonly used in the literature), the current computations were carried out in a fictitious upper medium with $\varepsilon_r = 5.3$. Then, the resultant current waveforms were used in Green function expressions to calculate remote field components above a perfectly conducting soil assuming $\varepsilon_r = 1$. Another adjustment parameter is the value of the distributed resistance R along the antenna. This value was selected by trial and error to provide an agreement between model-predicted and measured electric fields at close distances. It was assumed for the adopted AT model that $R = 0.1\ \Omega/\text{m}$. It should be noted that the use of this model is restricted only to the electromagnetic analysis of a single RSC without the presence of nearby structures.

In the second adopted AT model [27], inductive energy-storing elements are included in the antenna representation of the lightning channel to control the return-stroke speed. Thus, the unrealistic assumption of higher permittivity of the surrounding medium is avoided for the control of the propagation speed. By introducing an additional uniform distributed inductance along the antenna, the resultant propagation speed at height z of the channel is given by (5.95).

Due to logarithmic (weak) dependencies of C_0 and L_0 on z, their values at height $z = 2{,}400$ m are applied to the entire channel. Accordingly, for a monopole antenna with length $L = 2{,}400$ m and radius $a = 0.05$ m representing the channel, the values of C_0 and L_0 are 4.84 pF/m and 2.29 μH/m, respectively. Therefore, $L^{add} = 10\ \mu\text{H/m}$ results in $v = 1.3 \times 10^8$ m/s along the channel. The value $R = 0.5\ \Omega/\text{m}$ was assumed for the distributed resistance of this AT model [27].

Figure 5.41 depicts the current distribution obtained using various AT models as a function of time at heights of 500 m, 1,000 m, and 1,500 m from the channel base. It should be noted that the plot for AT 1 described in [27] shows oscillations around the current peak. It also exhibits a very large current dispersion mainly due to the low-pass behavior of the distributed inductance in terms of frequency.

The spatial–temporal distribution of current along the channel predicted by the new model is very similar to the one predicted by AT 2 [21]. However, the slower decay of currents for AT 2 compared to the new model is mainly due to the value of distributed resistance in AT 2 which was set by trial and error. Interestingly, the new model provides lightning currents similar to the MTLE model with fewer adjustment parameters compared to conventional AT models.

Figures 5.42(a) and (b) and 5.43(a) and (b) depict the variation of the electric and magnetic fields versus time, respectively, at distances of 50 m and 500 m from the lightning channel base computed using various AT models. As seen in these figures, the electromagnetic fields predicted by the three models are similar in terms of amplitude and wave-shape which shows that the radiating mode of the monopole antenna is independent of its excitation. Once again, the slower rise time seen in the AT 1 model is due to the low-pass behavior of the distributed reactance. The slight amplitude discrepancy observed in the three models is mainly due to the difference between current waveforms of the three models. Indeed, the three models are generated by the same numerical technique but with different types of adjustment parameters.

Comparison with an electromagnetic model based on the FEM technique

In this section, we compare the underground electromagnetic fields predicted by the proposed model with those predicted by an electromagnetic model of the lightning channel based on FEM. The effects of soil stratification in the characterization of the underground fields due to lightning were first investigated by Paknahad *et al.* [93] using a full-wave finite-element-based solution of Maxwell's

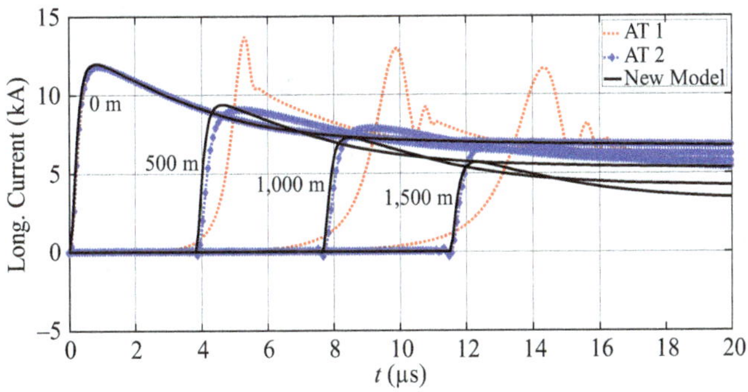

Figure 5.41 Channel current as a function of time at different heights above perfect ground as predicted by models. Adapted from Moini et al. *[83]*

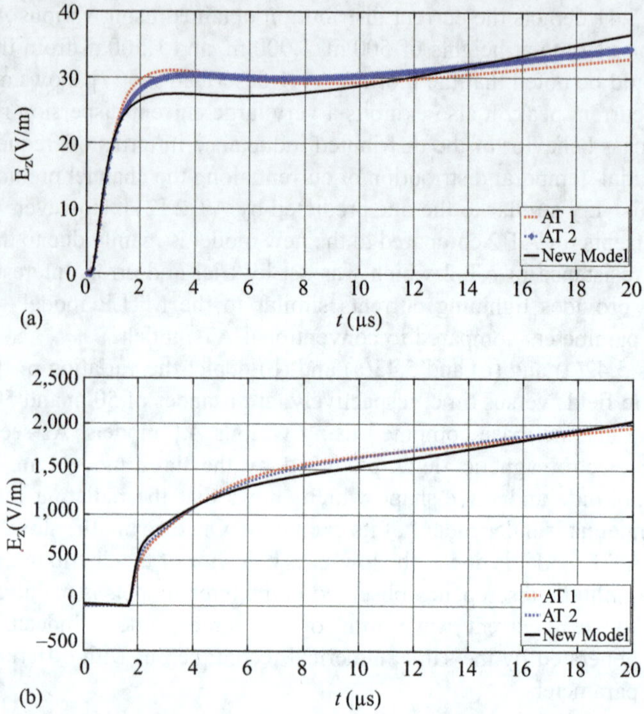

Figure 5.42 Vertical component of the electric field above the perfect ground at different distances (r) from the channel. (a) r = 50 m. (b) r = 500 m. Adapted from Moini et al. *[83]*

equations. Thus, all numerical results obtained by the FEM technique are adapted from [93].

Two different soil models are considered: a uniform soil and a two-layer soil. The upper medium for both cases is free space. The electromagnetic fields due to lightning at depths of 1 m and 5 m inside the soil are investigated. The calculations are performed for three observation points located at horizontal distances of 50 m, 300 m, and 1,000 m from the lightning channel.

In the first configuration, a uniform soil with a resistivity $\rho = 100\ \Omega$.m and a relative permittivity $\varepsilon_r = 10$ is considered. Figures 5.44–5.46 depict, respectively, the time-domain variation of the horizontal electric field, vertical electric field, and horizontal magnetic field for this configuration. As seen in these figures, the results predicted by the new model are similar to those obtained by the FEM technique in terms of amplitude and wave shape. However, for observation points closer to the air-soil interface and at distances farther from the channel, the electromagnetic fields predicted by the FEM technique feature an oscillatory behavior in their early times. This may lead to electromagnetic fields with a larger peak magnitude and a slower rise time compared to the proposed model. For the FEM technique, a better

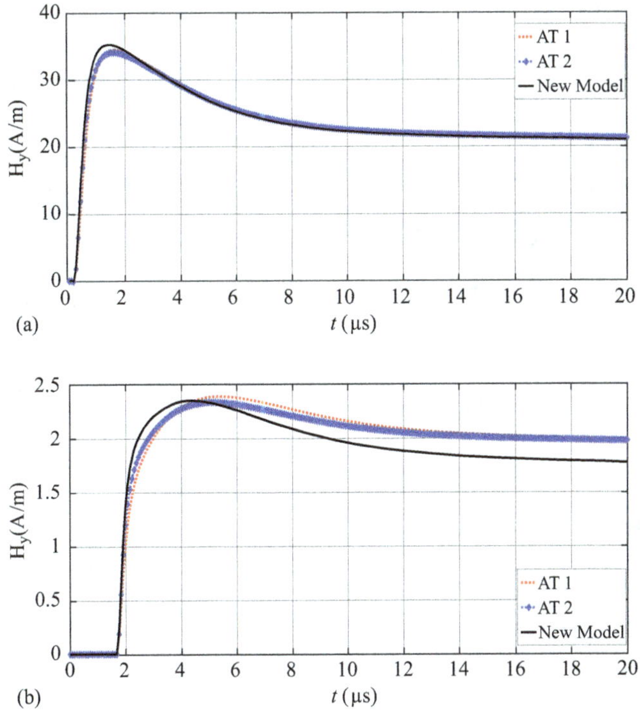

Figure 5.43 Horizontal component of the magnetic field above the perfect ground at different distances (r) from the channel. (a) r = 50 m. (b) r = 500 m. Adapted from Moini et al. *[83]*

accuracy for the electromagnetic fields estimates may be obtained by using a finer mesh, but with a much higher computation cost. Evidently, this is more pronounced when the analysis has to be extended to the lightning channel in the presence of a complex electromagnetic environment with electronic equipment susceptible to being destroyed by low threshold energy. The total simulation time for this transient study with 8,192 frequency samples is about 19 min. For the methodology described in this section, this mainly consists of three stages for each frequency: a matrix-filling phase of MoM for the computation of the mutual interactions, inversion of equation for the excitation vectors (solution phase), and computation of the resulting electromagnetic fields for the air–soil environment.

A two-layer soil is considered for the second configuration. The upper and lower layers of the soil are characterized by soil resistivities $\rho_u = 100\ \Omega$m and $\rho_d = 1{,}000\ \Omega$m, respectively. A relative permittivity of 10 is selected for both layers. The computations are performed considering a thickness of 2 m for the upper medium. Figures 5.47–5.49 depict, respectively, the time-domain variation of the horizontal electric field, vertical electric field, and horizontal magnetic field. As seen in these figures, the results predicted by the new model are similar to those

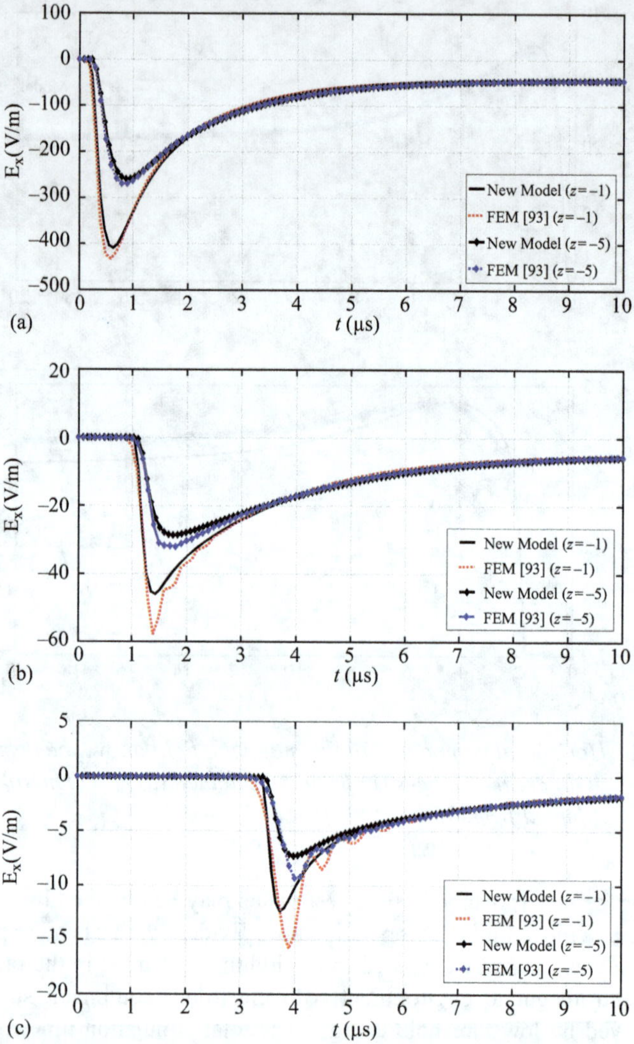

Figure 5.44 Horizontal component of the electric field at depths of 1 m and 5 m in uniform soil (ρ = 100 Ω.m , ε_r = 10) at different distances (r) from the channel. (a) r = 50 m. (b) r = 300 m. (c) r = 1,000 m. Adapted from Moini et al. *[83]*

obtained by the FEM technique in terms of amplitude and wave shape. However, the following remarks may be considered:

- Underground electric fields are more affected than their magnetic counterparts by the soil stratification.
- The horizontal component of the underground electric fields, which has a noticeable inducing effect on buried cables, is less affected by observation depths.

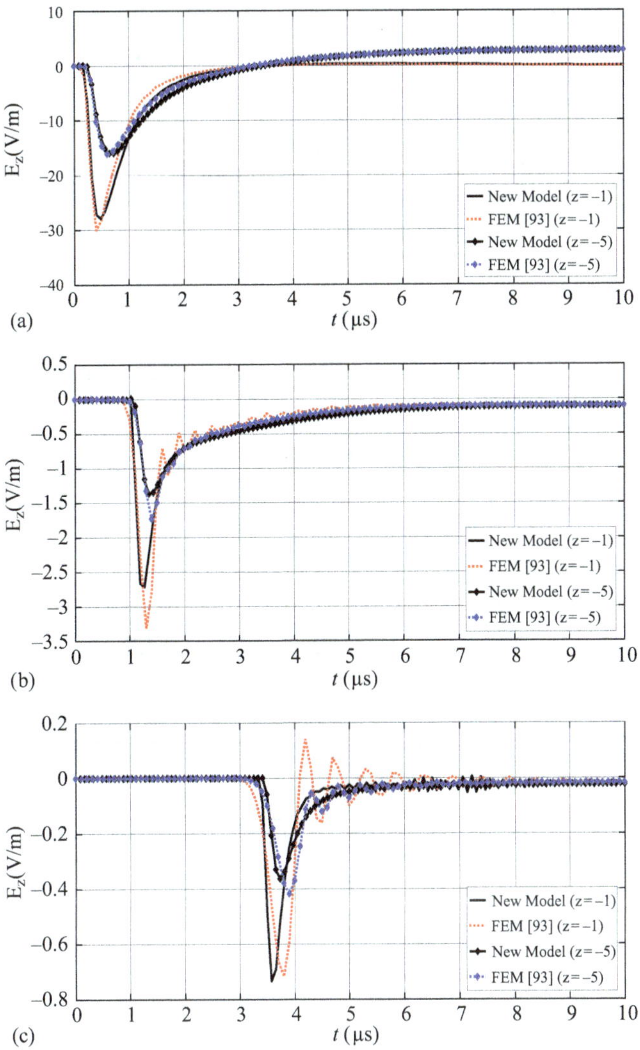

Figure 5.45 Vertical component of the electric field at depths of 1 m and 5 m in uniform soil ($\rho = 100\ \Omega$.m , $\varepsilon_r = 10$) at different distances (r) from the channel. (a) $r = 50$ m. (b) $r = 300$ m. (c) $r = 1{,}000$ m. Adapted from Moini et al. *[83]*

- The oscillatory behavior exhibited in the FEM technique is not seen in the adopted model presented in this section.
- Signal dispersion at far distances is more pronounced in the FEM technique than in the model presented in this section.

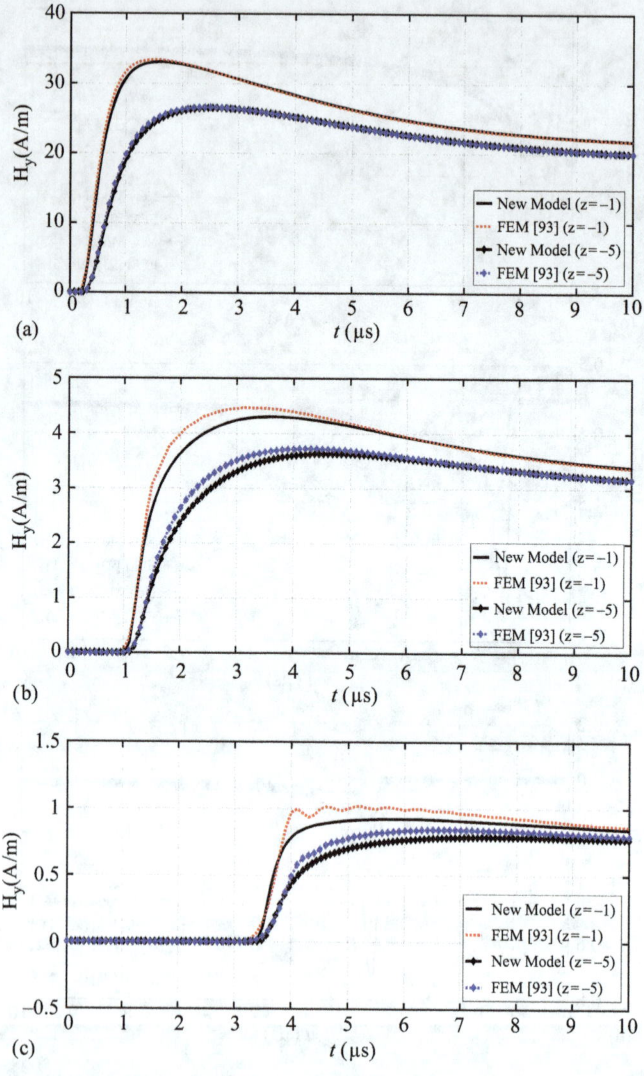

Figure 5.46 Horizontal component of the magnetic field at depths of 1 m and 5 m in uniform soil (ρ = 100 Ω.m, $\varepsilon_r = 10$) at different distances (r) from the channel. (a) r = 50 m. (b) r = 300 m. (c) r = 1,000 m. Adapted from Moini et al. *[83]*

Comparison of lightning-induced voltages on overhead lines

Figure 5.50 depicts the configuration of the lightning channel and its position relative to a single-phase transmission line terminated on each end with a load impedance. The length of the transmission line, located at a height of 10 m above ground, is 1 km. The lightning channel is positioned at a distance of 50 m from the

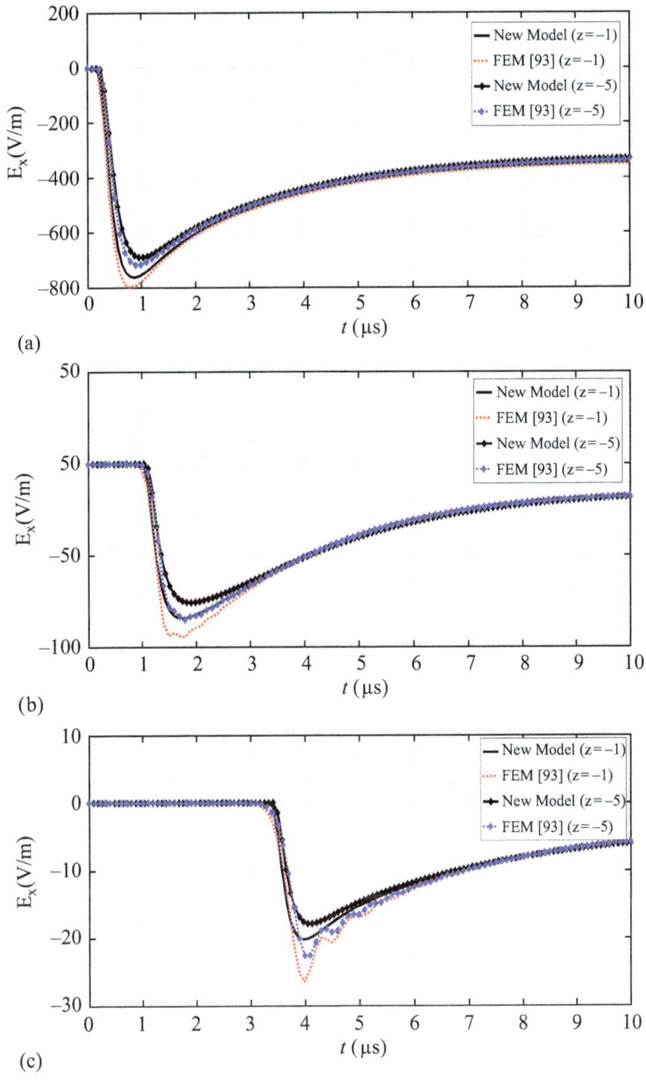

Figure 5.47 Horizontal component of the electric field at depths of 1 m and 5 m in a two-layer soil ($\rho_u = 100\ \Omega$.m, $\rho_d = 1,000\ \Omega$.m) at different distances (r) from the channel. (a) r = 50 m. (b) r = 300 m. (c) r = 1,000 m. Adapted from Moini et al. *[83]*

center of the transmission line. For simulation purposes, two different soil models are considered: one with a uniform soil with perfect conductivity, and the other a two-layer soil. In order to avoid any reflection at the line terminals, we consider the overhead line as being matched at both ends. Note that the geometry of the line is adapted from [94].

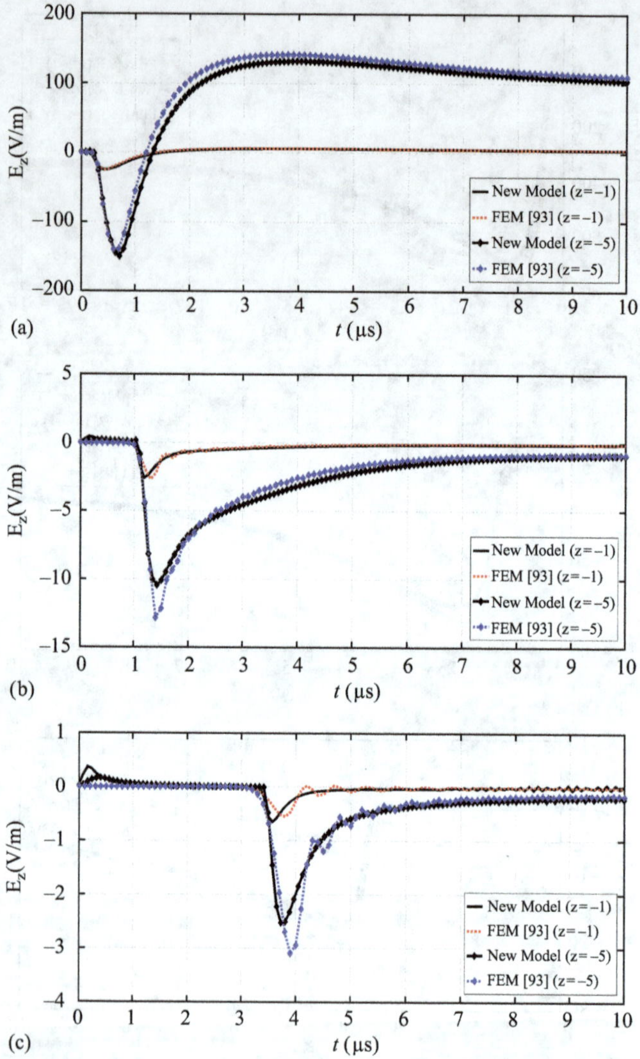

Figure 5.48 Vertical component of the electric field at depths of 1 m and 5 m in a two-layer soil ($\rho_u = 100\ \Omega$.m, $\rho_d = 1{,}000\ \Omega$.m) at different distances (r) from the channel. (a) r = 50 m. (b) r = 300 m. (c) r = 1,000 m. Adapted from Moini et al. *[83]*

The lightning return-stroke channel is represented by a monopole antenna with a length of 1,300 m. The return stroke channel is divided into 130 segments. For each segment a current source excitation is considered. To define a spatial–temporal distribution of current along the channel similar to the modified transmission line with exponential decay (MTLE) model [90] the amplitude and phase

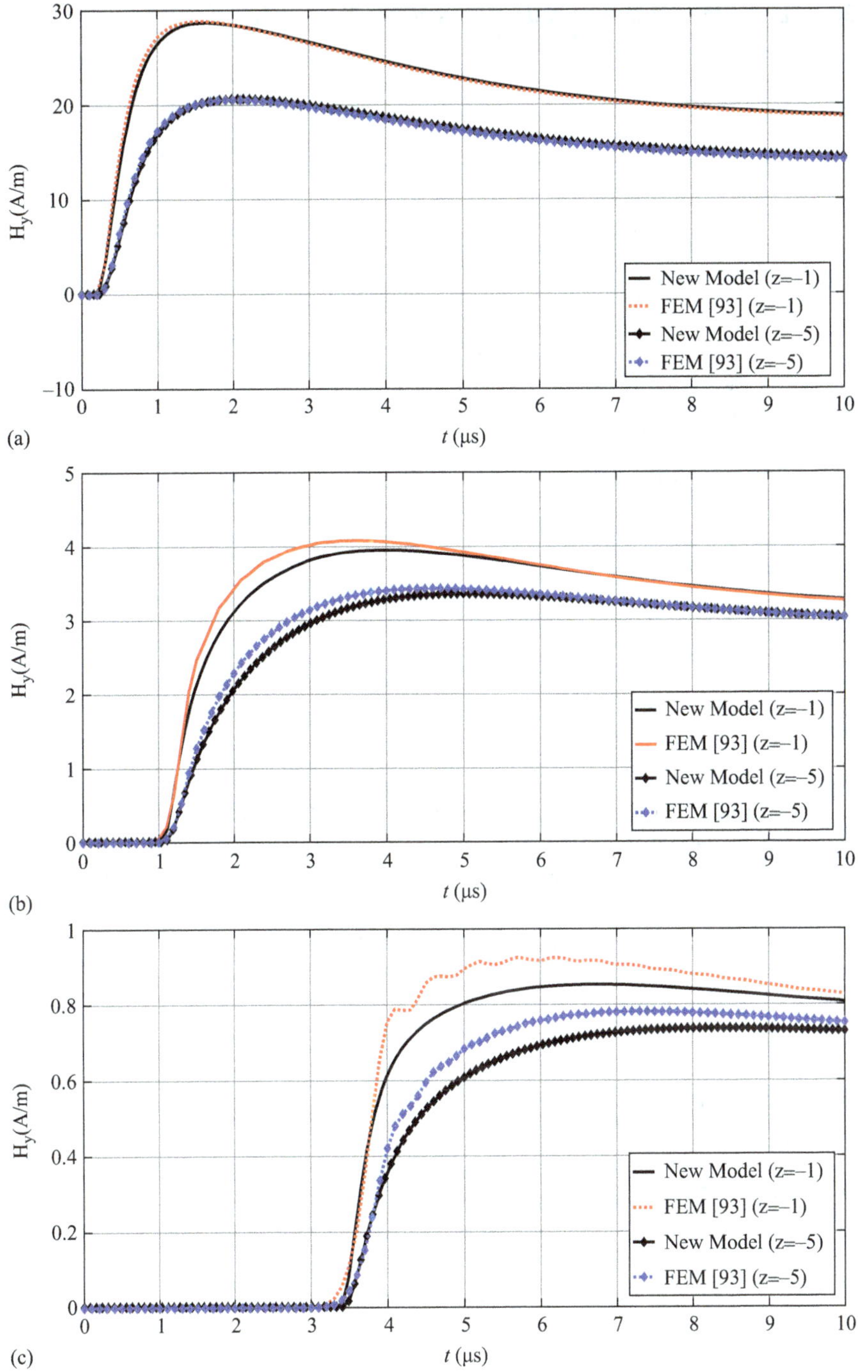

Figure 5.49 Horizontal component of the magnetic field at depths of 1 m and 5 m in a two-layer soil ($\rho_u = 100\ \Omega$.m, $\rho_d = 1,000\ \Omega$.m) at different distances (r) from the channel. (a) r = 50 m. (b) r = 300 m. (c) r = 1,000 m. Adapted from Moini et al. *[83]*

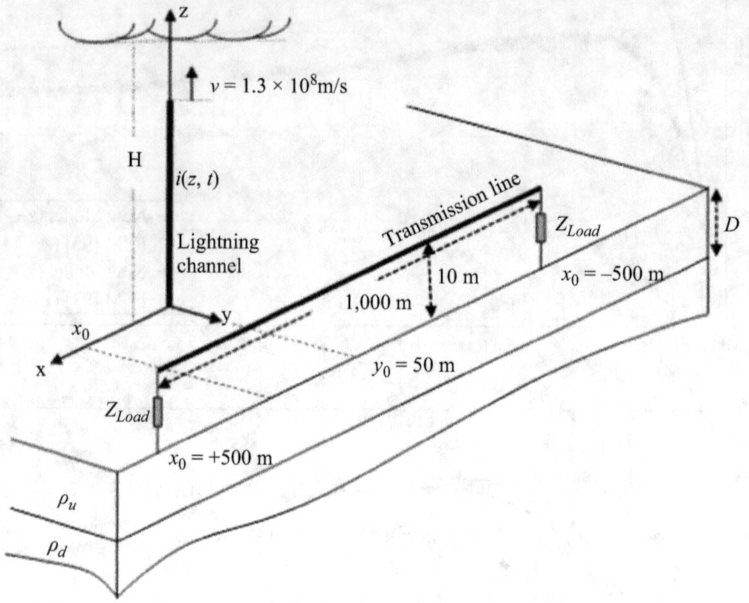

Figure 5.50 Configuration of the lightning channel and its position relative to a single-phase transmission line. Adapted from Moini et al. *[84]*

of each source are obtained by choosing $v = 1.3 \times 10^8$ m/s and $\lambda = 2$ km. For the evaluation of the lightning-induced voltage level at a specified location on the line, a vertical conductor with a large resistance (e.g. 1 MΩ) is used to connect that location to the ground. The induced voltage is obtained by multiplying the current carried by that conductor by the resistance value. Details on the technique used to model the lightning channel and the parameters used for the lightning return-stroke current are given in the previous sections. Figure 5.51(a), (b), and (c) shows the level of the induced voltages at points $x_0 = 0$ m , $x_0 = 250$ m, and $x_0 = 500$ m of the line, respectively. It is assumed that the line is matched at both ends. The soil is considered homogeneous with a resistivity of 10^{-4} Ω·m (essentially, a perfect ground plane) and a dielectric constant of 1. The obtained results are compared to those of [94], which are based on a combination of the MTLE model of lightning channel and a field-to-transmission line coupling model. The current distribution used in this example is similar to that used in [94], implemented using the concept of distributed sources. Also, in the method used in this work, all electromagnetic coupling mechanisms between the RSC and the overhead line and the presence of a multilayer soil are inherently taken into account (i.e., all inductive, capacitive, and conductive effects).

The effect of a multilayer soil on lightning-induced voltages is also investigated. First, cases of uniform soils characterized by a low resistivity ($\rho = 333$ Ω.m) or a high resistivity ($\rho = 3{,}333$ Ω.m), are considered. Then, these soil resistivity

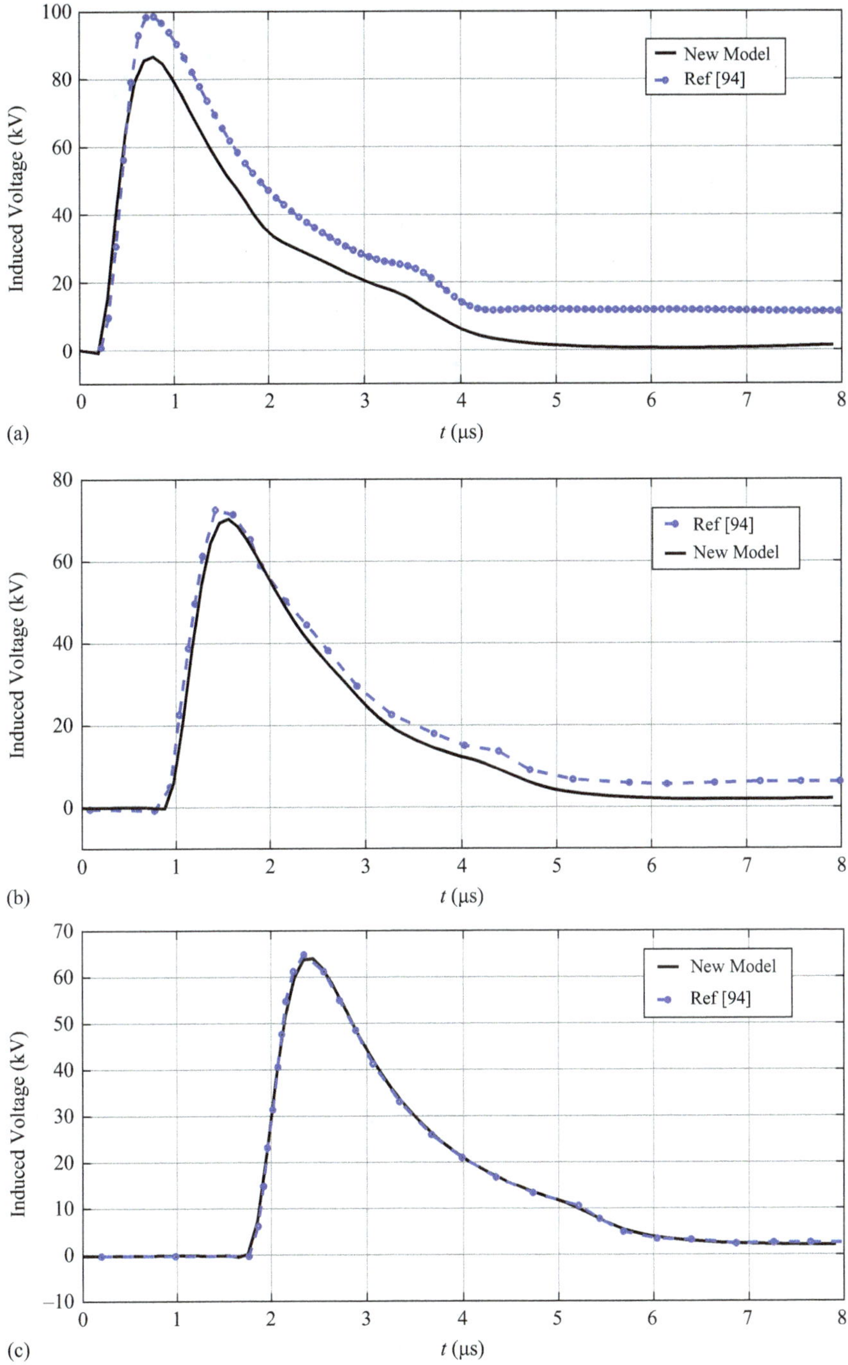

Figure 5.51 Voltage caused by lightning channel at different positions along the overhead line for a perfectly conductive ground. (a) $x_0 = 0$ m. *(b)* $x_0 = 250$ m. *(c)* $x_0 = 500$ m. *Adapted from Moini* et al. *[84]*

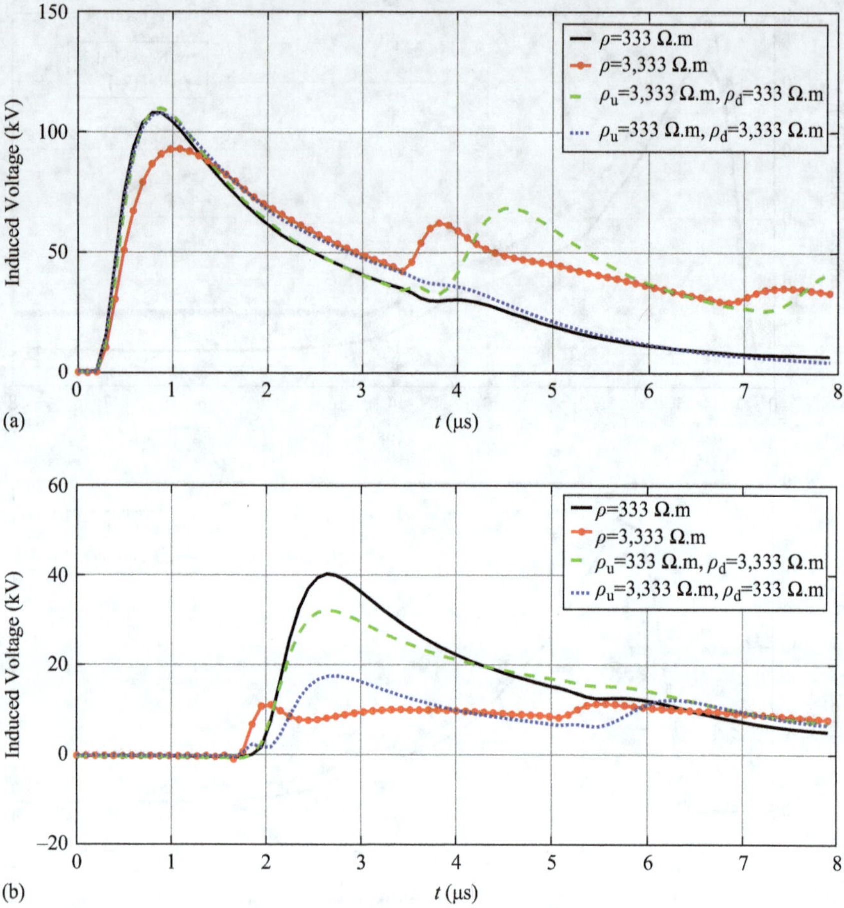

Figure 5.52 Effects of soil stratification on voltage caused by lightning channel at the center and at the end of the transmission line. (a) $x_0 = 0$ m. *(b)* $x_0 = 500$ m. *Adapted from Moini* et al. *[84]*

values are used to investigate the effects of a two-layer soil on the induced voltage. The upper layer thickness of the soil is $D = 5$ m. Figure 5.52(a), (b), and (c) shows the effects of various realistic soil models on the lightning-induced voltages at points $x_0 = 0$ m, $x_0 = 250$ m, and $x_0 = 500$ m of the line, respectively. The value of the termination load impedances remains unchanged compared to the case of a perfectly conducting ground. It should be noted that, normally, the value of the characteristic impedance of the line changes with the electrical parameters of the soil and its stratification. Therefore, multiple reflections from both ends of the line cause secondary peaks after a few microseconds as seen in Figure 5.52. Indeed, this is less pronounced for a uniform soil with a low resistivity value ($\rho = 333$ Ω.m)

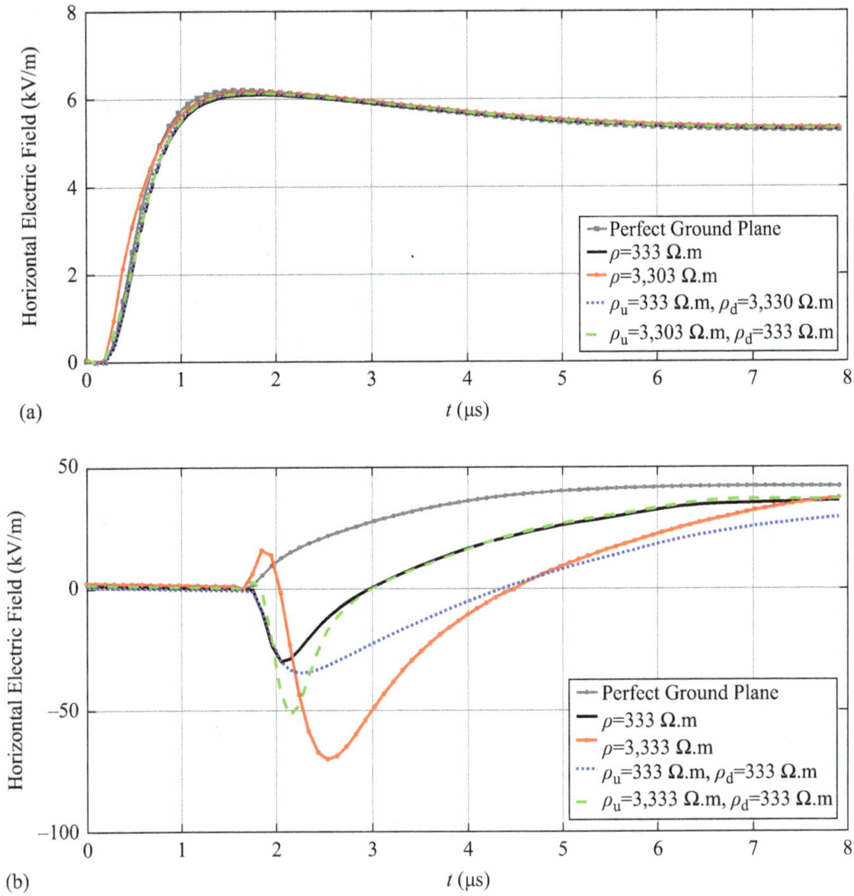

Figure 5.53 Effects of soil stratification on the horizontal component of the electric field of lightning channel at the center and at the end of the transmission line. (a) $x_0 = 0$ m. *(b)* $x_0 = 500$ m. *Adapted from Moini et al. [84]*

compared to the case of high resistivity uniform soil ($\rho = 3{,}333$ Ω.m). For a two-layer soil, the resistivity of the upper layer of the soil has a dominant effect. Thus, the induced voltages along the line when the upper soil layer is less resistive than the lower layer exhibit a similar behavior to the case of uniform soil with the lower resistivity.

The secondary peaks representing the transmission line reflections appear after a few microseconds. Since secondary peaks are related to the line termination mismatch, it can be concluded that the amplitude and rise time of the generated voltages for the first peak are similar to the case of the matched line for the first few microseconds. The center point of the transmission line ($x_0 = 0$ m) is

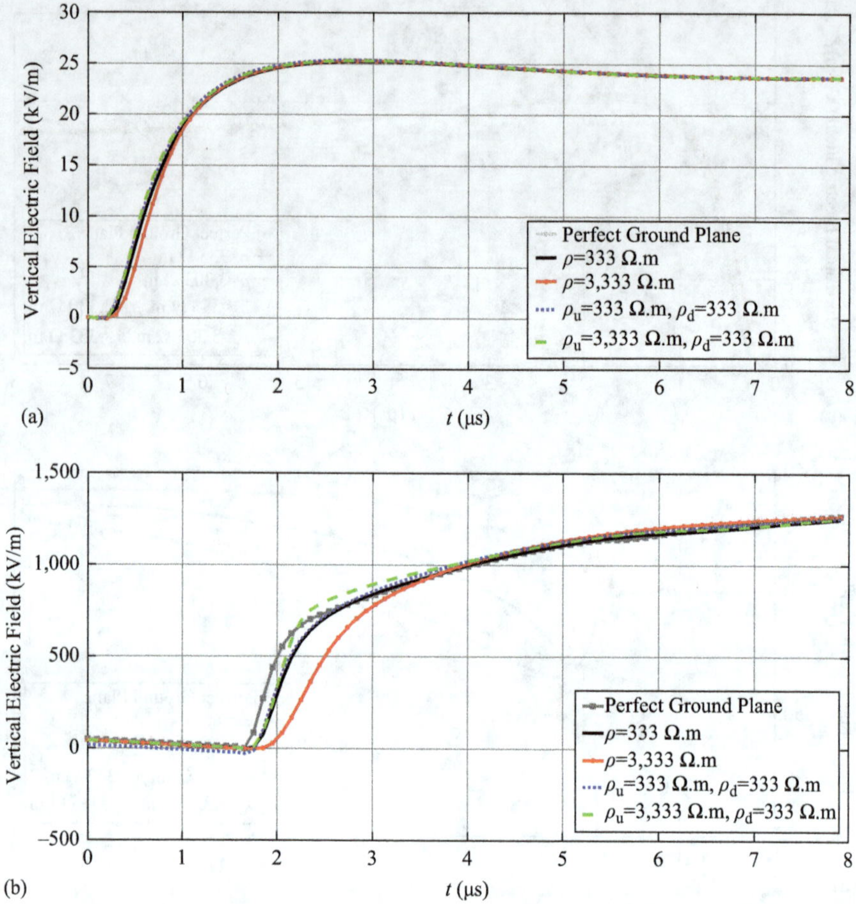

Figure 5.54 Effects of soil stratification on the vertical component of the electric field of lightning channel at the center and at the end of the transmission line. (a) $x_0 = 0$ m. *(b)* $x_0 = 500$ m. *Adapted from Moini* et al. *[84]*

located very close to the lightning channel, at a distance of 50 m. A comparison of Figures 5.51(a) and 5.52(a) shows that the first peak of the voltage at the center point is not affected by the soil stratification. However, a comparison of Figures 5.51(c) and 5.52(c) reveals that at a larger distance from the channel ($x_0 = 500$ m), the soil stratification can affect the amplitude and the rise time of the induced voltage waveform. This is more pronounced when the upper layer of the soil has a high resistivity value.

The overvoltage induced on the transmission line is due to the contribution of the electromagnetic field from the lightning channel in the presence of a stratified soil. Accordingly, the effects of soil stratification on the electric field components

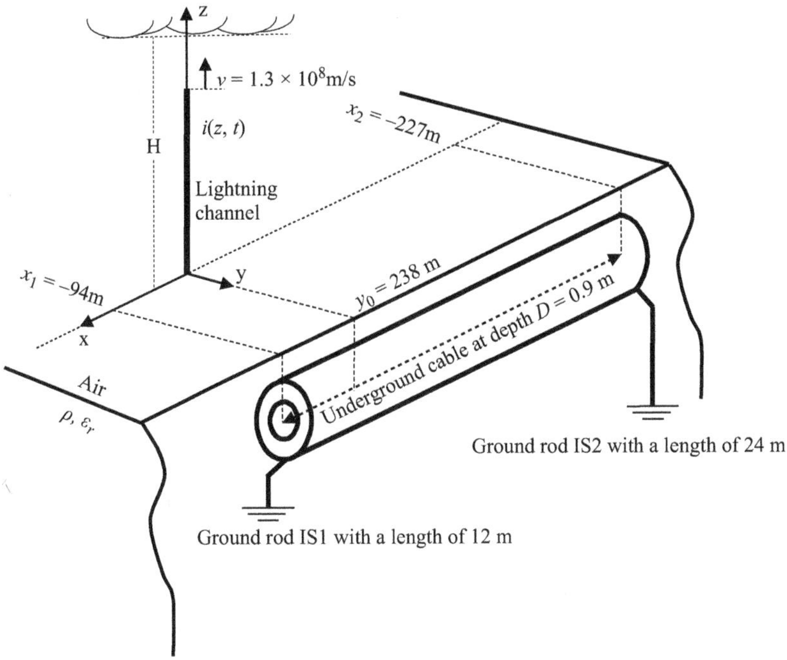

Figure 5.55 Configuration of the lightning channel and its position relative to an underground cable. Adapted from Moini et al. *[84]*

at the center and at the end of the transmission line are also investigated. Figures 5.53(a) and 5.54(a) depict the horizontal and vertical component of the electric field due to the lightning channel at the center of the transmission line. As seen in these figures, none of the components of the electric field are affected by the soil stratification at a very close distance to the channel ($x_0 = 0$ m). On the contrary, when the distance increases ($x_0 = 500$ m) the horizontal component of the electric field is highly perturbed by the soil stratification (Figure 5.53(b)), while the vertical component is less affected (Figure 5.54(b)). As the horizontal component of the electric field has a major impact on the overvoltages induced by lightning on overhead transmission lines, the treatment of a multilayer soil has to be performed by accurate numerical solutions instead of simplified formulas.

Comparison of lightning-induced voltages on buried cables

In this section, the proposed methodology, including the improved lightning channel model, is used to analyze the lightning-induced disturbance on buried cables. Thus, the experimental setup presented by Paolone *et al.* [95] for such an analysis, is selected for the numerical simulation. In [95], the current generated by indirect lightning events were measured at the ends of a buried cable, both in the cable sheath and in the inner conductor. The experimental results obtained for the cable sheath were also compared to numerical simulations based on the FDTD

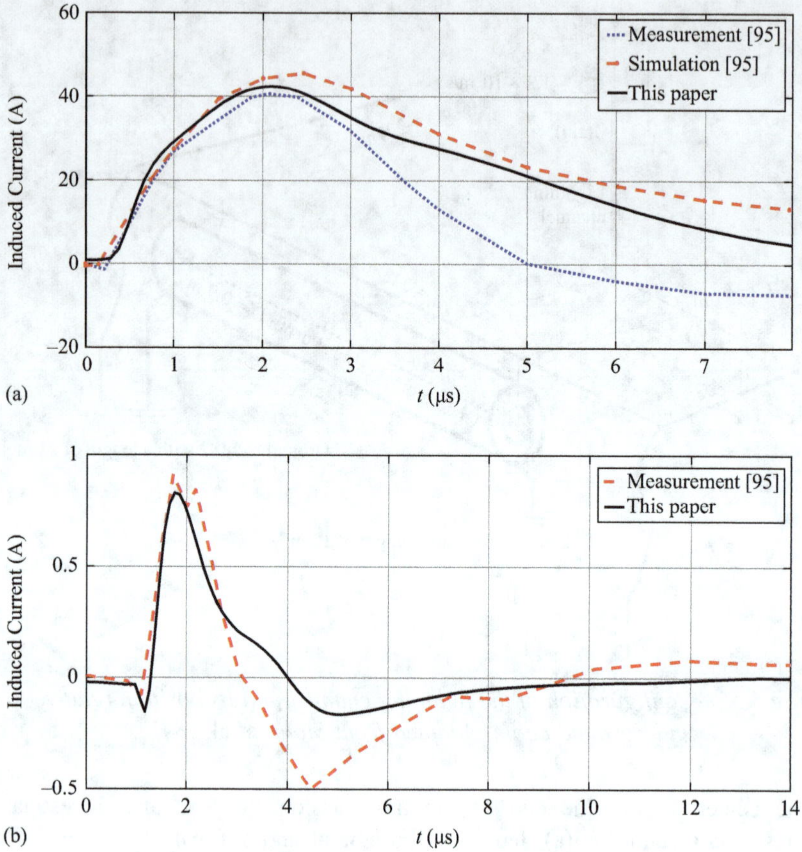

Figure 5.56 Lightning generated currents al location IS1. (a) Cable sheath. (b) Inner conductor. Adapted from Moini et al. *[84]*

technique [96]. Figure 5.55 depicts the lightning channel and its position relative to the underground cable for the selected event. The length of the cable, which is buried at 0.9 m, is assumed to be 133 m. The cable sheath is connected to vertical ground electrodes at both ends. The lengths of the grounding rods at location IS1 and IS2 are 12 m and 24 m, respectively. A complete description of the cable, its configuration, and its location relative to the lightning channel is presented in [95]. The lightning current at the base of the channel is presented by a sum of two Heidler's functions with the parameters [83]: $A = 23.1$ kA, $\tau_{11} = 0.28$ μs, $\tau_{12} = 4.74$ μs, $n_1 = 5$, $B = 9.7$ kA, $\tau_{21} = 5$ μs, $\tau_{22} = 100$ μs, $n_2 = 5$. A homogeneous soil characterized by a resistivity $\rho = 1{,}000$ Ω.m and relative permittivity $\varepsilon_r = 10$ is selected for simulation purposes.

The currents on the sheath and core of the buried cable at locations IS1 and IS2 are depicted in Figures 5.56 and 5.57, respectively. Figures 5.56(a) and 5.57(a) also present the experimental data and simulation results for the lightning-induced

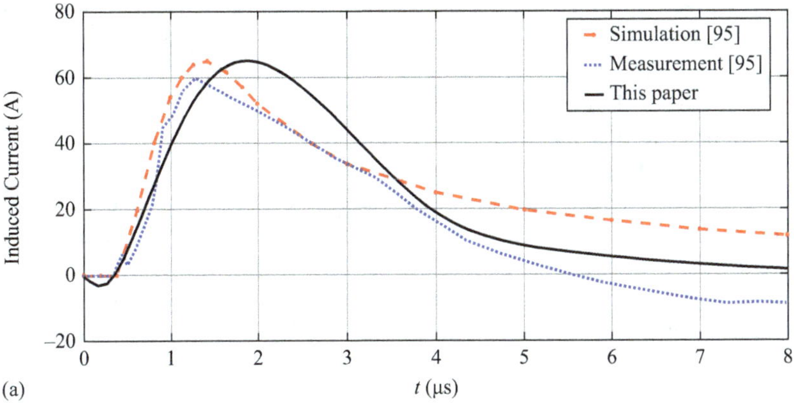

(a)

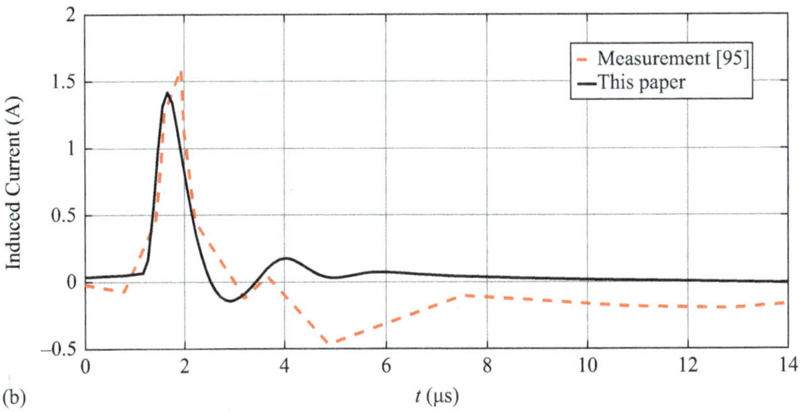

(b)

Figure 5.57 Lightning generated currents al location IS2. (a) Cable sheath. (b) Inner conductor. Adapted from Moini et al. *[84]*

currents in the sheath of the cable, adapted from [95]. A good agreement between the proposed technique, the measurements and the FDTD technique is observed for the early time response of the cable, especially for the peak value of the induced current. Differences in the late time response, more pronounced in the FDTD results, may be related to the choice of soil resistivity value and its stratification [93]. As observed in Figures 5.56(b) and 5.57(b), the transient currents in the cable core predicted by the proposed method, are in agreement with experimental data. In addition, the low peak value is due to the attenuation provided by the cable sheath.

5.6 Summary

In this chapter, we have presented the antenna theory-based electromagnetic models of the lightning return stroke where the lightning return-stroke channel (RSC) is considered as a monopole wire antenna above a conducting ground.

We have also described the time- and frequency-domain solutions of the governing electric field integral equation (EFIE), utilizing the method of moments. The EFIE is used to determine the current distribution along the channel from which remote electromagnetic fields are readily computed.

In the original antenna theory (AT) model in the time domain, the lightning RSC is represented by a lossy vertical monopole antenna, which is fed at its lower end by a voltage source. The voltage waveform is specified on the basis of the assumed input current of the antenna and the antenna resistance per unit length. There are only two adjustable parameters in this model, namely, the wave propagation speed for a nonresistive channel and the value of the distributed channel resistance. Once these two parameters are specified, the spatial and temporal distributions of the current along the channel are found by solving the governing EFIE, using the method of moments. The time-domain AT model has been compared with other lightning return-stroke models in terms of current and line charge density distributions along the channel, and the predicted remote electromagnetic fields. The primary features of the time-domain AT model are as follows: (1) the current amplitude decreases and current rise time increases as the current wave propagates along the channel, in agreement with optical observations, (2) the current wave propagates along the channel at a speed lower than the speed of light due to corona effects and ohmic losses in the channel, and (3) the model-predicted electric and magnetic fields are reasonably consistent with typically measured fields.

In the ATIL model, the lightning RSC is represented by a lossy vertical monopole antenna above a perfectly conducting ground loaded by a set of constant (ATIL-F) or height-variable (ATIL-V) distributed inductances, fed at its lower end by a voltage source. The ATIL-V model allows one to have more control over the variation of the propagation speed along the channel and also over the current distribution and resultant electromagnetic field waveforms, without artificially changing the relative permittivity of the surrounding medium, as done in the original AT model. The ATIL-F and ATIL-V models are compared to the original AT model in terms of the current distribution along the channel and remote electromagnetic fields. The current dispersion predicted by both the ATIL-F and ATIL-V models is more consistent (relative to the original AT model) with optical observations of lightning. It has been shown that adjusting the relative permittivity of the surrounding medium, as done in the AT model, does not affect the current dispersion along the RSC, while in both versions of the ATIL model, the current dispersion increases with increasing propagation speed.

We have also described a nonlinear AT model. In this model, the lightning return-stroke channel is modeled with fixed inductive loading while resistive elements are considered as nonlinear distributed loads whose resistance is a function of both current and time. Characteristics of the resistive loads are drawn from physical models and observations of the channel. The main feature of the proposed model is its ability to predict the zero crossing in both electric and magnetic field waveforms at far distances, as observed in all measurement data. It has been shown that the model can successfully predict the most well-known features of a lightning return-stroke channel, including EM fields at intermediate and far distances, the current distribution along the channel, the current wave propagation speed profile, and the line charge density distribution, although it fails to predict the electric field flattening in the near distance.

In the frequency-domain AT model, the RSC is represented by a lossy vertical straight wire antenna above ground. Since a current source is used for the excitation of the wire, there is no requirement to compute the input impedance calculations for the antenna in the time-domain AT model. The model can reproduce all electromagnetic field benchmark features used for model evaluation, except for the hump following the initial peak in far magnetic fields and the zero crossing in both electric and magnetic fields at far ranges. The finite conductivity of the lower half-space is accounted for in computing lightning-radiated electromagnetic fields by including a surface wave component. Complete electromagnetic fields over lossy ground are analyzed, and it has been shown that the rise times and peak values of both electric and magnetic fields are significantly influenced by propagation over a poorly conducting ground. Decreasing the ground's conductivity results in the rise time increasing and the peak value decreasing. The related results also suggest that neglecting the influence of the ground's finite conductivity is not acceptable in most practical situations.

A new type of AT model based on a thin-wire antenna with distributed current sources is also presented for the electromagnetic analysis of lightning. The required propagation speed of the current wave along the channel was simulated by the implementation of an appropriate phase shift between consecutive current sources. Contrary to conventional AT models, the new model does not require additional adjustment parameters to tune the wave propagation along the channel. Furthermore, any type of spatial–temporal distribution of current along the channel can be obtained by defining an adequate attenuation for the distributed sources. A solution of the EFIE by the MoM was adopted for modeling the lightning channel allowing the treatment of multilayer soil by Sommerfeld integrals. The latter has permitted a direct computation of the electromagnetic fields due to lightning for complex electromagnetic problems without using approximate or simplified formulas.

Overall, various AT models provide an appropriate representation of the lightning RSC in terms of predicted currents and electromagnetic fields. AT models provide several degrees of freedom to simulate all known behaviors of the lightning RSC consistent with measured characteristics. The main feature of the approach described here is its ability to include nearby metallic structures in the modeling stage of the RSC. This feature enables more effective studies of the RSC on overhead and buried lines with complex geometries, the current distribution along tall structures struck by lightning, and multiple other lightning-related phenomena.

References

[1] V. A. Rakov and M. A. Uman, "Review and evaluation of lightning return-stroke models including some aspects of their application," *IEEE Trans. Electromagn. Compat.*, vol. 40, no. 4, pp. 403–426, 1998.

[2] S. I. Braginskii, "Theory of developing of a spark channel," *Sov. Phys. JETP*, vol. 34, pp. 1068–1074, 1958.

[3] E. I. Dubovoy, M. S. Mikhailov, A. L. Ogonkov, and V. I. Pryazhinsky, "Measurement and numerical modeling of radio sounding reflection from a lightning channel," *J. Geophys. Res.*, vol. 100, pp. 1497–502, 1995.

[4] M. N. Plooster, "Numerical model of the return-stroke of the lightning discharge," *Phys. Floyds*, vol. 14, pp. 2124–2133, 1971.

[5] V. A. Rakov, "Characterization of lightning electromagnetic fields and their modeling," in *Proc. 14th Int. Zurich Symp. EMC*, Supplement, Zurich, Switzerland, February 20–22, 2001, pp. 3–16.

[6] A. S. Bizjaev, V. P. Larionov, and E. H. Prokhorov, "Energetic characteristics of lightning channel," *in Proc. 20th Int. Conf. on Lightning Protection*, Interlaken, Switzerland, 1990, pp. 1.1/1–3.

[7] P. F. Little, "Transmission line representation of lightning return-stroke," *J. Phys.*, vol. 11, pp. 1893–1910, 1978.

[8] Y. Baba and V. Rakov, "On the mechanism of attenuation of current waves propagating along a vertical perfectly conducting wire above ground: application to lightning," *IEEE Trans. Electromagn. Compat.*, vol. 47, no. 3, pp. 521–532, 2005.

[9] N. Theethayi and V. Cooray, "On the representation of the lightning return-stroke as a current pulse propagating along a transmission line", *IEEE Trans. Power Del.*, vol. 20, no. 2, pp. 823–837, 2005.

[10] S. Visacro and A. De Conti, "A distributed-circuit return-stroke model allowing time and height parameter variation to match lightning electromagnetic field waveform signatures," *Geophys. Res. Lett.*, vol. 32, p. L23805, 2005, doi: 10.1029/2005GLO24336.

[11] V. Cooray and N. Theethayi, "Pulse propagation along transmission lines in the presence of corona and their implication to lightning return-strokes," *IEEE Trans. Antennas and Propagat.*, vol. 56, no. 7, pp. 1948–1959, 2008, Digital Object Identifier 10.1109/TAP.2008.924678.

[12] V. A. Rakov, "Lightning return-stroke modeling: Recent developments," in *Proc. 3rd Brazilian Workshop Atmospheric Electricity/Int. Conf. Grounding and Earthing*, Rio de Janeiro, Brazil, Nov. 4–7, pp. 85–96, 2002.

[13] R. Moini, B. Kordi, G. Z. Rafi, and V. A. Rakov, "A new lightning return-stroke model based on antenna theory model," *J. Geophys. Res.*, vol. 105, no. D24, pp. 29,693–29,702, 2000.

[14] B. Kordi, R. Moini, W. Janischewskyj, A. M. Hussein, V. Shostak, and V. A. Rakov, "Application of the antenna theory model to a tall tower struck by lightning," *J. Geophys. Res.*, vol. 108, no. D17, p. 4542, 2003, doi: 10.1029/2003JD003398.

[15] Y. Baba and V.A. Rakov, "Electromagnetic models of the lightning return-stroke," *J. Geophys. Res.*, vol. 112, p. D04102, 2007, doi:10.1029/2006JD007222.

[16] A. Taflove, *Advances in Computational Electrodynamics: The Finite-Difference Time-Domain Method*, Norwood, MA: Artech House, 1998.

[17] Y. Baba and V. A. Rakov, "On the transmission line model for lightning return-stroke representation," *Geophys. Res. Lett.*, vol. 30, no. 24, p. 2294, 2003, doi:10.1029/2003GL018407.

[18] E. K. Miller, A. J. Poggio, and G. J. Burke, "An integrodifferential equation for time-domain analysis of thin wire structure, Part I," *J. Comput. Phys.*, vol. 12, pp. 24–48, 1973.

[19] G. J. Burke and A. J. Poggio, "Numerical Electromagnetic Code (NEC) – method of moments," *Naval Ocean System Center, San Diego, CA, Tech.* Document 116, 1980.

[20] Y. Baba and M. Ishii, "Numerical electromagnetic field analysis of lightning current in tall structure," *IEEE Trans. Power Del.*, vol. 16, no. 2, pp. 324–328, 2001.

[21] Shoory, R. Moini, S. H. H. Sadeghi, and V. A. Rakov, "Analysis of lightning radiated electromagnetic fields in the vicinity of lossy ground," *IEEE Trans. Electromagn. Compat.*, vol. 47, no. 1, pp. 131–145, 2005.

[22] A. S. Podgorski and J. A. Landt, "Numerical analysis of the lightning CN-tower interaction," *Presented at the Int. Zurich Symp. Electromagn. Compat.*, Zurich, Switzerland, 1985.

[23] M. Van Blaricum and E. K. Miller, "TWTD—a computer program for time-domain analysis of thin-wire structures," *Lawrence Livermore Laboratory, Rept.* UCRL-51 277, 1972.

[24] A. S. Podgorski and J. A. Landt, "Three dimensional time domain modeling of lightning," *IEEE Trans. Power Del.*, vol. PWRD-2, no. 3, pp. 931–938, 1987.

[25] Y. M. Lin, M. A. Uman, J. A. Tiller, *et al.*, "Characterization of lightning return-stroke electric and magnetic fields from simulation two-station measurements," *J. Geophys. Res.*, vol. 84, pp. 6307–6314, 1979.

[26] S. Bonyadi-Ram, R. Moini and S. H. H. Sadeghi, " Incorporation of distributed inductive loads in the antenna theory model of lightning return-stroke channel," *Presented at the Int. Conf. Lightning Protection (ICLP)*, Avignon, France, 2004.

[27] S. Bonyadi-Ram, R. Moini, S. H. H. Sadeghi, and V. A. Rakov, "On representation of lightning return-stroke as a lossy monopole antenna with inductive loading," *IEEE Trans. Electromagn. Compat..*, vol. 50, no. 1, pp. 118–127, 2008.

[28] D. M. Jordan and M. A. Uman, " Variation in light intensity with height and time from subsequent lightning return-stroke," *J. Geophys. Res.*, vol. 88, pp. 6555–6562, 1983.

[29] D. Wang, N. Takagi, T. Watanabe, V. A. Rakove, and M. A. Uman, "Observed leader and return-stroke propagation characteristic in the buttom 400 m of a rocket triggered lightning channel," *J. Geophys. Res.*, vol. 104, no. D12, pp. 14369–14376, 1999.

[30] S. Bonyadi-Ram, R. Moini, S. H. H. Sadeghi, and V. A. Rakov, "Incorporation of capacitive loads in the antenna theory model of lightning

return-stroke channel," *in Proc. Int. Symp. Electromagn. Compat. (EMC)*, Zurich, Switzerland, 2005.

[31] S. H. Seyed-Moosavi, R. Moini, and S. H. H. Sadeghi, "Representation of lightning return-stroke channel as a nonlinearly loaded thin-wire antenna," *IEEE Trans. Electromagn. Compat.*, vol. 51, no. 3, pp. 488–498, 2009.

[32] S. I. Drabkina, "The theory of the development of the spark channel," *Sov. Phys. JETP*, vol. 21, pp. 473–83, 1951.

[33] K. Sheshyekani, S. H. H. Sadeghi, and R. Moini, "A combined MoM-AOM approach for frequency domain analysis of nonlinearly loaded antennas in the presence of a lossy ground," *IEEE Trans. Antennas Propagat.*, vol. 56, no. 6, pp. 1717–1724, 2008.

[34] R. Moini, S. H. H. Sadeghi and B. Kordi, "An electromagnetic model of lightning return-stroke channel using electric field integral equation in time Domain," *J. Eng. Anal. Bound. Elem*, vol. 27, pp. 305–314, 2003.

[35] J. A. Stratton, *Electromagnetic Theory*, New York, NY: McGraw-Hill, 1941.

[36] A. Reineix, "Analyse theorique de la diffaraction d'ondes electromagnetiques impulsionnells," *Doctorat, Univ*. de Limoges, France, 1986.

[37] J. Herault, R. Moini, A. Reineix, and B. Jecko, "A new approach to microstrip antenna using a mixed analysis: transient-frequency," *IEEE Trans. Antennas Propagat.*, *38*, 1166–1175, 1990.

[38] R. Moini, B. Kordi, and M. Abedi, "Evaluation of LEMP effects on complex wire structures located above a perfectly conducting ground using electric field integral equation in time domain," *IEEE Trans. Electromagn. Compat.*, vol. 40, no. 2, pp.154–162, 1998.

[39] C. T. Tai, *Dyadic Green Functions in Electromagnetic Theory*, New York, NY: IEEE Press, 1993.

[40] J. H. Richmond, "Radiation and scattering by thin-wire structures in the complex frequency domain," in E. K. Miller (Ed.), (eds.) *Computational Electromagnetics*, New York, NY: IEEE Press, 1992.

[41] H. C. Pocklington, "Electrical oscillations in wire," *Cambr. Philos. Soc. Proc.*, vol. 9, pp. 324–332, 1897.

[42] T. K. Sarkar, "A study of various methods for computing electromagnetic field utilizing thin wire integral equation," *Radio Sci.*, vol. 18, pp. 29–38, 1983.

[43] A.Ishimaru, *Electromagnetic Wave Propagation, Radiation, and Scattering*, Hoboken, NJ: Prentice-Hall, 1991.

[44] J. A. Landt, E. K. Miller, and F. J. Deadrick, "Time-domain computer models of thin-wire antennas and scatterers," *Interaction Notes, Note 206, Lawrence Livermore Laboratory*, November 1973.

[45] E.K. Miller and J. A. Landt, "Direct time-domain techniques for transient radiation and scattering," *Interaction Notes, Note 334, Lawrence Livermore Laboratory, and Los Alamos Scientific Laboratory*, July 1976.

[46] G. J. Burke and A. J. Poggio, *Numerical Electromagnetic Code (NEC) – Method of Moments*, *Lawrence Livermore Laboratory*, 1981.

[47] J. H. Richmond, "Digital computer solutions of the rigorous equations for scattering problems," *Proc. IRE*, vol. 53, pp. 796–804, 1965.

[48] R. F. Harrington, "Matrix methods for field problems," *Proc. IEEE*, vol. 55. no. 2, pp. 136–149, 1967.

[49] S. H.Seyed-Moosavi, R. Moini, S. H. H. Sadeghi, and B. Kordi, "Application of the nonlinear antenna theory model to a tall tower struck by lightning for the evaluation of return stroke channel current and radiated electromagnetic fields," *J. Geophys. Res.*, vol. 116, D11118, 2011, doi:10.1029/2010 JD014684.

[50] L. Grcev, "Computer analysis of transient voltages in large grounding systems," *IEEE Trans. Power Del.*, vol. 11, no. 2, pp. 815–823, 1996.

[51] A. Sommerfeld, "Uber die ausbreitung der Wellen in der drahtlosen Telegraphie," *Ann. Phys.*, vol. 28, p. 665, 1909.

[52] T. Takashima, T. Nakae, and R. Ishibashi, "Calculation of complex fields in conducting media," *IEEE Trans. Elect. Ins.*, vol. EI-15, pp. 1–7, 1980.

[53] R. W. P. King, "Electromagnetic field of a vertical electric dipole over an imperfectly conducting half-space," *Radio Sci.*, vol. 25, pp. 149–160, 1990.

[54] R. W. P. King and S. S. Sandler, "The electromagnetic field of a vertical electric dipole over the earth or sea," *IEEE Trans. Antennas Propagat.*, vol. 42, no. 3, pp. 382–389, 1994.

[55] M. Abramowitz and I. Stegun, *Handbook of Mathematical Functions,* New York, NY: Dover, 1970.

[56] J. R. Wait, "On the wavetilt at high frequencies – a personal view," *IEEE Trans. Electromagn. Compat.*, vol. 39, no. 1, p. 65, 1997.

[57] J. R. Wait, "The ancient and modern history of EM ground-wave propagation," *IEEE Antennas Propagat. Mag.*, vol. 40, no. 5, pp. 7–24, 1998.

[58] J. R. Wait, *Electromagnetic Waves in Stratified Media*, New York, NY: IEEE Press, 1996.

[59] V. Cooray, "Some considerations on the Cooray–Rubinstein formulation used in deriving the horizontal electric field of lightning return-strokes over finitely conducting ground," *IEEE Trans. Electromagn. Compat.*, vol. 44, no. 4, pp. 560–566, 2002.

[60] V. Cooray and S. Lundquist "Effects of propagation on the rise times and the initial peaks of radiation fields from return-strokes," *Radio Sci.*, vol. 22, pp. 409–415, 1983.

[61] V. A. Rakov, "Some inferences on the propagation mechanisms of dart leaders and return-strokes," *J. Geophys. Res.*, vol. 103, pp. 1879–1887, 1998.

[62] C. A. Nucci, G. Diendorfer, M. A. Uman, F. Rachidi, M. Ianoz, and C. Mazzetti, "Lightning return-stroke current models with specified channel base current: A review and comparison," *J. Geophys. Res.*, vol. 95, pp. 20,395–20,408, 1990.

[63] V. A. Rakov and A. Dulzon, "A modified transmission line model for lightning return-stroke field calculations," in *Proc. Int. Zurich Symp. on Electromagn. Compat.*, vol. 9, pp. 229–235, 1991.

[64] R. Thottappillil, V. A. Rakov, and M. A. Uman, "Distribution of charge along the lightning channel: relation to remote electric and magnetic fields and to return-stroke models," *J. Geophys. Res.*, vol. 102, pp. 6987–7006, 1997.

[65] F. Bost, L. Nikolas, and G. Rojat, "Study of conduction and induction phenomena in electric circuits using a time-domain Integral formulation," *IEEE Trans. Magnetics*, vol. 36, no. 4, pp. 960–963, 2000.

[66] M. R. Guedira, "Etude en regime transitoire du comportement de structures filaires presentant de charges lineraires ou non lineraires," *These pour l'obtention du doctorat, Universite de Limoges, France*, 1983.

[67] C. A. Nucci, S. Guerrieri, M. T. Correia de Barros, and F. Rachidi, "Influence of corona on the voltages induced by nearby lightning on overhead distribution lines," *IEEE Trans. Power Del.*, vol. 15, no. 4, pp. 1265–1273, 2000.

[68] R. C. Olsen III, D. M. Jordan, V. A. Rakov, M. A. Uman, and N. Grimes, "Observed one-dimensional return-stroke propagation speeds in the bottom 170 m of a rocket triggered lightning channel," *Geophys. Res. Lett.*, vol. 31, L16107, 2004, doi: 10.1029/2004GL020187.

[69] V. Kodali, V. A. Rakov, M. A. Uman, *et al.*, "Triggered-lightning properties inferred from measured currents and very close electric fields," *Atmos. Res.*, vol. 76, pp. 355–376, 2005.

[70] V. P. Idone and R. E. Orville, "Lightning return-stroke velocities in the Thunderstorm Research International Program (TRIP)," *J. Geophys. Res.*, vol. 87, pp. 4903–4915, 1982.

[71] R. Thottappillil and M. Uman, "Comparison of lightning return-stroke models," *J. Geophys. Res.*, vol. 98, no. D12, pp. 22,903–22,914, 1993.

[72] C. E. Baum, "Return-stroke initiation," in *Proc. EMC Symposium, Zurich*, 1989, pp. 383–388.

[73] M. N. Plooster, "Numerical simulation of spark discharges in air," *Phys. Fluids*, vol. 14, pp. 2111–2123, 1971.

[74] M. N. Plooster, "Numerical model of the return-stroke of the lightning discharge," *Phys. Floyds*, vol. 14, pp. 2124–2133, 1971.

[75] A. De Conti, S. Visacro, N. Theethayi, and V. Cooray, "A comparison of different approaches to simulate a nonlinear channel resistance in lightning return-stroke models," *J. Geophys. Res.*, 113, D14129, 2008, doi:10.1029/2007JD009395.

[76] R. H. Golde, *Lightning, vol.1- Physics of Lightning, A.P.*, 1977.

[77] R. C. Jones, "Return-stroke core diameter," *J. Geophys. Res.*, vol. 73, no. 2, pp. 809–814, 1968.

[78] C. R. Paul, *Analysis of Multi-conductor Transmission Lines*, New York, NY: Wiley, 1994.

[79] L. Grcev, F. Rachidi, and V. A. Rakov, "Comparison of electromagnetic models of lightning return-strokes using current and voltage sources," *in Proc. 12th Int. Conf. Atmospheric Electricity*, Versailles, France, 2003, pp. 593–596.

[80] F. Rachidi, C. A. Nucci, M. Ianoz, and C. Mazzetti, "Influence of a lossy ground on lightning-induced voltages on overhead lines," *IEEE Trans. Electromagn. Compat.*, vol. 38, no. 3, pp. 250–264, 1996.

[81] F. Rachidi, "Formulation of the field-to-transmission line coupling equations in terms of magnetic excitation fields," *IEEE Trans. Electromagn. Compat.*, vol. 35, no. 3, 1993.

[82] M. Rubinstein, "An approximate formula for the calculation of the horizontal electric field from lightning at close, intermediate, and long ranges," *IEEE Trans. Electromagn. Compat.*, vol. 38, no. 3, pp. 531–535, 1996.

[83] R. Moini, A. Aghabarati, S. Fortin, and F.P. Dawalibi, "An improved antenna theory model of lightning return stroke using a distributed current source. Part I: theory and implementation," *IEEE Trans. Electromagn. Compat.*, vol. 61, no. 2, pp. 381–390, 2019.

[84] R. Moini, M. Nazari, A. Aghabarati, S. Fortin and F. P. Dawalibi, "An improved antenna theory model of lightning return stroke using a distributed current source – Part II: applications," *IEEE Trans. Electromagn. Compat.*, vol. 61, no. 2, pp. 391–399, 2019.

[85] "CDEGS CDEGS 17.0.8018," *Safe Engineering Services & technologies ltd. Montreal, Canada, 2021 [Online]. Available:* http://www.sestech.com/.

[86] A. Aghabarati, R. Moini, S. Fortin, and F. P. Dawalibi, "Electromagnetic shielding properties of spherical polyhedral structures generated by conducting wires and metallic surfaces," *IEEE Trans. Electromagn. Compat.*, vol. 59, no. 4, pp. 1285–1293, 2017.

[87] K. A. Michalski and D. Zheng, "Electromagnetic scattering and radiation by surfaces of arbitrary shape in layered media, Part I: theory," *IEEE Trans. Antennas Propagat.*, vol. 38, no. 3, pp. 335–344, 1990.

[88] S. Fortin, Y. Yang, J. Ma, and F. P. Dawalibi, "Electromagnetic fields of energized conductors in multilayer medium with recursive methodology," *in Asia-Pacific Power and Energy Engineering Conference*, Wuhan, China, 2009.

[89] F. P. Dawalibi and A. Selby, "Electromagnetic fields of energized conductors," *IEEE Trans. on Power Delivery*, vol. 8, no. 3, pp. 1275–1283, 1993.

[90] F. Rachidi and C. A. Nucci, "On the Master, Lin, Uman, standler and the modified transmision line lightning return stroke current models," *J. Geophys. Res.*, vol. 95, pp. 20,389–20394, 1990.

[91] F. Heidler, J. Cvetic, and B. V. Stanic, "Calculation of lightning current parameters," *IEEE Trans. Power Del.*, vol. 14, no. 3, pp. 399–404, 1999.

[92] K. Berger, R. B. Anderson, and H. Kroninger, "Parameters of lightning flashes," *Electra*, vol. 41, pp. 23–37, 1975.

[93] J. Paknahad, K. Sheshyekani, F. Rachidi, and M. Paolone, "Lightning electromagnetic fields and their induced currents on burried cables. Part II: the effect of a horizontally stratified ground," *IEEE Trans. Electromagn. Compat.*, vol. 56, no. 5, pp. 1146–1154, 2014.

[94] C.A. Nucci, F. Rachidi, M.V. Ianoz, and C.A. Mazzetti, “Lightning-induced voltages on overhead lines,” *IEEE Trans. Electromagn. Compat.*, vol. 35, no. 1, pp. 75–86, 1993.

[95] E. Paolone, E. Petrache, F. Rachidi, *et al.*, “Lightning-induced voltages on buried cables. Part II: experiment and model validation,” *IEEE Trans. Electromagn. Compat.*, vol. 47, no. 3, pp. 498–508, 2005.

[96] E. Petrache, F. Rachidi, M. Paolone, C. A. Nucci, V. A. Rakov, and M. A. Uman, “Lightning induced disturbances in buried cables—Part I: theory,” *IEEE Trans. Electromagn. Compat.*, vol. 3, p. 47, 2005.

[97] R. Moini, M. Nazari, S. Fortin, and F. P. Dawalibi, “On the use of the antenna theory model of lightning return stroke with distributed current source for EMC analysis,” *IEEE Trans. Electromagn. Compat.*, vol. 61, no. 3, pp. 674–680, 2019.

[98] M. Nazari, R. Moini, S. Fortin, F. P. Dawalibi and F. Rachidi, "Impact of frequency-dependent soil models on grounding system performance for direct and indirect lightning strikes," *IEEE Trans. Electromagn. Compat.*, vol. 63, no. 1, pp. 134–144, 2021.

Chapter 6

Transmission line models of the lightning return stroke

Alberto De Conti[1], *Fernando H. Silveira*[1] *and Silverio Visacro*[1]

6.1 Introduction

A typical return stroke in downward negative lightning is initiated when the bottom of a negatively charged leader propagating downward from the cloud is brought to ground potential. In first strokes, this happens when the stepped-leader encounters, at some tens of meters above the ground, one of the upward-propagating, positively charged leaders induced by the downward leader. The negative charges deposited by the stepped-leader are then gradually neutralized by voltage and current waves that propagate along the leader path, while the positive charges deposited by the upward connecting leader are neutralized by downward-propagating voltage and current waves. This is somehow analogous to a negatively charged transmission line that is switched to ground potential through a short transmission line grounded at its bottom, or, alternatively, to the excitation of an initially uncharged transmission line by a series lumped voltage source located somewhere down the line, this transmission line being also grounded at its bottom. A similar description can be applied to subsequent strokes, except that now the return stroke is initiated when the dart leader reaches the ground. The resulting transient process is, in this case, analogous to either a negatively charged, lossy transmission line that is suddenly switched to ground potential through a lumped grounding impedance, or the injection of a positive current pulse at one end of an initially neutral transmission line by a lumped current or voltage source.

Transmission line models of the return stroke use the analogy above to represent the propagation of the return stroke as being equivalent to the propagation of current and voltage pulses along a transmission line. In these models, the propagating pulse is initiated either with the discharge to ground of a previously charged transmission line or with the use of a lumped source to excite the channel. Currents and voltages along the lightning channel are then found from the solution of the

[1]LRC/UFMG – Lightning Research Center/Federal University of Minas Gerais, Brazil

well-known telegrapher's equations, which for constant, linear, and uniform per-unit-length parameters can be written in time domain as

$$-\frac{\partial V(z,t)}{\partial z} = RI(z,t) + L\frac{\partial I(z,t)}{\partial t} \tag{6.1}$$

$$-\frac{\partial I(z,t)}{\partial z} = GV(z,t) + C\frac{\partial V(z,t)}{\partial t} \tag{6.2}$$

where L, C, R, and G are, respectively, the per-unit-length inductance, capacitance, resistance, and conductance of the transmission line assumed to represent the lightning channel, z is the axial coordinate specifying position on the line, and t is time. Equations (6.1) and (6.2) are derived from Maxwell's equations under the assumption of a quasi-transverse electromagnetic (TEM) field structure, which means, for a two-conductor transmission line with both conductors parallel to the axial coordinate z, that the z-directed electric field component associated with the propagating current and voltage pulses is negligible compared to the electric field components laying on a plane transverse to z [1]. By using Kirchhoffs laws, (6.1) and (6.2) can also be derived from the equivalent circuit illustrated in Figure 6.1, which uses a combination of lumped circuit elements to represent a short line section of length Δz [1]. For this reason, the transmission line models of the return stroke are also called distributed-circuit return-stroke models [2].

If the per-unit-length parameters R, L, C, and G are known, telegrapher's equations (6.1) and (6.2) can be solved for any boundary conditions and any type of excitation by using either a numerical or analytical approach depending on the complexity of the problem. If the cross section and the properties of the dielectric between the line conductors are independent of z, as well as the cross section and the properties of the line conductors themselves, the per-unit-length parameters do not vary with the position on the line and the transmission line is said to be uniform [1]. However, if any of these conditions is violated, the transmission line is said to be nonuniform. This is exactly the case if the lightning channel is assumed to be represented as a long vertical wire positioned over a perfectly conducting plane, which leads to a variation of L and C with height [3]. The channel capacitance is also affected by the presence of the cloud at the top of the leader channel [4], by the existence of a corona sheath surrounding the channel core whose radius is believed

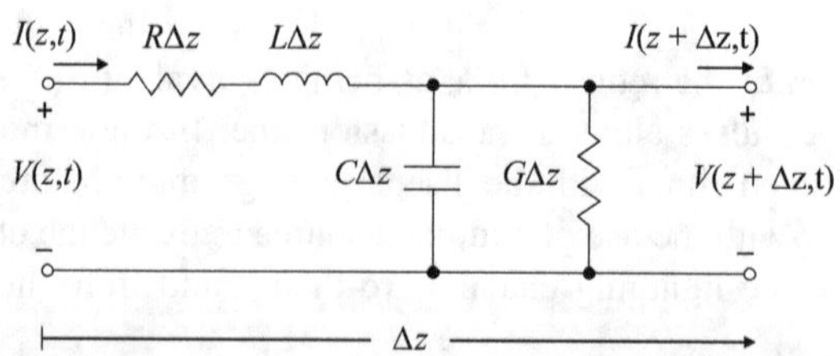

Figure 6.1 Lumped-circuit model of a transmission line segment with length Δz

to vary with height [5], and, to a lesser extent, by possible variations in the leader core radius with height, the latter also affecting the channel inductance.

In addition to L and C, the channel resistance and conductance per unit length are also expected to be dependent on z, the former mostly due to variations in leader temperature, radius, and conductivity [6,7], the latter due to variations in the radius and conductivity of the ionized region existing in the vicinity of the channel core [8]. Other factors, such as channel tortuosity and the existence of branches, can also be considered by attributing to specific channel sections different per-unit-length parameters, or different charge distributions. It is thus apparent that if the lightning channel is to be represented as a transmission line, this transmission line should be nonuniform.

The complexity of the problem increases if the time-varying and nonlinear nature of the channel parameters is also considered. It is known that the leader channel is composed of a thin, negatively charged core of few millimeters surrounded by a radially formed corona sheath which presumably contains the bulk of the charges deposited by the leader and whose external radius is believed to range from some centimeters to tens of meters [7]. When the return stroke is initiated, the gradual neutralization of the charges deposited by the leader is accompanied by the heating and expansion of the channel core and by the collapse of the corona sheath that surrounds it. If the lightning channel is represented as a transmission line, one can therefore expect such a transmission line to have different per-unit-length parameters ahead and behind the return stroke front. Indeed, in a subsequent stroke, the resistance of the leader channel is believed to decay about two orders of magnitude with the passage of the return stroke current [7], and to vary nonlinearly with this current [9]. In addition, the gradual neutralization of the corona sheath that surrounds the channel core results in a nonlinear variation of the channel capacitance and conductance [3,5,10,11]. Finally, the channel inductance and capacitance are also affected by the expansion of the channel core, although this variation is not significant [12]. It is thus apparent that, in addition to being nonuniform, the transmission line representing the lightning channel is also nonlinear.

Several transmission line models have been proposed to represent the lightning return stroke. In general, different assumptions are made in the derivation of the channel parameters per unit length and in the way these parameters are assumed to vary with position and time. Differences are also found in the form of excitation of the transmission line representing the channel, and in the method applied to solve telegrapher's equations (6.1) and (6.2). Section 6.2 presents a brief review of these models, dividing them into discharge-type transmission line models and lumped-excitation transmission line models. Section 6.3 is dedicated to the formulation of a simplified transmission line model of a subsequent stroke and calculation of the corresponding per-unit-length parameters. Section 6.4 presents computed results in which the effect of various channel parameters on predicted lightning currents and remote electromagnetic fields is discussed based on the simplified return stroke model formulated in Section 6.3. The obtained results suggest that the consideration of nonuniform and nonlinear channel parameters change the model predictions in such a way that they come closer to characteristics typically observed in actual lightning. Section 6.5 presents summary and conclusions.

6.2 Review of transmission line models of the lightning return stroke

Transmission line models of lightning return strokes can be roughly classified in two categories, namely as discharge-type models or lumped-excitation models [13]. Discharge-type models represent the leader channel as a transmission line charged to potential U, the return stroke being the result of the line discharge to ground (e.g., [4,5,14–17]). Lumped-excitation models represent the leader channel as an initially neutral transmission line that is excited at its bottom by a lumped source (e.g., [3,11,18,19]). One could possibly include a third category of transmission line models that uses the analogy between lightning and transmission lines to derive, based on expected lightning properties, relevant quantities such as the per-unit-length channel resistance and the channel radius (e.g., [7]). In some cases, one model can fall into more than one category, such as the model of Oetzel [20], which is primarily intended to estimate lightning parameters but can also be viewed as a discharge-type model.

A brief review of transmission line models of the return stroke pertaining to the discharge-type and lumped-excitation categories is presented throughout this section. This review is focused on the main assumptions adopted in the representation of the lightning channel and on the overall agreement between model (mostly current and remote electromagnetic field) predictions with experimental data. Models dedicated to the estimation of lightning parameters are not discussed in detail here because they usually present neither channel currents nor remote electromagnetic field predictions. Their estimates for the channel diameter and for the per-unit-length resistance will be however useful in verifying the validity of assumptions adopted in other models. A more detailed model review can be found in [13].

6.2.1 Discharge-type models

Most of the existing transmission line models of the return stroke are of discharge type. The main idea behind these models is illustrated in Figure 6.2. It consists of representing the leader channel as a charged transmission line that is suddenly brought to ground potential through the closing of a switch. The return stroke is assumed to be equivalent to the transient process resulting from the line discharge. Examples of models pertaining to this category are the models of Oetzel [20], Price and Pierce [14], Little [4], Strawe [12], Takagi and Takeuti [15], Gorin [5,21], Mattos and Christopoulos [16,22], Baker [23], Baum and Baker [10], Hoole and Hoole [17], and Bazelyan and Raizer [24].

In general, return stroke models of discharge type assume the channel to be straight and vertical, although exceptions are found (e.g., [25,26]). A voltage difference in the range of 10^7–10^8 V based on the estimates of Uman [27] is usually assumed to exist between the tip of the leader and the ground plane in the

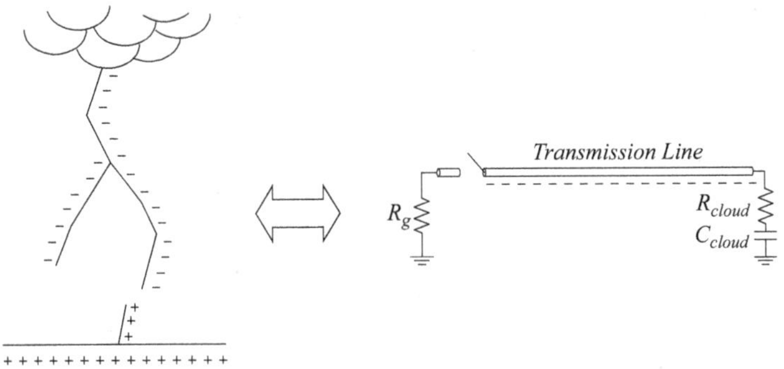

Figure 6.2 Typical representation adopted in discharge-type, transmission line models of the return stroke

simulation of first strokes. Channel lengths varying from 1.5 km [23] to 9.6 km [12] are found in the literature, values between 3 and 6 km being typical [4,15,16,20,24]. In some cases, infinite channels are also assumed for allowing an analytical solution for the transients in the line [10,14]. In general, the cloud termination of the leader channel is assumed to be open-circuited [12,23,24], purely capacitive [4,15,20], or resistive-capacitive [16,22], with capacitance values ranging from 0.01 to 1 μF and resistance values of the order of 1 kΩ. The selection of these parameters is based on assumptions that are often not completely justified and need further experimental validation. The bottom of the channel is usually short-circuited [20] or terminated in a lumped ground resistance with values ranging from 50 Ω to 200 Ω (e.g., [4,12,15,16,23]). The use of a matching resistance at the channel bottom, although unrealistic, was assumed by Price and Pierce [14] in order to let the solution of the line transients analytically tractable. In some cases, as in the model of Baum and Baker [10], the boundary conditions at the bottom of the channel were not clearly specified. Most of the models consider attachment points ranging from 0 to 100 m above ground. They are therefore able to represent both first and subsequent strokes.

Although the channel inductance L and capacitance C per unit length are expected to vary with position and time, as discussed in Section 6.1, the models of Oetzel [20], Price and Pierce [14], Strawe [12], Takagi and Takeuti [15], and Bazelyan and Raizer [24] assumed these parameters to be constant and uniform. Little [4] and later Mattos and Christopoulos [16] assumed the channel capacitance to vary with height due to the influence of both ground plane and cloud, but the obtained values were assumed to remain constant during the return stroke. The influence of a corona sheath surrounding the channel core was considered by Strawe [12], Gorin [5], Mattos and Christopoulos [22], Bazelyan and Raizer [24], Baker [23], and Baum and Baker [10] in the form of an increase in the capacitance associated with the channel core. The models of Gorin [5], Baker [23], and Baum and Baker [10] also considered this parameter to vary nonlinearly due to the neutralization of the corona sheath.

Despite the nonlinear nature of channel losses, the assumption of linear, constant and uniformly distributed series channel resistance is frequent in discharge-type models. This is the case, for instance, of the models of Little [4] ($R = 1$ Ω/m), Price and Pierce [14] ($R \approx 0.06$ Ω/m if $L = 2$ μH/m is assumed in their paper), and Takagi and Takeuti [15] ($R = 0.08$ Ω/m). Although sometimes selected without a clear physical justification, the values of R indicated above are in the range of channel resistances estimated by Rakov [7] for the leader ($R = 3.5$ Ω/m) and return stroke ($R = 0.035$ Ω/m) channels. A more complex representation of channel losses was considered by Strawe [12], Baker [23], Mattos and Christopoulos [16,22], Gorin [5,21], Bazelyan and Raizer [24], and Beroual *et al.* [25,26], who applied either hydrodynamic models or arc resistance models with different simplifying assumptions to simulate the nonlinear variation of the channel resistance.

Currents predicted by models of discharge type usually present decay in magnitude and increase in rise time with increasing channel height, which is consistent with return stroke luminosity profiles [28]. However, the agreement between predicted channel-base currents and measured current waveforms is often poor (see, e.g., Figure 3 in [22]) or not clearly specified (e.g., discussion of Figures 4 and 5 in [23]). Disagreements have also been found between estimated current parameters at the channel base and measured data (e.g., [4,14]).

Strawe [12], by defining the speed of current waves propagating along the channel in terms of the arrival of 63% of the peak current at a given channel height, obtained a current speed ranging from $0.2c$ to $0.3c$ in his model predictions, where c is the speed of light. These values are in the range of return stroke speeds inferred from optical measurements of lightning luminosity profiles, which are seen to vary from about $0.1c$ to $2c/3$ [27]. Other authors have also obtained average propagation speeds in the first kilometer or so of the lightning channel that fall into the limits above, namely Bazelyan and Raizer [24] ($0.4c$), Gorin [5] (0.1–$0.6c$), Little [4] ($0.5c$), Takagi and Takeuti [15] ($2c/3$), and Mattos and Christopoulos [22] (about $0.5c$), but it is not clear which criteria they considered to estimate this parameter. In general, the models above predict a decay of the propagation speed with increasing height that is consistent with measured lightning luminosity profiles [28].

Remote electromagnetic fields predicted by discharge type models are usually inconsistent with measured data. This is the case of the models of Price and Pierce (e.g., Figure 4 in [14]), Takagi and Takeuti (e.g., Figures 12 and 13 in [15]), Baker (Figures 3 and 6 in [23]), and Mattos and Christopoulos (Figures 7–9 in [22]). Other investigators such as Little [4], Strawe [12], Baum and Baker [10], Gorin [5,21], Bazelyan and Raizer [24] and Beroual *et al.* [25,26] did not calculate remote electromagnetic fields with their models. Since the agreement between measured and predicted field waveforms is probably the most important factor for assessing the validity of return stroke models, the overall validity of discharge-type models can be considered limited. More recently, a discharge-type model considering a nonlinear channel resistance and constant shunt parameters representing the corona sheath was proposed to investigate the propagation of M-components, which at ground level present much lower peak amplitudes and longer rise times than return-stroke currents [29]. The model was shown to present a good agreement with currents and electromagnetic fields measured in triggered lightning experiments.

6.2.2 Lumped excitation models

In the transmission line models of the return stroke that consider lumped excitation, the leader channel is represented as an initially uncharged transmission line that has one of its terminals excited by a lumped current or voltage source (see Figure 6.3). Examples of models pertaining to this category are the models of Uman and McLain [30], Amoruso and Lattarulo [18], Rondón *et al.* [31], Theethayi and Cooray [3,32], Visacro and De Conti [11], and De Conti *et al.* [19,33].

The simplest transmission line model of the return stroke considering lumped excitation is the well-known TL model [30]. This model assumes that the current $i(z,t)$ at any point z of the channel at time t is a delayed replica of the current $I(0,t)$ injected at the channel base by a lumped current source, the delay being dependent on the assumed wave speed v. Since the propagating current pulse is neither attenuated nor distorted in the TL model, the return stroke channel is analogous to a lossless uniform two-conductor transmission line surrounded by a medium with permittivity ε and permeability μ, where $v = (\mu\varepsilon)^{-0.5}$. The solution of telegrapher's equations leads in this case to a simple closed-form expression relating the current at any position of the channel to the current injected at the channel base. For this reason, the TL model is usually classified as an engineering return stroke model. Since a whole chapter is dedicated in this book to the engineering return stroke models, a detailed discussion of the TL model is omitted in favor of other transmission line models of the return stroke. Some of its characteristics and predictions will be however used for comparison purposes whenever necessary.

In the models of Amoruso and Lattarulo [18], Rondón *et al.* [31], Theethayi and Cooray [3], Visacro and De Conti [11], and De Conti *et al.* [19,33], many of the

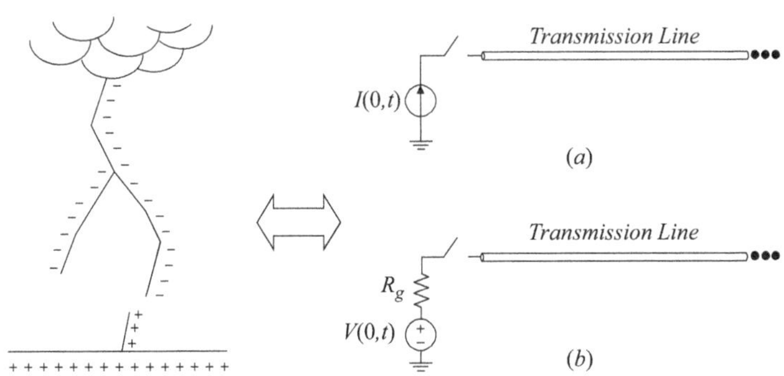

Figure 6.3 Typical representation of a return stroke in a transmission line return stroke model considering lumped excitation: (a) use of an ideal current source at the bottom of the channel; (b) use of a voltage source with internal resistance R_g at the bottom of the channel

simplifications adopted in the TL model are removed to let the modeling of the lightning channel more realistic, mostly with the consideration of nonuniform and/or nonlinear per-unit-length parameters. However, several simplifying assumptions are still adopted in the formulation of these models. For instance, channel branches and tortuosity are usually neglected, and therefore the channel is modeled as a straight vertical transmission line. Also, an ideal lumped current source is assumed to inject the lightning current at the bottom of the channel, even though a lumped voltage source with internal impedance representing the equivalent grounding impedance seen by the leader would be more adequate for assuring the appropriate boundary conditions at the lightning attachment point. This however does not pose a problem since identical model predictions can be obtained with either approach even if nonlinear and nonuniform channel parameters are considered in the simulation of lightning strikes to flat ground [34].

As opposed to the TL model, in which the channel inductance L and capacitance C per unit length are assumed to be independent of the z coordinate along the channel, the approximations adopted in the models of Rondón *et al.* [31], Theethayi and Cooray [3], Visacro and De Conti [11], and De Conti *et al.* [19,33] result in a logarithmic variation of these parameters with height. Amoruso and Lattarulo [18] represented the lightning channel as a lossless transmission line in which an exponential variation was attributed to L and C with respect to z. Such an exponential variation was justified as being the result of a channel core radius exponentially expanding with decreasing channel height. In their model and in the model of Theethayi and Cooray [3], L and C were assumed to be constant and linear, and the effect of the corona sheath surrounding the channel core was neglected. Visacro and De Conti [11] considered a time-varying capacitance to represent the corona sheath surrounding the channel core, with L assumed to be constant and linear. De Conti *et al.* [33] included the variation of L and C with time due to the radial expansion of the channel core, the presence of a radially formed corona sheath surrounding the channel core being neglected. In [19], a coaxial corona model is assumed to govern the nonlinear variation of the channel capacitance. None of the models above considered the effect of the charges accumulated in the cloud in the calculation of the channel capacitance, although it is not clear to what extent such inclusion would be relevant in terms of predicted channel currents and remote electromagnetic fields.

Except for the models of Amoruso and Lattarulo [18] and Rondón *et al.* [31], and for the simple TL model, which assume a lossless channel, the effect of nonlinear channel losses is usually included in the transmission line models of the return stroke that consider lumped excitation. Theethayi and Cooray [3] and also Visacro and De Conti [11] assumed the series resistance of the leader channel to decay exponentially with time after the arrival of the lightning current at a given point along the channel. Initial resistance values of the order of a few ohms to tens of ohms were assumed by both authors, with decay time constants varying from a fraction of microseconds to tens of microseconds. De Conti *et al.* [33] evaluated different approaches to simulate the nonlinear channel resistance in a transmission line model of the return stroke, namely an exponential decay of this parameter with time, the use of a simplified hydrodynamic model for describing the radial expansion of the channel core with time, and several arc resistance models. The latter two approaches were shown to be more consistent than the simple exponential approach for incorporating, according to different simplifications, the dependence that exists between the shape of the propagating current and the rate of change of the channel resistance. In addition to representing series losses in the channel core, Theethayi and Cooray [3,32] and De Conti *et al.* [19] introduced in their model a shunt conductance representing corona losses.

The lossy behavior of the lightning channel and the nonuniform variation of L and C lead to the attenuation and distortion of the return stroke currents predicted by the models of Theethayi and Cooray [3,32], Visacro and De Conti [11], and De Conti *et al.* [33], which agree with luminosity profiles for return strokes [28]. The models above also predict a decay of the propagation speed with height that is consistent with available experimental data [35]. De Conti and Visacro [36] obtained speed profiles presenting in some cases an increase with increasing height at the bottom of the channel followed by a decrease with increasing height, which is consistent with speed profiles observed in some rocket-triggered lightning experiments [37]. In the model of Amoruso and Lattarulo [18], the propagation speed was set as $c/3$. The time to peak current predicted by their model at various channel heights seem to be independent of the channel height, which is unrealistic. This is also the case of the simple TL model, which assumes the lightning current to propagate undistorted and unattenuated.

As opposed to the discharge-type models discussed in Section 6.2.1, most of the lumped-excitation models discussed in this section predict remote electromagnetic fields in overall good agreement with experimental data. This is the case of the models of Visacro and De Conti [11], Theethayi and Cooray [3], and De Conti *et al.* [19,33], although some of the characteristics observed in the electromagnetic field predictions of Visacro and De Conti (see Figure 2 in [11]) and Theethayi and Cooray (see Figures 9(c), 10, 20, and 21 in [3]) were later shown by De Conti *et al.* [33] to be artifacts of assuming the channel resistance to decay exponentially with time without any interaction with the shape of the lightning current. Remote field predictions of Amoruso and Lattarulo [18] were shown to present features that are in some instances consistent with measured data (see, e.g., Figure 3 in their paper), but the impedance mismatch at the upper end of the channel is seemingly responsible for unrealistic discontinuities at the tail of predicted field waveforms depending on selected model parameters (see Figures 4 and 5 in their paper).

In [19], it was shown that a return-stroke model considering corona and nonlinear losses lead to close vertical electric fields presenting waveforms, amplitudes, and decay with distance in good agreement with dart-leader electric-field changes measured in triggered-lightning experiments. The same model was used in [38] to investigate lightning strikes to towers, with focus on wave interactions occurring at the return-stroke front due to the arrival of current pulses that propagate upward on the channel after being transmitted from the tower. It was also considered in [39] to calculate close electric fields and lightning-induced voltages on a nearby overhead line, leading to good agreement with measured data.

6.3 Return-stroke model and calculation of channel parameters per unit length

As discussed in Section 6.1, if the per-unit-length parameters R, L, C, and G are known, the spatial and temporal distribution of the channel current can be found from the solution of telegrapher's equations (6.1) and (6.2). From Section 6.1, it is also known that a more realistic representation of the lightning channel should include time-varying (nonlinear) and nonuniform channel parameters per unit length. However, it is clear from the overview presented in Section 6.2 the great diversity of assumptions found in the conception of the transmission line models of the return stroke, which makes it difficult not only a direct comparison between model predictions but also a comparison between the techniques used by each author for deriving the channel parameters. For this reason, a simplified model is formulated in this chapter for representing a subsequent stroke, based on which a few possibilities are discussed for determining the channel parameters per unit length. This model assumes a lumped current source to inject the desired channel-base current at the bottom of a straight vertical transmission line positioned over a perfectly conducting ground, and considers the existence of two different regions in the lightning channel, which are described as follows:

1. *Leader Channel*: The leader channel is the region ahead of the return-stroke front, which is formed by the dart leader while propagating downward from the cloud. It consists of a lossy core of a few millimeters surrounded by a corona sheath whose radius can reach several meters containing the bulk of negative charges deposited by the leader.
2. *Return-Stroke Channel*: The return-stroke channel is the region behind the return-stroke front, where the negative charges deposited by the leader are gradually neutralized by the return stroke and the channel core radius expands due to an increase in channel temperature and pressure. Due to the gradual neutralization of the corona sheath, part of the charge injected at the channel base is subtracted from the upward-propagating current at each channel section. In addition, the expansion of the channel core and the variation of its physical properties lead to a progressive decay of the channel resistance with time.

The description above is used in the next subsections as a reference for calculating the per-unit-length parameters to be used in the modeling of a typical subsequent stroke.

6.3.1 Channel inductance and capacitance

If corona, channel tortuosity, losses, and the influence of the cloud termination are neglected, the return stroke channel can be represented as an infinitely long vertical wire positioned over a ground plane. This is equivalent to a two-conductor transmission line in which one of the conductors is the vertical wire and the return conductor is the ground. Assuming the line excitation to consist of a lumped source inserted between the ground plane and the bottom of the vertical conductor, the cross section of the dielectric between the two transmission line conductors increases as the observer moves along the z coordinate from the ground to the cloud. This leads to a decrease in the channel per-unit-length capacitance C and an increase in the channel per-unit-length inductance L with increasing height, so both quantities are a function of z, $L(z)$ and $C(z)$.

The derivation of $L(z)$ and $C(z)$ is relatively simple if the cylindrical conductor representing the vertical transmission line is treated as a conical antenna of infinite length whose cone angle is small [40]. If losses are neglected, the characteristic impedance of such a vertical transmission line is given by

$$Z(z) = 60 \ln \frac{2z}{r_a} \tag{6.3}$$

where z is the distance from the observation point on the line to the ground plane (or, strictly speaking, to the lumped source inserted between the ground plane and the bottom of the vertical conductor), and r_a is the conductor radius. The characteristic impedance given by (6.3) relates voltage and current at any point of the vertical transmission line, yielding larger values as z increases. The variation of $Z(z)$ with position gives support to the idea that if the propagation of voltage and current waves along a vertical conductor is to be modeled using transmission line theory, a nonuniform transmission line model is required.

The characteristic impedance is related to the line inductance and capacitance per unit length as

$$Z(z) = \sqrt{\frac{L(z)}{C(z)}} \tag{6.4}$$

Also, if the medium surrounding the vertical transmission line is air and losses are neglected, $L(z)$ and $C(z)$ are simply related as

$$v(z) = \frac{1}{\sqrt{L(z)C(z)}} = \frac{1}{\sqrt{\mu_0 \varepsilon_0}} = c \tag{6.5}$$

where $v(z)$ is the propagation speed of the current and voltage waves along the line, $\varepsilon_0 = 8.85 \times 10^{-12}$ F/m, $\mu_0 = 4\pi \times 10^{-7}$ H/m, and c is the speed of light. The relation expressed by (6.5) is independent of the position z on the vertical line even though the inductance and capacitance of the line vary with height.

From (6.3) to (6.5), one can straightforwardly obtain the following equations expressing the per-unit-length inductance and capacitance of a vertical transmission line

$$L(z) = \frac{\mu_0}{2\pi} \ln \frac{2z}{r_a} \tag{6.6}$$

$$C(z) = \frac{2\pi\varepsilon_0}{\ln \frac{2z}{r_a}} \tag{6.7}$$

Equations (6.6) and (6.7) were used by Gorin [5] and Rakov [7] in their return stroke models. A logarithmic variation of $L(z)$ and $C(z)$ with height was also considered, under different approximations, in the models of Rondón *et al.* [31], Theethayi and Cooray [3], Visacro and De Conti [11], and De Conti *et al.* [33]. Differences in the values of $L(z)$ and $C(z)$ obtained with their models are not likely to significantly affect the currents that propagate along the vertical conductor provided a lumped current source is used to excite the channel base and corona is neglected. In this case, it can be shown that if $L(z)$ and $C(z)$ vary logarithmically with increasing height, their logarithmic variation (and implicitly the logarithmic variation of $Z(z)$) is more important for the characterization of the return stroke current than the absolute values these parameters may assume [36].

The variation of the channel inductance and capacitance with height as predicted by (6.6) and (6.7) is illustrated in Figure 6.4 for channel radii of 1 mm and 1 cm. Both values are within the limits suggested by Oetzel [20] and Rakov [7] for the leader and return stroke channels. If a channel radius of 1 mm is assumed, an increase of 44% is observed in the per-unit-length inductance if an observer moves from a point on the channel located 25 m above the ground plane to a point 3,000-m high. As expected, the per-unit-length capacitance is reduced by the same amount. In the case of a channel radius of 1 cm, the variation of the channel parameters in the considered range is larger, reaching 56%. Finally, due to the logarithmic nature of the channel parameters, a tenfold increase in the channel radius from 1 mm to 1 cm leads to a variation in $L(z)$ and $C(z)$ of less than 30%. This suggests that the radial expansion of the channel core during the leader-return stroke transition is not likely to affect the channel inductance and capacitance per unit length significantly.

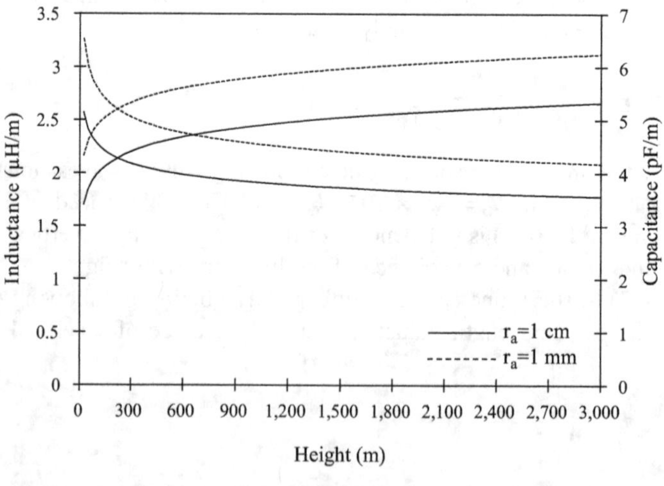

Figure 6.4 Per-unit length inductance (monotonically increasing curves) and capacitance (monotonically decreasing curves) obtained from (6.6) and (6.7) for two different radii

6.3.2 *Effect of corona on the calculation of channel parameters*

The equations presented in Section 6.3.1 for calculating the per-unit-length inductance and capacitance of a vertical conductor representing the lightning channel are derived neglecting corona. However, if the magnitude of the current propagating along the vertical conductor is such that the associated voltage exceeds a certain critical level, streamers will radiate from it. Consequently, the air in the vicinity of the conductor will be filled with either positive or negative charges depending on the polarity of the propagating pulse. The accumulation of corona charges at a given channel section can be interpreted as a dynamic increase of the capacitance of that section, while the associated corona losses can be interpreted as the effect of a dynamic conductance [41]. The equations presented in Section 6.3.1 for calculating the channel capacitance must therefore be modified to include corona. In addition, suitable equations must be proposed for representing the shunt conductance of the vertical conductor representing the lightning channel. The accumulation of corona charges is not expected to affect the channel inductance because the current propagating in the z direction is essentially confined to the channel core. For this reason, the calculation of the channel inductance as outlined in Section 6.3.1 remains unaltered if corona is considered.

Different theories have been proposed to explain the mechanisms of formation and neutralization of corona in lightning discharges. It is not the objective of this section to present a thorough account of existing corona theories, but instead to concentrate on the effect of corona on the calculation of the channel parameters, namely the capacitance and conductance per unit length. The discussion presented here is based essentially on the corona models of Gorin [5], Cooray [42], Cooray and Theethayi [41], and Maslowski and Rakov [8]. More details on corona mechanisms and corona modeling in lightning discharges, as well as corona interpretations that differ from the one presented here, can be found in [10,43–45].

6.3.2.1 Leader and return stroke channels in the presence of corona

Let one assume for the sake of simplicity that the leader channel can be represented as a coaxial line whose lateral view is shown in Figure 6.5(a). The central conductor with radius r_a represents the leader core, which is isolated from the ground by the switch S, r^- represents the external radius of the corona sheath created by the leader, which is assumed to have cylindrical symmetry, and finally an external conductor with radius $r_b > r^-$ not shown in the figure represents an effectively grounded return conductor. The distribution of negative charges in the region $r_a \leq r < r^-$ is assumed to decay inversely with the radial distance from the central core, which leads to a uniform voltage gradient E^- in the corona sheath [5,42]. The voltage gradient E^- is assumed as equal to the critical electric field necessary for the stable propagation of negative streamers, which is of the order of 1–1.5 MV/m [42].

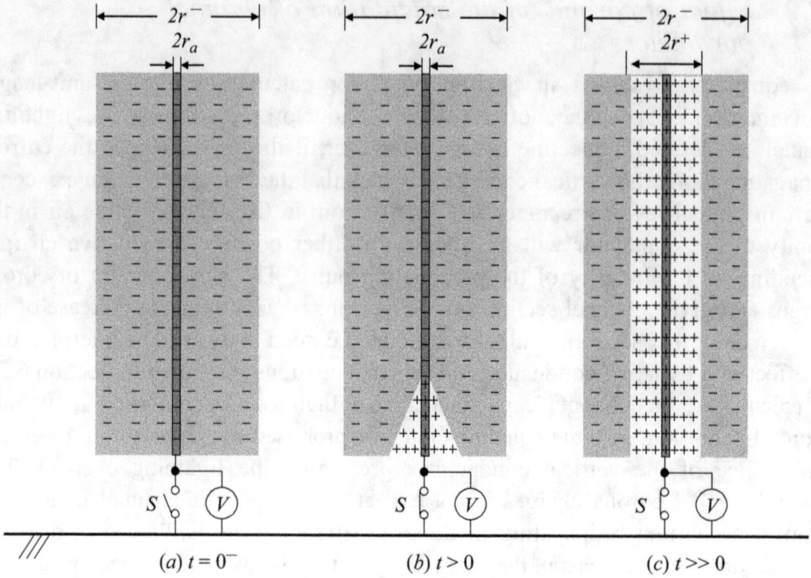

Figure 6.5 Idealization of the reverse corona mechanism in lightning assuming a coaxial structure to describe both the channel and corona geometry. The central conductor with radius r_a represents the leader/return stroke core, r^- represents the outer radius of the negative corona sheath created by the leader, and r^+ represents the outer radius of the positive corona sheath created by the return stroke. The return conductor with radius $r_b > r^-$ is not shown in the figure for clarity

The amount of negative charges deposited by the leader in a subsequent stroke must be such that the voltmeter placed across the switch S in Figure 6.5(a) will measure a negative voltage $V_0 = V(z = 0, t = 0^-)$ of the order of several MV. When the switch is closed at $t = 0$, the voltage difference between the central conductor and the return conductor will drop to zero. Let this transition be such that the voltage $V(z = 0,t)$ across the switch decays exponentially from its initial value V_0 to a final value $V_\infty = V(z = 0, t \rightarrow \infty) = 0$. Since the negative space charges deposited by the leader within r^- will not flow back instantaneously to the central conductor after the switching, a certain amount of positive charges will be induced in the central conductor in order to assure the expected voltage decay across the switch [42]. The positive charges at first cancel out the negative charges deposited by the leader in the central conductor and then start to accumulate thus forming an excess of bounded positive charges within r_a. Consequently, an inversion of polarity will be observed in the electric field at the surface of the central conductor. As soon as the positive electric field at $r = r_a$ attains a certain critical level E^+, positive streamers will radiate from it. Because of this, positive charges will start to be deposited in the region formerly occupied by the negative leader charges.

The positive streamers will propagate until they reach a radius r^+ beyond which the electric field due to the net positive charge within that region is less than E^+ [5]. With the continuous deposition of positive charges in the region formerly filled with negative charges, the voltage read by the voltmeter gradually approaches zero. The net charge per unit length also decays progressively due to the superposition of positive and negative charges until complete (or nearly complete) neutralization is attained [46]. A pictorial description of the progressive deposition of positive charges within the radius r^+ after the return stroke is initiated is illustrated in Figure 6.5(b) for times just after the closing of the switch, and in Figure 6.5(c) for longer times.

The reverse corona mechanism described above is supported by the experiments performed by Cabrera and Cooray [47] and Hermosillo and Cooray [46], who showed that, during the application of a negative voltage pulse on a cylindrical structure, streamers were initiated from the central conductor and the region between the coaxial conductors was filled with negative space charges. Then, if the central conductor was suddenly grounded, the space charge was neutralized by streamers of opposite polarity initiated from the central conductor. The obtained results formed the basis of the corona model proposed by Cooray [42]. Similar considerations were made in the development of the theoretical model of Gorin [5], whose argument was later used by Maslowski and Rakov [8] to infer corona properties using engineering return stroke models.

If the theories developed to explain the accumulation and neutralization of space charges in coaxial structures can be extended to lightning, the neutralization of the negative corona charges deposited by the leader can be thought of as the result of the propagation through the leader channel of the voltage change $\Delta V(0,t)$ created at the channel base after the attachment process is completed. For example, in order to model an exponential decay of the leader voltage V_0 at the base of a channel surrounded by negative corona charges, one can simply assume the application of an exponentially increasing voltage change $\Delta V(0,t)$ with final value $|V_0|$ at the bottom of an electrically neutral channel. When the electric field due to the voltage change $\Delta V(0,t)$ exceeds $E^\Delta = E^+ + E^-$, reverse corona will be initiated and positive corona charges will be deposited around the channel [5]. The total voltage at the channel base will then be the superposition of the effects due to $\Delta V(0,t)$ and V_0. This idea is depicted in Figure 6.6.

With the propagation of the voltage change $\Delta V(0,t)$ along the channel, the occurrence of reverse corona at each channel section will determine the deposition of a certain amount of positive charges to neutralize the negative charge previously deposited by the leader. Consequently, there will be positive corona currents leaving the channel transversely along the whole channel extension, the superposition of which will constitute both the current that propagates longitudinally along the channel and the resulting current at the channel base provided the appropriate time delays are taken into account. Alternately, if one thinks of a lumped current source injecting the desired current waveform at the bottom of an initially neutral conductor intended to represent the channel core, the resulting voltage waveform at the channel base will correspond to the voltage change

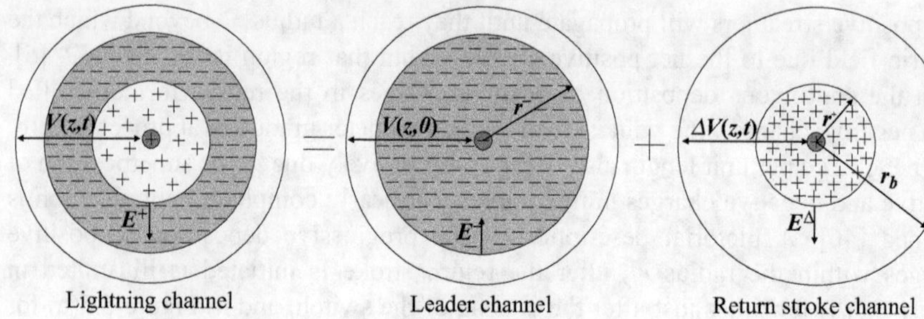

Figure 6.6 Idealization of the charge distribution in the vicinity of a channel section located at vertical coordinate z as the superposition of the negative charge deposited by the leader (initial condition at $t = 0$) with the positive charge deposited by the return stroke due the voltage change $\Delta(z,t)$ [5]

$\Delta V(0,t)$. The deposition of positive corona charges at each channel section will determine the attenuation and delay of the return stroke current as it propagates along the channel [41].

Although strictly valid for coaxial structures, an approximate account of the reverse corona currents leaving the lightning channel can be made with the model proposed by Cooray [42]. The input parameters of his model are the instantaneous voltage at a given channel section, the critical breakdown electric field obtained from Peek's formula, the inner (r_a) and outer (r_b) radius of the coaxial structure assumed to represent the lightning channel, the critical field E^+ necessary for the stable propagation of positive streamers during the reverse corona mechanism, and the critical field E^- necessary for the stable propagation of negative streamers. Once the total return stroke voltage (leader voltage plus voltage change) at a given channel section is calculated, both the amount of positive charges deposited by the reverse corona and the radius r^+ can be calculated iteratively at that channel section. The corona current can then be obtained as the time derivative of the positive corona charge. Details of such formulation can be found in [42]. However, if (a) $\Delta V(0,t)$ is much larger than the voltage necessary for the initiation and stabilization of the reverse corona mechanism, (b) $\Delta V(0,t)$ is so large that the voltage due to the charges deposited in the channel core can be neglected, and finally (c) it is assumed that $r^+ \gg r_a$, the corona model proposed by Cooray [42] reduces to the model of Gorin [5]. The resulting equations, which are given below as (6.8) and (6.9), allow the calculation of the positive charge $q^+(z,t)$ deposited at each channel section as a function of the voltage change $\Delta V(z,t)$, the critical electric field E^Δ, and the outer radius of the coaxial structure,

$$r^+\left[1 + \ln \frac{r_b}{r^+}\right]E^\Delta \approx \Delta V(z,t) \tag{6.8}$$

$$q^+(z,t) \approx 2\pi\varepsilon_0 r^+ E^\Delta \tag{6.9}$$

where the dependence of r^+ on time and on the vertical coordinate z were omitted for convenience. From (6.9), the corona current per unit length $i_c(z,t)$ can be calculated as

$$i_c(z,t) = \frac{dq^+(z,t)}{dt} \tag{6.10}$$

Equations (6.8) and (6.9) are valid for $\Delta V(z,t) \geq V^\Delta = r_a \ln(r_b/r_a)E^\Delta$, where V^Δ is the corona voltage threshold associated with E^Δ. If such condition is not met, no reverse corona will develop and r^+ should be assumed as equal to r_a. In equations (6.8) and (6.9), it is further assumed that r^+ does not contract after reaching its maximal distance from the channel core.

Examples of reverse corona currents calculated with (6.8), (6.9), and (6.10) are illustrated in Figure 6.7 for $r_a = 1$ cm, $r_b = 10$ m, $E^\Delta = 2.5$ MV/m ($E^+ = 1$ MV/m, $E^- = 1.5$ MV/m), and $\Delta V(z,t) = |V_0|[1 - \exp(-5 \times 10^6 t)]$. As explained before, V_0 can be thought of as the voltage associated with the negative charges deposited by the leader at a given channel section, while $\Delta V(z,t)$ is the voltage change necessary for bringing that channel section to ground potential. Of course, an increase in the absolute value of the leader potential is associated with a larger amount of negative charges deposited by the leader. Consequently, larger leader potentials will require a larger amount of positive charges to be deposited by the return stroke. Thus, as expected, the reverse corona currents increase with increasing $|V_0|$, which is shown in Figure 6.7 for values of $|V_0|$ ranging from 5 to 10 MV. An increase in the outer radius r_b, on the other hand, will lead to a decrease in the corona currents. This is illustrated in Figure 6.8 for $|V_0| = 10$ MV and three different values of r_b.

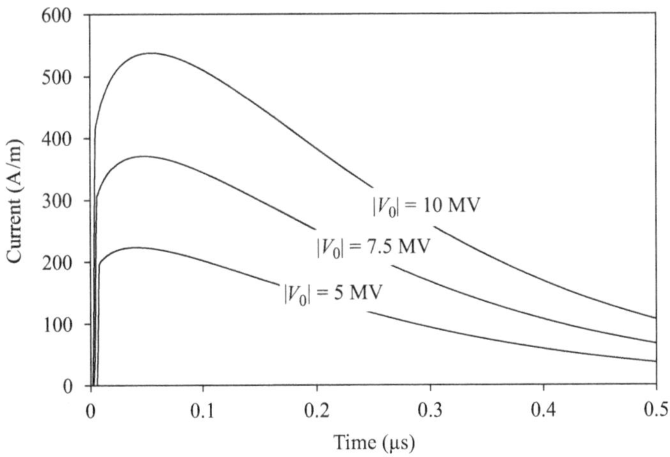

Figure 6.7 Reverse corona currents calculated with (6.8), (6.9), and (6.10) for a coaxial structure with $r_a = 1$ cm, $r_b = 10$ m, $E^\Delta = 2.5$ MV/m, and an applied voltage change $\Delta V(z,t) = |V_0|[1 - \exp(-5 \times 10^6 t)]$, considering different values of V_0

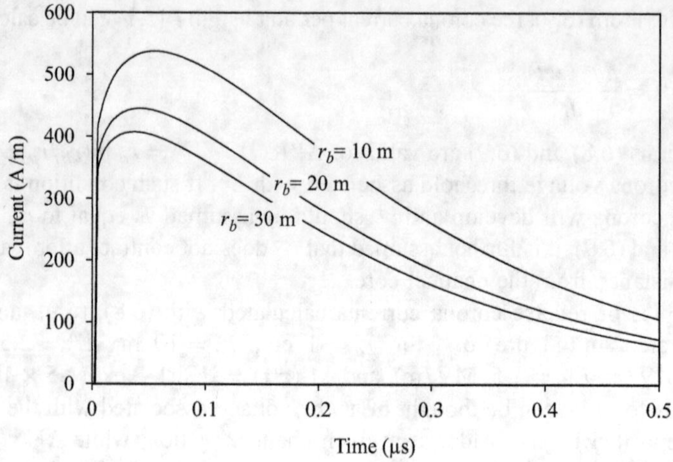

Figure 6.8 Same as Figure 6.7 but considering $|V_0| = 10$ MV and assuming different values for r_b

Since (6.8) and (6.9) are in principle valid only for coaxial structures, its extension for calculating corona currents in a vertical conductor requires further approximations and therefore must be viewed with caution. One of the possibilities is to assume that the radius r_b of the return conductor pertaining to the coaxial structure representing the channel increases with increasing height according to some law, as suggested by Gorin [5], Cooray and Theethayi [41], and De Conti *et al.* [19].

6.3.2.2 Transmission line equations in the presence of corona

In order to analyze the effect of corona on the calculation of the capacitance and conductance of the return-stroke channel, it is useful to rewrite telegrapher's equations (6.1) and (6.2) as

$$\frac{\partial V(z,t)}{\partial z} + RI(z,t) + L(z)\frac{\partial I(z,t)}{\partial t} = 0 \qquad (6.11)$$

$$\frac{\partial I(z,t)}{\partial z} + G(z,t)V(z,t) + C(z,t)\frac{\partial V(z,t)}{\partial t} = 0 \qquad (6.12)$$

Equation (6.11) is identical to (6.1), except that now the dependence of the inductance on position is explicitly shown. This is necessary because the inductance of the vertical conductor intended to represent the lightning channel increases with increasing height. Equation (6.12), on the other hand, differs from (6.2) by the consideration of per-unit-length parameters that present variation with both time and position. This modification is necessary for including the effect of corona charges on voltage and current pulses that propagate along the vertical conductor

[41]. If the voltage at point z is below the corona voltage threshold, only the variation of the channel capacitance with height is needed in (6.12). In this case, $C(z,t)$ reduces to $C(z)$, with $G(z,t)$ becoming null. If the voltage pulse exceeds the corona inception voltage V^{Δ}, corona charges will be deposited around the vertical conductor, determining a nonlinear increase in the capacitance, which will now be given by

$$C(z,t) = C(z) + C_{dyn}(z,t) \tag{6.13}$$

where $C_{dyn}(z,t)$ is the dynamic capacitance introduced by the corona charges. For reverse corona of positive polarity, which is the case for return strokes in negative lightning, $C_{dyn}(z,t)$ is given by

$$C_{dyn}(z,t) = \begin{cases} 0, & \text{if } V(z,t) < V^{\Delta} \\ \dfrac{q^{+}(z,t)}{V(z,t)}, & \text{if } V(z,t) \geq V^{\Delta} \text{ and } \dfrac{\partial V(z,t)}{\partial t} > 0 \\ 0, & \text{if } V(z,t) \geq V^{\Delta} \text{ and } \dfrac{\partial V(z,t)}{\partial t} \leq 0 \end{cases} \tag{6.14}$$

Also, it can be shown that

$$G_{dyn}(z,t) = \frac{\partial C_{dyn}(z,t)}{\partial t} \tag{6.15}$$

The derivation of (6.11)–(6.15) is presented in detail in [41]. In the same reference it is shown that (6.12) can be equivalently written as

$$\frac{\partial I(z,t)}{\partial z} + C(z)\frac{\partial V(z,t)}{\partial t} = -i_c(z,t) \tag{6.16}$$

where $i_c(z,t)$ is the corona current given by (6.10) and $C(z)$ is the capacitance of the vertical conductor in the absence of corona, considering $C_{dyn}(z,t) = 0$. The pair of (6.11) and (6.16) describe a nonuniform transmission line embedded in air in which distributed current sources inject corona currents. If the series resistance is neglected in (6.11), the speed of the propagating voltage and current pulses will still be given by (6.5). In other words, current pulses generated by the corona sources will travel along the line with the speed of light. The superposition of the voltage and current pulses due to the distributed corona sources, however, will lead to waveforms that appear to travel slower than the speed of light [41].

The modification of the channel capacitance $C(z,t) = C(z) + C_{dyn}(z,t)$ given by (6.13) and (6.14) due to the corona currents shown in Figure 6.7 can be seen in Figure 6.9. For the assumed values of $r_a = 1$ cm and $r_b = 10$ m, the geometric capacitance $C(z)$ can be calculated as $2\pi\varepsilon_0/\ln(r_b/r_a) = 8.0$ pF. As shown in the figure, after corona inception the total capacitance rises monotonically until reaching a final value that depends on the amplitude of the applied voltage change. In the illustrated cases, the final values of $C(z,t)$ are 2.5–3.3 times larger than $C(z)$.

The variation of the per-unit-length conductance due to the corona currents shown in Figure 6.7 is illustrated in Figure 6.10. As expected from (6.15), the conductance presents an initial rise that is related to the initial expansion of the corona sheath and the associated increase in the channel capacitance. After

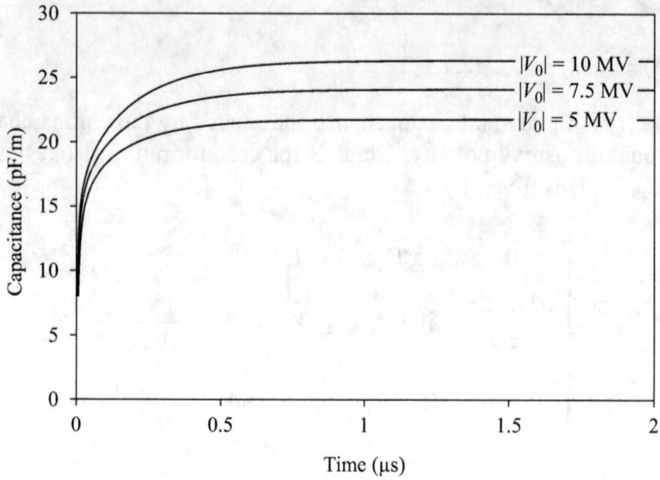

Figure 6.9 Variation of the total capacitance $C(z,t) = C(z) + C_{dyn}(z,t)$ of a coaxial structure with $r_a = 1$ cm and $r_b = 10$ m associated with the corona currents shown in Figure 6.7

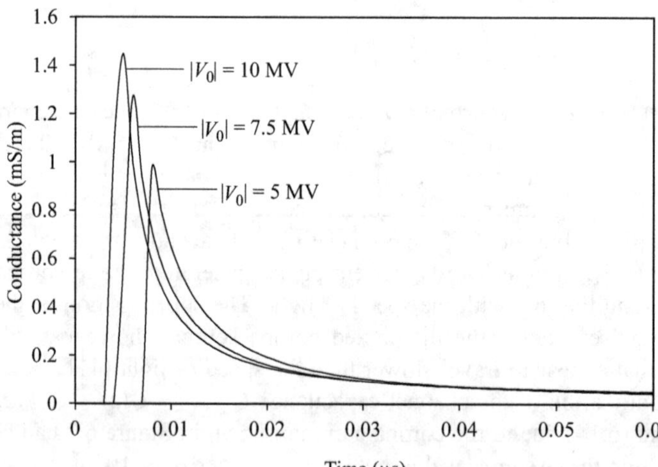

Figure 6.10 Variation of the per-unit-length conductance with time in a coaxial structure with $r_a = 1$ cm and $r_b = 10$ m associated with the corona currents shown in Figure 6.7

reaching a peak, it decays monotonically to zero as the corona sheath reduces its expansion rate. Both the amplitude and the inception time of the conductance curves are dependent on the value of V_0.

If all uncertainties involved in the modeling of the return stroke as a transmission line are taken into consideration, a simplified account of corona could possibly consider only the dynamic capacitance. One should keep in mind, however, that if one neglects the contribution of the dynamic conductance in (6.12) the associated corona currents will lose the fast rise observed in the waveforms shown in Figure 6.7 and reach different peak values. Also, there will be no losses due to charge leak. This will downplay the role of corona in the calculation of voltage and current pulses propagating along the line, and therefore should be viewed with caution.

6.3.3 Calculation of the channel resistance

The calculation of the channel resistance requires that account be made of physical phenomena occurring in the transition from the leader to the return-stroke phase of lightning. In the case of a subsequent stroke, the channel temperature rises rapidly from about 20,000 K to 30,000 K or more as the leader channel is gradually brought to ground potential, with the channel pressure rising to values of several atmospheres. As the channel pressure exceeds the pressure of the surrounding ambient, a shock wave is created, and the channel core expands until its pressure reduces to the atmospheric pressure. The channel expansion takes place initially at a faster rate due to the significant channel overpressure. Then, as a state of pressure equilibrium is gradually reached in tens of microseconds after the return stroke initiation and the channel temperature cools down to about 20,000 K or below, the channel expansion becomes gradually slower. The variation of channel temperature and pressure also leads to variations of electron and heavy particle densities in the channel core. Consequently, the channel conductivity is expected to change in the various stages of the return stroke. This variation ranges from about $1 \times 10^4 - 5 \times 10^4$ S/m in the initial phase of channel expansion (the strong-shock region) to values below 10^4 S/m in the final stage of channel expansion (the weak-shock region), when the channel is cooled down to temperatures below 15,000 K [7,48–50].

If the lightning channel is represented as a vertical transmission line, the gradual expansion of the channel core and the continuous change of its physical properties result in the time variation of the per-unit-length channel resistance. To some extent, the inductance and capacitance per unit length are also affected by the radial expansion of the core, so (6.11) and (6.12) should be rewritten as

$$\frac{\partial V(z,t)}{\partial z} + R(z,t)I(z,t) + L(z,t)\frac{\partial I(z,t)}{\partial t} = 0 \tag{6.17}$$

$$\frac{\partial I(z,t)}{\partial z} + G(z,t)V(z,t) + C(z,t)\frac{\partial V(z,t)}{\partial t} = 0 \tag{6.18}$$

where the time dependence of $R(z,t)$ and $L(z,t)$ is now explicitly shown (neglecting skin effect), and $C(z,t)$ is supposed to include not only the dynamic nature of corona but also the channel core expansion. The channel resistance is also allowed to vary with position in (6.17) in order to accommodate not only the local variation of this parameter during the return stroke but also a presumed increase of its initial value at the upper part of the leader channel due to channel cooling.

In practice, the channel inductance and capacitance are much less affected by the expansion of the central core than the channel resistance due to their logarithmic dependence on the core radius. It is therefore reasonable to assume that the channel core expansion and the continuous change of its physical properties will be significant only for the resistance calculation. This is confirmed by numerical results presented in Section 6.4.1.2.

A rigorous procedure for determining $R(z,t)$ requires the solution of three hydrodynamic equations representing the conservation of mass, momentum and energy applied to a short cylindrical channel segment, with the assumed lightning current as input parameter [2]. Examples of such hydrodynamic (or gas-dynamic) models are the models of Braginskii [9], Plooster [48,49], Paxton [50], Ripoll [51,52], and da Silva [53], which usually have as output parameters the channel temperature, pressure, energy, and electrical conductivity as a function of radial coordinate and time. The models of Paxton [50], Plooster [48,49], Ripoll [51,52], and da Silva [53] are perhaps the most complete among the gas-dynamic models, but their extension for use in transmission line models of the return stroke is not straightforward because the simultaneous solution of hydrodynamic and voltage-current propagation equations is needed [25,26,51,52]. Given the difficulties imposed by a rigorous solution, approximations are usually required for representing the nonlinear channel resistance. A detailed analysis of different approximations suitable for representing the nonlinear (time-varying) nature of the channel resistance is presented in [33]. A brief account of two such approximations is presented in Sections 6.3.3.1 and 6.3.3.2, which deal, respectively, with the strong-shock approximation proposed by Braginskii [9] for representing the radial expansion of spark channels and the arc resistance equation of Toepler [54].

6.3.3.1 Strong-shock approximation

The strong-shock approximation consists in solving the hydrodynamic equations describing the radial expansion of a spark channel by considering the pressure in the channel to be much higher than the pressure in the surrounding ambient. According to the strong-shock approximation, three radial regions can be defined [12]: (i) a central core containing the hot, low-density plasma, (ii) an outward expanding shell with negligible thickness containing the mass excluded from the channel, and (iii) an undisturbed region located beyond the shell radius. By assuming uniform temperature, pressure, electrical conductivity and mass density in the central core, Braginskii [9] used the strong-shock approximation to obtain a self-similar solution describing the dependence of the channel radius on the current traversing a short channel segment. His solution in integral form can be written as [55]

$$r_a(z,t) \approx \sqrt{\left(\frac{4}{\rho_0 \pi^2}\right)^{1/3} \int_0^t [\sigma(z)\xi]^{-1/3} i(z,t)^{2/3}\, dt + r_i(z)^2} \tag{6.19}$$

where $r_a(z,t)$ is the arc channel radius, in meters, $i(z,t)$ is the current, in amperes, $r_i(z)$ is the initial radius, $\sigma(z)$ is the channel conductivity, in S/m, $\rho_0 = 1.29\ \mathrm{kg/m^3}$ is the ambient atmospheric density, and ξ is a factor describing the rate of radial expansion of the arc channel. Although the strong-shock approximation is strictly valid during the initial phase of channel expansion, De Conti *et al.* [33] have shown that channel radii predicted by (6.19) compares well with predictions of the more accurate hydrodynamic model of Plooster [49] even in the weak-shock regime. Assuming that (6.19) is also valid for representing the radial expansion of the lightning channel and neglecting skin effect, the per-unit-length channel resistance can be obtained from

$$R(z,t) = \frac{1}{\sigma(z)\pi r_a(z,t)^2} \tag{6.20}$$

It is to be noted that a constant channel conductivity is assumed in (6.19) and (6.20), which is reasonable in the initial stage of the return stroke since this parameter is not expected to vary significantly for plasma temperatures above 15,000 K and channel pressures ranging from 1 to 100 atm [48]. However, this assumption introduces an error in (6.20) by specifying a monotonic resistance decay that is not valid at later times. This is explained as follows. If the channel conductivity is allowed to vary with time in (6.19) and (6.20), the overall variation of $R(z,t)$ with time can be roughly separated in three different stages: (i) in the strong-shock regime, the fast expansion of the channel core and the relative increase of the channel conductivity leads to a sharp resistance decay from an initial value ranging from 1 to 50 Ω/m associated with the characteristics of the leader channel (core radius of few millimeters and conductivity of about 10^4 S/m) to a value about two orders of magnitude lower [7,20]; (ii) as the channel core continues to expand and the core conductivity presents a relative decay due to gradual channel cooling, the resistance tends to reduce its decay rate, eventually reaching some stability because conductivity and radius will have opposite effects in (6.20); (iii) in the final stage of the return stroke, as the channel expands at reduced rate and the channel conductivity decays more significantly due to channel cooling, the counteracting effect of both parameters in (6.20) will eventually lead to a relative increase of the channel resistance [33]. This is exactly the behavior observed in experiments with spark channels [56]. Since the relative increase of the channel resistance cannot be predicted if a constant channel conductivity is assumed, (6.20) will inevitably underestimate the channel resistance in the weak-shock regime at the current tail.

6.3.3.2 Toepler's arc-resistance equation

Toepler's [54] empirical formula is an arc-resistance equation that assumes the per-unit-length resistance of the spark to be inversely proportional to the time integral of the input current according to

$$R(z,t) = \frac{\alpha}{q(0) + \int_0^t i(t)\,dt} \qquad (6.21)$$

where α is an empirical constant and $q(0)$ is the initial charge in the spark channel, which is equal to zero prior to arc initiation. The validity of (6.21) for simulating the nonlinear resistance of the lightning channel is, in principle, limited. However, it seems reasonable, at least in an engineering perspective, to assume that the leader channel behaves similarly as a long stationary arc that is subjected to a sudden expansion during the transition to the return stroke. Since the leader channel is believed to conduct currents of the order of 10^2 A [7], $q(0)$ can be reasonably assumed as different from zero at the time when the return stroke is initiated. In such perspective, if $q(0)$ is known, α can be suitably selected for adjusting the initial channel resistance. However, similarly as with the strong-shock approximation, (6.21) leads to a monotonic resistance decay. For this reason, the resistance values predicted by (6.21) in the final stage of the return stroke are probably underestimated.

6.3.3.3 Computation of the channel resistance

For illustrating the use of the strong-shock approximation and Toepler's equation in the modeling of the nonlinear channel resistance, the channel-base current proposed by Nucci *et al.* [57] was assumed as the input current in (6.19) and (6.21). It has a peak value of 11 kA, a front time of about 0.5 μs, and a maximum steepness of 105 kA/μs. The obtained results are shown in Figure 6.11, with the solid line corresponding to the resistance calculated with the strong-shock approximation and the dashed lines corresponding to the resistances calculated with Toepler's equation for three different sets of parameters. In the simulation with the strong-shock approximation, $\sigma(z) = 2.2 \times 10^4$ S/m, $\xi = 4.5$, and $r_i(z) = 1.7$ mm were considered in (6.19) and (6.20). The values assumed for $\sigma(z)$ and ξ are the same used by Braginskii [9]. The value assumed for $r_i(z)$ is within the limits suggested by Rakov [7] for the leader channel. The combination of $\sigma(z) = 2.2 \times 10^4$ S/m and $r_i(z) = 1.7$ mm in (6.20) leads to an initial channel resistance of 5 Ω/m. The parameters α and $q(0)$ used in Toepler's equation were selected in order to match this initial resistance value and to produce different decay rates for this parameter [58].

Figure 6.11 shows that the channel resistance initially presents a faster decay, which is associated with the rise of the input current, and then starts to decay more slowly. As expected, an increase in $q(0)$ determines a slower decay rate for the channel resistance predicted by Toepler's equation. This happens because in this case the integral term plays a role that is comparatively less important in the denominator of (6.21). After 40 μs (not shown in Figure 6.11), the channel

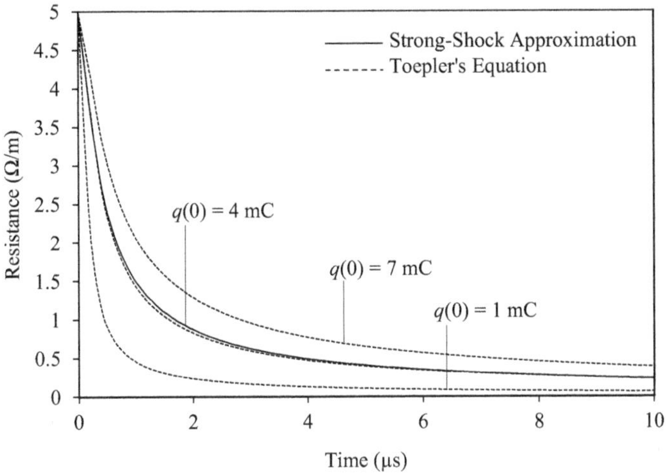

Figure 6.11 *Channel resistance predicted by (6.19) and (6.20) with* $\xi = 4.5$, $\sigma(z) = 2.2 \times 10^4$ *S/m and* $r_i(z) = 1.7$ *mm (strong-shock approximation), and by (6.21) with* $q(0) = 1$, *4 or 7 mC and* $\alpha = q(0)R_i$ *(Toepler's law). The current waveform of Nucci* et al. *[57] was assumed as input current.* R_i *is the initial resistance [58]*

resistance reaches either 0.07 Ω/m (strong-shock approximation and Toepler's equation with $q(0) = 4$ mC), 0.02 Ω/m ($q(0) = 1$ mC) or 0.13 Ω/m ($q(0) = 7$ mC). The value of 0.07 Ωm obtained with the strong-shock approximation after 40 μs corresponds to a channel core radius of 1.44 cm. Such values of channel radius and channel resistance are consistent with the predictions of Rakov [7] for the return-stroke channel.

It is apparent from Figure 6.11 that the resistance decay predicted by Toepler's equation agrees well with that obtained with the strong-shock approximation if $q(0) = 4$ mC. Indeed, $q(0)$ and α can always be adjusted to provide a good match with resistance curves obtained with the strong-shock approximation [33]. This suggests that the theories presented in Sections 6.3.3.1 and 6.3.3.2 can be equivalently used in the modeling of the channel resistance. Consistent results can be obtained with the strong-shock approximation provided initial channel radii of the order of few millimeters, channel conductivities of the order of 10^4 S/m, and ξ ranging from 1 to 10 are assumed [33,58,59]. The selection of the constants to be used in Toepler's equation can be made after a preliminary analysis with (6.19) and (6.20), or simply by selecting a desired initial channel resistance and a certain decay rate. In any case, regardless of the adopted calculation procedure, one should bear in mind that (6.19)–(6.21) have limitations especially in the weak-shock regime, and that a more accurate description of the nonlinear behavior of the channel resistance requires the simultaneous solution of hydrodynamic and voltage–current equations.

6.4 Computed results

In this section, the return-stroke model outlined in Section 6.3 is used to simulate a subsequent stroke to flat ground. First, the influence of various channel parameters on the return stroke current is investigated in Section 6.4.1. The analysis is presented in such way as to indicate how channel nonuniformity, losses, and corona let the characteristics of the calculated currents closer to observed lightning properties. A suitable combination of channel parameters is then selected for the calculation of lightning electromagnetic fields. The results are presented in Section 6.4.2.

6.4.1 Channel currents

The calculation of channel currents with the return stroke model described in Section 6.3 requires the numerical solution of telegrapher's equations (6.17) and (6.18) in time domain. The numerical results presented in this section were obtained with a first-order Finite-Difference Time-Domain (FDTD) scheme [1] assuming an ideal lumped current source to inject the current waveform of Nucci *et al.* [57] at the bottom of the channel.

6.4.1.1 Lossless nonuniform channel

To evaluate the modeling of the return stroke as a transmission line according to the discussion presented in Section 6.3, it is initially assumed that the channel is lossless and corona free. The obtained results are illustrated in Figure 6.12 in terms of channel currents calculated at heights of 0, 100, 300, 600, and 900 m above the

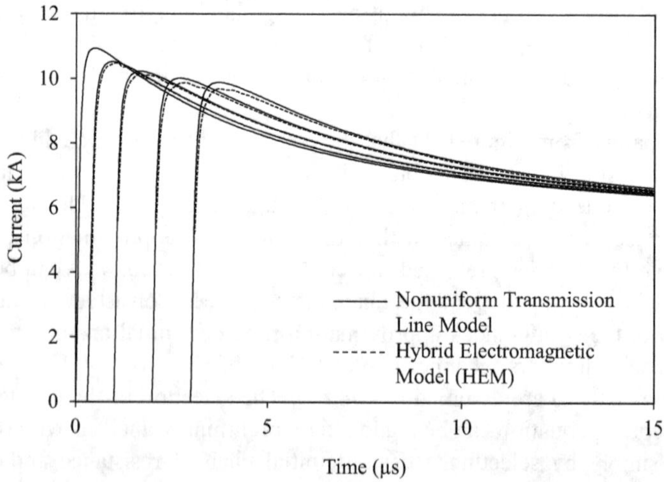

Figure 6.12 Currents calculated at heights of 0, 100, 300, 600, and 900 m (waveforms from left to right in the figure) with two different models considering an ideal current source to excite the bottom of a lossless channel with core radius of 1 mm in the absence of corona

ground surface. A radius $r_a = 1$ mm was considered for the calculation of the channel inductance and capacitance with (6.6) and (6.7). As discussed in Section 6.3.1, this leads to the representation of the lightning channel as a nonuniform transmission line with inductance and capacitance varying with the vertical coordinate z. For comparison purposes, channel currents were also calculated with the Hybrid Electromagnetic Model (HEM) [60–62] assuming the channel to be represented as an infinitely long vertical monopole excited by an ideal current source at its lower end. HEM solves Maxwell's equations numerically in the frequency domain by formulating the problem in terms of the vector magnetic and scalar electric potentials, with time domain solutions being obtained through the inverse Fourier Transform.

The waveforms illustrated in Figure 6.12 clearly indicate that the injected current propagates with the speed of light along the channel. This is expected because the medium surrounding the lossless vertical wire representing the channel is air. It is also seen in Figure 6.12 that the propagating current is attenuated and distorted even in the absence of losses and corona. This is explained by the non-Transverse Electromagnetic (TEM) field structure associated with the propagation of current waves along a vertical wire [63]. In HEM, the non-TEM field structure is implicitly considered in the numerical solution of Maxwell's equations. In the transmission line model of the channel, however, the attenuation and distortion of the propagating current are a result of nonuniform channel parameters, which determine the occurrence of multiple current reflections at each channel section [3,63]. The good agreement seen in Figure 6.12 between currents calculated with the transmission line model of the vertical wire and currents calculated with the numerical solution of Maxwell's equations suggests that the representation of the lightning channel as a nonuniform transmission line is able to emulate satisfactorily the non-TEM field structure associated with this problem. This result is in line with the analysis of Baba and Rakov [63], who studied this problem in detail.

Interestingly, a lossless vertical monopole with infinitesimally small radius excited by a point current source can be shown to support a spherical TEM field structure, which means that the injected current will in this case propagate unattenuated and undistorted with the speed of light [64]. This somehow provides a physical justification for the popular TL model [30], which assumes the lightning current to propagate unattenuated and undistorted along a uniform transmission line. The hypothesis of an undisturbed lightning current is however not supported by measured lightning luminosity profiles, which indicate that the return stroke current should reduce its amplitude and increase its rise time while propagating up to the cloud [28]. This suggests that the representation of the lightning channel as a nonuniform transmission line and its effect on the channel currents as seen in Figure 6.12 change the model predictions in such way that they will come closer to the experimental observations, at least to some extent.

6.4.1.2 Lossy nonuniform channel

The variation of the channel inductance and capacitance with the vertical coordinate z is important for the modeling of the lightning channel as a transmission line, but the results presented in Figure 6.12 were obtained for a lossless channel, which is unrealistic. In this section, this simplifying assumption is removed, and the return stroke current is calculated for a lossy nonuniform channel. The channel resistance is assumed to be governed by the strong-shock approximation described in Section 6.3.3.1, with $\sigma(z) = 2.2 \times 10^4$ S/m, $r_i(z) = 1$ mm, and $\xi = 4.5$. This set of parameters leads to an initial channel resistance of 14.5 Ω/m that is representative of the pre-return-stroke channel [33]. At the channel base, the resistance decays from its initial value to 1.88, 0.43, and 0.13 Ω/m after 1, 5, and 20 μs of the return stroke initiation, while the core radius expands from 1 mm to 1.4 cm in 40 μs. Resulting channel currents calculated at heights of 0, 100, 300, 600, and 900 m above ground are shown in Figure 6.13. The dashed lines were obtained considering the radial core expansion predicted by (6.19) to affect the inductance and capacitance, whereas the solid lines were obtained neglecting this effect and considering $r_a = 1$ mm in (6.6) and (6.7).

It is apparent from Figure 6.13 that, compared to the lossless case illustrated in Figure 6.12, the consideration of a nonlinear channel resistance increases the attenuation and distortion of the propagating current. This feature, which is consistent with measured lightning luminosity profiles [28,35], is seen in Figure 6.13 to remain nearly unaffected if inductance and capacitance are allowed to vary with

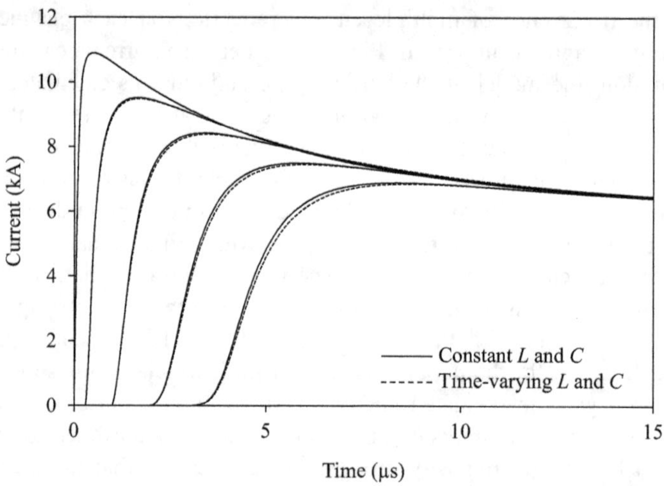

Figure 6.13 Currents calculated at heights of 0, 100, 300, 600, and 900 m (waveforms from left to right in the figure) considering an ideal current source to excite the bottom of a lossy channel with $r_i = 1$ mm, $\sigma(z) = 2.2 \times 10^4$ S/m, and $\xi = 4.5$ in the absence of corona. The channel inductance and capacitance were assumed to either change their values following the radial expansion of the channel core or to keep a constant value calculated with $r_a = 1$ mm in (6.6) and (6.7)

time due to channel core expansion. This confirms the discussion of Section 6.3.3 that the core expansion is likely to be significant only for the resistance calculation, with the assumption of a constant core radius being acceptable for the calculation of the channel inductance and capacitance.

It can also be inferred from the current waveforms shown in Figure 6.13 that the consideration of a nonlinear channel resistance leads to a reduction of the return stroke speed to values below the speed of light and also to a monotonic decay of this parameter with increasing height, both features supported by experimental data [35,65]. One possible approach to evaluate the return stroke speed is the use of the apparent speed $v(z)$, defined as the ratio between the distance z propagated by the return stroke current and the time required for this current to reach a certain threshold level that can be related to the minimum energy necessary to excite an optical instrument used in measuring lightning luminosity profiles [3]. If the threshold level is assumed to be 2.5% of the current peak at the channel base, a monotonic decay of $v(z)$ from about 2.9×10^8 m/s near the channel base to 2.0×10^8 m/s at a height of 3,000 m is obtained for the set of parameters considered in Figure 6.13. If the channel conductivity is reduced from 2.2×10^4 S/m to 1.0×10^4 S/m, $v(z)$ will vary from 2.6×10^8 m/s near the channel base to 1.5×10^8 m/s at a height of 3,000 m [33]. These values approach the estimates of Mach and Rust [35] for the return stroke speed, in which upper limits of the order of 2.8×10^8 m/s and 2.0×10^8 m/s were respectively obtained for natural and triggered negative lightning.

Overall, the results presented in this section show that currents calculated with the lightning channel represented as a lossy nonuniform transmission line (considering the nonlinear variation of the channel resistance) behave closer to measured data than if channel losses are neglected.

6.4.1.3 Lossy nonuniform channel with corona

In this section, in order to include more of lightning physics into the lossy nonuniform transmission line model used so far, the corona model of Section 6.3.2 is assumed to govern the radial diffusion of electric charges from the central core. The outward expansion of the corona radius and the deposited corona charge are calculated with (6.8) and (6.9) assuming $r_b = 2z$ and $E^{\Delta} = 2$ MV/m as in [5]. The dynamic capacitance and conductance associated with the corona currents are obtained from (6.14) and (6.15), respectively. A core radius of 1 cm is used for calculating V^{Δ}, the corona voltage threshold associated with E^{Δ}. The same radius is used for calculating the channel inductance and geometrical capacitance. The obtained results are shown in Figure 6.14 in terms of currents calculated at heights of 0, 100, 300, 600, and 900 m above ground level. The solid lines shown in the figure were calculated assuming a lossless channel. The dashed lines were obtained assuming the channel resistance to be governed by the strong-shock approximation with the same parameters of Section 6.4.1.2. To avoid a premature resistance decay, a current threshold $I^{\Delta} = V^{\Delta}/Z(z)$ was defined at each channel section below which a constant resistance is assumed, where $Z(z)$ is given by (6.3).

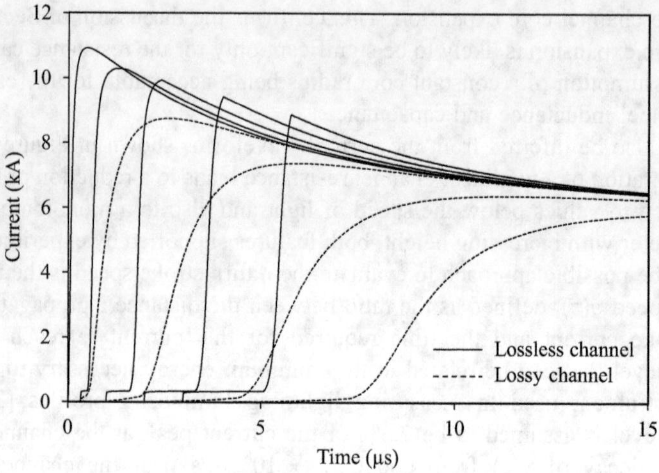

Figure 6.14 Currents calculated at heights of 0, 100, 300, 600, and 900 m (waveforms from left to right in the figure) considering an ideal current source to excite the bottom of the channel in the presence of corona. The solid lines were obtained for a lossless channel. The dashed lines were calculated assuming the channel resistance to decay according to the strong-shock approximation, with r_i = 1 mm, $\sigma(z) = 2.2 \times 10^4$ S/m, and $\xi = 4.5$

Figure 6.14 shows that currents below the corona threshold propagate unaffected by corona. This is clearer in the lossless case, in which the leading tip of the current waveforms is well defined and propagates with the speed of light. In the lossy case, the leading tip is severely attenuated by the channel resistance, resulting in slower propagation and in a concave profile in the rising part of the return stroke current.

Above the corona threshold, the apparent speed of the upward moving front is significantly reduced due to the superimposition of the injected current with the forward- and backward-moving corona currents, all of which propagating with the speed of light [41]. In the lossless case, for example, an average value of 1.9×10^8 m/s can be inferred for $v(z)$ in the first kilometer or so of the channel if the lower edge of the slowly moving part of the current is taken as reference. In the lossy case, as a result of attenuation and distortion, the apparent propagation speed decays monotonically from 1.8×10^8 m/s near the channel base to 1.5×10^8 m/s, 1.3×10^8 m/s, and 1.2×10^8 m/s at heights of 300, 600, and 900 m if a threshold of 2.5% of the peak current at the channel base is considered. These values are in the range of the average propagation speed of negative return strokes in natural and triggered lightning, which is typically between one-third and one-half of the speed of light [37]. The monotonic decrease of the return stroke speed with increasing height inferred from the calculated current waveforms is also supported by measured data (e.g., [65]), although there is experimental evidence that the return stroke speed may vary non-monotonically along the lightning channel, initially increasing in the first several tens of meters of the channel and then decreasing with increasing height [37]. Overall, it is seen that the inclusion of corona let the model predictions closer to experimental observations.

6.4.2 Predicted electromagnetic fields

The results presented in Section 6.4.1.3 show that the modeling of the lightning channel as a lossy nonuniform transmission line in the presence of corona is able to reproduce the most important characteristics of subsequent return-stroke currents in negative lightning. However, a more complete assessment of the validity of a return stroke model requires the computation of remote electromagnetic fields and the comparison of predicted field waveforms with signatures observed in measured lightning electromagnetic fields. This task is performed here by assuming the lightning channel to consist of an array of short vertical dipoles excited by the currents calculated at each channel section with the considered transmission line model. The remote electromagnetic fields are then calculated in time domain using the formulation of Uman *et al.* [66]. The results are illustrated in Figure 6.15, which shows electromagnetic fields calculated at distances of 50 m, 5 km, and 100 km from the channel base. In the simulations, the lightning channel was modeled as a lossy nonuniform transmission line in the presence of corona. The same parameters of Section 6.4.1.3 were considered, except that a lower limit of 0.145 Ω/m was assumed for the channel resistance. This value is exactly two orders of magnitude lower than the initial resistance of 14.5 Ω/m associated with the leader channel in the simulated case, which is in line with the discussion presented in [7]. The use of

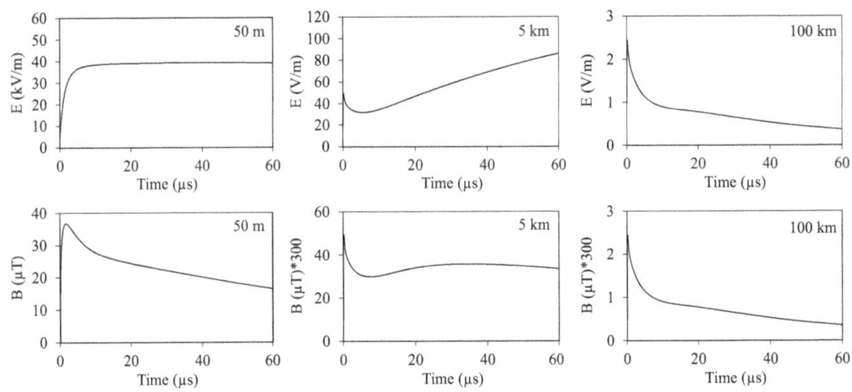

Figure 6.15 Remote electric and magnetic fields associated with the currents shown as dashed lines in Figure 6.14. The channel resistance is assumed to decay according to the strong-shock approximation, with r_i = 1 mm, $\sigma(z) = 2.2 \times 10^4$ S/m, and $\xi = 4.5$, with a final resistance value of 0.0145 Ω/m. The channel is modeled considering corona, with r_a = 1 cm, $r_b = 2z$, and E^{Δ} = 2 MV/m

such a limiting value was necessary for avoiding a monotonic decay of the channel resistance, which is unrealistic as discussed in Section 6.3.3, and for leading to more realistic electric field waveforms in the very close range at later times. Despite the use of a lower limit for the channel resistance, the resulting channel currents are essentially the same as those illustrated as dashed lines in Figure 6.14.

For analyzing the field waveforms illustrated in Figure 6.15, it is useful to resort to Table 6.1, which summarizes the signatures typically observed in measured lightning electromagnetic fields [2,67]. The ability of the evaluated model to reproduce each of the listed signatures is also shown in the table. It is seen that most of the field signatures are reproduced by the model, namely a sharp initial peak in electric and magnetic fields at 5 km and 100 km, a slow ramp after the initial peak in electric fields at 5 km, a hump after the initial peak in magnetic fields at 5 km, and the flattening of electric fields measured at 50 m. The only feature that is not predicted by the evaluated model is the zero-crossing usually observed at the tail of remote field measurements. In [11], it is argued that an increase of the initial channel resistance with increasing height due to the cooling of the upper part of the leader channel could be one of the factors determining the zero-crossing. This could be easily included in the evaluated model by assuming non-uniformly distributed values for both channel conductivity and radius. Another possible reason for the lack of zero crossing in the far field waveforms is that the strong-shock approximation used in the evaluated model is not able to accurately predict the dynamics of the channel resistance in the weak-shock regime, for which a more complete hydrodynamic model would be necessary.

In any case, the lack of zero crossing in the far field region is not a problem if one is primarily interested in using the return stroke model discussed in this chapter for evaluating the interaction of lightning with nearby electrical systems (e.g., calculating lightning-induced voltages on overhead lines [39,68]). This conclusion is supported by the fact that the considered model satisfactorily reproduces all signatures observed in lightning electromagnetic fields measured at distances below 5 km. This conclusion remains essentially unaffected if a more accurate resistance model is used, since modifications in the final value of channel resistance are unlikely to affect the peak values and shape of predicted field waveforms in the first tens of microseconds, which are of more importance is this type of analysis.

Table 6.1 Main signatures observed in measured lightning electromagnetic fields

	Field signature	**Model prediction**
(i)	Sharp initial peak at 5 km and 100 km	Yes
(ii)	Slow ramp after the initial peak in electric fields at 5 km	Yes
(iii)	Hump after the initial peak in magnetic fields at 5 km	Yes
(iv)	Zero crossing in the tail of field waveforms at 100 km	No
(v)	Flattening of electric fields measured at 50 m	Yes

6.5 Summary and conclusion

This chapter discusses the modeling of the return stroke channel as a transmission line. A brief review of the pertinent literature indicates that the transmission line models of the return stroke can be roughly classified as either discharge-type models, which assume the return stroke to correspond to the discharge of a previously charged transmission line to ground through a closing switch, lumped-excitation models, which assume the return stroke to correspond to an initially neutral transmission line that is fed by a lumped voltage or current source at one of its terminations, or finally as models that use transmission-line or distributed-circuit theory to infer relevant lightning properties.

A discussion is presented on the calculation of the per-unit-length parameters necessary for simulating the return stroke channel as a transmission line. It is shown that if a transmission line is intended to represent the lightning channel, it must be nonuniform (in order to accommodate the variation of the channel parameters with position) and nonlinear (in order to accommodate the temporal variation of the channel resistance as a function of the channel current and the gradual neutralization of the corona sheath surrounding the channel core). Engineering equations are presented for estimating the per-unit-length parameters associated with a vertical transmission line in the presence of nonlinear losses and corona. The proposed equations are suitable to computer simulation and can be easily implemented using a first-order FDTD scheme.

Computed results show that the modeling of the lightning channel as a lossy nonuniform transmission line in the presence of corona is able to reproduce the most important characteristics of subsequent return strokes of negative lightning, including the reduction in amplitude and increase in front time of the lightning current with increasing height, realistic return stroke speeds, realistic speed profiles, and, if a suitable set of parameters is selected, most signatures typically observed in measured electromagnetic fields. Besides confirming the consistency of modeling the return stroke channel using transmission line theory, the obtained results suggest that the consideration of relevant lightning properties let model predictions closer to measured data. It can be expected, therefore, that model predictions are likely to consistently improve if better models become available for representing the various physical processes in the lightning discharge. A transmission line model of the return stroke is flexible enough to accommodate virtually any model improvement provided it can be written, analytically or numerically, in the form of per-unit-length parameters.

References

[1] Paul CR. *Analysis of Multiconductor Transmission Lines*. 1st ed. New York, NY: Wiley; 1994.

[2] Rakov VA, Uman MA. Review and evaluation of lightning return stroke models including some aspects of their application. *IEEE Transactions on Electromagnetic Compatibility*. 1998;40(4):403–426.

[3] Theethayi N, Cooray V. On the representation of the lightning return stroke as a current pulse propagating along a transmission line. *IEEE Transactions on Power Delivery.* 2005;20(2):823–837.

[4] Little PF. Transmission line representation of a lightning return stroke. *Journal of Physics D: Applied Physics.* 1978;11(13):1893–1910.

[5] Gorin BN. Mathematical modeling of the lightning return stroke (in Russian). *Elektrichestvo.* 1985;4:10–16.

[6] Borovsky JE. An electrodynamic description of lightning return strokes and dart leaders: guided wave propagation along conducting cylindrical channels. *Journal of Geophysical Research: Atmospheres.* 1995;100(D2):2697–2726.

[7] Rakov VA. Some inferences on the propagation mechanisms of dart leaders and return strokes. *Journal of Geophysical Research: Atmospheres.* 1998;103(D2):1879–1887.

[8] Maslowski G, Rakov VA. A study of the lightning channel corona sheath. *Journal of Geophysical Research: Atmospheres.* 2006;111(D14):1–16.

[9] Braginskii SI. Theory of the development of a spark channel. *Soviet Physics JETP.* 1958;34:1068–1074.

[10] Baum CE, Baker L. Analytic return-stroke transmission-line model. In: Gardner RL, editor. *Lightning Electromagnetics.* New York: Hemisphere; 1990. p. 17–40.

[11] Visacro S, De Conti A. A distributed-circuit return-stroke model allowing time and height parameter variation to match lightning electromagnetic field waveform signatures. *Geophysical Research Letters.* 2005;32(23):1–5.

[12] Strawe DF. Non-linear modeling of lightning return strokes, Report FAA-RD-79–6. In: *Proc. Federal Aviation Administration/Florida Institute of Technology Workshop on Grounding and Lightning Technology*; Melbourne, Florida, USA; 1979. p. 1–8.

[13] De Conti A, Silveira FH, Visacro S, *et al.* A review of return-stroke models based on transmission line theory. *Journal of Atmospheric and Solar-Terrestrial Physics.* 2015;136:52–60.

[14] Price GH, Pierce ET. The modeling of channel current in the lightning return stroke. *Radio Science.* 1977;12(3):381–388.

[15] Takagi N, Takeuti T. Oscillating bipolar electric field changes due to close lightning return strokes. *Radio Science.* 1983;18(3):391–398.

[16] da Frota Mattos MA, Christopoulos C. A nonlinear transmission line model of the lightning return stroke. *IEEE Transactions on Electromagnetic Compatibility.* 1988;30(3):401–406.

[17] Hoole PRP, Hoole SRH. Simulation of lightning attachment to open ground, tall towers and aircraft. *IEEE Transactions on Power Delivery.* 1993;8 (2):732–740.

[18] Amoruso V, Lattarulo F. The electromagnetic field of an improved lightning return-stroke representation. *IEEE Transactions on Electromagnetic Compatibility.* 1993;35(3):317–328.

[19] De Conti A, Silveira FH, Visacro S. A study on the influence of corona on currents and electromagnetic fields predicted by a nonlinear lightning

return-stroke model. *Journal of Geophysical Research: Atmospheres*. 2014;119(9):5142–5156.

[20] Oetzel GN. Computation of the diameter of a lightning return stroke. *Journal of Geophysical Research (1896-1977)*. 1968;73(6):1889–1896.

[21] Gorin BN. The closing-switch model of lightning return-stroke. In: *Proc. 9-th Int. Conf. Atmospheric Electricity; St. Petersburg*, Russia; 1992. p. 206–211.

[22] da F Mattos MA, Christopoulos C. A model of the lightning channel, including corona, and prediction of the generated electromagnetic fields. *Journal of Physics D: Applied Physics*. 1990;23(1):40–46.

[23] Baker L. Return-stroke transmission line model. In: Gardner RL, editor. *Lightning Electromagnetics*. New York, NY: Hemisphere; 1990. p. 63–74.

[24] Bazelyan EM, Raizer YP. *Lightning Physics and Lightning Protection*. Bristol: IOP Publishinge; 2000.

[25] Beroual A, Rakotonandrasana JH, Fofana I. Predictive dynamic model of the negative lightning discharge based on similarity with long laboratory sparks – Part 1: physical process and modeling. *IEEE Transactions on Dielectrics and Electrical Insulation*. 2010;17(5):1551–1561.

[26] Beroual A, Rakotonandrasana JH, Fofana I. Predictive dynamic model of the negative lightning discharge based on similarity with long laboratory sparks Part 2: validation. *IEEE Transactions on Dielectrics and Electrical Insulation*. 2010;17(5):1562–1568.

[27] Uman MA. *Lightning*. 1st ed. New York, NY: Dover Publications, Inc.; 1984.

[28] Jordan DM, Uman MA. Variation in light intensity with height and time from subsequent lightning return strokes. *Journal of Geophysical Research*. 1983;88(C11):6555–6562.

[29] Tran MD, Rakov VA. An advanced model of lightning M-component. *Journal of Geophysical Research: Atmospheres*. 2019;124(4):2296–2317.

[30] Uman MA, McLain DK. Magnetic field of lightning return stroke. *Journal of Geophysical Research*. 1969;74(28):6899–6910.

[31] Rondon DA, Silveira FH, Visacro S. A varying surge impedance transmission line model for simulation of lightning return current in time domain. In: *Proc. Int. Conf. Grounding and Earthing (GROUND)*; Rio de Janeiro, Brazil; 2002. p. 101–104.

[32] Theethayi N, Cooray V. Representation of the return stroke as a transmission line—The apparent return stroke velocity. In: *Proc. 27th Int. Conf. Lightning Protection (ICLP)*; Avignon, France; 2004. p. 110–115.

[33] De Conti A, Visacro S, Theethayi N, *et al.* A comparison of different approaches to simulate a nonlinear channel resistance in lightning return stroke models. *Journal of Geophysical Research: Atmospheres*. 2008;113 (D14).

[34] De Conti A, Visacro S. On the use of lumped sources in a nonlinear lightning return stroke model and extension for evaluating strikes to tall objects. *Journal of Geophysical Research: Atmospheres*. 2009;114(D11).

[35] Mach DM, Rust WD. Photoelectric return-stroke velocity and peak current estimates in natural and triggered lightning. *Journal of Geophysical Research: Atmospheres.* 1989;94(D11):13237–13247.

[36] De Conti A, Visacro S. Reproducing lightning electromagnetic field signatures with a new current return-stroke model based on a distributed-circuit approach. In: *Proc. 28th Int. Conf. Lightning Protection (ICLP)*; Kanazawa, Japan; 2006. p. 169–174.

[37] Rakov VA. Lightning return stroke speed: a review of experimental data. In: *Proc. 27th Int. Conf. Lightning Protection (ICLP)*; Avignon, France; 2004. p. 139–144.

[38] De Conti A, Silveira FH, Visacro S. Lightning strikes to tall objects: A study of wave interactions at the return-stroke front using a nonlinear transmission line model. *Journal of Geophysical Research: Atmospheres.* 2015;120 (13):6331–6345.

[39] De Conti A, Silveira FH, Visacro S. Close electric fields and lightning-induced voltages predicted by a return-stroke model including corona and nonlinear channel resistance. *Electric Power Systems Research.* 2015; 118:8–14.

[40] Jordan EC, Balmain KG. *Electromagnetic Waves and Radiating Systems.* 1st ed. Englewood Cliffs, NJ: Prentice-Hall, Inc.; 1968.

[41] Cooray V, Theethayi N. Pulse propagation along transmission lines in the presence of corona and their implication to lightning return strokes. *IEEE Transactions on Antennas and Propagation.* 2008;56(7):1948–1959.

[42] Cooray V. Charge and voltage characteristics of corona discharges in a coaxial geometry. *IEEE Transactions on Dielectrics and Electrical Insulation.* 2000;7(6):734–743.

[43] Heckman SJ, Williams ER. Corona envelopes and lightning currents. *Journal of Geophysical Research: Atmospheres.* 1989;94(D11):13287–13294.

[44] Rao M, Bhattacharya H. Lateral corona currents from the return stroke channel and the slow field change after the return stroke in a lightning discharge. *Journal of Geophysical Research.* 1966;71(11):2811–2814.

[45] Ignjatovic M, Cvetic J. Drift-diffusion model of corona discharge in coaxial geometry due to negative lightning impulse voltage. *International Journal of Electrical Power & Energy Systems*. 2021;129:106815.

[46] Hermosillo VF, Cooray V. Space-charge generation and neutralisation in a coaxial cylindrical configuration in air under a negative voltage impulse. *Journal of Electrostatics.* 1996;37(3):139–149.

[47] Cabrera VM, Cooray V. On the mechanism of space charge generation and neutralization in a coaxial cylindrical configuration in air. *Journal of Electrostatics.* 1992;28(2):187–196.

[48] Plooster MN. Numerical simulation of spark discharges in air. *The Physics of Fluids.* 1971;14(10):2111–2123.

[49] Plooster MN. Numerical model of the return stroke of the lightning discharge. *The Physics of Fluids.* 1971;14(10):2124–2133.

[50] Paxton AH, Gardner RL, Baker L. Lightning return stroke. A numerical calculation of the optical radiation. *The Physics of Fluids*. 1986;29(8):2736–2741.

[51] Ripoll JF, Zinn J, Jeffery CA, *et al.* On the dynamics of hot air plasmas related to lightning discharges: 1. Gas dynamics. *Journal of Geophysical Research: Atmospheres*. 2014;119(15):9196–9217.

[52] Ripoll JF, Zinn J, Colestock PL, *et al.* On the dynamics of hot air plasmas related to lightning discharges: 2. Electrodynamics. *Journal of Geophysical Research: Atmospheres*. 2014;119(15):9218–9235.

[53] da Silva CL, Sonnenfeld RG, Edens HE, *et al.* The plasma nature of lightning channels and the resulting nonlinear resistance. *Journal of Geophysical Research: Atmospheres*. 2019;124(16):9442–9463.

[54] Toepler M. Zur kenntnis der gesetze der gleitfunkenbildung. *Annalen der Physik*. 1906;326(12):193–222.

[55] Barannik SI, Vasserman SB, Lukin AN. Resistance and inductance of a gas arc. *Soviet Physics. Technical Physics* 1975;19(11):1449–1453.

[56] Engel TG, Donaldson AL, Kristiansen M. The pulsed discharge arc resistance and its functional behavior. *IEEE Transactions on Plasma Science*. 1989;17(2):323–329.

[57] Nucci CA, Diendorfer G, Uman MA, *et al.* Lightning return stroke current models with specified channel-base current: a review and comparison. *Journal of Geophysical Research: Atmospheres*. 1990;95(D12):20395–20408.

[58] De Conti A, Visacro S, Theethayi N, *et al.* Modeling of the nonlinear channel resistance using some simplified theories. In: *Proc. Int. Conf. on Grounding and Earthing (GROUND)*; Florianopolis, Brazil; 2008. p. 69–74.

[59] De Conti A, Visacro S, Theethayi N, *et al.* Simulation of the time varying channel resistance: exponential decay versus strong-shock approximation. In: *Proc. 29th Int. Conf. Lightning Protection (ICLP)*; Uppsala, Sweden; 2008. p. 1–9.

[60] Visacro S, Soares A. HEM: a model for simulation of lightning-related engineering problems. *IEEE Transactions on Power Delivery*. 2005;20 (2):1206–1208.

[61] Visacro S, Silveira FH. Evaluation of current distribution along the lightning discharge channel by a hybrid electromagnetic model. *Journal of Electrostatics*. 2004;60(2):111–120.

[62] CIGRE Working Group C4-47. Electromagnetic computation methods for lightning surge studies with emphasis on the FDTD method. *CIGRE*; 2019. CIGRE Brochure 785.

[63] Baba Y, Rakov VA. On the mechanism of attenuation of current waves propagating along a vertical perfectly conducting wire above ground: application to lightning. *IEEE Transactions on Electromagnetic Compatibility*. 2005;47(3):521–532.

[64] Thottappillil R, Uman MA and Theethayi N. Electric and magnetic fields from a semi-infinite antenna above a conducting plane. *Journal of Electrostatics*. 2004;61(3):209–221.

[65] Idone VP, Orville RE. Lightning return stroke velocities in the thunderstorm research international program (TRIP). *Journal of Geophysical Research: Oceans.* 1982;87(C7):4903–4916.
[66] Uman MA, McLain DK, Krider EP. The electromagnetic radiation from a finite antenna. *American Journal of Physics.* 1975;43(1):33–38.
[67] Lin YT, Uman MA, Tiller JA, *et al.* Characterization of lightning return stroke electric and magnetic fields from simultaneous two-station measurements. *Journal of Geophysical Research: Oceans.* 1979;84(C10):6307–6314.
[68] De Conti A, Silveira FH, Visacro S. Influence of a nonlinear channel resistance on lightning-induced voltages on overhead lines. *IEEE Transactions on Electromagnetic Compatibility.* 2010;52(3):676–683.

Chapter 7

Measurements of lightning-generated electromagnetic fields

Mahendra Fernando[1], Lasitha Gunasekara[1] and Vernon Cooray[2]

7.1 Introduction

Lightning-generated electromagnetic fields consist of vertical and horizontal components. The horizontal field consists of two components depending on the alignment of the antenna. The vertical component of the electric field generated by lightning flashes can be measured either using a field mill [1] or using a flat plate (or a vertical whip) antenna [2,3]; each method has its advantages and disadvantages. The electric field's three components can be measured using specially adapted spherical antennas [4].

The conventional method used to measure the magnetic field is the crossed loop antenna [5–7], but in some studies, magnetometers have been used to measure this field component. The frequency spectrum of the electromagnetic fields generated by lightning flashes can be estimated by either Fourier transforming the broadband signal or using antenna systems tuned to the desired frequencies [8–13].

A brief description of the theory pertinent to these measuring techniques is given in this chapter. Some parts of this chapter are adapted from Refs. [14,15].

7.2 Electric field mill or generating voltmeter

The principle of operation of the field mill is illustrated in Figure 7.1. The plate marked S is the detector placed in a background electric field assumed to be uniform and steady for the moment. The plate marked M is a movable electrode that is at ground potential. This electrode can be moved back and forth in front of the sensing plate, either exposing it to or screening it from the background electric field. Consider the situation shown in Figure 7.1. In which the sensing plate is completely exposed to the electric field. The electric field line sends to the plate,

[1]Department of Physics, University of Colombo, Sri Lanka
[2]Department of Electrical Engineering, Uppsala University, Sweden

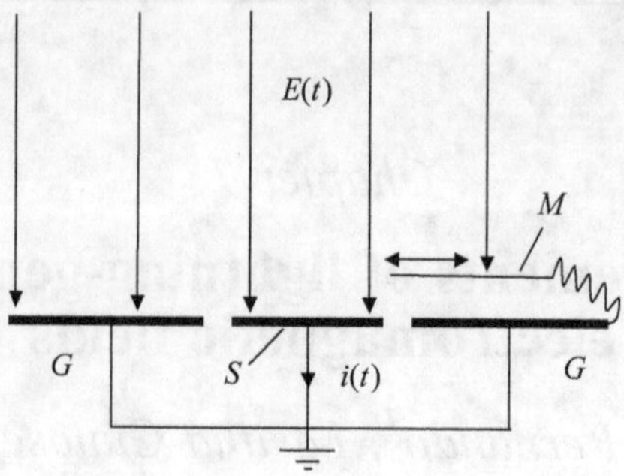

Figure 7.1 Diagram illustrates the basic principle of the electric field mill

and the total charge induced on the sensing plate is $A\epsilon_0 E$, where A is the area of the plate. Assume that plate M is moved back and forth in front of the sensing plate. This will change the exposed area of the sensing plate as a function of time, and since the charge induced on the plate is a function of the exposed area of the sensing plate, and a current will flow between the sensing plate and the ground. If the background electric field is constant, this current is given by

$$i(t) = \frac{da(t)}{dt}\epsilon_0 E \tag{7.1}$$

where $a(t)$ is the instantaneous exposed area of the sensing plate. Thus, knowing how the sensing plate's exposed area varies in time, the background electric field can be obtained by measuring the current flowing between the sensing plate and the ground.

Assume that M moves periodically back and forth over the sensing plate, thus alternatively shielding and un-shielding the sensing plate from the background electric field. If M moves back and forth n times per second, then the output current oscillates with a period, T_p equal to $1/n$ and the peak amplitude of the output is proportional to the background electric field. If the background electric field varies with the time, then the envelope of the oscillating output voltage follows the background electric field. However, the field mill cannot measure any rapid variation in the background electric field faster than the period Tp. In other words, the time resolution of the field mill is on the order of T_p. Thus, the rate of the periodic motion of M gives an upper limit to the resolution of the field that a field mill can measure. In modern field mills, the time resolution is increased by utilizing a rotating vaned wheel that alternatively shields or un-shields the sensing electrode from the electric field as each vane rotates over it. If the rotational speed of the metal vane is n revolutions per second, and if it has m vanes, then the time resolution of the field mill will decrease to $1/mn$. Such field mills can measure faster variations in the background electric field than described above. In general, the upper-frequency limit of the modern field mills may range from 1 to 10 kHz. The main advantage of the field mill is that it can be used to measure the absolute value of the background field.

7.3 Plate or whip antenna

7.3.1 Measurement of electric field

The physical configuration of the plate (or whip) antenna is shown in Figure 7.2. In principle, an antenna is a metal object connected to the ground through an electric circuitry. As the background electric field varies in time, the charge induced on the antenna also varies in time, generating a current in the electrical circuitry.

Suppose the dimension of the antenna is much smaller than the minimum wavelength of interest in the time-varying electric field. In that case, the antenna will act as a capacitive voltage source with the voltage proportional to the background electric field, $e(t)$. The constant of proportionality between the background electric field and the output voltage is called the effective height of the antenna.

The equivalent circuit of the antenna shown in Figure 7.2 is depicted in Figure 7.3, where C_a is the capacitance of the antenna to the ground and C_c is the

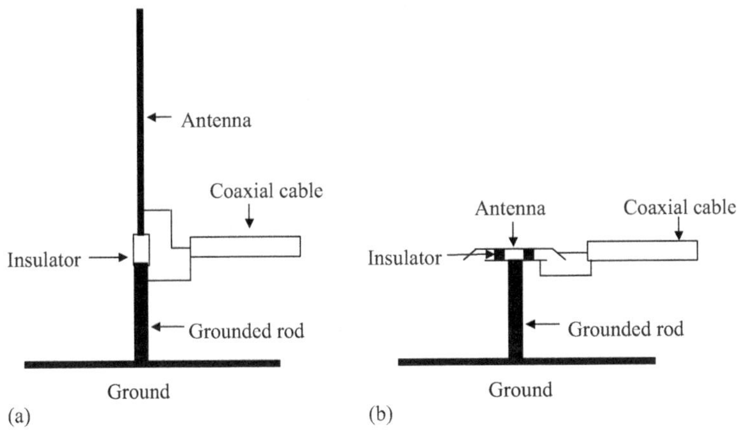

Figure 7.2 Antennas for the measurement of lightning-generated fields. (a) Whip antenna. (b) Plate antenna

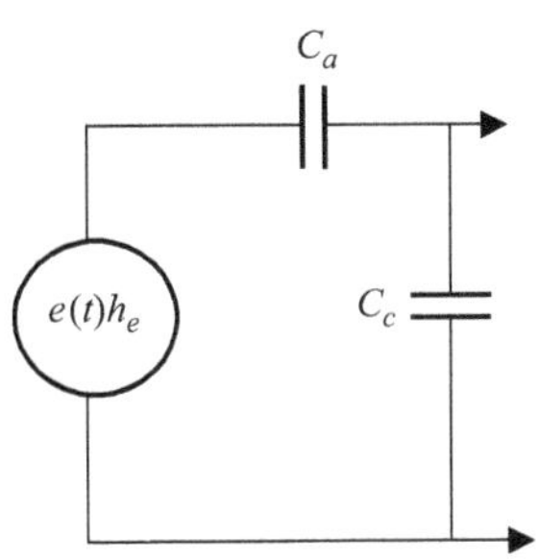

Figure 7.3 Equivalent circuit of the antenna and the attached cable. C_a is the capacitance of the antenna, and C_c gives the capacitance of the cable connected to the antenna

capacitance of the cable connected to the antenna. The effective height of the antenna can be either calculated from theory or measured by applying a known electric field to the antenna and measuring the output voltage.

The electronic circuitry that can be used to obtain the background electric field from a plate or whip antenna is given in Figure 7.4(a) [14,15]. The equivalent circuit of the electronic circuitry, when connected to the antenna, is given in Figure 7.4(b).

The relationship between the output signal, V_m of the circuit and the incident electric field E on the antenna is given in the frequency domain by,

$$V_m(s) = E(s)h_e\left(\frac{sC_aR}{1 + sRC + s^2RR_1C(C_a + C_c) + s(R_1 + R)(C_a + C_c)}\right) \tag{7.2}$$

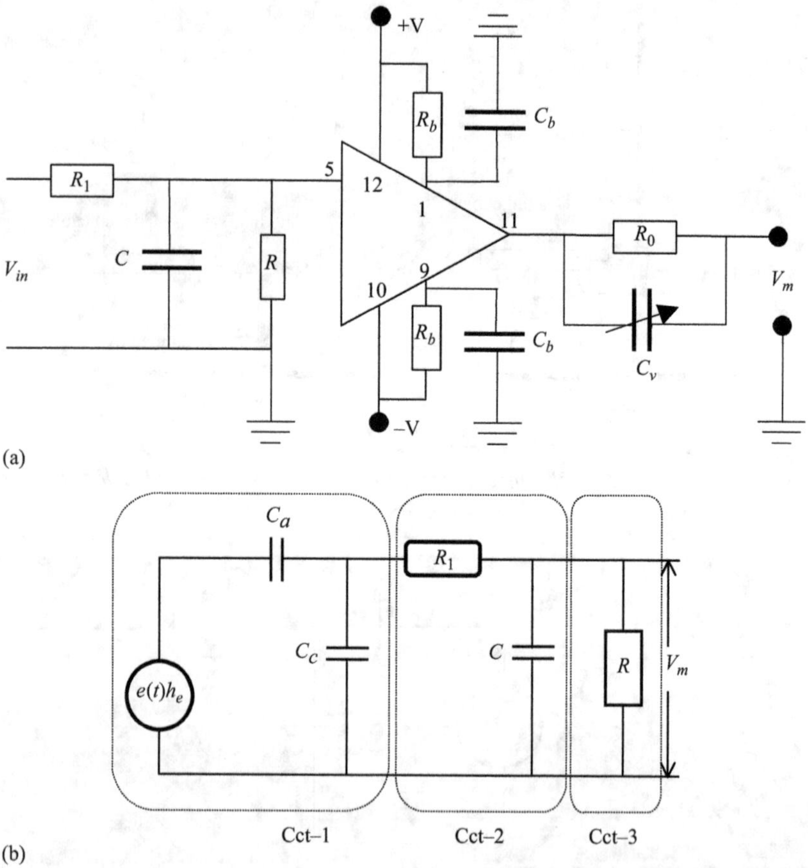

Figure 7.4 (a) Electronic circuit of the buffer amplifier system. (b) Equivalent circuit of the antenna system together with the attached electronics

here s is the Laplace variable, h_e is the effective height of the antenna and $E(s)$ is the Laplace transform of the background electric field, $e(t)$. In general, the resistance R_1 (equal to the cable impedance) is only a few tens of ohms, while the R is selected of about a few hundred mega ohms. Therefore, R_1 can be neglected in comparison to R in the term $(R_1 + R)$ in the denominator, and with that, (7.2) reduces to

$$V_m(s) = E(s)h_e\left(\frac{sC_aR}{1 + s^2RR_1C(C_a + C_c) + sR(C + C_a + C_c)}\right) \tag{7.3}$$

When $RR_1C(C_a + C_c) \ll R(C_a + C_c + C)$, which is true in practice, the time-domain solution of the above equation when the incident field is a Heaviside step function can be written as

$$V_m(t) = e(t)h_e\frac{C_a}{C_a + C_c + C} \quad e^{-\frac{t}{\tau_d}} \tag{7.4}$$

where

$$\tau_d = R(C_a + C_c + C) \tag{7.5}$$

To obtain an accurate measurement of the time-varying background electric field, the rise time constant of the antenna and the measuring system should be much shorter than the rising edge of the background field, and the decay time constant of the antenna system should be much longer than the total duration of the time-varying field. This point is illustrated further in the waveforms shown in Figure 7.5(a) and (b), where the effect of the time constants on the measurement of distant radiation field generated by a return stroke is illustrated. The data show that in order to obtain an accurate measurement of the distant radiation field of a return stroke, the rise time, the duration of which is about 0.1 and 100 ms, the rise time and decay time constants of the measuring system should be about 0.01 ms and 1 ms, respectively. They translate to 3 dB frequency limits of 1.6×10^2 Hz and 1.6×10^7 Hz. At close distances, due to the influence of the static electric field, the electric field's duration increases; therefore, the decay time constant should also be increased to get a faithful representation of the close field. In general, a decay time constant of about 1 sec is needed to faithfully measure the close electric fields produced by lightning flashes.

It is important to mention here that the upper-frequency limit of the bandwidth of the measuring system is also influenced by the physical dimension of the antenna, the electronic components used in the circuitry and the recording system used to record the output of the measuring system. If l is the length or the diameter of the antenna, it is necessary that $l \ll \lambda_m/4$, where 1_m is the minimum wavelength of interest in the electric field measurements. If this condition is not satisfied, the current induced in different parts of the antenna will reach the electronic circuitry at different times, thus invalidating the abovementioned theory. However, in many practical applications, l may not exceed a few metres in the case of whip antenna and a few tens of centimetres in the case of plate antenna. Thus, the

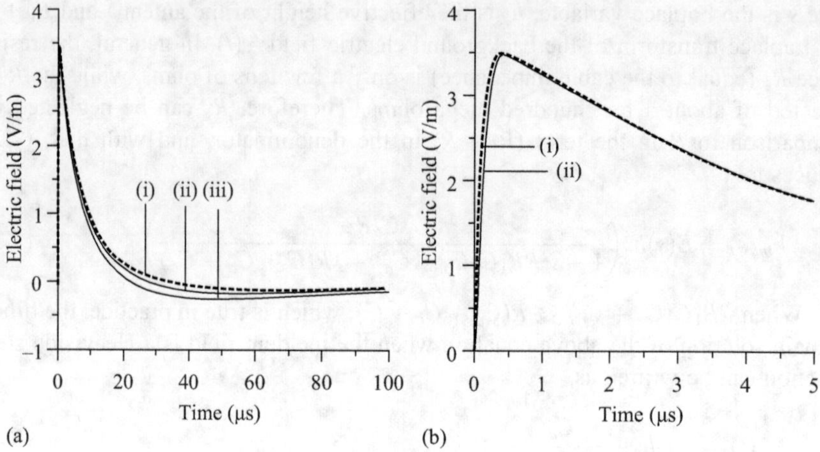

Figure 7.5 The influence of the rise time constant and the decay time constant of the antenna system for measuring the electric radiation field of a return stroke. The dashed line shows the actual electric field at 100 km and the solid lines show the antenna's output. (a) Effect of the decay time constant: (i) $t_r = 5$ *ns*, $t_d = 1$ *ms, (ii)* $t_r = 5$ *ns*, $t_d = 250$ μs, *(iii)* $t_r = 5$ *ns*, $t_d = 100$ μs. *(b) Effect of the rise time constant: (i)* $t_r = 10$ *ns*, $t_d = 1$ ms, *(ii)* $t_r = 50$ *ns*, $t_d = 1$ ms. *The electric field is calculated using the modified transmission line model MTLE (see Chapter 6). Note that the curves corresponding to (i) coincide with the actual electric field*

upper-frequency limit of the bandwidth is determined mainly by the electronic circuitry and the recording system.

This antenna system has an advantage over the field mill in providing a higher time resolution in the measurements. However, due to the effect of leakage currents across insulators (limiting the highest value of R) and the drastic reduction in the output amplitude (limiting the highest value of C), there is a limit to which the decay time constant could be increased. Thus, it is difficult to use this antenna system and the associated electronics to measure the electric fields' very low-frequency components, including DC fields.

7.3.2 Measurement of the derivative of the electric field

When both t_r and t_d are much smaller than the time variations in the signal under consideration, the output signal of the antenna system follows the temporal behaviour of the electric field derivative. For example, when these conditions are satisfied, the terms containing s (i.e. frequency) in the denominator of (7.3) can be neglected in comparison to unity leading to

$$V_m(s) = E(s) h_e s C_a R \tag{7.6}$$

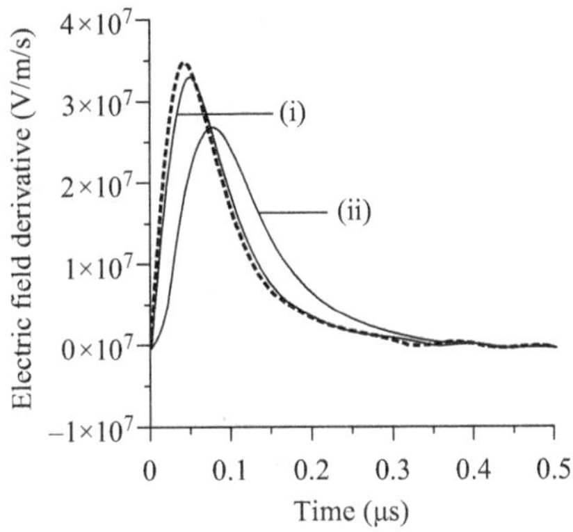

Figure 7.6 Diagram shows that the antenna system's output approaches that of the electric field derivative when the rise time constant and the decay time constant are much less than the rise time of the radiation field. The dashed line shows the derivative of the electric radiation field at 100 km. The solid lines show the antenna system's output after considering the changes in the output when the circuit parameters are changed. (i) $\tau_r = 2$ *ns,* $\tau_d = 5$ *ns, (ii)* $\tau_r = 5$ *ns,* $\tau_d = 30$ *ns. The electric field is calculated using the modified transmission line model MTLE (see Chapter 6)*

If the incident background electric field is a Heaviside step function, the output of the circuit in the time domain, $v_m(t)$, is given by

$$V_m(t) = h_e C_a R \delta(t) \tag{7.7}$$

where $\delta(t)$ is the Dirac delta function, which is the derivative of the Heaviside step function (i.e. the input electric field). Figure 7.6 depicts the output of the antenna system for several values of rise and decay time constants. The output signal is normalized by removing the effect of changing circuit parameters as t_r and t_d are varied. Note that in order to get the electric field derivative faithfully, t_r and t_d have to be much less than about 0.01 μs, which approximately is the rise time of the electric radiation field under consideration.

7.4 Measurements of the three electric field components in space

Almost all the studies reported in the literature deal with the vertical component of the electric field at ground level. If the ground is a perfect conductor, then the electric field at ground level has only the vertical component. However, at points above the ground, the electric field has three components one vertical (z-direction) and two horizontal (x- and y-direction). If the ground is finitely conducting, the electric field at ground level too will have a horizontal component. The horizontal component (or x- and y-component) of the lightning-generated electric field is smaller than the vertical component for the ground conductivities of practical interest. However, it plays a significant role in creating lightning-induced voltages on long horizontal conductors.

Thomson and co-workers measured the three components of the lightning electric fields using a spherical antenna system [4]. A schematic diagram of the antenna system used by these authors is shown in Figure 7.7(a).

The sensor is a metallic sphere with three caps placed perpendicular to each other. The diameter of the sphere was about 45 cm, and the cap radius was set to 7.6 cm. All the detecting and controlling electronics were placed at the centre of the sphere. The sensor is supported by an insulating structure and is placed about 1.5 m (about three times the diameter of the sphere) above the ground level to minimize any perturbation on the ambient surface charge density. By measuring the charge induced on the three caps in the presence of a background electric field, the three components of the field in the x, y, and z-directions can be calculated. Note that only the field component perpendicular to the cap will induce a net charge on the cap. In order to illustrate the theory, assume that the electric field component is directed in the z-direction (see Figure 7.7(b)). This field component will induce a net charge only in the cap perpendicular to the z-direction, i.e. bottom cap in the figure. The total charge induced on this cap due to an electric field E_z (directed

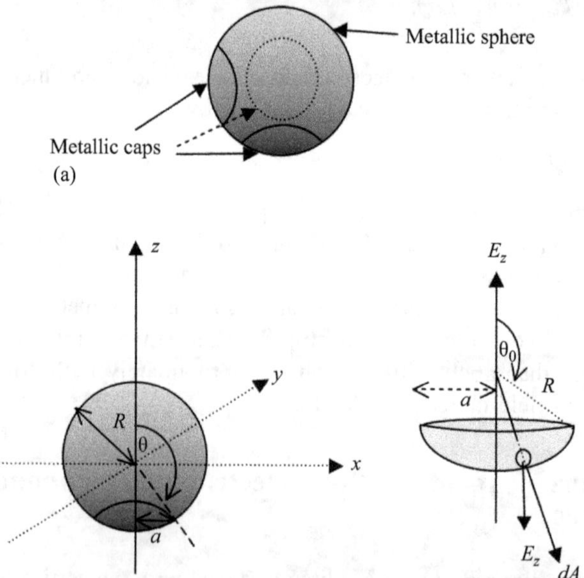

Figure 7.7 (a) Block diagram of the three-dimensional electric field sensor. The sensor consists of a metallic sphere with three caps placed perpendicular to each other. (b) A schematic diagram of the antenna system and a cap

along the z-axis) is given by

$$Q = \int \sigma(\theta) dA \tag{7.8}$$

where $\sigma(\theta)$ is the angular-dependent surface charge density on the cap. Now, the angular distribution of charge density induced on the sphere by E_z is given by

$$\sigma(\theta) = 3\epsilon_0 E_z \cos\theta \tag{7.9}$$

Moreover

$$dA = R^2 \sin\theta \, d\theta \, d\phi$$

In the above equation, ε_o is the permittivity of air and ϕ is the azimuthal angle of the area element dA.

Then the total charge induced on the cap is

$$Q_T = 3\epsilon_0 E_z R^2 \int_0^{2\pi} d\phi \int_{\theta_0}^{\pi} \sin\theta \, d\theta \tag{7.10}$$

where,

$$\theta_0 = \pi - \sin^{-1}\left(\frac{a}{R}\right) \tag{7.11}$$

After performing the integration, we obtain

$$Q_T = -3\epsilon_0 \pi a^2 E_z \tag{7.12}$$

The induced charge on each cap will induce an equal and opposite charge on the rest of the body. This induced charge would be evenly distributed along the rest of the spherical body. If this charge were assumed to be distributed on an imaginary cap of the same dimensions in the opposite location to the actual one, these two caps would constitute a capacitor with a separation equal to the sphere diameter.

By using Gauss's law, the following relations can be derived:

$$\int E.ds = \frac{Q}{\varepsilon_0} \tag{7.13}$$

$$\int_0^{\theta}\int_0^{2\pi} E \, R^2 \sin\theta \, d\theta \, d\varphi = \frac{Q}{\varepsilon_0} \tag{7.14}$$

$$E = \frac{Q}{2\pi\varepsilon_0 R\left[R - \sqrt{R^2 - a^2}\right]} \tag{7.15}$$

$$V = -\int_0^{2R} E.dr = -\frac{Q}{\pi\varepsilon_0\left[R - \sqrt{R^2 - a^2}\right]} \tag{7.16}$$

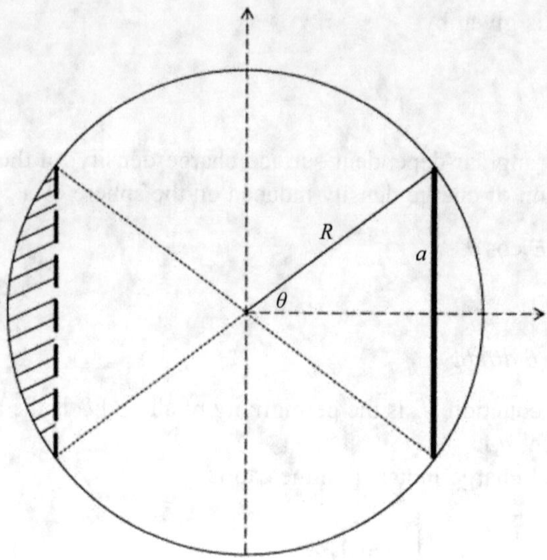

Figure 7.8 The imaginary caps scenario

By applying (7.12),

$$V_g = \frac{3a^2 E_z}{\left[R - \sqrt{R^2 - a^2}\right]} \tag{7.17}$$

From (7.17), the measured voltage (V_g) is directly proportional to the E field component perpendicular to each cap by which, the respective E field can be derived.

Thomson *et al.* [4] also presented some of the known errors associated with their spherical antenna. The errors are as follows:

- Sphere tilt (sensor tilt)

This is the most notable error source of this antenna type. As per Figure 7.9, the measured ($E_{v\ meas}$, $E_{h\ meas}$) and actual ($E_{v\ act}$, $E_{h\ act}$) E field values are related as follows.

$$E_{v\ act} = E_{v\ meas} \cos\theta - E_{h\ meas} \sin\theta \tag{7.18}$$

$$E_{h\ act} = E_{h\ meas} \cos\theta + E_{v\ meas} \sin\theta \tag{7.19}$$

The error due to the sphere tilt on the $E_{v\ act}$ is less since the $E_{h\ meas}$ is a comparatively small value. The opposite holds for the $E_{h\ act}$ since the contribution from $E_{v\ meas}$ is quite large. Physically, by adjusting the antenna caps' directional alignment, the error can be minimized to some extent, and in addition, an adjustment to the recorded data can be made to rectify the remaining error values. If θ is quite small, we can find it by using (7.19) and then use the same to calculate the

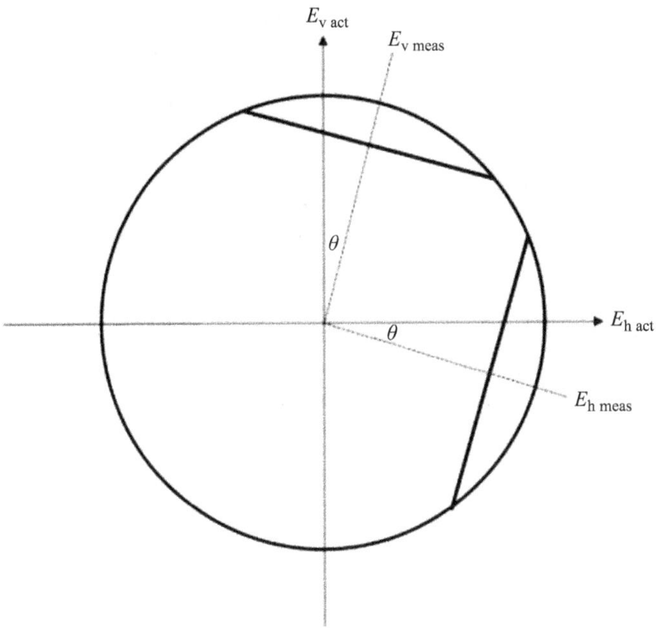

Figure 7.9 The sensor tilt scenario

corrected E values as well. Thus if θ is quite small and the actual Eh is also small, (7.19) becomes,

$$0 \sim E_{h\ meas} + E_{v\ meas}\theta$$

$$\theta \sim -\frac{E_{h\ meas}}{E_{v\ meas}} \tag{7.20}$$

By using (7.20), the tilt angle can be approximated, and thus the adjustments can be made accordingly.

- The proximity of the antenna to the ground

  The antenna's placement (around three diameters above the earth's surface) creates an electric field distortion in it which is caused by the rearrangement of surface charge on the ground, which was initially self-induced by the antenna. Simply said, it is a reciprocal effect where it is very existence above the ground influences its charge density. Using the method of images, where the spherical antenna is considered to be one end of the dipole, and the ground is taken as the other end, the calculations yielded that the measured result was within 0.1% deviation of the actual ones.
- The permittivity of the dielectric support and ropes

  The sensor was supported by PVC pipes and Nylon ropes, which are considered as dielectric cylinders that are in an external E field ($\boldsymbol{E_0}$). The maximum field perturbation at a distance $\boldsymbol{r}$ from the axis of the cylinder with a dielectric constant ***K***, inner radius ***a*** and an outer radius ***b*** is given by (7.21):

$$|\Delta E_r| = E_0 \frac{(K-1)(K+1)(b+a)(b-a)b^2}{[K(b+a)+b-a][K(b-a)+b+a]} \tag{7.21}$$

- Conductance of the support

As mentioned in Thomson *et al.* [4], the conductance of the support has a major impact, especially in wet conditions. Therefore, special care should be taken to prevent the sensor from getting wet during data acquisition.

Nearly a quarter of a century later, in 2018, Gunasekara *et al.* [16,17], revisited the three-dimensional field measurements by utilizing an identical setup to that of Thomson and coworkers [4]. This setup was deployed on two occasions starting from 2013 and in 2014 in two coastal regions of Sri Lanka. A data sample consisting of 65 Return Strokes (RS), 50 Negative Narrow Bipolar Pulses (NNBP) and 40 Positive Narrow Bipolar Pulses (PNBP) was selected and analyzed. Gunasekara *et al.* [16] took a step further by cross verifying the physically measured horizontal *E* field data against theoretically derived horizontal *E* field data.

This verification was achieved using the Cooary formula [14] which output theoretical horizontal *E* field values based on measured vertical *E* field input. This approach was implemented to all types of aforementioned events and the resultant *E* field signatures tallied with minute differences (see Figure 7.10).

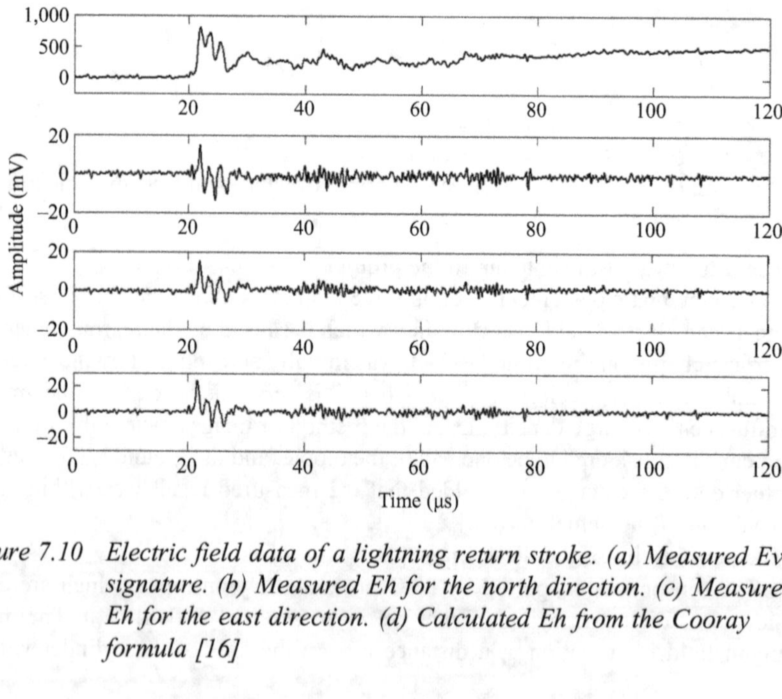

Figure 7.10 Electric field data of a lightning return stroke. (a) Measured Ev signature. (b) Measured Eh for the north direction. (c) Measured Eh for the east direction. (d) Calculated Eh from the Cooray formula [16]

7.5 Crossed loop antennas to measure the magnetic field

The voltage induced in a loop antenna due to an incoming magnetic field is proportional to the area of the loop multiplied by the derivative of the magnetic field component perpendicular to the loop. By measuring the voltage induced in two magnetic loops placed orthogonal to each other, the component of the magnetic field parallel to the plane containing the two axes of the loops can be obtained. The magnetic field generated by a vertical lightning channel is parallel to the ground plane and is directed perpendicular to the line joining the point of observation and the lightning channel. Therefore, the direction of the lightning flash from a given point can be obtained by measuring the ratio of the voltages induced in two orthogonal magnetic loops.

The equivalent circuit of the loop antenna is shown in Figure 7.11. Here L is the inductance of the loop, and R is its resistance. In practice, the resistance of the loop can be neglected. The voltage, V, induced in the loop, assumed to be electrically small, is given by

$$V = -n\frac{d\phi}{dt} \tag{7.21}$$

where n is the number of turns in the loop and j is the magnetic flux threading the loop. The flux is given by

$$\phi = B(t)A\cos\theta \tag{7.22}$$

where $B(t)$ is the time-varying magnetic field, A is the area of the loop and q is the angle between the axis of the loop and the magnetic vector. Since the antenna's output is proportional to the derivative of the magnetic field, it has to be integrated to obtain a signal that is proportional to the magnetic field. An antenna and the corresponding electronics suitable for this purpose, as developed by Krider and Noggle [6], are shown in Figure 7.12.

The intermediate integration time constant of the integrator determines the lower bandwidth limit of the magnetic field measuring system. Thus, the integration time constant of the system should be much longer than the duration of the waveforms of interest. In the circuit shown in Figure 7.12, with C_1 = 1,000 pF, the

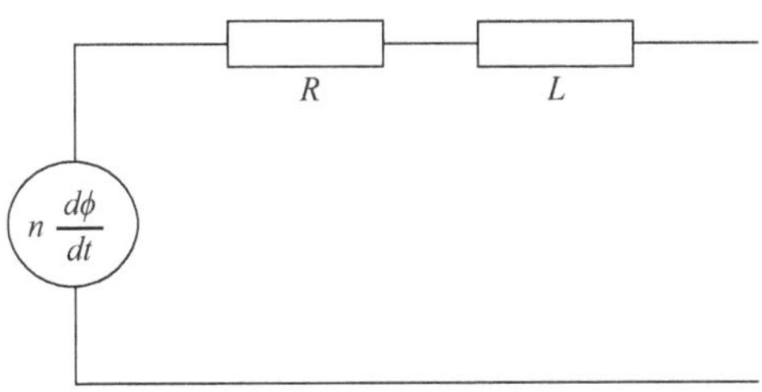

Figure 7.11 Equivalent circuit of the magnetic loop

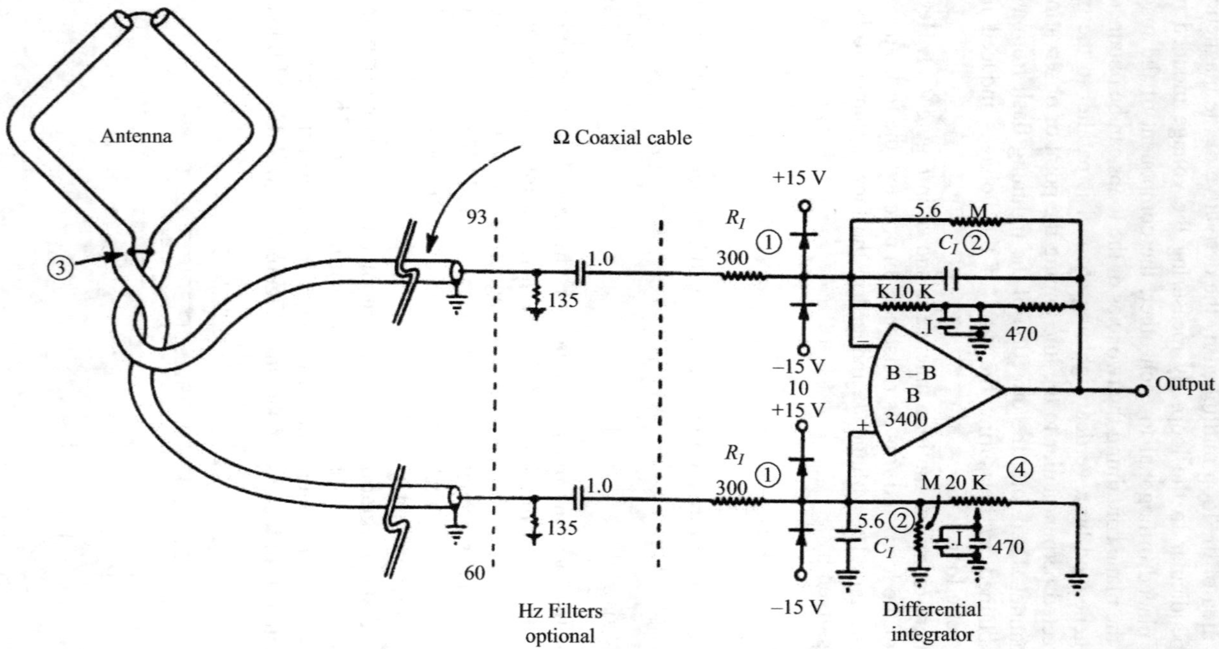

Figure 7.12 Antenna and the corresponding electronics suitable for the measurement of the magnetic field generated by a lightning return stroke. From Krider and Noggle [5]

time constant of the integrator was 4 ms without the 60 Hz filters and 1.3 ms when they were included. As in the electric field measuring system, the upper limit of the bandwidth is determined by the electronic circuitry and the recording system.

In measuring magnetic fields, it is also necessary to avoid any contamination of the measurements due to the electric fields. In the measuring system shown in Figure 7.12, this is achieved: (a) by shielding the magnetic field sensor with an outer screen that is broken at the top to avoid any circulating currents and (b) by measuring the difference in the voltages induced at the two ends of the antenna thus cancelling out any contribution from the electric field.

7.6 Magnetic field measurements using anisotropic magnetoresistive (AMR) sensors

Even though the loop antenna systems are widely used in magnetic field measurements as well as in direction finding systems, one problem in using loop antenna systems is the necessity of integrating the measured voltage signal. This can be avoided by using magnetometers [e.g. 18,19,20,22]. The fluxgate magnetometer used by Williams and Brook [15] was sensitive to changes as low as 10 μG. Apart from the sensors used in the above studies, Hall effect sensors and anisotropic magnetoresistive sensors (or AMR sensors) can also be used for magnetic field measurements. Fernando *et al.* [21] used Honeywell's HMC1022 AMR sensor to measure the magnetic field generated by lightning. Their results suggest that AMR sensors can be used successfully for lightning magnetic field measurements. Honeywell's HMC1022 sensor has two sensors oriented perpendicularly to each other on a single chip. It has a resolution of as low as 85 mG through 6 G, a typical sensitivity of 1 mV/V/G and a bandwidth of 5 MHz with angular accuracy of $\pm 1^\circ$. The sensitivity of the measuring setup was increased using the LMH6624 op-amp. The measuring setup can measure magnetic fields as low as 300 mG in the range of DC to 4 MHz. A block diagram of the sensor setup used by Fernando *et al.* [21] is shown in Figure 7.13.

7.7 Narrowband measurements

The frequency content of lightning electric fields spans from VLF to the UHF range. Basically, two techniques were used in different studies to obtain the frequency spectrum. In one technique, the measured (as described in Section 7.2) broadband signal is Fourier transformed to obtain the frequency spectrum [see, e.g. 9, 10,11,17]. The other technique used narrowband oscillators to sample the electric field at the desired frequency [see, e.g. 11,12,13].

Narrowband oscillators in the VLF and HF range mainly consist of passive devices such as inductors and capacitors. Since the antenna (flat plate or whip) acts as a capacitor (say C_a), one can design a narrowband oscillator in principle by connecting an inductor L and a resistor R in series with the antenna [13,23]. The equivalent circuit of such a system is given in Figure 7.14.

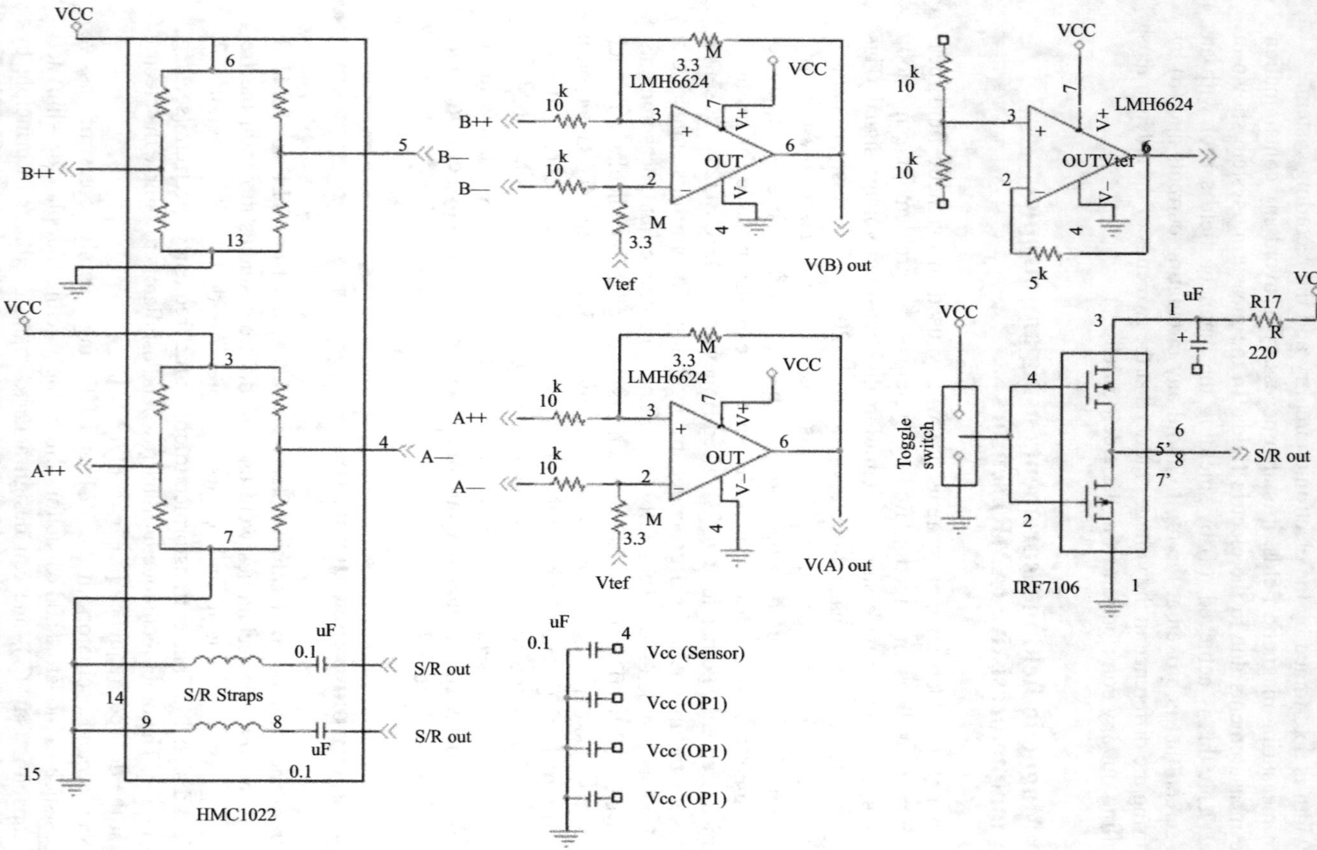

Figure 7.13 Block diagram of the electronic circuit for HMC1022 sensor as given in [21]

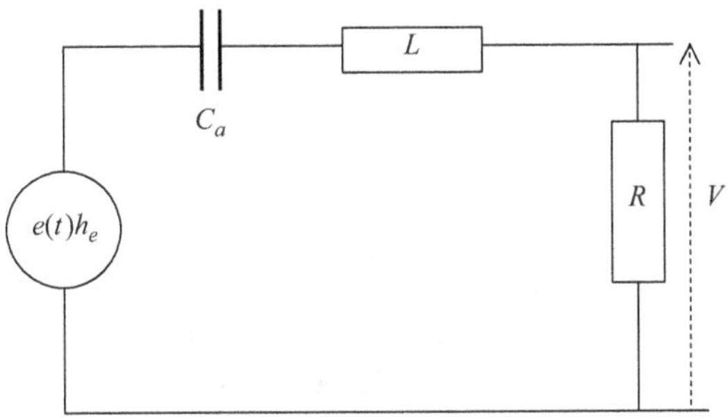

Figure 7.14 Block diagram of the narrowband oscillator circuit. Here, C_a is the antenna capacitance, L is the inductor, and R is the resistor connected between the antenna and ground

Let the measured voltage across the resistor is V. Then, it is a simple matter to show that

$$V(s) = E(s)h_e \frac{sR}{s^2L + sR + \frac{1}{C_a}} \tag{7.23}$$

where, as before, h_e is the effective height of the antenna, s is the Laplace variable, and $E(s)$ is the Laplace transform of the background electric field, $e(t)$. Assume that the incident electric field is a Heaviside step function. Then the antenna output in the time domain is given by

$$v(t) = h_e\sqrt{LC_a}e^{\left(-\frac{R}{2L}\right)t}\sin\frac{t}{\sqrt{LC_a}} \tag{7.24}$$

The output oscillates with a frequency f_o given by $1/2\pi\sqrt{LC_a}$, and it decays exponentially with a time constant of $2L/R$. The larger the R, the smaller the time constant, and the signal decays to zero rapidly. Thus, by selecting the values of R and L, one can confine the oscillations of the circuit created by a given impulse in the electric field to the desired time duration.

References

[1] Malan, D. J. and B. F. J. Schonland, An electrostatic fluxmeter of short response time for use in studies of transient field changes, *Proc. Phys. Soc. London Ser. B*, 63, 402–408, 1950.

[2] Kitagawa, N. and M. Brook, A comparison of intracloud and cloud-to-ground lightning discharges, *J. Geophys. Res.*, 65, 1189–1201, 1960.

[3] Cooray, V. and S. Lundquist, On the characteristics of some radiation fields from lightning and their possible origin in positive ground flashes, *J. Geophys. Res.*, 87, 11203–11214, 1982.

[4] Tomson, E. M., E. J. Medelius, and M. A. Uman, A remote sensor for the three components of transient electric field, *IEEE Trans. Indus. Electro.*, 35, 426–433, 1988.

[5] Norinder, H. and O. Dahle, Measurements by frame areas of current variations in lightning discharges, *Ark. Mat. Astron. Fys.*, 32A, 1–70, 1945.

[6] Krider, E. P. and R. C. Noggle, Broadband antenna system for lightning magnetic fields, *J. Appl. Meterolog.*, 14, 252–256, 1975.

[7] Krider, E. P., R. C. Noggle, A. E. Pifer, and D. L. Vance, Lightning directionfinding system for forest fire detection, *Bull. Am. Meteorol. Soc.*, 61(9), 980–986, 1980.

[8] Willet, J. C., J. C. Bailey, C. Leteinturier, and E. P. Krider, Lightning electromagnetic radiation field spectra in the interval from 0.2 to 20 MHz, *J. Geophys. Res.*, 95(D12), 20367–20387, 1990.

[9] Weidman, C. D. and E. P. Krider, The amplitude spectra of lightning radiation fields in the interval from 1 to 20 MHz, *Radio Sci.*, 21(6), 964–970, 1986.

[10] Sonnadara, U., V. Cooray, and M. Fernando, The lightning radiation field spectra of cloud flashes in the interval from 20 kHz to 20 MHz, *IEEE Trans. Electromag. Compat.*, 48(1), 234–239, 2006.

[11] Malan, D. J., Radiation from lightning discharges and its relation to the discharge process, in *Recent Advances in Atmospheric Electricity, Proceedings of the 2nd Conference on Atmospheric Electricity*, pp. 557–563, 1958.

[12] LeVine, D. M. and E. P. Krider, The temporal structure of HF and VHF radiation during Florida lightning return strokes, *Geophys. Res. Lett.*, 4, 13–16, 1977.

[13] Cooray, V., Temporal behaviour of lightning HF radiation at 3 MHz near the time of return strokes, *J. Atmos. Terr. Phys.*, 48, 73–78, 1986.

[14] Cooray, V., Mechanism of the lightning flash, in: Cooray, V. (Ed.), *The Lightning Flash*, The Institution of Electrical Engineers, London, UK, 2003.

[15] Galvan, A. and M. Fernando, Operative characteristics of a parallel-plate antenna to measure vertical electric fields from lightning fields from lightning flashes. *Report UURIE*, pp. 285–300, 2000.

[16] Gunasekara, T. A. L. N., M. Fernando, U. Sonnadara, and V. Cooray. Horizontal electric fields of lightning return strokes and narrow bipolar pulses observed in Sri Lanka. *J. Atmos. Solar-Terrest. Phys.*, 173, pp. 57–65, 2018.

[17] Gunasekara, T. A. L. N., S. N. Jayalal, M. Fernando, U. Sonnadara, and V. Cooray, Time-frequency analysis of vertical and horizontal electric field changes of lightning negative return strokes observed in Sri Lanka, *J. Atmos. Solar-Terrestrial Phys.*, 179, pp. 34–39, 2018.

[18] Williams, D. P. and M. Brook, Magnetic measurement of thunderstorms currents, 1. *Continuing currents in lightning, J. Geophys. Res.*, 68, 3243–3247, 1963.

[19] Meese, A. D. and W. H. Evans, Charge transfer in the lightning stroke as determined by the magnetograph, *J. Franklin Inst.*, 273, 375–382, 1962.
[20] Nelson L. D., Magnetographic measurements of charge transfer in the lightning flash, *J. Geophys. Res.*, 73, 5967–5972, 1968.
[21] Fernando, M., S. R. Sharma, P. Hettiarachchi, and N. Jayawantha, Use of AMR sensors for lightning magnetic field measurement, *Proc. Tech. Sessions, IPSL.*, 22, 37–45, 2006.
[22] Pierce E. T., The charge transferred to earth by a lightning flash, *J. Franklin Inst.*, 268, 353–354, 1968
[23] Jayaratne K. P. S. C. and V. Cooray, Lightning HF radiation at 3 MHz during leader and return stroke processes, *J. Atmos. Terr. Phys.*, 56, 493–501, 1994.

Chapter 8

HF and VHF electromagnetic radiation from lightning

Chandima Gomes[1]

8.1 Introduction

In any electrical process where charges undergo a rate of change with respect to time, electromagnetic fields are generated. Accelerating charge produces radiation fields. Depending on the rate of change of spatial positioning of charge, radiation fields created will have a spectrum of frequencies. The International Telecommunication Union (ITU) has labelled these frequencies for the convenience of identification (Table 8.1). In this chapter, we adhere to this ITU definitions of radio frequencies.

Lightning is a complex electrical phenomenon where charge motion takes place at various spatial elevations under various ambient electric field conditions. The radiated energy due to this varying spatio-temporal behaviour of current in the discharge channels undergoes further changes as they reach receiver ends due to propagation effects such as pulse broadening due to scattering. Thus, as per the information that we have today, from the cloud electrification process to the completion of a ground flash lightning discharge, lightning radiates electromagnetic fields of a wide range of frequencies and spatial energy distribution. Out of this broad spectrum, in this chapter, we investigate a small window, 3–300 MHz, known as HF and VHF bands (Table 8.1). These bands play a key role in several aspects of lightning sciences as they contain vital information on various electrification processes that gave rise to their emission. Basically, lightning activities that radiate at the HF–VHF band produce sferics of rising or trailing times in the sub-microsecond scale. Once detected and the temporal informations are extracted, these radiation fields could lead to the reconstruction of the maps of the discharge process with finer details. Furthermore, the radio frequency receivers in satellites can also use the HF and VHF signals that they receive for real-time tracking of various tropospheric dynamics that give rise to the discharge processes, thus, leading to advanced weather nowcasting and forecasting abilities. However, it should be noted that satellite systems, such as the US Department of Defence sponsored Fast On-orbit Recording of Transient Events (FORTE) satellite [1], pick a larger content of earth noise at HF and VHF bands, both natural and man-made, than that is by their earth-based counterparts.

[1]School of Electrical and Information Engineering, University of the Witwatersrand, South Africa

Table 8.1 The frequency spectrum of radio waves as per the ITU definition

Frequency range (Hz)	Frequency band	Abbreviation
3–30 Hz	Extremely low frequency	ELF
30–300 Hz	Super low frequency	SLF
300–3,000 Hz	Ultra low frequency	ULF
3–30 kHz	Very low frequency	VLF
30–300 kHz	Low frequency	LF
300–3,000 kHz	Medium frequency	MF
3–30 MHz	High frequency	HF
30–300 MHz	Very high frequency	VHF
300–3,000 MHz	Ultra high frequency	UHF
3–30 GHz	Super high frequency	SHF
30–300 GHz	Extremely high frequency	EHF
300–3,000 GHz	Tremendously high frequency	THF

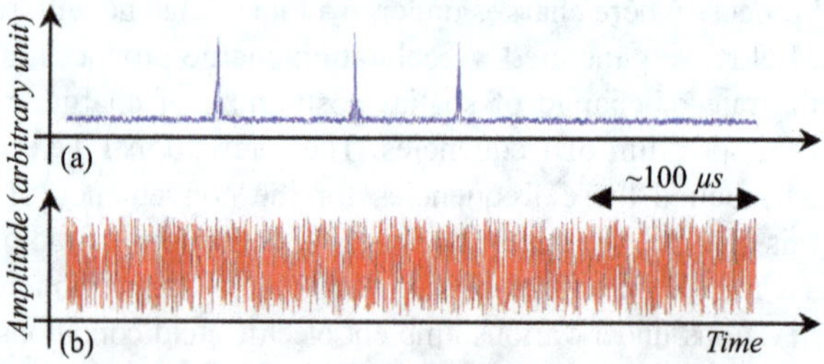

Figure 8.1 Lightning generated VHF signals. (a) Individual pulses of microsecond-scale duration (b) continuous emission of radiation that lasts for a few milliseconds. Adapted from [5]

Additionally, the HF and VHF radiation may provide crucial information to understand the physics of the discharge phenomenon.

Even though it is not as straightforward as in the clarity of information from optical sensors, many atmospheric environments are transparent to HF and VHF radio waves, whereas they are opaque or translucent to optical radiation [2–4]. In the discharge events of lightning, HF–VHF signals are emitted both as intermittent pulses of microsecond-scale width (Figure 8.1a) or continuous pulse trains of millisecond scale duration (Figure 8.1b) [5].

8.2 Information analysis and discussion

8.2.1 Significance of lightning-related HF–VHF emission

VLF and LF radiation sensors employed in some lightning detection networks such as NLDN, capture horizontally propagating EM fields ducted through the earth-ionosphere waveguide, generated mostly by vertical or vertically slated discharges.

These discharges are usually involved with cloud-to-ground lightning. On the other hand, HF and VHF radiation are generated by a much diverse spectrum of events, which include intra-cloud flashes, partial breakdowns of inter-cloud flashes and return strokes of ground flashes as well [6].

Early work on the detection of broadband electric fields reported that both LF–VLF and HF–VHF emissions are prevalent in the leading and trailing ends of the lifetime of a thunderstorm [7,8]. They found that the two spectrums are sometimes interrelated, however, the correlation is complex and a liaison recognition is not straightforward. In a few of these studies [6,9,10], HF–VHF radiation has been classified as major and minor based on their pattern of occurrence. The HF–VHF radiation that occurs in a time-bound pulse group is termed major and those that occur in isolation are termed minor. Major radiation is considered associated with LF–VLF whereas minor is assumed to be emitted at higher altitudes and not directly associated with LF–VLF band emissions.

The long vertical displacement of charge in a ground flash is the main source of LF–VLF radiation. On the other hand, in both inter- and intra-cloud lightning flashes, the short branches and segmentations at regular intervals give rise to short-distance breakdown processes. Such brief discharges provide weak LF–VLF signals, usually at the initial stage of the discharge. In contrast, HF–VHF radiation is emitted in the neutralization process of all branches along the total length of the cloud flash [11,12]. Thus, HF–VHF radiation bands could be treated as having much higher efficiency in mapping cloud flash domains.

Confirming the above argument, a white paper produced by VAISALA [13] claims that their cloud lightning mapping network Vaisala's TLS200/LS8000 could reconstruct the map of all branches in an inter- or intra-cloud lightning flash with a detection efficiency of over 90% of the cloud flashes whereas their VLF–LF based detection system has less than 50% efficiency. Furthermore, they state that the LF–VLF system could not map 90% of the cloud flashes covering their entire spatial volume. This significantly increased efficiency of the HF–VHF detection system enhances the accuracy of nowcasting both severe and non-severe thunderstorms by a large margin over that of the LF–VLF system.

8.2.2 Preliminary breakdown pulse trains

Pulse trains consisting of microsecond scale pulses of bipolar nature, preceding negative first return strokes, have been observed in electric and magnetic field records of ground flashes. They are referred to as preliminary breakdown pulses of negative return strokes (NPBP). These bipolar pulses have an initial polarity similar to that of the return stroke succeeding them. The first detailed analysis of NPBP was published by Weidman and Krider [14], where a large sample recorded in Florida has been taken into account. Seventy-seven per cent of their pulse trains, with pulses of positive initial polarity, have preceded ground strokes. However, they have not specified the polarity of these ground flashes, which may be negative. The rest 23% of pulse trains have not preceded any return stroke events. They claim that the characteristics of these pulse trains (the rest 23%) are similar to that of those preceding return strokes. The initial polarity of each pulse in a given sequence tends to be the same. A few fast narrow pulses have been superimposed on the initial half cycle. The bipolar overshoot is smaller in amplitude compared with the initial peak, in most cases. The frequency spectrum of a few pulses of NPBP trains has been given in Weidman [15] and Weidman and Krider [16]. The time duration of the preliminary activity is given by Beasley [17]. In the last 25 years, a significant number of works have been published that report the characteristics of NPBP on various geographical location [18–22]. Modelling of the discharge process that leads to NPBP emission has also been attempted in a few studies [23–27].

A more recent study by Kolmašová [28] presents information on magnetic fields of NPBP that sheds some light on the RF emissions associated with the discharge process. They have recorded simultaneous measurements of the broadband NPBP trains and RF emission at the 60–66 MHz band which belongs to VHF. In most cases analyzed, except for a very few, the sensors could not pick any geolocated VHF emission sources that occur time-synchronized with the train of NPBP. Thus, it was confirmed in previous observations that the discharge phenomenon that gives rise to the VHF emission and that generates NPBP are only weakly related [29–31]. Interestingly, Kolmašová *et al.* [28] also report that a more rigorous analysis delineates that at individual pulse levels, the most intense VHF emissions are coinciding with the occurrence of NPBPs. Furthermore, the starting time of the first NPBP overlaps with an abrupt initiation of the VHF emission which then continues throughout the NPBP train. The same team reconfirmed their observations later [32,33]. Similar observations have been reported in by thunderstorm recorded in Sweden at the lower HF range [34] as it is depicted in Figure 8.2.

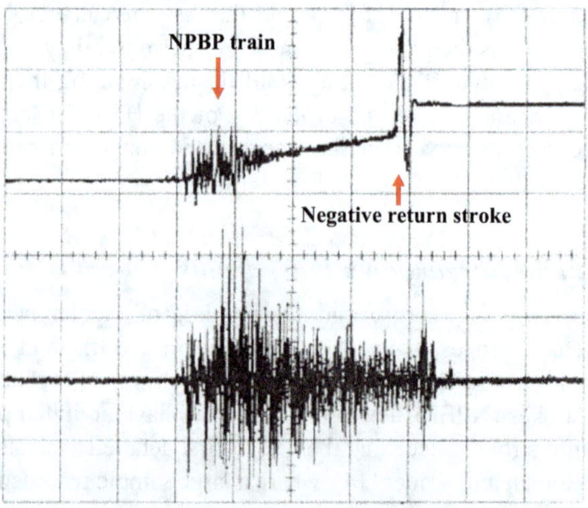

Figure 8.2 The radiation field (upper trace) and the simultaneously measured 3 MHz emission (lower trace) of an NPBP train recorded in Uppsala, Sweden. The X-axis is time and the Y-axis is the electric field. Units: X-axis: 0.5 ms/square, Y-axis: arbitrary units (adopted from [34])

In line with such observations, it can be concluded that the discharge phenomenon that produces NPBP trains and that gives rise to HF–VHF emission may be the same, although, there is no concrete evidence to deduce the geolocation of such discharge occurrence.

Similar to NPBP trains that are associated with negative lightning, positive ground flashes are also preceded by a train of pulses, in some cases (Figure 8.3). These may be referred to as PPBP trains. PPBPs are not as prevalent as NPBPs. Several studies report the observation of PPBPs, most often pertinent to a certain geographical location [34–40]. However, none of these studies, except Gomes [34] has made attempts to measure simultaneous HF-VHF emission in conjunction with PPBP trains. The few cases presented in Gomes [34] show that similar to NPBP, in PPBP too, 3 MHz radiation starts with the onset of the PPBP and continues with the pulse train (Figure 8.3). The most intense HF emission coincides with the peak amplitude of pulses in the broadband PPBP record. Thus, one may conclude that there is not much difference between the characteristics of PPBP and NPBP concerning HF emission.

In both NPBP and PPBP trains recorded in Sweden [34], the 3 MHz radiation, once initiated with the onset of the broadband pulses, continues right up to the return stroke. The highest intensity of the HF radiation usually coincides with the breakdown pulses of the highest amplitude. In almost all cases of positive lightning with PPBP, the 3 MHz signal continues even after the return stroke for tens of milliseconds with varying intensity (Figure 8.3). In negative strokes, the 3 MHz continuous after the return stroke only on a few occasions and in those cases also for less than half a millisecond. This phenomenon is yet to be explored in detail.

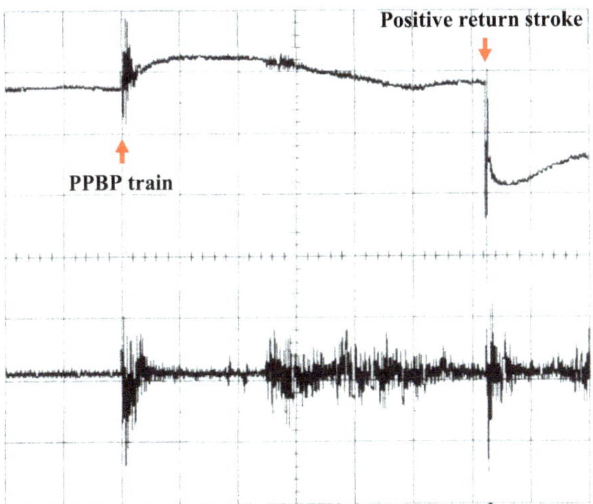

Figure 8.3 The radiation field (upper trace) and the simultaneously measured 3 MHz emission (lower trace) of PPBP train recorded in Uppsala, Sweden. The X-axis is the time and the Y-axis is the electric field. Units: X-axis: 5 ms/square, Y-axis: arbitrary units (adapted from [34])

Preliminary breakdown behaviour could be observed in some intra-cloud flashes as well [41]. However, there are no studies on the HF–VHF emission related to such breakdown phenomena.

8.2.3 Return stroke

Emission of RF radiation from the return stroke and its temporal neighborhood is well studied for the last 40 years [42–47].

Cooray [42] reported that once the propagation delays are eliminated, the broadband negative return stroke waveform coincides with the inception of the emission of HF at 3 MHz. The initiation of the return stroke typically marks the strongest emission of 3 MHz radiation. Similar observations have been reported in a few other studies as well [46,47]. Research outcomes presented by Gomes [34], which are pertinent to measurement at the same geographical location, done about 12 years after the studies reported in Cooray [42], partially confirm this observation for negative return strokes that succeed NPBP trains. In approximately 50% of the records, the intensity of 3 MHz radiation at the return stroke was insignificant compared with that associated with NPBPs (Figure 8.4). In the other 50%, in most cases, there was a time gap in the order of a few 10 s to a few 100 s of microseconds between the intense 3 MHz emission and the return stroke (Figure 8.4).

The US Department of Energy sponsored program, Fast on-Orbit Recording of Transient Events (FORTE) spacecraft, launched in the 1990s, provided a significantly large volume of information related to lightning and thunderstorm for the following couple of decades. Detection of the emission of RF radiation during

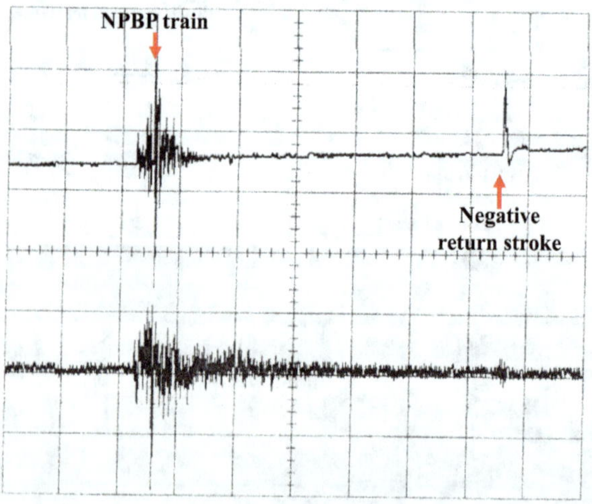

Figure 8.4 The radiation field (upper trace) and the simultaneously measured 3 MHz emission (lower trace) of NPBP train and the following return stroke recorded in Uppsala, Sweden. Units: X*-axis: 1 ms/square,* Y*-axis: arbitrary units (adapted from [34])*

discharge activities is one of the key observations of the project [48]. As per the observation in the FORTE program, Suszcynsky *et al.* [45] report that usually the HF–VHF emission from the temporal location of a return stroke is a blend of RF signals from the propagation of the final part of the stepped leader, the charge neutralization that takes place during the attachment process and the return stroke current itself. For subsequent return strokes, the dominant RF source is the dart leader.

Nakano *et al.* [44] have reported that there is a contrasting difference in the intensity of VHF–SHF emission around return stroke in summer and winter lightning. The RF emission is quite strong and detectable in almost all summer lightning whereas in winter lightning over 50% of the cases such emission is undetectable. This is not well explained so far.

Research done over the last few decades shows that RF emission from positive lightning events is significantly less than that from their negative counterpart [49,50]. Despite this weak emission of RF radiation, scientists pay high attention to the detection of emissions from positive lightning due to several reasons. Among them, one of the most significant reasons is the intensification of positive lightning at the trailing age of thunderstorms prior to flash floods, hailstorms and tornadoes, in supercells and strati-form sections of mesoscale convective storms. Thus, detection of VHF emission from positive lightning could be an accurate indicator in forecasting extreme weather events mentioned earlier [46].

There are a few studies in the past that have been successful in recording the VHF signals from positive lightning [46,51–53]. In the last few years, several techniques were developed where weak VHF signals emitted by lightning discharges could be detected and discriminated from the noise and stray signals [54,55]. Thus several studies could be conducted to find finer details of VHF emitted from positive lightning [50].

Zhang *et al.* [49] deployed the Lightning Mapping Array (LMA) system that is based on the reception of lightning-generated VHF signals combined with GPS technology, to track the discharge events of positive lightning. The system could reconstruct the discharge event in 3-D by detecting the radiation at 50 ns resolution and spatial precision of 50–100 m.

As per their observations on the pattern of VHF emission, Zhang *et al.* [49], identified three distinct phases of positive lightning (Figure 8.5). They classified them into three stages; discharge, continuing current and final. The process that takes place inside the cloud was considered the first or discharge stage. The flow of the continuing current following the return stroke was identified as the second stage. And the post-continuing-current period was referred to as the final stage.

The charge neutralization in the positive charge center gave rise to the emission of radiation in all stages. The velocity of the charge transfer has been detected to be directly proportional to the intensity of the RF signal. Thus, as is expected, in the post-return stroke phase during which the channel conductivity is maximum, the FR emission was detected to be the most intense. Zhang *et al.* [49] have also observed that the inception of a positive leader in a cloud is possible only if the ambient potential exceeds a certain threshold. The VHF emission prior to the return stroke is negligibly small.

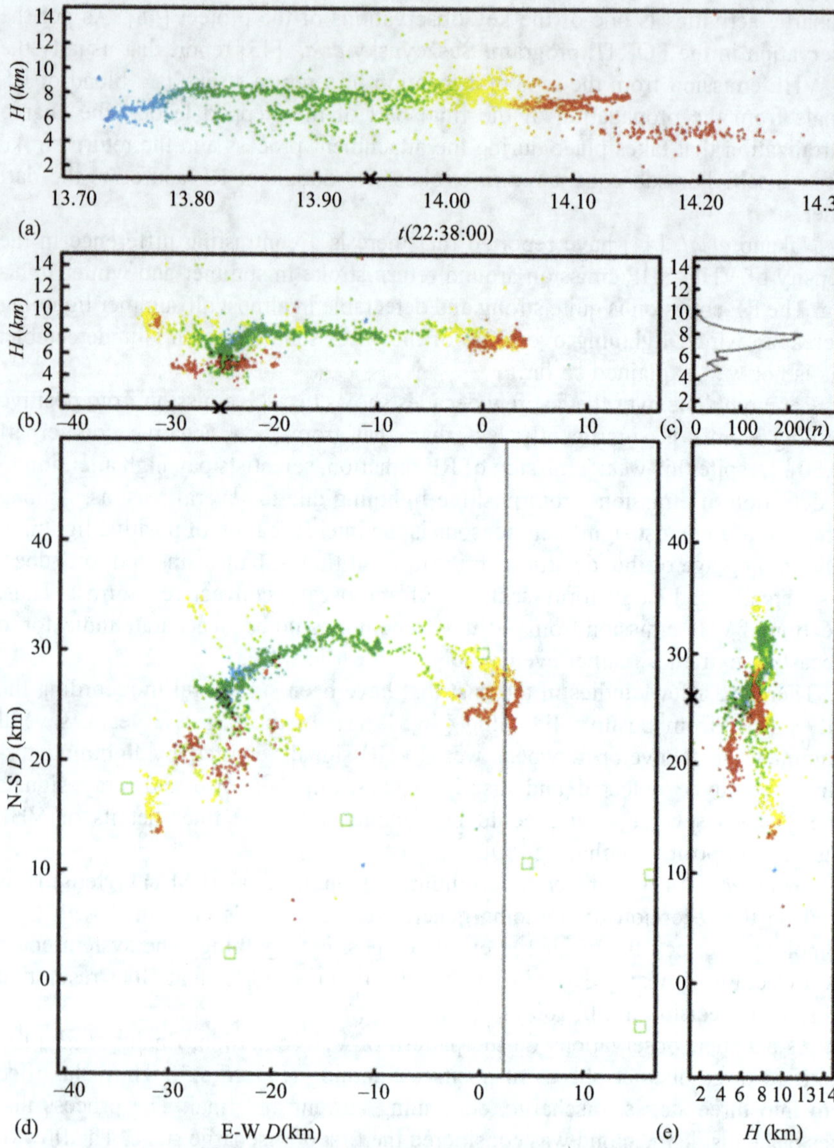

Figure 8.5 The space–time variation of radiation of positive lightning. The color scheme that varies from blue, green, and yellow to red represents the temporal distribution of VHF radiation. (a) Height-time plots; (b) north–south vertical projections; (c) height distribution of number (n) of radiation events; (d) plan views; (e) east–west vertical projections of the lightning radiation sources adapted from [49]

Recent studies have revealed that although the VHF emission of positive lightning is weak, the LF–VLF band emission, especially at low power is more dominant in positive lightning than that in their negative counterpart Hare *et al.* [56]. VHF and high-speed camera records reveal that short-burst negative recoils propagate both downward [57,58] and upward [59,60] in positive leaders. Integration of information from VHF sensors and high-speed cameras could provide many other important details on the characteristics of positive lightning such as the effects of side bi-directional leaders on the branching of the channel [61,62].

8.2.4 Cloud flash pulse trains

Electric and magnetic fields due to cloud flashes, which are not presumably associated with any ground flash activity, have been reported in many studies [63,64]. They have commonly been termed cloud flash pulses (CFP), intra-cloud pulses, etc. and the process that gives rise to such pulse trains is termed a cloud discharge. A typical cloud flash pulse train starts with a large static field change on which bipolar radiation pulses are superimposed. The pulse shapes of CFP are characterized by a few tens of microsecond long pulse width and pulse separation. The pulse trains may last for a few milliseconds to a few tens of milliseconds [65,66]. There are many pulses with multiple initial peaks and some of them are bipolar with a single initial peak, with several sharp unipolar pulses embedded in the leading edge [67].

HF–VHF emissions from CFP trains have been observed for a few decades [11,12,63,68]. Cloud flashes radiate intermittently at HF (10 MHz). The amplitude of the radiation is maximum at the beginning of the flashes. However, there are several occasions when the amplitude of HF radiation appears high in the later stages [63]. Betz *et al.* [12] report that the optical emission of cloud flashes is closely correlated to the intensity of the VHF radiation emitted. Most cloud flash pulse trains are not very intense HF–VHF emitters unless there are special events (narrow bipolar pulses or TIPPS) involved [68]. These events are discussed in the next section.

8.2.5 Trans-ionospheric pulse pairs (TIPPs)

An interesting VHF radiation phenomenon was discovered in the 1990s which is known today as Trans-ionospheric Pulse Pairs (TIPPs) [6,9,10]. Over the years, many other observations have been done to shed more light on this phenomenon [69–71].

TIPPs were first reported by the VHF sensor system, nicknamed Blackbeard, which was mounted in the ALEXIS satellite. TIPPs are energy-intense RF emissions in the VHF band (in a narrow spectrum; 25–200 MH) that occur in pairs. In the time domain, these RF bursts are a few microseconds in pulse width and a few tens of microseconds in pulse separation. They are known to be associated with lightning, however, their energy content is much larger than that of other lightning-related RF emissions. Wu [72] states that there are TIPPs, although not common, of which the power content may be in the order of 1 MW. Despite their origin at cloud height, they could penetrate the ionosphere to reach the height of satellites, thus, they were given the name TIPPs. In a single thunderstorm, there may be tens of

TIPPs generated, and by the time of publishing the work, Wu [72] reports that there are over 500,000 records of TIPPs available in global databases.

Since its discovery, there were several models proposed to explain their origin and behaviour, with special attention to the existence of the second pulse in the pair [69–75]. Models proposed for TIPPs have the challenge of giving reasonable explanations to its features such as pulse duration and inter-pulse separation, polarization and coherence of pulses, Ionospheric dispersion (group delay caused by the plasma of ionosphere on RF signals imposing a group delay that which has an approximate inverse square relation to the frequency), energy content ratio of the two pulses, correlation between pulse profiles, and bimodal characteristics.

Consequent to the earliest detection of TIPPs, the occurrence has been attributed to the RF emission of certain in-cloud discharges of which the direct reception and reflection from the ground surface give rise to the pulse pair [10,48,76–78].

As it was described in Jacobson and Light [74], the ground reflection model could provide information to estimate various spatial geometries of the discharge process. It is simply understood that the pulse separation of the pair reaching the satellite is proportional to the height from ground level to the emission source, leading to the computation of relative heights of various RF emitters in the cloud. If the information on the horizontal positioning of the discharge is known (thus, the observation angle is inferred), then the above calculation could be extended to infer the absolute height of the RF emission source. The RF emission process, of which the location has been computed in this method, was later referred to as the compact intra-cloud discharge [79].

The ground-reflection model of TIPPs provided a reasonable explanation for most of the observations, however, there were a few serious mismatches between the model and the observed pulse amplitudes [75]:

(a) If the second pulse is due to the reflected signal from the earth surface its amplitude should be less than that of the first due to the long distance it has travelled in air and the loss of energy at the reflection surface.
   (i) However, many studies confirm that the pulses have similar contents of mean energy [10,77,78].
   (ii) Below 100 MHz, in some cases, the second pulse of TPPIs have amplitudes greater than twice [76] or occasionally even ten times [80] that of the first.
   (iii) Above 110 MHz, in more than 50% of the cases, the amplitude of the second pulse exceeds that of the first [69].
(b) If the two pulses are of the same origin (both space and time) their wave profiles should be correlated apart from the dispersion and scattering effects. However, Jacobson *et al.* [69] report that above 110 MHz, in all cases the two pulses are grossly uncorrelated.
(c) In contrast to the previous conclusion, about 80% of the TPPIs occur independently of compact intra-cloud discharges which are considered the origin of TPPIs [70].

Citing the above drawbacks of the ground-reflection model, Wu [75] proposed a model where they ascribe TIPPs to relativistic electrons emitted in ground flashes which upon incident on earth surface will emit the bursts of RF signals into space within a narrow conical volume. Wu [72] claims that the outcomes of the model are in agreement with the electron parameters computed from the detection of X-ray emission of lightning by satellites. Wu [75] model predicts that the TIPP could even be detected on the ground (although they are much prominently being observed in space) and attribute the observation of Smith and Holden [81] to the TIPPs rather than that is due to an anthropogenic source as it was claimed in Smith *et al.* [82].

During the last few decades, scientific observations have revealed that there are clear indications of intensified discharge activities in thunderclouds, a few to few tens of minutes prior to extreme weather events such as severe storms that lead to flash floods, hurricanes etc. [39,83–89]. As lightning-related high energetic phenomena such as TPPI can be observed by satellite-mounted radiation sensors, both detection and modelling of such phenomena could lead to accurate extreme weather prediction models.

8.2.6 Narrow bipolar events (NBEs)

Since the 1980s, lightning researchers have observed a very strong VHF radiation phenomenon associated with intra-cloud discharges or unrelated to any other known discharge phenomena of a lightning flash, and characterized by bipolar pulses of a few tens of microsecond width and typically with positive initial peak (similar to a positive return stroke generated electric field) [82,90,91]. They were later termed narrow bipolar pulses (NBPs) [92,93]. NBEs are not frequently observed in cloud flashes, however, once occurred they make a distinct impact on the recording systems due to their high intensities that are comparable with fields due to return strokes of ground flashes [93]. Despite the return-stroke-like intensity, NBPs are not known to be associated with detectable optical emission [94,95].

The observations of Wu *et al.* [96] reveal that NBEs are typically involved with the deepest convection in a thundercloud, however, there are cases where the event height exceeds 15 km, instead of being at the center of the deep convection. The polarity of the NBEs also shows height dependence where the negative events occur at higher elevations of the storm with a threshold height of about 14 m [97]. Positive events are usually located in the middle parts of the storm. Wu *et al.* [96] report that negative NBEs are possibly detected only in very intense thunderstorms.

Analysis of NBP characteristics concludes that the length of the channel of an NBE could not be more than about 1 km [98]. Thus, NBEs are often termed compact intra-cloud discharges.

Similar to TIPPs, the highly energetic nature of NBPs makes them unable to penetrate the ionosphere to reach the heights of satellites. Thus, they are also potential candidates for predicting weather events. However, the unexplained relation between NBEs and other known physical events of a lightning flash, so far, pauses a barrier to utilizing the detection of NBPs for forecasting or nowcasting the succeeding extreme weather events.

8.2.7 Applications in lightning detection and mapping

Based on the type of pulses or pulse trains emitted in the VHF band by lightning events, different types of detection and mapping techniques could be adopted to make precise reconstruction of the discharge channel. Emission of intermittent pulses (Figure 8.1a) is well-fitting into the time-of-arrival type detection techniques where the minimum requirement of the number of ground-based detectors is four (at individual separations of a few tens of kilometers for best accuracy) for 3-D mapping [51,52,99].

Pulse trains with millisecond scale duration are usually employed in the interferometric technique, where the radiation sensors are closely spaced dipole arrays oriented to detect the azimuth angle of arriving signal. The interferometric method could also reconstruct the 3-D map of the discharge event with a suitable sensor arrangement [100–102]. Such 3-D reconstruction of the discharge channel requires a minimum of three sensors typically separated by 50–150 km from one another [102].

Lojou and Cummins [5] compared the performance of the two techniques and reported that they are equally good in producing 2-D mapping of both space and time features of both storms and storm cells. They are also equally accurate in the flash count during a thunderstorm. However, in 3-D mapping, the interferometric method could have a larger footprint with a lesser number of antennas than the time-of-arrival technique. The larger sensor spacing of the interferometric method reduces its location accuracy over the time of arrival method. As it is expected HF–VHF-based techniques could cover a much larger discharge activity volume than LF–VLF-based detection methods.

It is reasonable to assume that a majority of discharge events in a thunderstorm give rise to a certain level of HF–VHF bands of radiation emission. However, the intensity of the RF emission has a wide range of levels depending on the nature of the charge movement involved with the event. There are several electromagnetic field phenomena detected through broadband sensors that are not investigated for their simultaneous narrow band RF emission. Chaotic pulse trains [103], H-bursts [104], and Q-bursts [105] are a few examples.

8.3 Conclusions

HF–VHF emission is extremely useful in developing prediction mechanisms for extreme weather events such as severe storms and supercells, flash floods, and hurricanes, as many such events are directly correlated to cloud electrification activities. However, to utilize the RF signals for weather forecasting, two important requirements should be fulfilled; the emitted RF from a discharge event should be strong enough for remote sensing (highly advantageous if they reach satellite altitudes) and the RF emission should be correlated with a known event of atmospheric dynamics. Once these two conditions are satisfied the information could be incorporated with smart algorithms to make precise forecasting and at least nowcasting. In this respect, TIPPs and narrow bipolar pulses could make a significant impact on weather forecasting methodologies in future. There are several other lightning activities of which the broadband existence is known but the HF–VHF emission behaviour is yet to be explored.

References

[1] Peterson M. 'FORTE measurements of global optical lightning waveforms and implications for optical lightning detection'. Earth and Space Science. 2022; 9 (6), e2022EA002280, https://doi.org/10.1029/2022EA002280

[2] Peterson M. 'Modelling the transmission of optical lightning signals through complex 3-D cloud scenes'. *Journal of Geophysical Research: Atmospheres*. 2020; 125(23), e2020JD033231. https://doi.org/10.1029/2020jd033231

[3] Peterson M. 'Holes in optical lightning flashes: identifying poorly transmissive clouds in lightning imager data'. *Earth and Space Science*. 2021; 8 (2), e2020EA001294. https://doi.org/10.1029/2020ea001294

[4] Brunner K.N. and Bitzer P.M. 'A first look at cloud inhomogeneity and its effect on lightning optical emission'. *Geophysical Research Letters*. 2020; 47(10), e2020GL087094. https://doi.org/10.1029/2020gl087094

[5] Lojou J.-Y. and Cummins K.L. 'Total lightning mapping using both VHF interferometry and time-of-arrival techniques'. In *19th International Lightning Detection Conference*; Tucson, AZ, USA, April 2006.

[6] Jacobson A.R., Cummins K.L., Carter M., Klingner P., Roussel-Dupre D., and Knox S.O. 'FORTE radio-frequency observations of lightning strokes detected by the national lightning detection network'. *Journal of Geophysical Research*. 2000; 105(D12), 15,653. https://doi.org/10.1029/2000JD900103

[7] Massey R.S. and Holden D.N. 'Phenomenology of transionospheric pulse pairs'. *Radio Science*. 1995; 30(5), 1645–1659. https://doi.org/10.1029/95RS01563

[8] Massey R.S., Holden D.N., and Shao X.-M. 'Phenomenology of transionospheric pulse pairs: further observations'. *Radio Science*. 1998; 33(6), 1755–1761. https://doi.org/10.1029/98RS02031

[9] Taylor W.L., Brandes E.A., Rust W.D., and MacGorman D.R. 'Lightning activity and severe storm structure'. *Geophysical Research Letters*. 1984; 11 (5), 545–548. https://doi.org/10.1029/GL011i005p00545

[10] Holden D.N., Munson C.P., and Devenport J.C. 'Satellite observations of transionospheric pulse pairs'. *Geophysical Research Letters*. 1995; 22(8), 889–892. https://doi.org/10.1029/95GL00432

[11] Proctor D.E. 'VHF radio pictures of cloud flashes'. *Journal of Geophysical Research: Oceans*. 1981; 86(5), 4041–4071. https://doi.org/10.1029/JC086iC05p04041

[12] Betz H.-D., Marshall T.C., Stolzenburg M., *et al.* 'Detection of in-cloud lightning with VLF/LF and VHF networks for studies of the initial discharge phase'. *Geophysical Research Letters*. 2008; 35, L23802. https://doi.org/10.1029/2008GL035820

[13] VAISALA, *WEA-MET-G-Benefits of VHF Total Lightning Mapping-white paper*. Ref Ref. B211153EN-A, Vaisala, 2012.

[14] Weidman C.D. and Krider E.P. 'The radiation field wave forms produced by intracloud lightning discharge processes'. *Journal of Geophysical Research: Oceans*. 1979; 84(C6), 3159–3164. https://doi.org/10.1029/JC084iC06p03159

[15] Weidman C.D., Krider E.P., and Uman M.A. 'Lightning amplitude spectra in the interval from 100 kHz to 20 MHz'. *Geophysical Research Letters*. 1981; 8(8), 931–934. https://doi.org/10.1029/GL008i008p00931
[16] Weidman C.D. and Krider E.P. 'The amplitude spectra of lightning radiation fields in the interval from 1 to 20 MHz'. *Radio Science*. 1986; 21(6), 964–970. https://doi.org/10.1029/RS021i006p00964
[17] Beasley W., Uman M.A., and Rustan P.L. Jr. 'Electric fields preceding cloud-to-ground lightning flashes'. *Journal of Geophysical Research: Oceans*. 1982; 87(C7), 4883–4902. https://doi.org/10.1029/JC087iC07p04883
[18] Gomes C., Cooray V., and Jayaratne C. 'Comparison of preliminary breakdown pulses observed in Sweden and in Sri Lanka'. *The Journal of Atmospheric and Terrestrial Physics*. 1998; 60, 975–979. https://doi.org/10.1016/S1364-6826(98)00007-8
[19] Nag A. and Rakov V.A. 'Electric field pulse trains occurring prior to the first stroke in negative cloud-to-ground lightning'. *Electromagnetic Compatibility, IEEE Transactions*. 2009; 51(1), 147–150. https://doi.org/10.1109/TEMC.2008.2005488
[20] Wu T., Takayanagi Y., Yoshida S., Funaki T., Ushio T., and Kawasaki Z. 'Spatial relationship between lightning narrow bipolar events and parent thunderstorms as revealed by phased array radar'. *Geophysical Research Letters*. 2013; 40(3), 618–623. https://doi.org/10.1002/grl.50112
[21] Baharudin Z.A., Ahmad A., Fernando M., Cooray V., and Mäkelä J.S. 'Comparative study on preliminary breakdown pulse trains observed in Johor, Malaysia and Florida, USA'. *Atmospheric Research*. 2012; 117, 111–121. https://doi.org/10.1016/j.atmosres.2012.01.012
[22] Baharudin Z.A., Fernando M., Ahmad A., Mäkelä J.S., Rahman M., and Cooray V. 'Electric field changes generated by the preliminary breakdown for the negative cloud-to-ground lightning flashes in Malaysia and Sweden'. *Journal of Atmospheric and Solar-Terrestrial Physics*. 2012; 84, 15–24. https://doi.org/10.1016/j.jastp.2012.04.009
[23] Carlson B., Liang C., Bitzer P., and Christian H. 'Time domain simulations of preliminary breakdown pulses in natural lightning'. *The Journal of Atmospheric and Terrestrial Physics*. 2015; 120, 5316–5333. https://doi.org/10.1002/2014JD022765
[24] da Silva C.L. and Pasko V.P. 'Physical mechanism of initial breakdown pulses and narrow bipolar events in lightning discharges'. *Journal of Geophysical Research-Atmospheres*. 2015; 120, 4989–5009. https://doi.org/10.1002/2015JD023209
[25] da Silva C.L., Merrill R. A., and Pasko V.P. 'Mathematical constraints on the use of transmission line models to investigate the preliminary breakdown stage of lightning flashes'. *Radio Science*. 2016; 51, 367–380. https://doi.org/10.1002/2015RS005853.
[26] Karunarathne S., Marshall T.C., Stolzenburg M., *et al.* 'Locating initial breakdown pulses using electric field change network'. *Journal of*

Geophysical Research-Atmospheres. 2013; 118, 7129–7141. https://doi.org/10.1002/jgrd.50441.

[27] Karunarathne S., Marshall T.C., Stolzenburg M., and Karunarathna N. 'Modeling initial breakdown pulses of CG lightning flashes'. *Journal of Geophysical Research-Atmospheres*. 2014; 119, 9003–9019. https://doi.org/10.1002/2014JD021553

[28] Kolmašová I., Santolík O., Defer E., *et al.* 'Lightning initiation: strong pulses of VHF radiation accompany preliminary breakdown'. *Scientific Reports*. 2018; 8, 3650. https://doi.org/10.1038/s41598-018-21972-z

[29] Stolzenburg M., Marshall T.C., Karunarathne S., *et al.* 'Luminosity of initial breakdown in lightning'. *Journal of Geophysical Research: Atmospheres*. 2013; 118(7), 2918–2937. https://doi.org/10.1002/jgrd.50276

[30] Wilkes R.A., Uman M.A., Pilkey J.T., and Jordan D.M. 'Luminosity in the initial breakdown stage of cloud-to-ground and intracloud lightning'. *Journal of Geophysical Research-Atmospheres*. 2016; 121(3), 1236–1247. https://doi.org/10.1002/2015JD024137

[31] Marshall T. C., Stolzenburg M., Maggio C. R., *et al.* 'Observed electric fields associated with lightning initiation'. *Geophysical Research Letters*. 2005; 32(3), 1–5. https://doi.org/10.1029/2004GL021802

[32] Kolmašová I., Marshall T., Bandara S., *et al.* 'Initial breakdown pulses accompanied by VHF pulses during negative cloud-to-ground lightning flashes'. *Geophysical Research Letters*. 2019; 46, 5592–5600. https://doi.org/10.1029/2019GL082488

[33] Karunarathne N., Karunarathne S., Marshall T.C., and Stolzenburg M. 'Modelling initial breakdown pulses of lightning flashes using a matrix inversion method'. *Radio Science*. 2019; 54(3), 268–280. https://doi.org/10.1029/2018RS006695

[34] Gomes C. *'On the nature of lightning flashes: with special attention to the initiation, modelling and remote sensing of return strokes'. PhD Thesis*, University of Colombo, Sri Lanka/Uppsala University, Sweden, 1999.

[35] Ushio T., Kawasaki Z., Matsura K., and Wang D. 'Electric fields of initial breakdown in positive ground flash'. *Journal of Geophysical Research: Atmospheres*. 1998; 103(D12), 14135–14139. https://doi.org/10.1029/97JD01975

[36] Gomes C. and Cooray V. 'Radiation field pulses associated with the initiation of positive cloud to ground lightning flashes'. *The Journal of Atmospheric and Solar-Terrestrial Physics*. 2004; 66(12), 1047–1055. https://doi.org/10.1016/j.jastp.2004.03.015

[37] Schumann C., Saba M.M.F., Da Silva R.B.G., and Schulz W. 'Electric fields changes produced by positives cloud-to-ground lightning flashes'. *The Journal of Atmospheric and Solar-Terrestrial Physics*. 2013; 92, 37–42. https://doi.org/10.1016/j.jastp.2012.09.008

[38] Qie X., Wang Z., Wang D., Liu M., and Xuan Y. 'Characteristics of positive cloud to-ground lightning in Da Hinggan Ling forest region at relatively high latitude. Northeastern China'. *Journal of Geophysical Research.*

Atmospheres. 2013; 118, 13,393–13,404. https://doi.org/10.1002/2013JD020093

[39] Zhang Y., Zhang Y., Lü W., and Zheng D. 'Analysis and comparison of initial breakdown pulses for positive cloud-to-ground flashes observed in Beijing and Guangzhou'. *Atmospheric Research*. 2013;129–130, 34–41. https://doi.org/10.1016/j. atmosres.2013.03.006

[40] Johari D., Cooray V., Rahman M., Hettiarachchi P., and Ismail M.M. 'Characteristics of preliminary breakdown pulses in positive ground flashes during summer thunderstorms in Sweden'. *Atmosphere*. 2016; 7(3), 39. https://doi.org/10.3390/atmos7030039

[41] Wu T., Yoshida S., Akiyama Y., Stock M., Ushio T., and Kawasaki Z. 'Preliminary breakdown of intracloud lightning: initiation altitude, propagation speed, pulse train characteristics, and step length estimation'. *Journal of Geophysical Research: Atmospheres*. 2015; 120, 9071–9086. https://doi.org/10.1002/2015JD023546

[42] Cooray V. 'Temporal behaviour of lightning HF radiation at 3 MHz near the time of first return strokes'. *Journal of Atmospheric and Terrestrial Physics*. 1986; 48(1), 73–78. https://doi.org/10.1016/0021-9169(86)90114-5

[43] Weidman C.D. and Krider E.P. 'The amplitude spectra of lightning radiation fields in the interval from 1 to 20 MHz'. *Radio Science*. 1986; 21(6), 964–970. https://doi.org/10.1029/RS021i006p00964

[44] Nakano M., Kawasaki Z.I., Takeuti T., Sirai T., and Kawamata Y. 'VHF/UHF radiation from positive and negative lightning'. In Kikuchi H. (eds), *Environmental and Space Electromagnetics*. Springer, Tokyo. 1991. https://doi.org/10.1007/978-4-431-68162-5_10

[45] Suszcynsky D.M., Kirkland M.W., Jacobson A.R., *et al.* 'FORTE observations of simultaneous VHF and optical emissions from lightning: basic phenomenology'. *Journal of Geophysical Research: Atmospheres*. 2000; 105(D2), 2191–2201, https://doi.org/10.1029/1999JD900993

[46] Chen Y., Zhang B., Wang Y., Wang Z., and Xu Y. 'The detailed time-frequency analysis of the HF/VHF emissions accompanied with CG lightning flashes'. In *11th Asia-Pacific International Conference on Lightning (APL)*, 2019; pp. 1–5, https://doi.org/10.1109/APL.2019.8815989

[47] Pathak P.P. and Rajput S. 'Signatures of radiation due to return stroke in LF-VHF range during ground discharge'. *Journal of Engineering and Technology Research*. 2021; 12(1), 15–18. https://doi.org/10.5958/2321-581X.2021.00003.9

[48] Jacobson A.R., Knox S.O., Franz R., and Enemark D.C. 'FORTE observations of lightning radio-frequency signatures: capabilities and basic results'. *Radio Science*. 1999; 34(2), 337–354. https://doi.org/10.1029/1998RS900043

[49] Zhang Y., Meng Q., Krehbiel P., Liu X., Yan M., and Zhou X. 'Spatiotemporal characteristics of positive cloud-to-ground lightning discharges and bidirectional leader of the lightning'. *Science in China Series D*. 2006; 49(2), 212–224. https://doi.org/10.1007/s11430-005-0054-1

[50] Li S., Qiu S., Shi L., and Li Y. 'Broadband VHF observations of two natural positive cloud-to-ground lightning flashes'. *Geophysical Research Letters*. 2020; 47, e2019GL086915. https://doi.org/10.1029/2019GL086915

[51] Zhang G., Zhao Y., Qie X., Zhang T., Wang Y-H., and Chen C-P. 'Observation and study on the whole process of cloud-to-ground lightning using narrowband radio interferometer'. *Science in China. Series D, Earth Sciences*. 2008; 51, 694–708. https://doi.org/10.1007/s11430-008-0049-9

[52] Zhang G., Wang Y., Qie X., *et al.* 'Using lightning locating system based on time-of-arrival technique to study three-dimensional lightning discharge processes'. *Science China Earth Sciences*. 2010; 53(4), 591–602. https://doi.org/10.1007/s11430-009-0116-x

[53] Akita M., Stock M., Kawasaki Z., Krehbiel P., Rison W., and Stanley M. 'Data processing procedure using distribution of slopes of phase differences for broadband VHF interferometer'. *Journal of Geophysical Research: Atmospheres*. 2014; 119, 6085–6104. https://doi.org/10.1002/2013JD020378

[54] Wang T., Qiu S., Shi L.H., and Li Y. 'Broadband VHF localization of lightning radiation sources by EMTR'. *IEEE Transaction on Electromagnetic Compatibility*. 2017; 59(9), 1949–1957. https://doi.org/10.1109/TEMC.2017.2651142

[55] Wang T., Li-Hua S., Shi Q., *et al.* 'Multiple-antennae observation and EMTR processing of lightning VHF radiations'. *IEEE Access*, 2018;6, 26558–26566. https://doi.org/10.1109/ACCESS.2018.2833115

[56] Hare B.M., Scholten O., Dwyer J., *et al.* 'Needle-like structures discovered on positively charged lightning branches'. *Nature*, 2019; 568(7752), 360–363. https://doi.org/10.1038/s41586-019-1086-6

[57] Kong X., Qie X., and Zhao Y. 'Characteristics of downward leader in a positive cloud-to-ground lightning flash observed by high-speed video camera and electric field changes'. *Geophysical Research Letters*. 2008; 35, L05816. https://doi.org/10.1029/2007GL032764

[58] Saba M.M.F., Cummins K.L., Warner T.A., *et al.* 'Positive leader characteristics from high-speed video observations'. *Geophysical Research Letters*. 2008; 35, L07802. https://doi.org/10.1029/2007GL033000

[59] Kotovsky D., Uman M.A., Wilkes R.A., and Jordan D.M. 'High-speed video and lightning mapping array observations of in-cloud lightning leaders and an M component to ground'. *Journal of Geophysical Research: Atmospheres*. 2019; 124, 1496–1513. https://doi.org/10.1029/2018JD029506

[60] Pu Y. and Cummer S.A. 'Needles and lightning leader dynamics imaged with 100–200 MHz broadband VHF interferometry'. *Geophysical Research Letters*. 2019; 46, 13,556–13,563. https://doi.org/10.1029/2019GL085635

[61] Yuan S., Jiang R., Qie X., Sun Z., Wang D., and Srivastava A. 'Development of side bidirectional leader and its effect on channel branching of the progressing positive leader of lightning'. *Geophysical Research Letters*. 2019; 46, 1746–1753. https://doi.org/10.1029/2018GL080718

[62] Qie X., Lu G., Liu D., *et al.* 'Study on dynamic-microphysical-electrical processes in severe thunderstorms and lightning hazards (STORM973)'. In

Proceedings in the 24th International Lightning Detection Conference & 6th International Lightning Meteorology; San Diego, CA, USA, April 2016, 18–21.

[63] Villanueva Y., Rakov V.A., Uman M.A., and Brook M. 'Microsecond-scale electric field pulses in cloud lightning discharges'. *Journal of Geophysical Research: Atmospheres*. 1994; 99(D7), 14353–14360. https://doi.org/10.1029/94JD01121

[64] Fernando M. and Cooray V. 'Sub microsecond structure of electric field pulses generated by cloud flashes'. In *24th International Conference on Lightning Protection*; UK, 1998.

[65] Sharma S.R., Fernando M., and Gomes C. 'Signatures of electric field pulses generated by cloud flashes'. *Journal of Atmospheric and Solar–Terrestrial Physics*. 2005; 67(4), 413–422. https://doi.org/10.1016/j.jastp.2004.09.007

[66] Bodhika J.P., Fernando M., and Cooray V. 'Characteristics of cloud flashes inferred from measurements of electric fields over Sri Lanka a tropical country'. In *Proceedings of the 28th International Conference on Lightning Protection;* Kanazawa, Japan, 2006.

[67] Ahmad N.A., Fernando M., Baharudin Z.A., *et al.* 'The first electric field pulse of cloud and cloud-to-ground lightning discharges'. *Journal of Atmospheric and Solar-Terrestrial Physics*. 2010;72(2–3), 143–150. https://doi.org/10.1016/j.jastp.2009.11.001

[68] Stolzenburg M., Marshall T.C., Bandara S., Hurley B., and Siedlecki R. 'Ultra-high speed video observations of intracloud lightning flash initiation'. *Meteorology and Atmospheric Physics*. 2021; 133, 1177–1202. https://doi.org/10.1007/s00703-021-00803-3

[69] Jacobson A.R., Holzworth R.H., and Shao X.-M. 'Observations of multi-microsecond VHF pulsetrains in energetic intracloud lightning discharges'. *Annals of Geophysics*. 2011; 29, 1587–1604. https://doi.org/10.5194/angeo-29-1587-2011

[70] Jacobson A.R. and Light T.E.L. 'Revisiting narrow bipolar event intracloud lightning using the FORTE satellite'. *Annals of Geophysics*. 2012; 30, 389–404. https://doi.org/10.5194/angeo-30-389-2012

[71] Dolgonosov M.S., Gotlib V.M., Vavilov D.I., Zelenyi L.M., and Klimov S.I. 'Solitary trans-ionospheric pulse pairs onboard of the microsatellite Chibis-M'. *Advances in Space Research*. 2015; 56(6), 1177–1184. https://doi.org/10.1016/j.asr.2015.06.009

[72] Wu HC. 'Relativistic-microwave theory of ball lightning'. *Scientific Reports*. 2016; 6, 28263. https://doi.org/10.1038/srep28263

[73] Jacobson A.R. 'How do the strongest radio pulses from thunderstorms relate to lightning flashes'. *Journal of Geophysical Research*. 2003; 108(D24), 4778. https://doi.org/10.1029/2003JD003936

[74] Jacobson A.R. and Light T.E.L. 'Bimodal radio frequency pulse distribution of intracloud-lightning signals recorded by the FORTEsatellite'. *Journal of Geophysical Research*. 2003;108(D9), 4266. https://doi.org/10.1029/2002JD002613

[75] Wu H.-C. 'Origin of transionospheric pulse pairs' *Geophysical Research Letters*. 2017; 44, 2597–2603. https://doi.org/10.1002/2017GL072741

[76] Massey R.S. and Holden D.N. 'Phenomenology of trans ionospheric pulse pairs'. *Radio Science*. 1995; 30(5), 1645–1659. https://doi.org/10.1029/95RS01563

[77] Massey R.S., Holden D.N., and Shao X.-M. 'Phenomenology of trans ionospheric pulse pairs: further observations'. *Radio Science*. 1998; 33(6), 1755–1761. https://doi.org/10.1029/98RS02031

[78] Massey R.S., Knox S.O., Franz R.C., Holden D.N., and Rhodes C.T. 'Measurements of trans ionospheric radio propagation parameters using the FORTE satellite'. *Radio Science*. 1998; 33(6), 1739–1753. https://doi.org/10.1029/98RS02032

[79] Nag A. and Rakov V.A. 'Compact intracloud lightning discharges: 1. Mechanism of electromagnetic radiation and modeling'. *Journal of Geophysical Research*. 2010; 115, D20102. https://doi.org/10.1029/2010JD014235

[80] Tierney H. E., Jacobson A.R., Roussel-Dupré R., and Beasley W.H. 'Transionospheric pulse pairs originating in maritime, continental, and coastal thunderstorms: pulse energy ratios'. *Radio Science*, 2002; 37(3), 11-1-11-7. https://doi.org/10.1029/2001RS002506

[81] Smith D.A. and Holden D.N. 'Ground-based observations of sub ionospheric pulse pairs'. *Radio Science*. 1996; 31(3), 553–571. https://doi.org/10.1029/96RS00007

[82] Smith D.A., Shao X.M., Holden D.N., *et al.* 'A distinct class of isolated intracloud lightning discharges and their associated radio emissions'. *Journal of Geophysical Research*. 1999; 104(D4), 4189–4212. https://doi.org/10.1029/1998JD200045

[83] Rigo T. and Llasat M.C. 'Forecasting hailfall using parameters for convective cells identified by radar'. *Atmospheric Research*. 2016; 169, Part A, 366–376. https://doi.org/10.1016/j.atmosres.2015.10.021

[84] Farnell C., Rigo T., and Pineda N. 'Lightning jump as a nowcast predictor: application to severe weather events in Catalonia'. *Atmospheric Research*. 2017; 183, 130–141. https://doi.org/10.1016/j.atmosres.2016.08.021

[85] Wapler K. 'The life-cycle of hailstorms: lightning, radar reflectivity and rotation characteristics'. *Journal of Atmospheric Research*. 2017; 193, 60–72. https://doi.org/10.1016/j.atmosres.2017.04.009

[86] Tian Y., Yao W., Sun Y., *et al.* 'A method for improving the performance of the 2σ lightning jump algorithm for nowcasting hail'. *Atmospheric Research*. 2022; 280, 106404. https://doi.org/10.1016/j.atmosres.2022.106404

[87] Carey L.D., Schultz E.V., Schultz C.J., *et al.* 'An evaluation of relationships between radar-inferred kinematic and microphysical parameters and lightning flash rates in Alabama Storms'. *Atmosphere*. 2019; 10, 796. https://doi.org/10.3390/atmos10120796

[88] Xu L., Chen S., and Yao W. 'Evaluation of lightning prediction by an electrification and discharge model in long-term forecasting experiments'. *Advances in Meteorology*. 2022; Article ID 4583030. https://doi.org/10.1155/2022/4583030

[89] Meng Q., Yao W., and Xu L. 'Development of lightning nowcasting and warning technique and its application'. *Advances in Meteorology*. 2019; Article ID 2405936, 9. https://doi.org/10.1155/2019/2405936

[90] Le Vine D.M. 'Sources of the strongest RF radiation from lightning'. *Journal of Geophysical Research*. 1980; 85(C7), 4091–4095. https://doi.org/10.1029/JC085iC07p04091

[91] Willett J.C., Bailey J.C., and Krider E.P. 'A class of unusual lightning electric field waveforms with very strong high-frequency radiation'. *Journal of Geophysical Research*. 1989; 94(D13), 16255–16267. https://doi.org/10.1029/JD094iD13p16255

[92] Eack K.B. 'Electrical characteristics of narrow bipolar events'. *Geophysical Research Letters*. 2004; 31(20), L20102. https://doi.org/10.1029/2004GL021117

[93] Rison W., Krehbiel P., Stock M., *et al.* 'Observations of narrow bipolar events reveal how lightning is initiated in thunderstorms'. *Nature Communications*. 2016; 7, 10721. https://doi.org/10.1038/ncomms10721

[94] Jacobson A.R., Light T.E.L., Hamlin T., and Nemzek R. 'Joint radio and optical observations of the most radio-powerful intra cloud lightning discharges'. *Annals of Geophysics*. 2013; 31(3), 563–580. https://doi.org/10.5194/angeo-31-563-2013

[95] Wu T., Yoshida S., Ushio T., Kawasaki Z., and Wang D. 'Lightning-initiator type of narrow bipolar events and their subsequent pulse trains'. *Journal of Geophysical Research: Atmospheres*. 2014; 119, 7425–7438. https://doi.org/10.1002/2014JD021842

[96] Wu T., Dong W., Zhang Y., *et al.* 'Discharge height of lightning narrow bipolar events'. *Journal of Geophysical Research*. 2012; 117, D05119. https://doi.org/10.1029/2011JD017054

[97] Wu T., Dong W., Zhang Y., and Wang T. 'Comparison of positive and negative compact intracloud discharges'. *Journal of Geophysical Research*. 2011; 116, D03111. https://doi.org/10.1029/2010JD015233

[98] Liu H., Dong W., Wu T., Zheng D., and Zhang Y. 'Observation of compact intracloud discharges using VHF broadband interferometers'. *Journal of Geophysical Research*. 2012; 117, D01203. https://doi.org/10.1029/2011JD016185

[99] Thomas R.J., Krehbiel P.R., Rison W., *et al.* 'Accuracy of the lightning mapping array'. *Journal of Geophysical Research*. 2004; 109, D14207. https://doi.org/10.1029/2004JD004549

[100] Richard P., Delannoy A., Labaune G., and Laroche P., 'Results of spatial and temporal characterization of the VHF-UHF radiation of lightning'. *Journal of Geophysical Research: Atmospheres*. 1986; 91(D1), 1248–1260. https://doi.org/10.1029/JD091iD01p01248

[101] Defer E., Blanchet P., Thery C., *et al.* 'Lightning activity for the July 10, 1996, storm during the Stratosphere-Troposphere Experiment: radiation, Aerosol, and Ozone-A (STERAO-A) experiment'. *Journal of Geophysical Research*. 2001; 106(D10), 10,151–10,172. https://doi.org/10.1029/2000JD900849

[102] Qing M., Yijun Z., He P., Xiaofeng X., *et al.* 'Lightning observation and field experiment with 3D SAFIR system during summer 2003 in Beijing-Hebei Area'. In *27th International Conference on Lightning Protection*; Avignon – France, September 2004, 13–16.

[103] Gomes C., Cooray V., Fernando M., Montano R., and Sonnadara U. 'Characteristics of chaotic pulse trains generated by lightning flashes'. *Journal of Atmospheric and Solar-Terrestrial Physics*. 2004; 66(18), 1733–1743. https://doi.org/10.1016/j.jastp.2004.07.036

[104] Rakov V.A., Uman M.A, Hoffman G.R., Masters M.W., and Brook M. 'Burst of pulses in lightning electromagnetic radiation: observations and implications for lightning test standards'. *IEEE Transactions on Electromagnetic Compatibility*. 1996; 38(2), 156–164. https://doi.org/10.1109/15.494618

[105] Bór J., Ludván B., Attila N., and Steinbach P. 'Systematic deviations in source direction estimates of Q-bursts recorded at Nagycenk, Hungary'. *Journal of Geophysical Research: Atmospheres*. 2016; 121, 5601–5619. https://doi.org/10.1002/2015JD024712

Chapter 9
Microwave radiation generated by lightning

Mohd Riduan Ahmad[1], *Joan Montanya*[2] *and Vernon Cooray*[3]

9.1 Introduction

The term microwave used in this chapter follows the general definition of microwave frequency band which falls between 300 MHz and 300 GHz. In the early days of the radio, electrical sparks were used to excite circuits of the transmitters. For example, H. Hertz built a spark-gap transmitter at 450 MHz to investigate the properties of electromagnetic waves. At higher frequencies, using spark-gap transmitters with resonators, J. C. Bose and O. Lodge generated 3 mm and 2.5 cm waves, respectively. Moreover, electrical sparks such as lightning occur naturally in thunderstorms. In lightning, a wide variety of discharges ranging from millimeter to kilometer scale with temporal characteristics from sub-nanosecond to a second produce wideband radio emissions. The research study on microwave radiation from lightning has been started decades ago with the first complete experimental work reported by Brook and Kitagawa in 1964 [1]. Since then, more research works on the microwave radiations generated by lightning flashes have been reported. Besides lightning, thunderclouds present sub-microsecond radiation extending to microwaves identified to be associated with micro discharges between cloud's hydrometeors (e.g. [2–4]). In a laboratory experiment, Keeney [5] showed the production of a microwave radiation by the neutralization of two drops of water of the same diameter (1.05 mm) carrying opposite charges. This chapter presents previous and current findings in the study of microwave radiations generated by lightning flashes.

9.2 Measurement of microwave radiation from lightning

The first complete description of the measurement of microwave radiation was reported by Brook and Kitagawa in 1964 [1]. The experiment used two helical

[1]Atmospheric Science and Disaster Management Research Group, Universiti Teknikal Malaysia Melaka, Malaysia
[2]Department of Electrical Engineering, Polytechnic University of Catalonia, Spain
[3]Department of Electrical Engineering, Uppsala University, Sweden

antennas to cover two resonance frequencies at 420 and 850 MHz. The bandwidth of the tuned receiver was 1.5 MHz. All results were taken from thunderstorms that located at distance between 10 and 30 km from the sensor with 92 samples from ten thunderstorms. It has been suggested that microwave radiations were associated primarily with the formation of streamers (breakdown process includes electron avalanches) based on the observation results where both stepped leader and dart leader were accompanied by strong impulsive microwave radiation. On the other hand, only some of return strokes (less than 50%) were accompanied by microwave radiations and in some cases the microwave radiations were detected 60–100 μs after the return stroke onset. Moreover, microwave radiations of cloud flash pulses were found to be much stronger than the microwave radiation of the stepped leader, dart leader and return stroke.

A study conducted in [6] measured microwave radiations emitted by natural lightning at Lenin Hills of Moscow during the summer of 1968. There were four microwave band receivers at 0.4, 0.7, 0.9 and 1.3 GHz. Vertical polarized antennas with 1 MHz bandwidth were used to measure the microwave radiations. The lightning sensing system (include those four microwave receivers) was triggered with a frame antenna operated from 2 to 60 kHz. A total of 200 records of microwave radiation bursts were obtained at 400, 700, and 900 MHz. Out of 200 records, only 20 microwave radiation bursts were detected at 1.3 GHz. Due to the low number of captured samples and low amplitude intensity, the microwave radiation at 1.3 GHz was not discussed by the authors. Based on Figures 9.4–9.7 in their paper, microwave bursts at 700 and 900 MHz were the strongest while microwave bursts of 1.3 GHz was very small.

An observation of microwave radiation at 2.2 GHz captured from cloud-to-ground (CG) and intra-cloud (IC) flashes has been reported in [7]. The flashes have been detected close to Kennedy Space Centre (KSC) in Florida in July 1977. A parabolic antenna with gain of 42 dB was used to capture the microwave radiation and the signal was down converted to 22 MHz with 3 MHz bandwidth by the receiver. The microwave radiation bursts were found to occur simultaneously with dart leaders and preliminary field changes prior to the first return stroke lasted between 5 and 10 ms. On the other hand, microwave bursts were detected few hundred microseconds after the first return stroke onset, similar to observation in [1]. It has been assumed that this delay was due to the high elevation angle that the parabolic antenna pointed toward the cloud while the microwave radiations from the return stroke were at low elevation angle.

In the preparation for the current space-based lightning imagers, simultaneous measurements of optical radiation from below and above thunderstorms by an instrumented U-2 aircraft were conducted in [8]. At ground, a 2.2 GHz receiver with a horn antenna complemented these observations. A record of a lightning flash with three return strokes showed a good agreement between the optical pulses detected at 20 km altitude and from below the thunderstorm. Although the authors did not analyze the measured microwave signals, they found coincidence between the optical pulses and the microwave radiation. The onsets of optical and microwave signals were coincident whereas not the times of their peaks.

A comprehensive review on very high frequency (VHF) and microwave radiations generated by lightning flashes has been provided in [9]. The experiments were conducted in New Mexico, USA in 1982 and France in 1983 and 1984 at seven frequency bands; 60 MHz, 100 MHz, 175 MHz, 300 MHz, 500 MHz, 900 MHz and 4.6 GHz. Vertically polarized dipole antennas have been used to capture the radiations and the bandwidth of the receivers was around 350 kHz. Intense VHF and microwave radiation bursts were observed to accompany preliminary breakdown pulses, stepped leaders, first return stroke and subsequent return strokes from 60 MHz to 500 MHz. Interestingly, microwave radiations at 900 MHz and 4.6 GHz were visible only during the first return stroke and third subsequent return stroke and absent from other return strokes. Thus, the authors suggested this radiation especially at 4.6 GHz was a result of micro breakdowns induced in the cloud between charged hydrometeors [2–5] through the electric field wave radiated by the return stroke [10,11].

Fedorov *et al.* [12] set up a measurement system (included radiometer) to measure millimeter wave radiations from lightning at 37.5 GHz (refer Figure 2.19 in their paper). The signals analyzed were limited to those with occurrence distances within 5 km as the authors would like to minimize the attenuation of the received signals. Only two recorded signals were analyzed and discussed by the authors. These two signals were found to be having complex shapes with a series of single pulses that exceed 5 μs, lasting about 60 μs. The authors suggested that the recorded signals were corresponding to the current pulses that arise from return stroke in the conducting channel. It only could be concluded that lightning flashes do emit millimeter wave radiation and the spectral intensity of that millimeter wave radiation is corresponding to the temperature of the antenna.

Yoshida [13] observed and studied microwave radiation of upward positive lightning (UPL) and initial continuous current (ICC) of the triggered lighting at 2.9 GHz. The author found that the amplitude change of the microwave burst is proportional to the burst of the UPL and ICC pulses. Most of the microwave radiation pulses appeared before the detection of ICC pulses with a period less than 1 μs. The negative leader development is found to be accompanied by strong microwave power but in short pulse train durations. The authors believed that these short pulses were the results of the charge movement in a short distance at the tips of negative leaders because of the short propagation of the charge would produce both VHF and 2.9 GHz impulsive radiations. Meanwhile, the long duration microwave pulses were believed to be the results of current pulse due to the absence of VHF radiations along the long duration microwave pulses. The authors suggested that these long duration 2.9 GHz radiations were generated by the thermal emissions from the channel of lightning current.

Ahmad *et al.* [14] studied about the microwave radiation associates with the narrow bipolar event (NBP) at 2.4 GHz frequency spectrum. The authors reported that there was more than 70% of the noise-liked burst microwave radiation of NBP started few microseconds before the onset of their corresponding electric field change. The range of microwave burst duration for NBP is between 2 μs and 8 μs and the average burst duration of negative narrow bipolar event and positive narrow bipolar event were 5.39 μs and 0.84 μs, respectively. The average peak amplitude of negative and positive NBPs was 1.46 and 1.80 V, respectively within a range of 0.28–3.78 V.

An experimental observation on microwave radiations emitted by natural lightning flashes has been reported by [15] where the authors used a 1.63 GHz ceramic patch antenna with 2 MHz bandwidth receiver (down converted to base-band with 64 MHz sampling rate) to capture the microwave radiations. The authors had observed and reported the microwave radiations from negative CG flashes. The microwave radiations were further analyzed for each individual pulse within a negative CG flash such as preliminary breakdown, step leaders, return stroke, and dart leaders. It was reported that preliminary breakdown, stepped leader and dart leader were associated with trains of individually resolvable pulses while the return stroke generated stronger continuous noise-like burst. The authors further classified the microwave burst into strong impulses and intermittent impulses. The preliminary breakdown process was found to be associated with strong isolated impulses while stepped and dart leaders were associated with intermittent impulses. There was no presence of burst during the interval of dart leader and return stroke. The pulse train duration burst of dart leader and return stroke were reported to be 1 ms and 0.2 ms, respectively. Therefore, the authors have suggested the preliminary breakdown process might be different from the development process of stepped and dart leaders due to this distinct characteristic. On the other hand, no impulsive radiations were detected just prior to the onset of all return strokes (first and subsequent return strokes).

Recent studies conducted in 2020 [16,17] analyzed NBPs accompanied by VHF and microwave radiations. The authors found that microwave radiations of 16 NBPs were found to lead both VHF and fast electric field changes waveforms with lead time of 125.53 ± 81.32 ns and 600.65 ± 222.34 ns, respectively. Both burst trains of VHF and microwave radiations were observed with rising phase (RP) and damping phase (DP) characteristic. A total of 21 microwave and 22 VHF waveforms were found to have initial stage (IS) at the earlier part of the RP with clear bipolar shape waveform. Another recent studies [18,19] analyzed stepped leaders of ten very close negative CG flashes (within reversal distance) accompanied by both VHF and microwave radiations. The authors found that all microwave radiations were observed to precede stepped leaders process with average leading time of 0.423 ± 0.378 μs. Around 80.41% of microwave radiation bursts preceded VHF radiation bursts with average leading time of 0.540 ± 0.596 μs.

9.3 The effect of microwave radiation from lightning

A study in [20] analyzed the interference of lightning flashes with multiple antennas wireless communication systems operating in the microwave band at 2.4 GHz and 5.8 GHz. A bit error rate (BER) measurement method was used to evaluate BER and packet error rate (PER) during five heavy thunderstorms on 25 January 2011 and 17–20 March 2011. The transmitter–receiver separation was fixed at 10 meters with line-of-sight (LOS) consideration. The authors found maximum recorded BER was 9.9×10^{-1} and the average recorded BER and PER were 2.07×10^{-2} and 2.44×10^{-2} respectively during the thunderstorms with the average fair weather BER and PER values under the influence of adjacent channel interference (ACI) and co-channel interference (CCI) being 1.75×10^{-5} and 7.35×10^{-6}, respectively. Thus, the authors concluded that multiple antennas wireless communication systems operating at the microwave frequency can be significantly interfered by lightning.

Continuation to the work done in [20, 21] studied the severity level of interference caused by natural lightning to wireless communication system at 2.4 GHz. The authors found that NBP was the strongest interferer to the bit's transmission than other processes of lightning flashes. Even though NBP consists of only a single bipolar pulse, it has been observed to interfere with the bit's transmission more severely than typical IC and CG flashes (see their Figures 9.2, 9.3, and 9.6). It is interesting how a single NBP pulse could produce very intense microwave radiation pulses that could interfere the bits transmission severely. Consider that if proposal put forwarded by [22] to be true that NBP discharge is the result of a high-energy electron avalanches mechanism then, the spectral amplitude of the resultant radiation field would be peaked at around 1 GHz as estimated in [23]. On top of that, based on VHF interferometer mapping, Rison *et al.* [24] and Tilles *et al.* [25] proposed that NBP is the result of fast breakdown mechanism (consists of only electron avalanches and fast propagating streamers) which possibly could emit intense microwave radiation. This might explain why a single NBP pulse could produce a very intense microwave burst at the 2.4 GHz band and cause the most severe burst error to the wireless communication network.

9.4 Sources generating microwave radiation

Brook and Kitagawa [1] and Le Boulch *et al.* [9] have strongly suggested that initial breakdown or ionization of the virgin air (electron avalanches and streamer) is the source that emit microwave radiations. As a matter of fact, all atmospheric discharges (electrostatic discharges and lightning discharges) are initiated by the same mechanism known as electron avalanches [26,27]. Electrostatic discharge is due to micro breakdowns induced in the cloud between charged hydrometeors through the electromagnetic field wave radiated by the return stroke [2–5,10,11]. Electron avalanches can take place either as in isolated discharge process (corona) or as initial breakdown process to initiate other processes such as streamers and leaders.

Cooray and Cooray [23] have simulated the electron avalanche process under various scenarios such as the tip of a streamer, and inside the model of coaxial and spherical geometries. The aim of the simulation was to evaluate the electromagnetic fields radiated by the electron avalanches by using an approach of accelerating charges. This is because electron avalanches consist of electrons accelerating and decelerating in given background electric field. The total radiation fields generated by the accelerating electron avalanche are given in [23]. The simulated radiation field has a duration of approximately 1–2 ns with a rise time in the range of 0.25 ns. The results of the radiated fields were taken into the calculations of the frequency spectrum analysis. The frequency spectrum of the electron avalanches was found to peak at around 100 MHz under coaxial and spherical geometries and around 1 GHz when it was simulating at the tip of a streamer. These simulation results were in good agreement with the experimental results in [28] where they carried out the experiment to study about the current produced by the spark by two spherical bodies inside the lab. The calculated frequency spectrum from the current produced typically peaked at two points, 100 MHz and 1 GHz. Therefore, Cooray and Cooray [23] postulated that the calculated 1 GHz frequency spectrum by [28] was due to the discharges of electron avalanche itself as the gap between the two spherical bodies was too short and the background potential inside the lab was not enough to sustain the development of the electron avalanche into streamer which led to the electron avalanche discharge alone.

Luque [29] simulated streamer collisions by using electromagnetic model of streamer propagation by solving Maxwell equations together with electron transport and reactions including photoionization. Streamer collisions commonly occur in a long air gap and during the progressing of a stepped leader. The simulation results showed that streamer collisions radiated electromagnetic field pulses with spectrum peaked at 300 MHz and had almost flat frequency spectrum response between 100 MHz and 1 GHz (see their Figure 9.3b). The radiated far-field pulse was found to decay only as $1/r$ as it moves away from the collision point. It has an asymmetric bipolar shape with sharp peak of opposite polarity to the field applied in the discharge followed by an overshoot and a slower decay. The total pulse durations calculated to be less than 5 ns (see their Figure 9.3a).

Furthermore, this microwave electromagnetic fields were observed to penetrate the streamer heads and capable to accelerate electrons up to about 100 keV. This mechanism is suggested to generate runaway electrons that possibly colliding with a nucleus and emit X-ray photons. In other word, streamer collisions are precursors of high-energy processes in electrical discharges. Recently, emission of X-rays in a long spark has been observed to associate with the emission of microwave radiation at 2.4 GHz [30].

Shi *et al.* [31] simulated two bidirectional propagating streamers and streamer collisions by solving Poisson's equation rather than solving full Maxwell's equations like [29] did. The simulation result showed that propagating streamers dominated the spectrum below 1 GHz while streamer collisions dominated spectrum at 1 GHz and above. Moreover, microwave radiation above 1 GHz became stronger as the total length of colliding streamers or the ambient field increased. When the collisions take place, the maximum localized electric field was greatly increased up to around $10E_k$ but collapsed very fast within 10 ps due to the large increase of the conductivity in the collision region. The E_k is the conventional breakdown field at ground estimated around 3.2 MV/m in the simulation.

Microwave radiation is related to the acceleration of electrons during the development of a discharge. In addition, electron–atom and electron–ion interactions (bremsstrahlung) result in another source of microwave radiation in plasmas [32]. This mechanism was proposed in [33] to be the source of microwave emissions from lightning. That was based on the measurements by [34] were they found that the measured power in the range from 100 MHz to 1.3 GHz resulted independent of the frequency. In addition, in [33] the calculations of the optical thickness for a fully ionized plasma corresponding to the return stroke channel demonstrate that it was completely opaque to microwave frequencies contrary to partially ionized plasmas of lightning leaders. So, the entire radiation was absorbed within the return stroke channels and not in the leader channels. In addition, the calculations showed how electron–ion interactions can be discarded because the calculated emitted power is four orders of magnitude lower than the electron–atom interactions. More recently, Moore *et al.* [35] calculated, that after certain time of the return stroke, some appreciable amount of microwave radiation can be detected.

Finally, the production of X-rays from lightning leaders [36] was assumed to result of bremsstrahlung radiation by runaway electrons accelerated in the intense fields present at lightning leader tips. So, it is expected some microwave radiation associated with the acceleration of runaway electrons as well with the bremsstrahlung electron–atom interactions that produce the source of the detected X-ray photons.

9.5 Method of experimentation

The experiments conducted in [1,6–15] had used receiver with down converting technique (not direct) where the microwave signal was down converted to much lower band in the baseband with narrow bandwidth. When signals undergo converter, there would be conversion loss and phase shifting (non-linear effects) as a downconverter includes a mixer for producing the intermediate frequency signal indicative of a frequency difference between the reference signal and the microwave signal. Unlike the amplifier which operates within its linear region even though it is one of the non-linear devices. Therefore, avoiding the usage of downconverter and usage of wide bandwidth are very crucial to accurately evaluate the temporal and spectral characteristics of microwave radiation signals.

Instead of using down converter, a direct time-domain measurement system without downconverter receiver and wide system bandwidth has been proposed and used in [16–20]. The system measured directly microwave signal at 0.961 GHz and the signal was sampled at 2.5 GS/s with bandwidth around 20 MHz. An air-gap parallel plate antenna has been chosen in [16–19] because of its omnidirectional receiving properties and the frequency independent of the air. The largest dimension of the top patch of the antenna was around 300 mm [37]. The simulations were conducted using parametric technique. Each simulation design was done with the aim to observe the return loss (S_{11}) changes with the corresponding changes of the distance between two plates, *d*, varied from 5 mm to 40 mm. The air gap separation cannot be too wide due to the consideration of the edging or fringing effect.

Figure 9.1 shows the variation of the S_{11} (which is the power loss in the signal transmitted to antenna port caused by the power reflected by that antenna port, so-called return loss) with the variations of *d*. The first simulation from 10 mm to 40 mm with 10 mm step carried out and the results as shown in Figure 9.1(a). Only

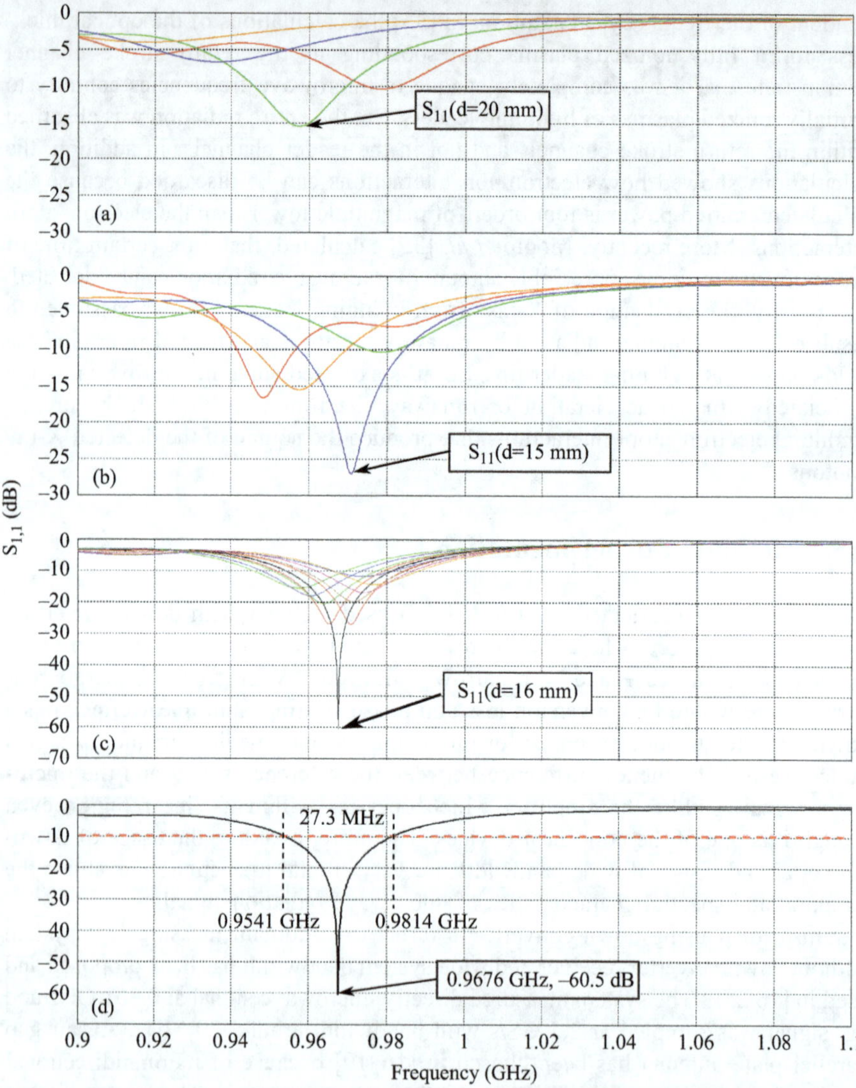

Figure 9.1 Return loss (S1,1) simulation results of air-gap parallel plate antenna for various air-gap separation (d). (a) 10–40 mm, (b) 5–20 mm, (c) 11–20 mm, and (d) the optimized S1,1 when d = *16 mm*

optimum result with S_{11} smaller than -10 dB was at 20 mm. The second simulation carried out with the air gap varied from 10 mm to 20 mm, with 5 mm step, and the results as shown in Figure 9.1(b). From the figure, the resonating frequency varied between 0.94 GHz and 0.98 GHz and the return loss varied significantly as d changes. The optimal design was at 15 mm with the lowest value of S_{11} at −26.5 dB. The third simulation was carried out with smaller simulation step 1 mm and the results are shown in Figure 9.1(c). The most optimal design was found at 16 mm with the lowest value of S_{11} at −60.5 dB resonated at center frequency around 0.9676 GHz with 27.3 MHz bandwidth as shown in Figure 9.1(d).

The result of simulated far-field radiation pattern of the most optimal air gap at 16 mm distance with antenna pattern was almost omnidirectional and vertically polarized with maximum power of 0 dB at 59 degrees of Theta (θ, azimuth) with 90 degrees of Phi (φ, elevation) and the minimum power was −40 dB at the position of 90° θ with 270° φ. The fabricated antenna was calibrated by using vector network analyzer (VNA) to evaluate the resonance frequency and return loss [16–19]. Figure 9.2 shows the calibrated resonance frequency at 0.961 GHz with 20 MHz bandwidth and the measured return loss of −29.31 dB. The calibration result of the fabricated antenna was comparable to the simulation. This antenna connected to commercial low noise amplifier (LNA) and bandpass filter (BPF) have been used by authors in [16–20] to conduct direct measurement of microwave radiation associated with NBPs, stepped leaders, and initial breakdown pulses.

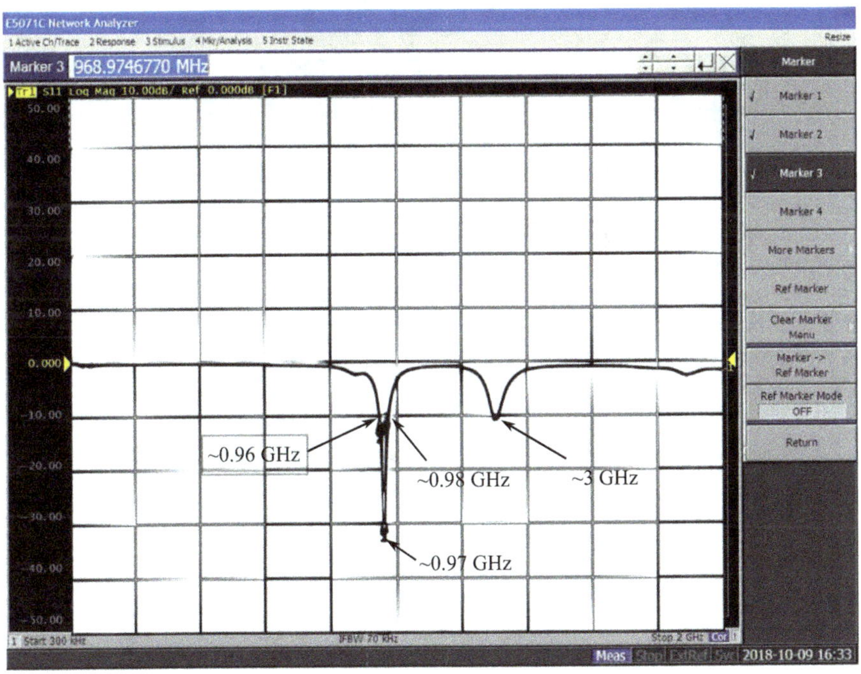

Figure 9.2 Calibrated return loss of the fabricated air-gap parallel plates antenna measured by using vector network analyser (VNA)

9.6 Microwave radiation associated with narrow bipolar pulses

The studies in [16,17] analyzed 74 NBPs (38 positive and 36 negative) that have been captured from tropical storms in Malacca Malaysia between December 2018 and January 2019. The authors found that microwave radiation pulses of NBPs preceded both VHF radiation pulses and fast electric field changes waveforms with lead time of 125.5 ± 81.3 ns and 600.7 ± 222.3 ns, respectively. The results obtained by [16,17] are similar to the finding in [18,19] where microwave radiation pulses of stepped leaders have been observed to precede both VHF and stepped leader pulses. The authors in [19] have suggested that the microwave radiation was emitted by electron avalanches or corona discharges while the VHF radiation was emitted by propagating streamers based on the fact that microwave radiation detected earlier than VHF radiation.

The microwave and VHF radiation bursts associated with NBPs are characterized into two main parts in [16,17], namely raising phase (RP) and damping phase (DP) as illustrated in Figures 9.3 and 9.4 for positive and negative NBP,

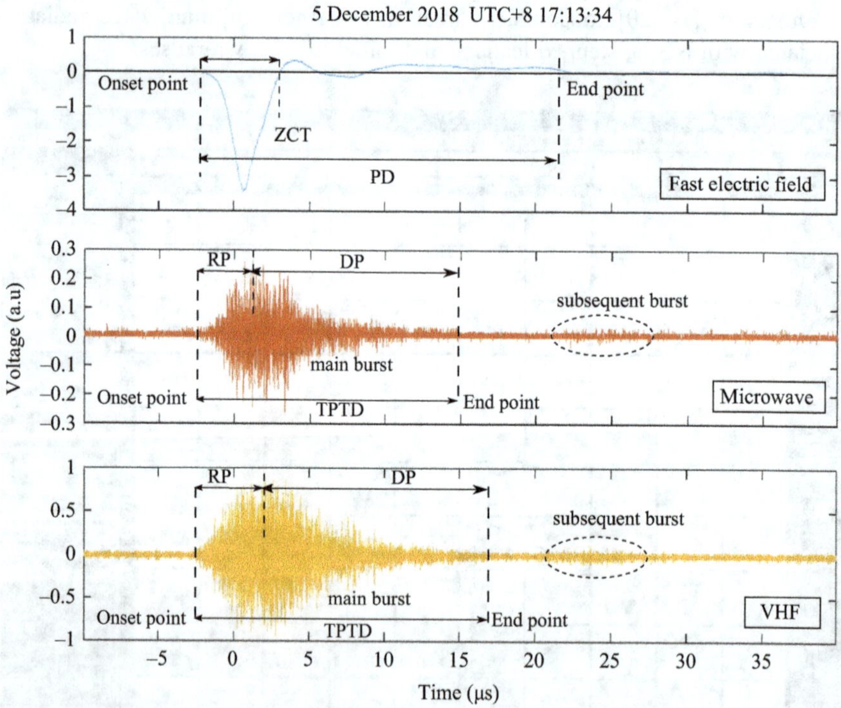

Figure 9.3 The microwave and VHF radiation pulses associated with a positive NBP. The total pulse duration of the positive NBP (atmospheric electricity sign convention) was 24 μs

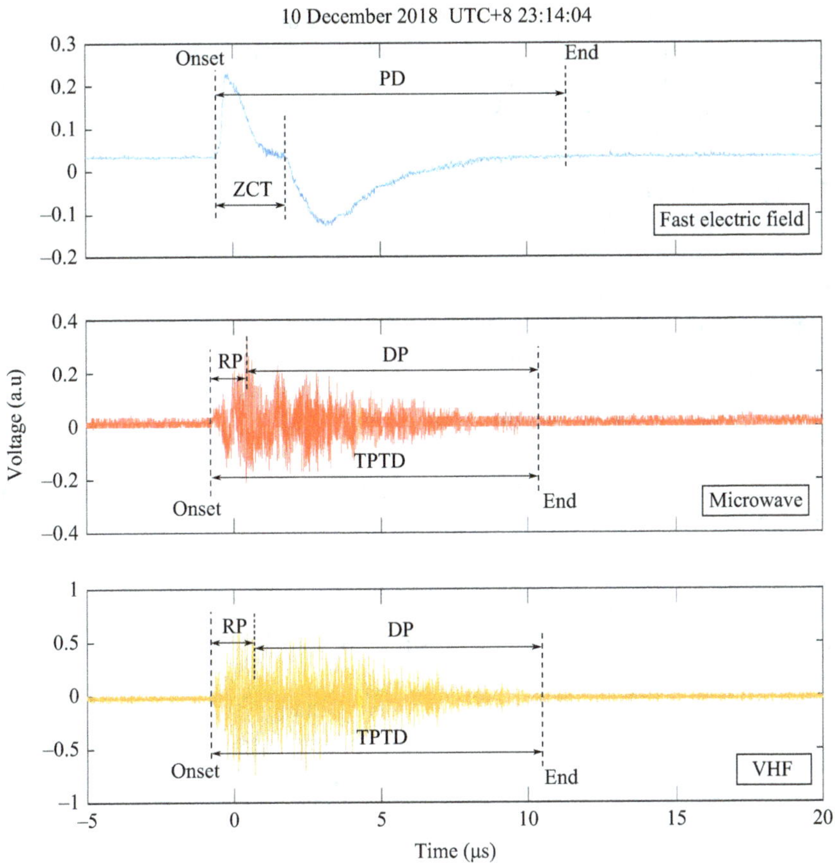

Figure 9.4 The microwave and VHF radiation pulses associated with a negative NBP. The total pulse duration of the negative NBP (atmospheric electricity sign convention) was 12 μs

respectively. The RP was defined as the period starting from its onset until it reaches its peak point while the DP was defined as the period from its peak point until the end point. The whole period of the burst occurrence was defined as total pulse trains duration (TPTD). In some cases, there would be a few clear individual bipolar pulses during the initial part of the burst train. The duration from the first individual bipolar pulse until the last individual pulse was defined as initial stage or IS. Around 30% of microwave and VHF radiation bursts have been observed to have IS at the earlier part of the RP with clear bipolar shape pulses.

The VHF radiation bursts were observed to have similar characteristics as microwave radiation bursts except the observation of a unique characteristic known as quiet phase. During quiet phase, VHF oscillation pulses were absent, but microwave pulses still can be detected. Moreover, a small number of VHF and microwave radiation bursts were found to have one or multiple subsequent bursts following the main burst after a short period. Figure 9.3 shows an example of subsequent burst following the main burst of the microwave and VHF radiation pulses associated with a positive NBP.

9.7 Microwave radiation associated with stepped leader and return stroke

A total of 100 stepped leaders of ten negative CG flashes accompanied by VHF and microwave radiations have been analyzed in [19]. All the stepped leaders were captured from two tropical storms in Malacca Malaysia on 12 and 24 November 2019. All first return strokes of the ten negative CG flashes were detected between 3.12 and 7.95 km from the measurement station.

Figure 9.5 shows an example of a stepped leader selected from a negative CG flash detected on 12 November 2019 at 15:37:45 local time (UTC+8) [19]. The location of the first return stroke was 3.46 km from the measurement station. Figure 9.6(A) and (B) shows an example of microwave and VHF radiation waveforms associated with a stepped leader, respectively. Both microwave and VHF radiation pulses were detected as individual oscillating pulses. The total pulse duration (TPD) of a radiation pulse was estimated from the onset to end time of one complete oscillation.

The study in [19] found that all microwave radiation pulses preceded stepped leader pulses with average leading time of 0.423 ± 0.378 μs. Around 80.41% of

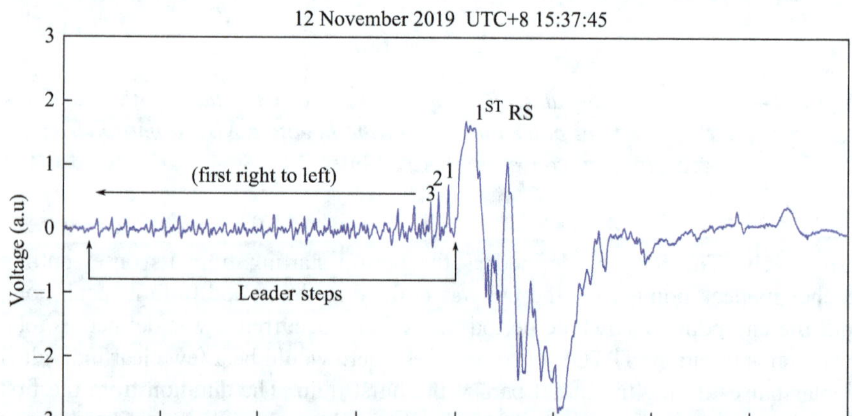

Figure 9.5 An example of the first return stroke of a negative cloud-to-ground (CG) flash preceded by a stepped leader. The first return stroke was located 3.46 km from the observation station. This waveform has been filtered to remove the static component

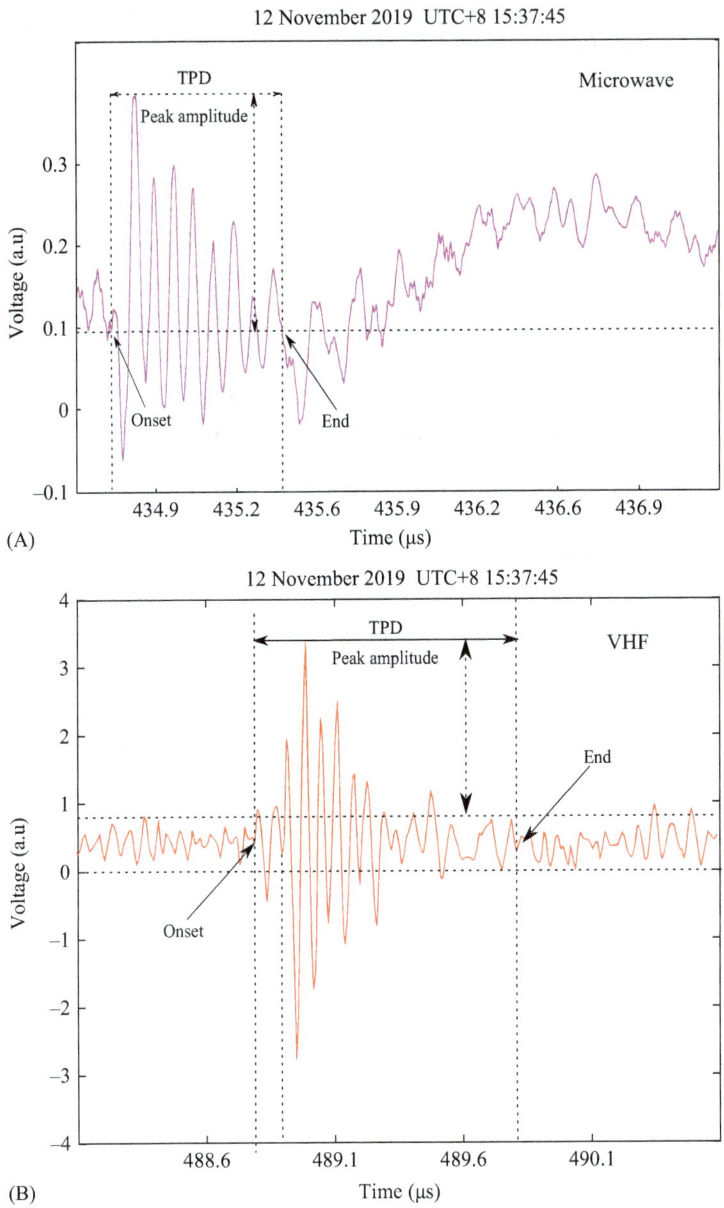

Figure 9.6 The microwave and VHF radiation pulses associated with one of the stepped leaders in Figure 9.5. (A) Individual microwave radiation oscillating pulse at 1 GHz, and (B) individual VHF radiation oscillating pulse at 60 MHz

microwave radiation bursts preceded VHF radiation bursts with average leading time of 0.540 ± 0.596 μs. The microwave radiation characteristics associated with stepped leaders have been classified into three categories [19]. The first category was the event of two microwave bursts detected during quiet period (QP) while second category was the event when one microwave burst only detected. For Category 3, no microwave burst detected at all during QP. The QP or quiet period was defined in [19] as the period between two stepped leader pulses where no fast electric field changes activity detected (see Figure 9.7).

The first and second microwave bursts of Category 1 have been suggested to be associated with electron avalanches and corona breakdown at the tip of negative leader and space stem, respectively. Microwave burst in Category 2 was suggested to be radiated by electron avalanches and corona breakdown at the tip of negative leader while space stem was absent. For Category 3, the detected VHF burst was suggested to be emitted by upward propagating positive streamers. It has been strongly suggested in [19] that microwave and VHF radiation pulses were emitted by different breakdown processes.

Interestingly, almost none or very weak microwave and VHF radiation pulses detected during the return stroke process particularly during the rising stage (slow front and rise time) [19]. This observation is similar to the previous findings in [1,7,9].

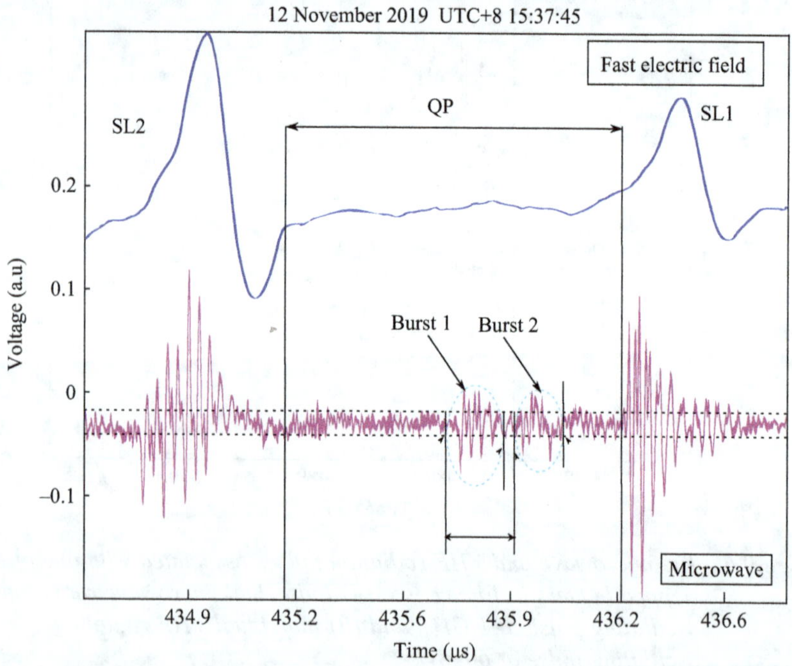

Figure 9.7 An example of two microwave pulses detected during quiet period (QP). Note that the QP is a duration between two SLs (SL1 and SL2)

9.8 Microwave radiation associated with initial breakdown process

Lightning flashes have been observed to be initiated by a series of bipolar electromagnetic field event that has several names known as breakdown pulses, preliminary breakdown pulses (PBP), initial breakdown (IB) pulse and characteristic pulse. Typically, the process is categorized into two types namely narrow IB and classical IB pulses [38–42]. The classical IB pulse is relatively larger than narrow IB pulse where the typical pulse duration is around 10 μs and longer and the main bipolar shape has two or three fast rising pulses or superimposed on the main bipolar pulse. The narrow IB pulse was often detected during initial stage of IB process, and it was a precursor event to the classical IB pulses.

Lightning initiation events of five negative CG flashes accompanied by VHF and microwave radiations have been analyzed in [38]. Figure 9.8 shows an example of lightning initiation event of a negative CG flash. The location of the first return stroke was 5.05 km from the measurement station. The figure shows four records of waveforms namely fast field (fast antenna system), slow field (slow antenna system), VHF (60 MHz), and microwave (~1 GHz). It has been found that first IB pulse has been accompanied by VHF, microwave, and initial electric field changes (IECs). The first microwave pulse has been observed to precede both first VHF and first IB pulses by average of 8.404 μs. Moreover, all first VHF pulses have been detected with positive polarity while the first microwave pulses were all detected

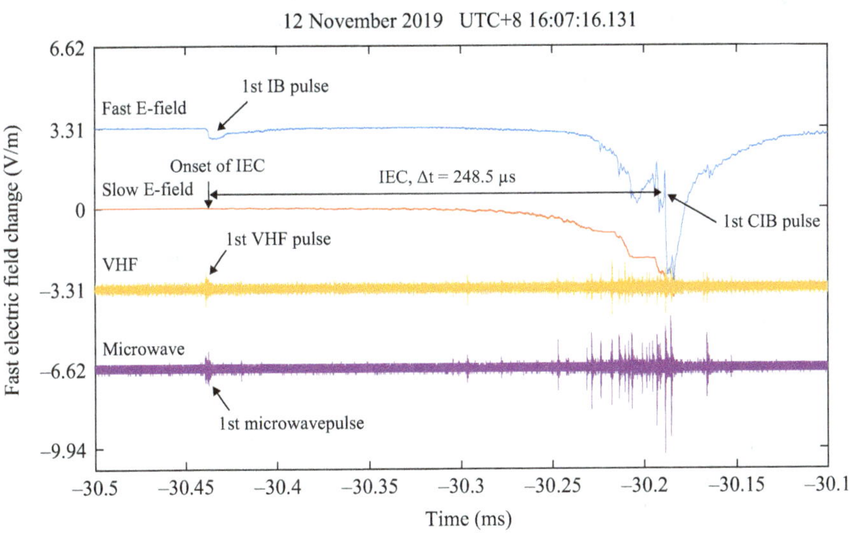

Figure 9.8 The detection of VHF and microwave pulses initiated the first IB pulse and IECs before the first classical initial breakdown (CIB) pulse. The duration from the onset of IECs to the peak of the first CIB was 248.5 μs

with negative polarity. From this observation, the authors have suggested that the VHF and microwave pulses might have been radiated by two different sources due to the opposite charge propagation [38].

Recent studies regarding lightning initiation have found that both CG and IC flashes have been initiated by a slowly varying electric field process known as initial electric field changes or IECs [39–43]. The studies suggested that IB was not the first process that initiates lightning. Instead, IEC was the first process that initiates lightning and then followed by the first IB pulse after sometimes (in the scale of milliseconds). The IEC is defined as a process of slow electric field changes that occurs (moving upward for CGs and downward for ICs) and ended just before the first classical IB pulse. This process is only possible to be observed when it happens very close to lightning sensors (within reversal distance [39–43]).

9.9 Conclusion

This chapter has presented a comprehensive review on the previous and current studies of microwave and VHF electromagnetic radiations generated by lightning flashes. Since the first experimental work done by Brook and Kitagawa in 1964, more research works have been conducted at various geographical regions to understand the underlying mechanism behind the emissions of VHF and microwave radiations by lightning flashes. Based on the findings of previous studies, most likely the microwave and VHF radiations have been generated by two different processes. The VHF radiation is generated by propagating streamers while microwave radiation is generated by electron avalanches and corona discharges.

References

[1] Brook M. and Kitagawa, N. 'Radiation from lightning discharges in the frequency range 400 to 1000 Mc/s'. *Journal of Geophysical Research.* 1964;**69**(12):2431–2434.

[2] Sartor, J. D. 'Radio observation of the electromagnetic emission from warm clouds'. *Science.* 1964;**143**(3609):948–950.

[3] Atkinson W.R. and Paluch, I. 'Electromagnetic emission from pairs of water drops exchanging charge'. *Journal of Geophysical Research*, 1966;**71**(16):3811–3816.

[4] Sartor J.D. and Atkinson, W.R. 'Charge transfer between raindrops: Microwave temperatures of thunderstorms can be deduced from radiation emitted by colliding drops'. *Science.* 1967;**157**(3794):1267–1269.

[5] Keeney J. *"The microwave emission spectrum of colliding charged water drops". Ph.D. dissertation*, New Mexico Institute of Mining and Technology, Socorro, NM, 1967.

[6] Kosarev E.L., Zatsepin V.G., and Mitrofanov, A.V. 'Ultrahigh frequency radiation from lightnings'. *Journal of Geophysical Research.* 1970;**75**(36):7524–7530.

[7] Rust W.D., Krehbiel P.R., and Shlanta, A. 'Measurements of radiation from lightning at 2200 MHz'. *Geophysical Research Letters*. 1979;**6**(2):85–88.

[8] Brook M., Tennis R., Rhodes C., Krehbiel P., Vonnegut B., and Vaughan Jr, O.H. 'Simultaneous observations of lightning radiations from above and below clouds'. *Geophysical Research Letters*. 1980;**7**(4):267–270.

[9] Le Boulch M., Hamelin J., and Weidman, C. 'UHF-VHF radiation from lightning'. *Electromagnetics*.1987;**7**(3–4):287–331.

[10] Atlas, D. 'Meterorological "Angel" echoes'. *Journal of Atmospheric Sciences*. 1959;**16**(1):6–11.

[11] Oetzel, G.N. 'Computation of the diameter of a lightning return stroke'. *Journal of Geophysical Research*. 1968;**73**(6):1889–1896.

[12] Fedorov V.F., Frolov Y.A., and Shishkov, P.O. 'Millimetric electromagnetic radiation of a lightning return stroke'. *Journal of Applied Mechanics and Technical Physics*. 2001;**42**(3):392–396.

[13] Yoshida S. 'Radiations in association with lightning discharges'. *Ph.D. dissertation, Department of Information and Communications Technology Division of Electrical Electronic and Information Engineering Graduate School of Engineering Osaka University*, 2008.

[14] Ahmad M.R., Esa M.M., and Cooray, V. 'Narrow bipolar pulses and associated microwave radiation'. In *Proceedings of Progress in Electromagnetics Research*; Stockholm, Sweden, August 2013. pp. 1087–1090.

[15] Petersen D. and Beasley, W. 'Microwave radio emissions of negative cloud-to-ground lightning flashes'. *Atmospheric Research*. 2014;**135**:314–321.

[16] Seah B.Y., Baharin S.A.S., Esa M.M., *et al.* 'Narrow bipolar events within reversal distance and associated UHF-VHF emissions'. In *Proceedings of IEEE International Conference on Electrical Engineering and Computer Science (ICECOS)*; Batam, Indonesia, October 2019. pp. 132–136.

[17] Seah B.Y. 'Temporal characteristics of microwave radiations emitted by narrow bipolar events in tropical thunderstorm'. *Master thesis, Faculty of Electronics and Computer Engineering, Graduate Study Centre of Universiti Teknikal Malaysia Melaka*, 2020.

[18] Baharin S.A.S., Ahmad M.R., AlKahtani A.A., Esa M.R.M., and Sidik, M.A. B. 'Temporal analysis of microwave radiation emitted by stepped leaders of a cloud-to-ground flash'. *Journal of Advanced Manufacturing Technology*. 2020;**142**(2):223–234.

[19] Baharin S. A. S., Ahmad M. R., Al-Shaikhli T. R. K., *et al.* 'Microwave radiation associated with stepped leaders of negative cloud-to-ground flashes'. *Atmospheric Research*. 2022;**270**:106091.

[20] Ahmad M.R., Esa M.M., Rahman M., Cooray V., and Dutkiewicz, E. 'Lightning interference in multiple antennas wireless communication systems'. *Journal of Lightning Research*. 2012;**4**(1):155–165.

[21] Ahmad M.R., Esa M.R.M., Cooray V., and Dutkiewicz, E. 'Interference from cloud-to-ground and cloud flashes in wireless communication system. *Electric Power Systems Research*. 2014;113:237–246.

[22] Cooray V., Cooray G., Marshall T., Arabshahi S., Dwyer J., and Rassoul, H. 'Electromagnetic fields of a relativistic electron avalanche with special attention to the origin of lightning signatures known as narrow bipolar pulses'. *Atmospheric Research*. 2014;**149**:346–358.

[23] Cooray V. and Cooray, G. 'Electromagnetic radiation field of an electron avalanche'. *Atmospheric Research*. 2012;**117**:18–27.

[24] Rison W., Krehbiel P.R., Stock M.G., *et al.* 'Observations of narrow bipolar events reveal how lightning is initiated in thunderstorms'. *Nature Communications.* 2016;**7**:1–12.

[25] Tilles J.N., Liu N., Stanley M.A., *et al.* 'Fast negative breakdown in thunderstorms'. *Nature Communications*. 2019;**10**(1):1648.

[26] Townsend, J.S. 'Electricity in gases'. Рипол Классик. 1915.

[27] Gallimberti, I. 'The mechanism of the long spark formation'. *Le Journal de Physique Colloques*. 1979;**40**(C7):C7–193.

[28] Chauzy S. and Kably, K. 'Electric discharges between hydrometeors'. *Journal of Geophysical Research: Atmospheres*. 1989;**94**(D11):13107–13114.

[29] Luque A. 'Radio frequency electromagnetic radiation from streamer collisions'. *Journal of Geophysical Research: Atmospheres*. 2017;**122**(19):97–104.

[30] Montanyà J., Fabró F., March V., *et al.* 'X-rays and microwave RF power from high voltage laboratory sparks'. *Journal of Atmospheric and Solar-Terrestrial Physics*. 2015;**136**:94–97.

[31] Shi F., Liu N., Dwyer J.R., and Ihaddadene, K.M. 'VHF and UHF electromagnetic radiation produced by streamers in lightning'. *Geophysical Research Letters*. 2019;**46**(1):443–451.

[32] Bekefi G., Hirsheld J.L., Sanborn, C. B. 'Incoherent microwave radiation from plasmas'. *Physical Review*. 1969;**116**(5):1051–1056.

[33] Rai J., Rao M., and Tantry, B.A.P. 'Bremsstrählung as a possible source of UHF emissions from lightning'. *Nature Physical Science*. 1972;**238**(82):59–60.

[34] Singh M. and Singh, M. 'UHF emission due to Bremsstrahlung from lightning return stroke'. *IOSR Journal of Applied Physics*. 2013;**3**:4–7.

[35] Moore C.B., Eack K.B., Aulich G.D., and Rison, W. 'Energetic radiation associated with lightning stepped-leaders'. *Geophysical Research Letters*. 2001;**28**(11):2141–2144.

[36] Montanya J., Oscar V.D.V., and Tapia, F.F. 'Simultaneous emissions of X-rays and microwaves from long laboratory sparks and downward lightning leaders'. In *Proceedings In AGU Fall Meeting Abstracts*; New Orleans, LO, December 2017. pp. AE32A–05.

[37] Balanis C.A. *Antenna Theory: Analysis and Design*, 3rd ed. New York, NY: John Wiley and Sons; 2005. pp. 151–205.

[38] Sabri M.H.M., Al-Kahtani A.A., Ali N.H.N., *et al.* 'Very high frequency and microwave radiation associated with initial breakdown process in CG lightning flashes from tropical storms'. In *Proceedings In 2021 35th*

International Conference on Lightning Protection (ICLP) and XVI International Symposium on Lightning Protection (SIPDA), September 2021. Sri Lanka: IEEE; 2021. pp. 1–4.

[39] Marshall T., Stolzenburg M., Karunarathna N., and Karunarathne, S. 'Electromagnetic activity before initial breakdown pulses of lightning'. *Journal of Geophysical Research: Atmospheres*. 2014;**119**(22):12–558.

[40] Chapman R., Marshall T., Karunarathne S., and Stolzenburg, M. 'Initial electric field changes of lightning flashes in two thunderstorms'. *Journal of Geophysical Research: Atmospheres*. 2017;**122**(7):3718–3732.

[41] Marshall T., Bandara S., Karunarathne N., *et al.* 'A study of lightning flash initiation prior to the first initial breakdown pulse'. *Atmospheric Research*. 2019;**217**:10–23.

[42] Sabri M.H.M., Ahmad M.R., Esa M.R.M., *et al.* 'Initial electric field changes of lightning flashes in tropical thunderstorms and their relationship to the lightning initiation mechanism'. *Atmospheric Research*. 2019;**226**:138–151.

[43] Sabri M.H.M., Ahmad M.R., Al-Kahtani A.A.N., *et al.* 'A study of cloud-to-ground lightning flashes initiated by fast positive breakdown'. *Atmospheric Research*. 2022;276:106260.

Chapter 10

The Schumann resonances

Colin Price[1]

10.1 Introduction

The Schumann resonances (SR) are global electromagnetic resonances excited within the Earth–ionosphere waveguide, primarily by lightning discharges. These resonances occur in the extremely low-frequency (ELF) range, with resonant frequencies around 8, 14, 20, 26, ... Hz. The history of the SR is an interesting story [1]. While Schumann [2,3] got most of the credit for the first prediction of the existence of the SR, the idea of natural global electromagnetic resonances was first presented by George F. Fitzgerald in 1893, and then again by Nikola Tesla in 1905 [4]. However, while others formulated the idea before Schumann, it was Schumann, together with König, who attempted to measure the resonant frequencies for the first time, unsuccessfully [5,6]. It was not until measurements made by [7,8] that adequate analysis techniques were available to extract the resonance information from the back-ground noise. Today, we know that we need 5–10 min of data to detect the SR clearly in the spectrum. For further insight into the history of the SR, the reader is pointed to the excellent review by Besser [1].

Following Schumann's landmark paper in 1952, there was an increasing interest in SR in a wide variety of fields. Because of the low attenuation of ELF waves in the SR band (~0.5 dB/Mm), it was discovered that not only lightning can produce SR but any large explosion in the atmosphere will also induce SR transients [9–11]. Hence, until the ban of atmospheric nuclear explosions in the 1960s, there was great interest in using the SR to monitor the enemy's nuclear explosions in remote parts of the globe. Another application of ELF waves related to the SR, due to the low attenuations of the ELF waves, was the man-made transmission of these waves for long-range communications with submarines [12,13]. However, due to the extremely long wavelengths at ELF, such transmitters need to be huge (>200 km length), with huge power outputs due to very low efficiencies of these transmitters. Nevertheless, since the signals propagate globally, the superpowers were still using these ELF transmitters until recently. The United States transmitter broadcasts at 76 Hz [14], while the Russian transmitter broadcasts at 82 Hz [15].

[1]Porter School of the Environment and Earth Sciences, Tel Aviv University, Israel

Besides the military uses of ELF resonances and propagation theory, from the very beginning of SR studies there was an interest to track global lightning activity using the SR [16–21]. It has also been suggested that extraterrestrial lightning may be detected and studied using SR [22,23]. However, the recent focus on SR research since the 1990s was a result of the connection between lightning activity and the Earth's climate. It was first suggested in 1990 that global warming may result in significant increases in lightning activity [24]. Since the SR is one way to monitor global lightning activity, Williams [25] suggested that the SR may be used to monitor global temperature variations, acting as a global thermometer. This started a new interest in SR research as related to global climate change that continues today.

Ever since the start of SR research in the 1950s, researchers were intrigued with the possible link between the environmental SR ELF fields, and biological systems. In the 1960s and 1970s, studies appeared showing links between the SR and the circadian rhythms in humans and birds [26] while similar results in flies were observed by [27]. More recently, new studies looking at the interaction between the SR and animal cells [28,29] and plant cells [30,31] have appeared.

Finally, with the discovery of transient luminous events (TLEs) such as sprites, elves, and jets, it was shown that SR transient pulses are closely linked to the occurrence of TLEs – sprites and elves [32–36]. Hence, SR research became also a major part of a relatively new field of research related to upper atmospheric discharges.

10.2 Theoretical background

Lightning discharges are considered as the primary natural source of SR. The vertical lightning channels behave like huge antennas that radiate electromagnetic energy at frequencies below about 100 kHz [37]. While the maximum radiated energy occurs around 10 kHz, the attenuation at these frequencies is about 10 dB/Mm. Hence, these frequencies can only be detected at a range of thousands of kilometers from the lightning discharge. While lightning signals below 100 Hz have very low amplitudes (magnetic fields of 10^{-12} T or pT), the attenuation is only 0.5 dB/Mm, and hence the electromagnetic waves from an individual discharge can propagate a number of times around the globe before decaying into the background noise. For this reason, the Earth–ionosphere waveguide behaves like a resonator at ELF frequencies and amplifies the spectral signals from lightning at the resonance frequencies due to constructive interference of electromagnetic waves propagating around the globe in opposite directions [37].

If the terrestrial waveguide was an ideal one, the resonant frequencies f_n would have been determined by the Earth's radius a and the speed of light c – (10.1) [2]. Even Schumann made these assumptions and arrived at the expected SR first mode of 10 Hz. However, the Earth–ionosphere waveguide is not a perfect electromagnetic cavity. Losses due to finite ionosphere conductivity make the system resonate at lower frequencies than would be expected in an ideal case (7.8 Hz), and the observed peaks are wider than expected. In addition, there are a number of horizontal asymmetries – day–night transition, latitudinal changes in the Earth magnetic field, sudden ionospheric disturbances, polar cap absorption, etc. – that complicate the SR power spectra:

$$fn = \left(\frac{c}{2\pi a}\right)\sqrt{n(n+1)} \quad n = 1, 2, 3, \ldots \tag{10.1}$$

The problem of wave propagation in the Earth–ionosphere cavity is most naturally formulated in spherical coordinates (r,θ,φ). The excitation source is represented by a vertical dipole with a current moment (Ids) located between two concentric spherical shells at $\theta = 0$. The radius of the inner shell – the Earth – is denoted by $r = a$, and the radius of the outer shell – the ionosphere – by $r = a + h$, assuming a sharp and frequency independent upper boundary. Both the observer and the source are assumed to be located on the Earth surface. Maxwell's equations are then solved assuming time dependence of $e^{i\omega t}$ and requiring continuity on the boundaries (ground–cavity transition at $r = a$, and cavity–ionosphere transition at $r = a + h$). The electric and magnetic components are then [38]:

$$E_r = i\frac{Ids}{8\pi a^2 \varepsilon_0 f}\frac{\nu(\nu+1)}{h}\frac{P_\nu^0(-\cos\theta)}{\sin\nu\pi}; \qquad H_\varphi = -\frac{Ids}{4a}\frac{1}{h}\frac{P_\nu^1(-\cos\theta)}{\sin\nu\pi} \tag{10.2}$$

In (10.2), ε_0 is a free space permittivity and P_ν^l are the associated Legendre functions. The complex parameter ν is calculated in terms of the complex sine of the wave incidence angle S via [39]:

$$S^2 = \frac{\nu(\nu+1)}{(\kappa_0 a)^2} \tag{10.3}$$

where k_0 is the free space wave number. The dimensionless quality factor Q of the resonant cavity may be determined as a ratio between the stored energy and the energy loss per cycle. Considering only the electrically stored energy [39]:

$$Q = \frac{\mathrm{Re}S}{2\,\mathrm{Im}\,S} \tag{10.4}$$

On Earth, the resonance is characterized by a quality factor Q ranging from 4 to 6 [40].

The resulting fields are shown in Figure 10.1 for the first three SR modes. For a single lightning discharge, the E-field always has a maximum at the location of the flash and the antipode, while the magnetic field (orthogonal to electric) has a minimum at the same locations, regardless of the mode. For other locations, the relative intensity of the electric and magnetic fields depends uniquely on the source–observer distance (SOD). Figure 10.2 shows the theoretical spectra for the vertical electric field as a function of different SODs. At a distance of 10 Mm from the source lightning (dotted curve), the electric field shows a minimum intensity at 8 and 20 Hz ($n = 1, 3$) while a maximum occurs at 14 and 26 Hz ($n = 2, 4$). Every distance has a specific spectral pattern in both the electric and magnetic fields, a

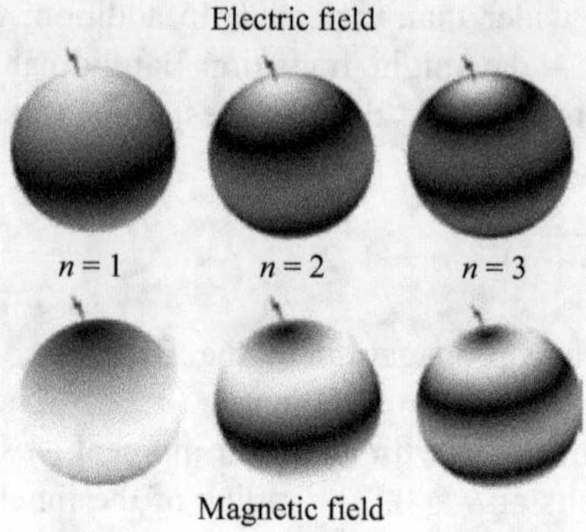

Figure 10.1 Electric and magnetic fields of the first three SR modes. White shading shows the maximum amplitudes in the fields and dark areas show the minimum in the field for the different modes and distances from the lightning discharge

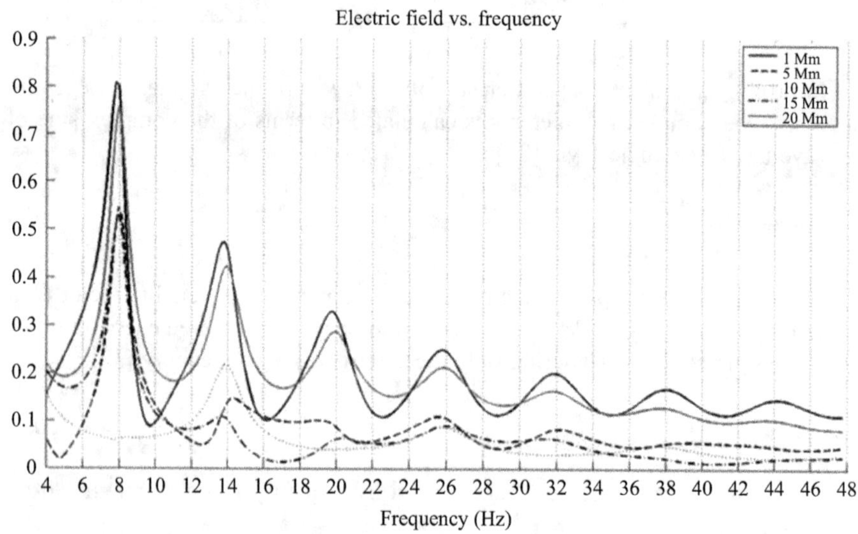

Figure 10.2 Electric field spectra as a function of SOD

characteristic often used in SR geolocation of intense lightning flashes using a single station [34,41–46]).

More realistic models are far more complex. Methods of introducing more complicated ionosphere structure include two-layer [47] and multi-layer models [48–50], and the more realistic two-exponential [51], 'knee' [52], and 'multi-knee' [23] profiles. Recently an open source python model for simulating the SR has also been developed [53,54].

10.3 SR measurements

The electromagnetic sensors used to measure SRs normally consist of two horizontal induction coils for detecting the horizontal magnetic field in the north-south (H_{NS}) and the east-west (H_{EW}) direction, and one vertical antenna for observing the vertical electric field, E_Z (Figure 10.3). The electric component is commonly measured with a ball antenna, suggested by Ogawa *et al.* [55], connected to a high-impedance amplifier. The magnetic induction coils consist of tens of thousands of turns around material with very high magnetic permeability. The measured ELF AC fields are very small (mV/m for the electric field, and pT for the magnetic fields) compared with the DC static electric and magnetic fields in the atmosphere that ranges from 100 V/m in the fair weather to kV/m on a stormy day, and a magnetic field of 50,000 nT.

Man-made noise produces various interferences in the ELF band, from high voltage power supply lines to traffic and pedestrians [40], forcing us to locate SR measuring stations in isolated rural areas, away from industrial activity. When choosing a site, the electromagnetic field sensors should be located as far away from power supply lines as possible. Complete battery power supply is preferable, but is expensive and limits long-term monitoring. Open spaces with uniform underlying geology and high soil conductivity should also be considered [40]. Since the sensors are exposed to external static electric and magnetic fields, even the slightest vibration of an antenna will result in huge signals induced at the input of the receiver. Hence, the horizontal magnetic antennas should be buried in the ground to avoid the signals induced by ground vibrations or wind. Ideally, electric and magnetic channels should be identical, being calibrated periodically, sampled using a 16 bit A/D (analog-to-digital) converter, equipped by a GPS clock for time stamping the data, and if necessary, a notch filter for reducing the anthropogenic

(a) (b)

Figure 10.3 *(a) Vertical electric field ball antenna with author; and (b) two horizontal induction coils at the Mitzpe Ramon, Israel SR site*

50 Hz (or 60 Hz) interference. The sampling frequency can vary from several tens of hertz to a few hundreds of hertz in order to cover the SR band without aliasing. It is advisable to save all raw data for later post-processing, although some groups use real-time analysis and save only the spectral parameters of the SR (peak frequency, peak amplitude and Q-factor) [56], together with short-time segments of ELF transients.

In the time domain, the electric and magnetic signals produce a constant background signal, which is a superposition of individual pulses arriving from about 50 random lightning flashes per second occurring all over the world [57]. Superimposed upon the background noise are intense transient pulses from individual powerful lightning discharges, with amplitudes often 10 times higher than that of the background noise (Figure 10.4(a)) [58]. After processing the time series by using the Fast Fourier Transform (FFT) algorithm, SR modes can usually be

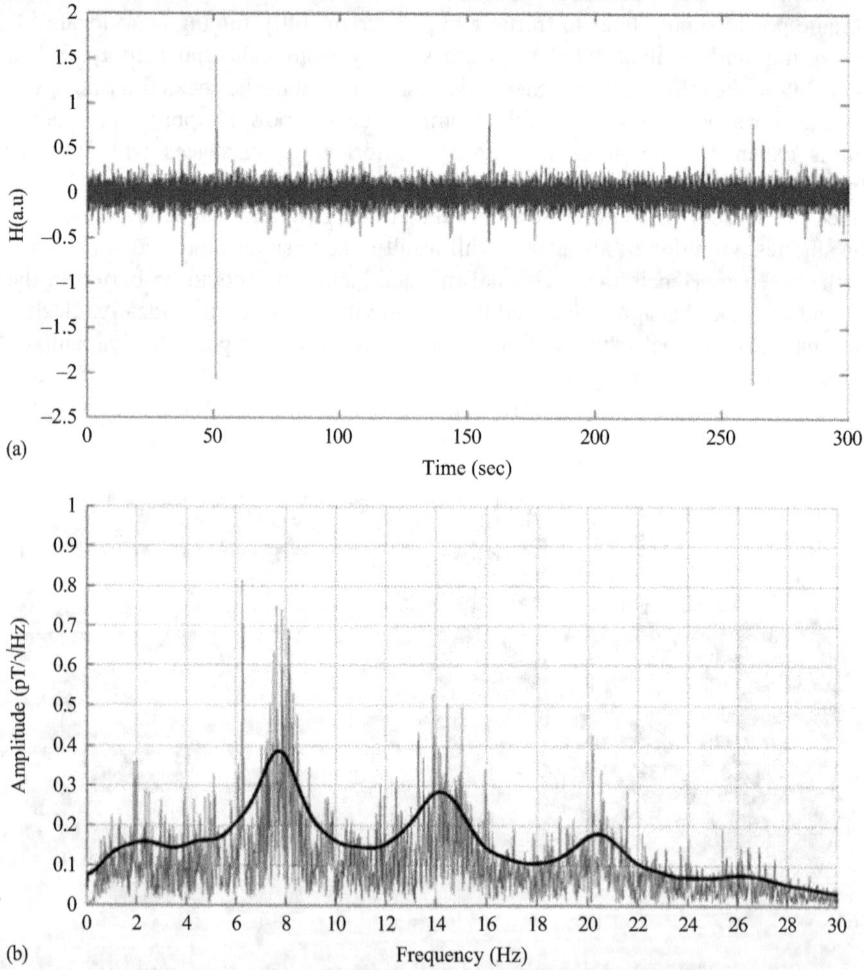

Figure 10.4 (a) Time series of 5 min of raw magnetic field data, (b) frequency spectrum of data in (a)

observed in the frequency domain at 8, 14, 20, 26 ... Hz (Figure 10.4(b)). For studies of global lightning activity, the SR spectra are normally fitted to a set of Lorentzian curves [40,59–62] where the curve for each mode is described by three parameters: peak amplitude, peak frequency, and the quality factor.

As mentioned above, the duration of data collection of up to 10 min is needed to obtain stable estimates of the SR spectrum. Nickolaenko and Hayakawa [40] suggest that this may explain the unsuccessful early experiments by Schumann and König [6] to detect the global resonances: the natural signal is actually random 'noise' and the resonance peaks become visible only after relatively long integration time. A 10 min interval was used in the first successful experiment by Balser and Wagner [63].

10.4 SR background observations of global lightning activity

At any given time, there are about 2,000 thunderstorms around the globe [17,19,25,55,64,65]. Producing ~50 lightning events per second [57], these thunderstorms create the background SR signal.

Determining the spatial lightning distribution from the background SR records is a complex problem: in order to properly estimate the lightning intensity from SR records, it is necessary to account for the distance to sources. The common approach to this problem is based on the preliminary assumption of the spatial lightning distribution. The most widely used approaches are the models of the three thunderstorm centers – Southeast Asia, Africa and South America [17,20,66–68], and a single thunderstorm center travelling around the globe showing the background ELF field [69], together with transient ELF pulses. An alternative approach is placing the receiver at the North or South Pole, which remains approximately equidistant from the main thunderstorm centers during the day [46]. A distinct method, not requiring preliminary assumptions on the lightning distribution [70–72], is based on the decomposition of the average background SR spectra, utilizing ratios between the average electric and magnetic spectra and between their linear combinations.

The best documented and the most debated features of the SR phenomenon are the diurnal variations in the background SR power spectrum. Some of the earliest studies were made by Holzer [18], Raemer [73], Balser and Wagner [16], and Polk and Fitchen [74]. The first investigators realized that the SR field power variations were related to global thunderstorm activity [16,18,73,75]. Thus, SR measurements became a convenient tool for studying global lightning activity [25,56,76–81]. Figure 10.5 shows the daily mean values of the first SR mode (8 Hz) measured simultaneously in Israel and California, over a 25-day period [82]. The agreement is quite remarkable given that the instruments, data acquisition and software algorithms were entirely independent of each other. This agreement is further evidence of the global nature of the SR, and its value of studying global lightning variability and trends.

Figure 10.6 shows the 4-year (1999–2002) mean diurnal and seasonal power variations in the first SR mode from the Mitzpe Ramon (MR), Israel ELF station, after fitting the data with Lorentzian curves. The geographical location of the MR site (32N,

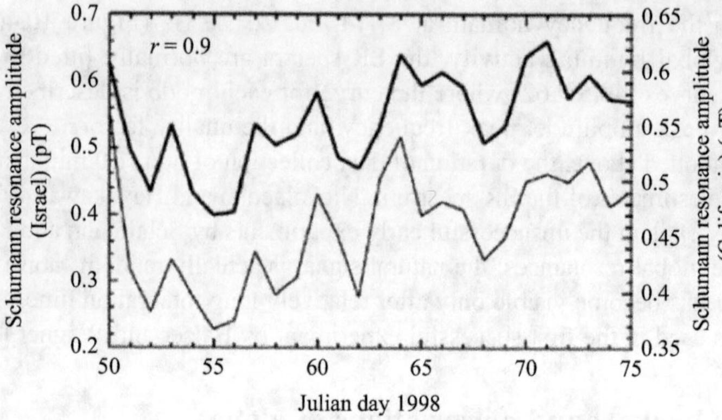

Figure 10.5 Comparison of the SR amplitude of the first mode (8 Hz) measured simultaneously in Israel and California (from [82])

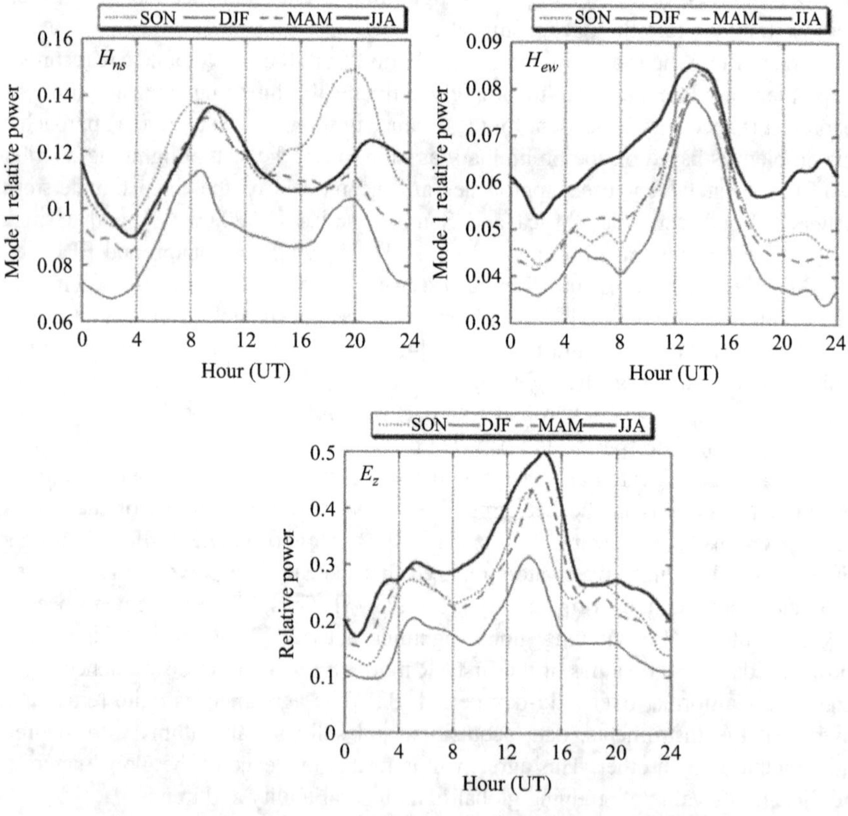

Figure 10.6 The 4-year mean diurnal and seasonal variations in the SR power for the first mode – individual electromagnetic components of the SR field (from [56])

34E) results in the clear spatial and temporal separation of the three main thunderstorm source regions. Two maxima in the H_{NS} component are easily identified around 9:00 and 20:00 UT and are associated with increased thunderstorm activity from Southeast Asia and South America in the late afternoon, local time. In the H_{EW} component, there is a strong maximum around 14:00 UT associated with the peak in afternoon African lightning activity. The three dominant maxima are clearly seen during all seasons, associated with the three 'hot spots' of planetary lightning activity. The time and amplitude of the peaks vary throughout the year, reflecting the seasonal changes in lightning activity. The electric field sensor is sensitive to lightning activity from all directions, and hence shows a combination of all three peaks in the diurnal variations.

10.5 SR transient measurements of global lightning activity

One of the most interesting problems in SR studies is determining the lightning source characteristics (the 'inverse problem'). Temporally resolving each individual flash in the background SR signal is impossible due to the overlapping of many different lightning waveforms at ELF frequencies. However, there are intense ELF transient events, also named 'Q-bursts', which appear as prominent excursions above the SR background signal (Figure 10.7). Q-bursts are triggered by intense lightning strikes, associated with a large charge transfer and often high peak current [44,55,83]. Amplitudes of Q-bursts can exceed the SR background level by a factor of 10 and they appear with intervals from ~10 s to a few minutes [71]. This separation in time allows us to consider the Q-bursts as isolated events and to determine their source lightning locations and charge moments [43,45,46,84–89].

The lightning location problem can be solved with either multi-station or single-station techniques. The multi-station techniques are the more accurate, but require more complicated and expensive facilities, involving a network of direction finders or

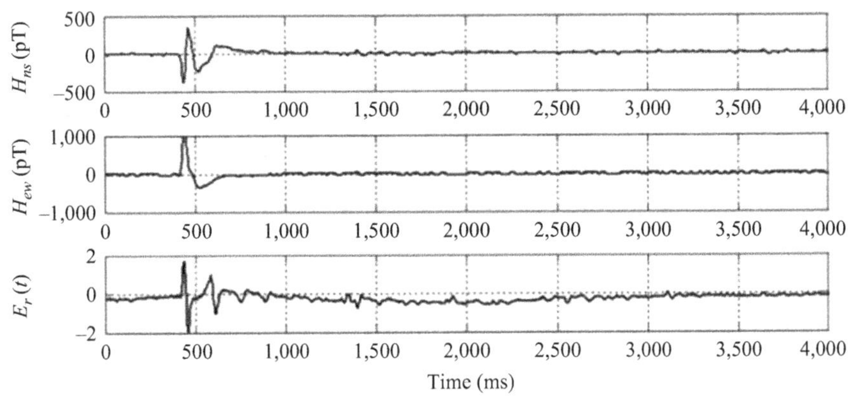

Figure 10.7 Example of ELF transient recorded at Mitzpe Ramon (32N, 34E) station, Israel [90]

time-of-arrival sensors [91]. Single-station systems usually combine a direction finder technique with a SOD estimation technique (Figure 10.2). The transients can be geo-located with SOD and/or source-bearing techniques, based on the relationship between the electric and the magnetic field components [34,41,43–45,92]. Source location techniques can be calibrated using the general location of flashes above continental regions [45,85], the proximity of cold cloud tops in visible and infrared satellite images [93], global lightning measurements from space by the optical transient detector (OTD) and the lightning imaging sensor (LIS) [41], and local measurements of lightning with ground networks, such as US National Lightning Detection Network in North America [34]. Geolocation of the source lightning using the single-station SR methodology can be identified with an accuracy of ~1 Mm anywhere on the globe.

10.6 Using SR as a climate research tool

The warming of the Earth has been the subject of intense debate and concern for many scientists for at least the past three decades. One of the important aspects in understanding global climate change is the development of tools and techniques that would allow continuous and long-term monitoring of processes affecting, and being affected by, the global climate. SRs are one of the very few tools that can provide such *global* information continuously, reliably, and cheaply.

Williams [25] suggested that global temperature may be monitored via the SR. The link between SR and temperature is lightning flash rate, which increases nonlinearly with temperature [24,94–96]. The nonlinearity of the lightning-to-temperature relation provides a natural amplifier of the subtle (several tenths of 1°C [97,98]) temperature changes and makes SR a sensitive 'thermometer'. Additional analysis using other SR data sets also shows strong positive correlations between surface temperatures and SR power on seasonal and daily timescales [99]. Figure 10.8 presents an example of daily observations of the 8 Hz magnetic field originating in South America, and recorded in Israel, and surface temperatures integrated over South America. Although the correlation coefficient is only 0.57, it is clear that on warmer days, there is more lightning activity than on cooler days.

Monitoring and predicting global climate change requires the understanding and modelling of factors that determine atmospheric concentrations of important greenhouse gases and feedbacks that determine the sensitivity of the climate system. Continental deep-convective thunderstorms produce most of the lightning discharges on Earth. In addition, they transport large amounts of water vapor into the upper troposphere, dominating the variability of global upper tropospheric water vapor (UTWV). UTWV is a key element of the Earth's climate, which has direct effects as a greenhouse gas, as well as indirect effect through interaction with clouds, aerosols and tropospheric chemistry. UTWV has a much greater impact on global warming than water vapor in the lower atmosphere [100], but whether this impact is a positive, or a negative feedback has been debated [101–105]. The main challenge in addressing this question is the difficulty in monitoring UTWV globally over long timescales. Price [82] and Price and Asfur [99] showed that changes in the UTWV can be monitored from records of the SR. Figure 10.9 shows an example of the connection between daily SR amplitudes and UTWV over Africa, the largest source of lightning and thunderstorms on the planet. It should be noted that the UTWV curve has been shifted one day to show the agreement between

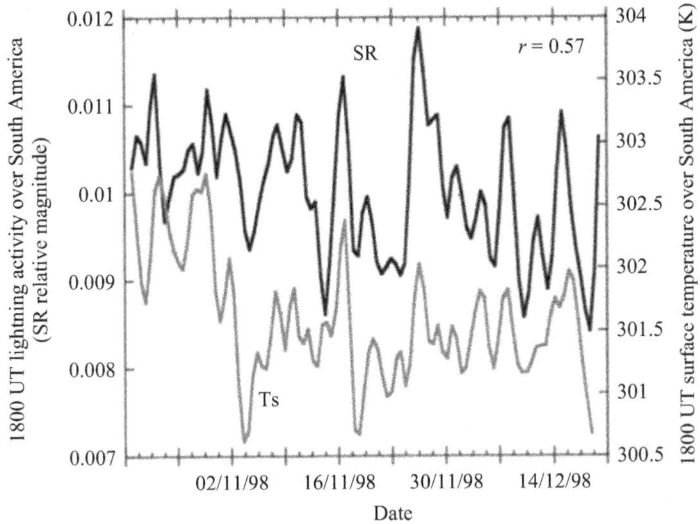

Figure 10.8 *Relative amplitude of the 8 Hz SR signal at 18:00 UT every day, arriving from South America, but detected in Israel (black), compared with the spatially averaged 18:00 UT tropical land surface temperatures over South America (grey)*

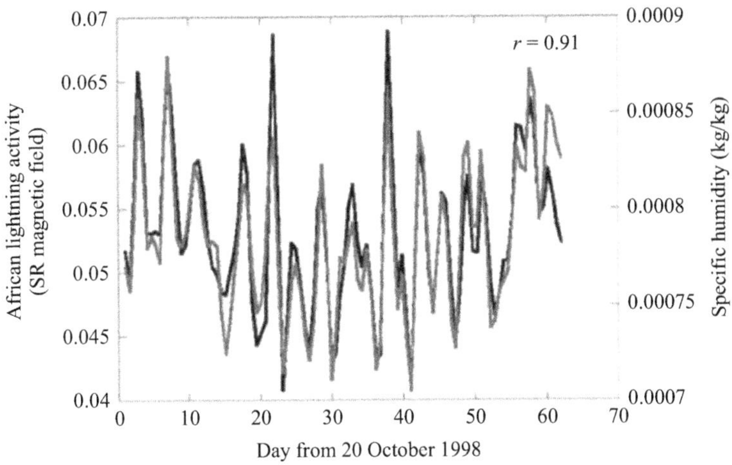

Figure 10.9 *Daily SR 8 Hz magnetic field records (black) and upper tropospheric water vapor (grey) over tropical Africa (from [99])*

the curves. However, the lightning activity peaks one day before the peak in the UTWV.

In addition to climate change, the SR has also been shown to be linked to the natural climate oscillation El Nino/Southern Oscillation (ENSO) that changes the Earth's climate every few year [106,107]. During ENSO years, the convection, and hence the lightning activity, shifts position relative to the fixed SR stations. Therefore, changes in SR parameters can be used to monitor the natural ENSO cycle, and the shifts in convection that occur during these events.

The above results show that two of the most important parameters of global climate change – surface temperature and UTWV – can be monitored via observations of the SR, utilizing its relation to worldwide thunderstorm activity. In addition, the SRs may also help us understand important feedback effects in the climate system, such as the water vapor feedback in the upper troposphere. One of the great advantages of this method is the availability of long-term calibrated data sets that can provide past and future records of global lightning variations on Earth.

10.7 SR in transient luminous events (TLE) research

It is now believed that many of the SR transients (Q-bursts) are related to TLEs, spectacular optical flashes in the upper atmosphere above active thunderstorms. The existence of TLEs was theoretically predicted by Wilson [108], but the official discovery came with the first image captured above a thundercloud by Franz *et al.* [109]. In the last 30 years, there has been an extensive hunt for TLEs using photography from ground stations, aircrafts, satellites, space shuttles and the International Space Station (ISS), leading to TLE documentation in different geographical locations all over the world [110–124].

TLEs can be classified into two main classes: sprites and elves [125], although there are also blue jets, gigantic jets, halos and trolls. Both elves and sprites are short-lived luminous events associated with active thunderstorms. Elves are dim doughnut-shaped glows of red light with a radius of a few hundred kilometers, lasting typically ~1 ms, and occurring at altitudes of ~90–100 km, located above the parent lightning discharge. Elves are produced by the electromagnetic pulse of the lightning, with the intensity of elves related to the lightning peak current [126]. Sprites are also red in color (due to the excitation of atmospheric nitrogen molecules [127]), while being a lot brighter than elves. Sprites have a much longer lifetime of tens of milliseconds and occur at lower altitudes in the atmosphere (40–90 km). Unlike the uniform featureless elves, sprites can be very varied in shapes, structure and size, with widths ranging to 50–100 km horizontally. Sprites are produced by the quasistatic electric field induced above thunderstorms immediately after large cloud-to-ground (CG) lightning [128]. In the case of sprites, the brightness appears to be related to the charge removed by the lightning, and not the peak current [129]. Since the SR transients are dominated by large charge moments, irrelevant to peak currents, the SR are better suited for studying sprites than elves [130].

The physical mechanisms responsible for sprites and elves initiation are independent of the polarity of the lightning flash [127,128,131–134]; however, the vast majority of sprites are initiated by positive CG flashes [32,91,135]. These powerful positive flashes emit strong electromagnetic energy in the ELF range, indicative of continuing currents lasting over timescales of at least a few milliseconds [91], and thus can be detected in the SR band. Boccippio *et al.* [32] suggested that sprites are produced by positive CG occurring in the stratiform region of a thunderstorm system, and are accompanied by large-amplitude transient pulses ('Q-burst') in the SR band. Observations [32–34,36,89,136] reveal that occurrences of sprites and transient SR are highly correlated (Figure 10.10).

SR records can be used to estimate the magnitude of the charge removed from CG [138,139], which appears to be one of the crucial parameters in determining which lightning discharge can produce sprites. A method of charge moment estimation of sprite-inducing CG discharges from SR data was developed by Huang *et al.* [33], who showed that the charge moments of sprite-inducing CG discharges range from 200 to 2,000 Ckm. Hu *et al.* [140] suggested a sprite initiation probability as a function of charge moments of positive CG discharges, and hence the charge moment estimation derived from SR data can possibly enable us to estimate

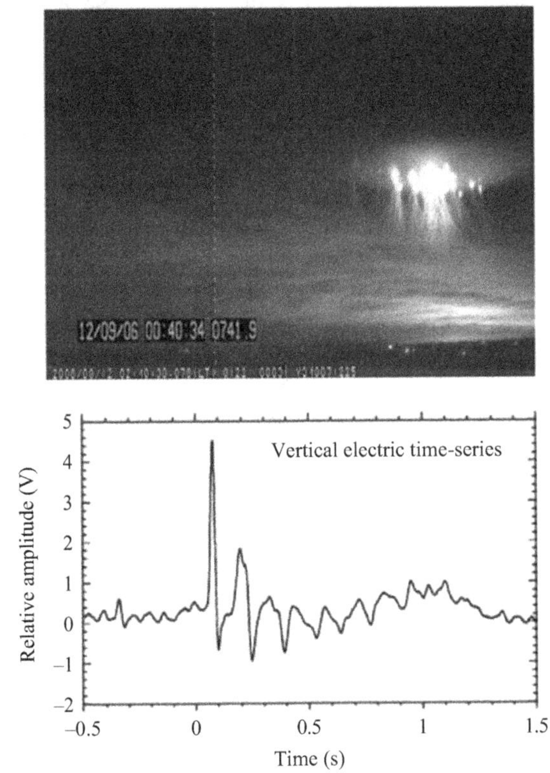

Figure 10.10 A sprite observed by the ILAN team [137] together with a typical ELF transient, showing clearly the 8 peaks per second (8 Hz) in the vertical electric field [34]

the global occurrence rate of sprites. However, it should be noted that not all sprites produce strong ELF transients [141].

Since sprites are rather rare, occurring at a rate of only a few per minute (while regular lightning occurs at a rate of 50–100 flashes per second around the globe) SR techniques appear to be one of the most convenient and low-cost tools for continuous TLE monitoring.

10.8 SR in extraterrestrial lightning research

Existence of SRs depends generally on two factors – presence of a substantial ionosphere with electric conductivity increasing with height from low values near the surface (or a high-conductivity layer, in case of gaseous planets) to form an ELF waveguide, and a source of excitation of electromagnetic waves in the ELF range. In the Solar System, there are a number of candidates for SR detection: Venus, Mars, Jupiter, Saturn, and its moon Titan [142].

The speculations that lightning occurs on Venus first arose about 45 years ago. The strongest evidence for lightning on Venus comes from the impulsive electro-magnetic waves seen by the Venera 11 and 12 landers [143–146] and the Pioneer Venus Orbiter [147,148]. On Mars lightning activity has not been detected, but charge separation and lightning strokes are considered possible in the Martian dust storms [149–152]. Jupiter and Saturn are the only planets where lightning activity is well established. Existence of lightning on Jupiter was predicted by Bar-Nun [153] and it is supported by data from Galileo, Voyagers 1 and 2, Pioneers 10 and 11 and Cassini [154,155]. Recently lightning on Saturn has also been confirmed by measurements from the Cassini spacecraft [156,157]. Although no lightning was observed during Voyager flybys of Titan in 1980 and 1981, it was long suggested that lightning dischargers do take place on this moon of Saturn [158,159]. However, recent data from Cassini/Huygens seems to indicate that there is no lightning activity on Titan [160,161].

Modelling of SR parameters on the planets and moons of the Solar System is complicated by the lack of knowledge of the waveguide parameters. SR frequencies depend on the structure of the lower part of the ionosphere, which is not sufficiently studied. On Jupiter and Saturn, the situation is yet more complicated. Little is known about the electrical parameters of the interior of Jupiter and Saturn. Even the question of what should serve as the lower waveguide boundary is a non-trivial one in the case of these gaseous planets. To our best knowledge, there are no works dedicated to SR on Saturn. There was only one attempt to model SRs on Jupiter – in the work by Sentman [162]. Sentman's calculations yielded resonant frequencies of ~0.76, 1.35, and 1.93 Hz with quality factors of roughly 7, predicting sharp, pronounced peaks.

The situation with other planets is a little better. SRs on Venus were studied by Nickolaenko and Rabinowicz [163], Pechony and Price [23], Yang *et al.* [164], and Simões *et al.* [165]. All studies, based on different conductivity profiles and with different models, yielded very close resonant frequencies: around 9, 16, and 23 Hz. The quality factors, though, differ substantially: Nickolaenko and Rabinowicz obtained Q-factors of ~5 while Pechony and Price acquired Q∼10. Such a difference – by a factor of two – was predicted by Nickolaenko and Rabinowicz for more sophisticated ionosphere representations.

Martian global resonances were modeled by Sukhorukov [166], Pechony and Price [23], Molina-Cuberos *et al.* [167], and Yang *et al.* [164]. The results of the studies are somewhat different. Sukhorukov obtained the resonant frequencies at about 13, 25, and 37 Hz with Q-factors around 3.5. The frequencies calculated by Pechony and Price are lower: 8.6, 16.3, and 24.4 Hz, with Q-factors of ∼2.4. The disparity can probably be explained by the different models of Martian lower ionosphere used in the two studies. Nevertheless, the low-quality factors obtained in both studies show that pronounced sharp peaks at resonance frequencies should not be expected for the Martian ELF waveguide. Significantly, different results were obtained by Molina-Cuberos *et al.* [167], where several ionosphere models were used. The first resonance occurred at 11–12 Hz (depending on ionosphere model), the second and third resonances interfered to form a single peak at 21–25 Hz and the fourth, fifth, and sixth modes produced a very smooth-shaped peak at around 60 Hz. The ionosphere of Titan is perhaps the most thoroughly modeled today. The recent interest in the largest satellite of Saturn is associated with the Cassini/Huygens Mission and the expectations of finding evidence of lightning activity on Titan. Consequently, SR on Titan received more attention than resonances on other celestial bodies. The resonant frequencies obtained for various ionospheric conductivity profiles tested in studies by Besser *et al.* [168], Morente *et al.* [169], Molina-Cuberos *et al.* [167], and Navarro *et al.* [170] range (for realistic models) from 11.0 to 15.0 Hz for the first mode, 21.2–27.8 Hz for the second and 35.6–41.6 for the third. Unfortunately, the quality factors were not calculated in these studies. Comparable results were obtained by other authors: resonant frequencies of 19.9, 35.8, and 51.8 Hz with Q-factors of 1–3 were obtained by Nickolaenko *et al.* [171], and 11.8, 22.5 and 34.1 Hz with Q-factors of ∼2 by Pechony and Price [23]. The low Q-factors acquired in these two studies show that the expected peaks, should lighting activity be found on Titan, are rather wide. After the Huygens probe entered Titan's atmosphere in 2005, a heated debate evolved over the existence of SR signals in the observed data [160,172,173].

Today there is no possibility to validate SR parameters calculated for other planets and moons. The values of the resonance frequencies and quality factors are sensitive to the ionospheric profile models. The accuracy of the latter is limited, and a deeper knowledge of planetary ionospheres would allow more precise predictions of SR parameters. On the other hand, experimental evaluation of SR parameters can aid in the elaboration of the effective model of the ionospheric conductivity profile and contribute substantially to the knowledge of lower ionospheres on planets of the Solar System.

10.9 SR and biology

One of the fundamental questions in biological sciences, and more specifically in brain research, is why organisms exhibit characteristic extremely low-frequency (ELF) oscillations in electric activity [29,174,175] that look so similar to the natural ELF SR oscillations in the atmosphere. It is surprising that many different types of species (vertebrates and invertebrates) exhibit similar low-frequency electrical wave activity, irrespective of their brain size, brain complexity, or even the existence of a cortex. What can it mean that most vertebrates, from fish and frogs to Homo sapiens, show essentially very similar electrical activity [174]. And why is the dominant peak frequency so close to 8 Hz in all examples?

In the 1960s and 1970s, research was done mainly on the circadian rhythms of people and birds showing that the ELF fields have a stabilizing effect on the biological clock [26,176]. Later experiments on flies showed similar effects of stabilizing the biological clock [27]. The interest in biological connections with the Schumann resonances has been rekindled in recent years with new laboratory experiments on spinal cord injuries [177], on cardiomyocytes (heart tissue) [28], as well as interactions with plants and impacts on photosynthesis [30,31].

Some of these studies show that the SR ELF fields allow the protection of cells (both animal and plant) from external stress factors. The fields appear to protect the cells. With further studies of the links between biology and the SR, we may see the developments of medical and agricultural treatments for cells under stress. This multi-disciplinary field has great potential for development in the future.

10.10 Summary

Being a global phenomenon, SRs have numerous applications in lightning research. Background SR records can serve as a convenient and a low-cost tool for global lightning activity monitoring. The SR can provide a global geoelectric index for monitoring climate changes. It provides one of the few tools that through variations in global lightning activity, can provide continuous and long-term monitoring of such important global climate change parameters as tropical land surface temperature and tropical UTWV.

SR transients (Q-bursts) can be used to geolocate intense lightning strikes anywhere on the planet. These large-amplitude pulses are apparently related to the occurrence of sprites and elves above thunderstorms, and therefore TLEs can be studied using SR observations. An additional application of SR is extraterrestrial lightning research. SRs may be used to detect and, if necessary, monitor lightning activity on the planets and moons of the Solar System.

There are still many open questions in SR research: importance of the day–night variation in the ionosphere conductivity profile [178,179], influence of the latitudinal changes in the Earth magnetic field, impacts of cosmic, solar and geomagnetic disturbances [180,181], polar cap absorption, accuracy of source geolocation and the determination of the spatial lightning distribution from the background records [43,89,182]. Despite these open problems, SR is one of the most promising tools in a variety of fields related to lightning electromagnetics.

In summary, the knowledge of the Schumann resonances in the atmosphere, their origins, propagation, variability and changes, may have implications for monitoring global climate changes, upper atmospheric TLEs, for space missions to other planets, for space weather research, and possibly even in the field of medicine and agriculture.

Acknowledgements

The authors would like to thank all of my graduate students and researchers who have been involved in our SR studies over the years, and who have contributed to the results in this chapter: Dr Mustafa Asfur, Dr Olga Pechony, Dr Eran Greenberg, Dr Michael Finkelstein, Dr Alex Melnikov, Mr. David Shtibelman, and Mr Boris Starobinets. In addition, many colleagues have contributed to the field over the years, as well as my personal education in the field of the SR: Sasha Nickolaenko, Earle Williams, Gabriella Satori, Jozsef Bor, Yasuhide Hobaro, Dave Sentman, Martin Fullekrug, Anirban Guha, and many more.

References

[1] Besser, B. P. (2007), Synopsis of the historical development of Schumann resonances, *Radio Sci.*, 42, RS2S02, doi:10.1029/2006RS003495.

[2] Schumann, W. O. (1952a), Über die strahlungslosen Eigenschwingungen einer leitenden Kugel, die von einer Luftschicht und einer Ionospha·renhu·lle umgeben ist, *Zeitschrift und Naturfirschung*, 7a, 149–154.

[3] Schumann, W. O. (1952b), Über die Dämpfung der elektromagnetischen Eigen-schwingnugen des Systems Erde – Luft – Ionosphäre, *Zeitschrift und Naturfirschung*, 7a, 250–252.

[4] Tesla, N. (1905), The transmission of electrical energy without wires as a means of furthering world peace, *Electrical World Engineer, January 7, 21–24.*

[5] Schumann, W. O. (1952c), Über die Ausbreitung sehr Langer elektriseher Wellen um die Signale des Blitzes, *Nuovo Cimento*, 9, 1116–1138.

[6] Schumann, W. O. and H. König (1954), Über die Beobactung von Atmospherics bei geringsten Frequenzen, *Naturwiss*, 41, 183–184.

[7] Balser, M. and C. Wagner (1962b), On frequency variations of the earth-ionosphere cavity modes, *J. Geophys. Res.*, 67, 4081–4083.

[8] Balser, M. and C. Wagner (1960a), Measurement of the spectrum of radio noise from 50 to 100 c/s, *J. Res. NBS*, 64D, 415–418.

[9] Balser, M. and C. Wagner (1963), Effect of a high-altitude nuclear detonation on the earth-ionosphere cavity, *J. Geophys. Res.*, 68, 4115–4118.

[10] Gendrin, R. and R. Stefant (1962b), Effet de l'explosion thermonucleaire a tres haute altitude du 9 juillet 1962 sur la resonance de la cavite Terreionosphere: Interpretation, *C.R. Acad. Sci. Paris*, 255, 2493–2495.

[11] Gendrin, R. and R. Stefant (1962a), Effet de l'explosion thermonucleaire a tres haute altitude du 9 juillet 1962 sur la resonance de la cavite Terre-ionosphere: Resultats experimentaux, *C.R. Acad. Sci. Paris*, 255, 2273–2275.

[12] Wait, J. R. (1977), Propagation of ELF electromagnetic waves and project Sanguine/Seafarer, *IEEE J. Oceanic Eng.*, 2(2), 161–172.

[13] Wait, J. R. (1974), Historical background and introduction to the special issue on extremely low frequency (ELF) communication, *IEEE Trans. Commun.*, 22(4), 353–354.

[14] Fraser-Smith, A. C. and P. R. Bannister (1998), Reception of ELF signals at antipodal distance, *Radio Sci.*, 33(1), 83–88, doi:10.1029/97RS01948.

[15] Yano, M., Y. Ida, Y. Hobara, M. Hayakawa, and A. P. Nickolaenko (2010), Reception of ELF transmitter signals at Moshiri, Japan, and their propagation characteristics, *Radio Sci.*, 45, RS1009, doi:10.1029/2009RS004224.

[16] Balser, M. and C. Wagner (1962a), Diurnal power variations of the earth-ionosphere cavity modes and their relationship to worldwide thunderstorm activity, *J. Geophys. Res.*, 67, 619–625.

[17] Heckman, S. J., E. Williams, and B. Boldi (1998), Total global lightning inferred from Schumann resonance measurements, *J. Geophys. Res.*, 103 (D24), 31775–31779.

[18] Holzer, R. E. (1958), *World thunderstorm activity and extremely low frequency spherics, in Recent Advances in Atmospheric Electricity*, L. G. Smith, ed., Pergamon Press, New York, NY, pp. 599–602.

[19] Nickolaenko, A. P. and L. M. Rabinowicz (1995) Study of the annual changes of global lightning distribution and frequency variations of the first Schumann resonance mode, *J. Atmos. Sol-Terr. Phys.*, 57(11), 1345–1348.

[20] Nickolaenko, A. P., G. Satori, B. Zieger, L. M. Rabinowicz, and L. G. Kudintseva (1998), Parameters of global thunderstorm activity deduced from the long-term Schumann resonance records, *J. Atmos. Sol-Terr. Phys.*, 60(3), 387–399.

[21] Yang, H., V. P. Pasko, and G. Sátori (2009), Seasonal variations of global lightning activity extracted from Schumann resonances using a genetic algorithm method, *J. Geophys. Res.*, 114, D01103, doi:10.1029/2008JD009961.

[22] Nickolaenko, A. P. and L. M. Rabinowicz (1987), On the applicability of extremely low frequency global resonances in the studies of lightning activity at Venus, *Space Res.*, 25, 301–308.

[23] Pechony, O. and C. Price (2004), Schumann resonance parameters calculated with a partially uniform knee model on Earth, Venus, Mars, and Titan, *Radio Sci.*, 39(5), RS5007, doi:10.1029/2004RS003056.

[24] Price, C. and D. Rind (1990), The effect of global warming on lightning frequencies, in *Proceedings of the AMS 16th Conf. on Severe Local Storms Conf. on Atmos. Electr.*, Kananaskis Park, Alberta, Canada, pp. 748–751.

[25] Williams, E. R. (1992), The Schumann resonance: a global tropical thermometer, *Science*, 256, 1184–1186.

[26] Wever, R. (1973), Human circadian rhythyms under the influence of weak electric fields and the different aspects of these studies, *Int. J. Biometeor.*, 17, 227–232.

[27] Engelmann, W., W. Hellrung, and A. Johnsson (1996), Recording method and effects of 10Hz square-wave electric fields, *Bioelectromagnetics*, 19, 100–110.

[28] Elhalel, G., C. Price, D. Fixler, and A. Shainberg (2019), Cardioprotection from stress conditions by weak magnetic fields in the Schumann Resonance band, *Sci. Rep.*, 9, 1645, https://doi.org/10.1038/s41598-018-36341-z

[29] Price, C., E. Williams, G. Elhalel, and D. Sentman (2020), Natural ELF fields in the atmosphere and in living organisms, *Int. J. Biometeorol.*, 65, 85–92 https://doi.org/10.1007/s00484-020-01864-6.

[30] Grinberg, M., M. Mudrilov, E. Kozlova, *et al.* (2021), Effect of extremely low-frequency magnetic fields on light-induced electric reactions in wheat, *Plant Signal. Behav.*, 17, 2021 664, https://doi.org/10.1080/15592324.2021.2021664.

[31] Sukhov, V., E. Sukhova, Y. Sinitsyna, *et al.* (2021), Influence of magnetic field with Schumann resonance frequencies on photosynthetic light reactions in wheat and pea, *Cells*, 10, 149, https://doi.org/10.3390/cells10010149.

[32] Boccippio, D. J., E. R. Williams, S. J. Heckman, W. A. Lyons, I. T. Baker, and R. Boldi (1995), Sprites, ELF transients, and positive ground strokes, *Science*, 269, 1088–1091.

[33] Huang, E., E. Williams, R. Boldi, *et al.* (1999), Criteria for sprites and elves based on Schumann resonance observations, *J. Geophys. Res.-Atmos.*, 104 (D14), 16943.

[34] Price, C., M. Asfur, W. Lyons, and T. Nelson (2002), An improved ELF/VLF method for globally geolocating sprite-produced lightning, *Geophys. Res. Lett.*, 29(3), 1.1–1.4.

[35] Sato, M. and H. Fukunishi (2003), Global sprite occurrence locations and rates derived from triangulation of transient Schumann resonance events, *Geophys. Res. Lett.*, 30(16), 1859, doi:10.1029/2003GL017291.

[36] Sato, M., H. Fukunishi, M. Kikuchi, H. Yamagishi, and W. A. Lyons (2003), Validation of sprite-inducing cloud-to-ground lightning based on ELF observations at Syowa station in Antarctica, *J. Atmos. Sol-Terr. Phys.*, 65, 607–614.

[37] Volland, H. (1984), *Atmospheric Electrodynamics*, Springer-Verlag, Berlin, 205 p.

[38] Wait, J. R. (1962), *Electromagnetic Waves in Stratified Media*, Pergamon Press, New York, NY.

[39] Galejs, J. (1972), *Terrestrial Propagation of Long Electromagnetic Waves*, Pergamon Press, New York, NY.

[40] Nickolaenko, A. P. and M. Hayakawa (2002), *Resonances in the Earth-Ionosphere Cavity*, Kluwer Academic Publishers, Dordrecht, Boston, London.

[41] Boccippio, D. J., C. Wong, E. Williams, R. Boldi, H. J. Christian, and S. J. Goodman (1998), Global validation of single-station Schumann resonance lightning location, *J. Atmos. Terr. Phys.*, 60, 701–712.

[42] Burke, C. P. and D. Ll. Jones (1992), An experimental investigation of ELF attenuation rates in the Earth-ionosphere duct, *J. Atmos. Terr. Phys.*, 54, 243–254.

[43] Greenberg, E. and C. Price, (2004), A global lightning location algorithm based on the electromagnetic signature in the Schumann resonance band, *J. Geophys. Res.*, 109, D21111, doi:10.1029/2004JD004845.

[44] Haung, H. G., E. Williams, R. Boldi, *et al.* (1999), Criteria for sprites and elves based on Schumann resonance observations, *J. Geophys. Res.*, 104, 16,943–16,964.

[45] Kemp, D. T. and D. Ll. Jones (1971), A new technique for the analysis of transient ELF electromagnetic disturbances within the Earth–ionosphere cavity, *J. Atmos. Terr. Phys.*, 33, 567–572.

[46] Nickolaenko, A. P. (1997), Modern aspects of Schumann resonance studies, *J. Atmos. Sol-Terr. Phys.*, 59, 806–816.

[47] Jones, D. L. (1974a), The calculations of the Q-factors and frequencies of Earth-ionosphere cavity resonances for a two-layer ionosphere model, *J. Geophys. Res.*, 69, 4037–14041.

[48] Hynninen, E. M. and Yu. P. Galyuck (1972), The field of a vertical electric dipole over the spherical Earth's surface below the vertically inhomogeneous ionosphere, *The Problems of Diffraction and Wave Propagation*, 11, 109–115 (in Russian).

[49] Jones, D. Ll. (1967), Schumann resonances and ELF propagation for inhomogeneous, isotropic ionosphere profiles, *J. Atmos. Terr. Phys.*, 29, 1037–1044.

[50] Yamashita, M. (1968), The propagation characteristics of ELF radio waves to great distances below the horizontally stratified ionosphere, *J. Atmos. Terr. Phys.*, 30, 1943–1953.

[51] Greifinger, C. and Ph. Greifinger (1978), Approximate method for determining ELF eigenvalues in the Earth-ionosphere waveguide, *Radio Sci.*, 13, 831–837.

[52] Mushtak, V. C. and E. R. Williams (2002), ELF propagation parameters for uniform models of the Earth-ionosphere waveguide, *J. Atmos. Sol-Terr. Phys.*, 64, 1989–2001.

[53] Bozoki, T., E. Prácser, G. Sátori, G. Dálya, K. Kapás, and J. Takátsy (2019), Modeling Schumann resonances with schupy, *JASTP*, 196, 105144, https://doi.org/10.1016/j.jastp.2019.105144.

[54] Prácser, E., T. Bozóki, G. Sátori, E. Williams, A. Guha, and H. Yu (2019), Reconstruction of global lightning activity based on Schumann resonance measurements: model description and synthetic tests, *Radio Sci.*, 54, 254–267, https://doi.org/10.1029/2018RS006772.

[55] Ogawa, T., Y. Tanka, T. Miura, and M. Yasuhara (1966), Observations of natural ELF electromagnetic noises by using the ball antennas, *J. Geomagn. Geoelectr.*, 18, 443–454.

[56] Price, C. and A. Melnikov (2004), Diurnal, seasonal and inter-annual variations of the Schumann resonance parameters, *J. Atmos. Sol-Terr. Phys.*, 66 (13–14), 1179.

[57] Christian, H. J., R. J. Blakeslee, D. J. Boccippio, *et al.* (2003), Global frequency and distribution of lightning as observed from space by the optical transient detector, *J. Geophys. Res.*, *108*(*D1*), 4005

[58] Greenberg, E. and C. Price (2007), Diurnal variations of ELF transients and background noise in the Schumann resonance band, *Radio Sci.*, 42, RS2S08, doi:10.1029/2006RS003477.

[59] Mushtak, V. C. and E. R. Williams (2009), An improved Lorentzian technique for evaluating resonance characteristics of the Earth-ionosphere cavity, *Atmos. Res.*, 91, 188–193.

[60] Sentman, D. D. (1987), *Magnetic elliptical polarization of Schumann resonances, Radio Sci.*, 22, 595–606.

[61] Poklad, Y. V., Ryakhovsky, I. A., Gavrilov, B. G., Ermak, V. M., Kozakova, E. N., and Achkasov, N. S. (2022), Investigation of the reaction of Schumann resonances to short transient geophysical events under the influence of atmospheric electromagnetic noise. *J. Geophys. Res.: Atmosp.*, 127, e2022JD036820, https://doi.org/10.1029/2022JD036820.

[62] Rodríguez-Camacho, J., J. Fornieles, M. C. Carrión, *et al.* (2018), On the need of a unified methodology for processing Schumann resonance measurements, *J. Geophys. Res.: Atmos.*, 123, 13277–13290, https://doi.org/10.1029/2018JD029462

[63] Balser, M. and C. Wagner (1960b), Observations of earth-ionosphere cavity resonances, *Nature*, 188, 638–641.

[64] Clayton, M. and C. Polk (1977), *Diurnal validation and absolute intensity of worldwide lightning activity, in Electrical Processes in Atmospheres*, H. Dolezalek and R. Reiter, eds., Steinkopff, Darmstadt, Germany, pp. 440–449.

[65] Sentman, D. D. and B. J. Fraser (1991), Simultaneous observation of Schumann resonances in California and Australia: evidence for intensity modulation by local height of the D region, *J. Geophys. Res.*, 96(9), 15973–15984.

[66] Belyaev, G. G., A. Yu. Schekotov, A. V. Shvets, and A. P. Nickolaenko (1999), Schumann resonances observed using Poynting vector spectra. *J. Atmos. Sol-Terr. Phys.*, 61, 751–763.

[67] Satori, G., J. Szendroi, and J. Vero, (1996), Monitoring Schumann resonances—I. Methodology. *J. Atmos. Terr. Phys.*, 58, 1475–1481.

[68] Shvets, A. V. (1999), Distance estimation to the world thunderstorm centers by measurement of the Schumann resonance background. *XXVI General Assembly URSI, University of Toronto*, Toronto, Ont., Canada, August 13–21, 1999, Abstracts, p. 297.

[69] Yatsevich, E. I., A. V. Shvets, L. M. Rabinowicz, A. P. Nickolaenko, G. G. Belyaev, and A. Yu. Schekotov (2005), Results of comparing Schumann resonance observations with a model of the single world thunderstorm center, *Izvestia Vuzov, Radiophysica*, 48(4), 283–298. English translation is published in *Radiophysics and Quantum Electronics*.

[70] Shvets, A. and M. Hayakawa (2011), Global lightning activity on the basis of inversions of natural ELF electromagnetic data observed at multiple stations around the world, *Surv. Geophys.*, 32, 705–732.

[71] Shvets, A. V. (2001), A technique for reconstruction of global lightning distance profile from background Schumann resonance signal, *J. Atmos. Sol-Terr. Phys.*, 63, 1061–1074.

[72] Shvets, A. V., M. Hayakawa, M. Sekiguchi, and Y. Ando (2009), Reconstruction of the global lightning distribution from ELF electromagnetic background signals, *J. Atmos. Solar Terr. Phys.*, 71, 1405–1412.

[73] Raemer, H. R. (1961), On the extremely low frequency spectrum of the earth-ionosphere cavity response to electrical storms, *J. Geophys. Res.*, 66, 1580–1583.

[74] Polk, C. and F. Fitchen (1962), Schumann resonances of the earth-ionosphere cavity – extremely low frequency reception at Kingston, R.I., *J. Res. NBS*, 66D, 313–318.

[75] Rycroft, M. J. (1963), *Low frequency disturbances of natural origin of the electric and magnetic fields of the earth*, Ph.D. thesis, University of Cambridge.

[76] Fullekrug, M. and A. C. Fraser-Smith (1997), Global lightning and climate variability inferred from ELF magnetic field variations, *Geophys. Res. Lett.*, 24, 2411–2414.

[77] Jones, D. L. (1974b), *ELF-VLF Radio Wave Propagation*, J. A. Holtet, ed., D. Reidel Publishing Company, Dordrecht, p. 207.

[78] Madden, T. and W. Thompson (1965), Low-frequency electromagnetic oscillations of the Earth-ionosphere cavity, *Rev. Geophys.*, 3(2), 211.

[79] Nickolaenko, A. P., M. Hayakawa, and Y. Hobara (1996), Temporal variations of the global lightning activity deduced from the Schumann resonance data, *J. Atmos. Terr. Phys.*, 58, 1699–1709.

[80] Polk, C. (1968), Relation of ELF noise and Schumann resonances to thunderstorm activity, in *Planetary Electrodynamics*, vol. 2, H. Volland, ed., CRC Press, Boca Raton, FL, pp. 55–83.

[81] Sátori, G. and B. Zieger (1999), El Ninõ-related meridional oscillation of global lightning activity. *Geophys. Res. Lett.*, 26, 1365–1368.

[82] Price, C. (2000), Evidence for a link between global lightning activity and upper tropospheric water vapor, *Nature*, 406, 290–293.

[83] Nickolaenko, A. P., M. Hayakawa, and Y. Hobara (2010), Q-bursts: Natural ELF radio transients, *Surv. Geophys.*, 31, 409–425.

[84] Bliokh, P. V., A. P. Nickolaenko, and Yu. F. Filippov (1980), *Schumann Resonances in the Earth-Ionosphere Cavity*, D. Ll. Jones, ed., Peter Peregrinus, Oxford.

[85] Burke, C. P. and D. Ll. Jones (1995), Global radiolocation in the lower ELF frequency band, *J. Geophys. Res.*, 100(D12), 26,263–26,272.

[86] Ishaq, M. and D. L. Jones (1977) Method of obtaining radiowave propagation parameters for the Earth–ionosphere duct at ELF, *Electron. Lett.*, 13, 254–255.

[87] Jones, D. Ll. and C. P. Burke (1992), An experimental investigation of ELF attenuation rates in the Earth–ionosphere cavity, *J. Atmos. Terr. Phys.*, 54, 243.

[88] Nakamura, T., M. Sekiguchi, Y. Hobara, and M. Hayakawa (2010), A comparison of different source location methods for ELF transients by using the parent lightning discharges with known positions, *J. Geophys. Res.*, 115, A00E39, doi:10.1029/2009JA014992.

[89] Price, C., E. Greenberg, Y. Yair, *et al.* (2004), Ground-based detection of TLE-producing intense lightning during the MEIDEX mission on board the Space Shuttle Columbia, *Geophys. Res. Lett.*, 31, L20107, doi:1029/2004GL020711.

[90] Greenberg, E., C. Price, Y. Yair, M. Ganot, J. Bór, and G. Sátori (2007), ELF transients associated with sprites and elves in eastern Mediterranean winter thunderstorms, *J. Atmos. Solar-Terr. Physics*, 69, 1569–1586.

[91] Reising, S. C., U. S. Inan, T. F. Bell, and W. A. Lyons (1996), Evidence for continuing current in sprite-producing cloud-to-ground lightning, *Geophys. Res. Lett.*, 23, 3639–3642.

[92] Rafalsky, V. A., A. V. Shvets, and M. Hayakawa (1995), One-site distance-finding technique for locating lightning discharges, *J. Atmos. Sol-Terr. Phys.*, 57(11), 1255–1261.

[93] Schmidt, C. T. (1993), *Detection of Distant Lightning Strikes from one Location Using Schumann Resonances*, M.Ph. thesis, Mich. Technol. Univ., Houghton.

[94] Price, C., E. Greenberg, Y. Yair, *et al.* (2004), Ground-based detection of TLE-producing intense lightning during the MEIDEX mission on board the Space Shuttle Columbia, *Geophys, Res. Lett.*, 31, L20107, doi:1029/2004GL020711.

[95] Price, C. and D. Rind (1994), Possible implications of global climate change on global lightning distributions and frequencies, *J. Geophys. Res.*, 99, 10823–10831.

[96] Williams, E. R. (2005), Lightning and climate: a review, *Atmos. Res.*, 76, 272–287.

[97] Angell, J. K. (1986), On the variation in period and amplitude of the quasi-biennial oscillation in the equatorial stratosphere, 1951–85, *Mon. Weather Rev.*, 114(11), 2272–2278.

[98] Jones, P. D., T. M. L. Wigley, and P. B. Wright (1986), Global Temperature-variations between 1861 and 1984, *Nature*, 322, 430–434.

[99] Price, C. and M. Asfur (2006), Can lightning observations be used as an indicator of upper-tropospheric water-vapor variability, *Bull. Am. Met. Soc.*, 87, 291, doi:10.1175/BAMS-87-3-xxx.

[100] Hansen, J., A. Lacis, D. Rind, *et al.* (1984), Climate sensitivity: Analysis of feedback mechanisms, in *Climate Processes and Climate Sensitivity*, J. E. Hansen and T. Takahashi, eds., *AGU Geophys. Monograph 29*, American Geophysical Union, Washington, D.C., pp. 130–163.

[101] Del Genio, A. D., W. Kovari Jr., and N. S. Yao (1994), Climatic implications of the seasonal variations of upper troposphere water vapour, *Geophys. Res. Lett.*, 21, 2701–2704.

[102] Lindzen, R. S. (1990), Some coolness concerning global warming, *Bull. Am. Meteorol. Soc.* 71, 288–299.

[103] Rind, D. (1998), Just add water vapor, *Science*, 28, 1152–1153.

[104] Rind, D., E. W. Chiou, W. Chu, *et al.* (1991), Positive water vapor feedback in climate models confirmed by satellite data, *Nature*, 349, 500–502.

[105] Sun, D. Z. and I. M. Held (1996), *A comparison of modeled and observed relationships between interannual variations of water vapor and temperature, J. Clim.*, 9, 665–675.

[106] Satori, G., E. R. Williams, and I. Lemperger (2009), Variability of global lightning activity on the ENSO time scale, *Atmos. Res.*, 91, 500–507.

[107] Satori, G. and B. Ziegler (1999), El Nino related meridional oscillations of global lightning activity, *Geophys. Res. Lett.*, 26, 1365–1368.

[108] Wilson, C. T. R. (1924), The electric field of a thundercloud and some of its effects, *Proc. Phys. Soc. London*, 37, 32D–37D, doi:10.1088/1478-7814/37/1/314.

[109] Franz, R. C., R. J. Nemzek, and J. R. Winckler (1990), Television image of a large upward electrical discharge above a thunderstorm system, *Science*, 249, 48.

[110] Boeck, W. L. and O. H. Vaughan Jr. (1990), Lightning observations from the STS-32 space shuttle mission, *EOS Trans. AGU*, 71, 1241.

[111] Boeck, W. L., O. H. Vaughan Jr., R. J. Blakeslee, B. Vonnegut, M. Brook, and J. McKune (1995), Observations of lightning in the stratosphere, *J. Geophys. Res.*, 100, 1465–1475.

[112] Chen, A. B., C. L. Kuo, Y. J. Lee, *et al.* (2008), Global distributions and occurrence rates of transient luminous events, *J. Geophys. Res.*, 113, A08306, doi:10.1029/2008JA013101.

[113] Dowden, R. L. and C. J. Rodger (1997), Decay of a vertical plasma column: a model to explain VLF sprites, *Geophys. Res. Lett.*, 24, 2765–2768.

[114] Fukunishi, H., Y. Takahashi, A. Uchida, M. Sera, K. Adachi, and R. Miyasato (1999), Occurrences of sprites and elves above the Sea of Japan near Hokuriku in winter, *Eos Trans. AGU*, 80(46), *Fall Meet. Suppl.*, F217.

[115] Hardman, S. F., R. L. Dowden, J. B. Brundell, J. L. Blahr, Z. I. Kawasaki, and C. J. Rodger (2000), Sprites observation in the northern territory of Australia, *J. Geophys. Res.* 105, 4689.

[116] Heavner, M. J., D. L. Hampton, D. D. Sentman, and E. M. Wescott (1995), Sprites over Central and South America (abstract), *Eos Trans. AGU*, 76(46), *Fall Meet. Suppl.*, 115.

[117] Lyons, W. A. (1994), Characteristics of luminous structures in the stratosphere above thunderstorms as imaged by low-light video, *Geophys. Res. Lett.*, 21, 875–878.

[118] Neubert, T., T. H. Allin, E. Blanc, *et al.* (2005), Co-ordinated observations of transient luminous events during the EuroSprite2003 campaign, *J. Atmos. Sol-Terr. Phys.*, 67, 807–820.

[119] Sentman, D. D. (1996), Observations of red sprites and blue jets, *Paper presented at 25th General Assembly, Union Radio Science International*, Lille, France, July 1996.

[120] Sentman, D. D. and E. M. Wescott (1993), Observations of upper atmosphere optical flashes recorded from an aircraft, *Geophys. Res. Lett.*, 20, 2857–2860.

[121] Su, H. T., R. R. Hsu, A. B. C. Chen, Y. J. Lee, and L. C. Lee (2002), Observation of sprites over the Asian continent and over oceans around Taiwan, *Geophys. Res. Lett.*, 29(*4*), 1044, doi:10.1029/2001GL013737.

[122] Yair, Y., C. Price, Z. Levin, *et al.* (2003), Sprite observations from the space shuttle during the Mediterranean Israeli Dust Experiment (MEIDEX), *J. Atmos. Sol-Terr. Phys.*, 65, 635–642.

[123] Neubert, T., N. Østgaard, V. Reglero, *et al.* (2021), Observations of the onset of a blue jet into the stratosphere, *Nature*,589, 371–375, https://doi.org/10.1038/s41586-020-03122-6

[124] Sato, M., T. Ushio, T. Morimoto, *et al.* (2015), Overview and early results of the global lightning and sprite measurements mission, *J. Geophys. Res. Atmos.*, 120, 3822–3851, doi:10.1002/2014JD022428.

[125] Williams, E. R. (2001), Sprites, elves, and glow discharge tubes, *Physics Today* 41, 41–47.

[126] Barrington-Leigh, C. P. and U. S. Inan (1999), Elves triggered by positive and negative lightning discharges, *Geophys. Res. Lett.*, 26, 683–686.

[127] Pasko, V. P., U. S. Inan, Y. N. Taranenko, and T. F. Bell (1995), Heating, ionization and upward discharges in the mesosphere due to intense quasi-electrostatic thundercloud fields, *Geophys. Res. Lett.*, 22, 365–368.

[128] Pasko, V. P., U. S. Inan, T. F. Bell, and Y. N. Taranenko (1997), Sprites produced by quasi-electrostatic heating and ionization in the lower ionosphere, *J. Geophys. Res.*, 102, 4529–4561.

[129] Takahashi, Y., A. Yoshida, M. Sato, *et al.* (2010), Absolute optical energy of sprites and its relationship to charge moment of parent lightning discharge based on measurement by ISUAL/AP, *J. Geophys. Res.*, 115, A00E55, doi:10.1029/2009JA014814

[130] Williams, E., E. Downes, R. Boldi, W. Lyons, and S. Heckman (2007a), Polarity asymmetry of sprite-producing lightning: a paradox?, *Radio Sci.*, 42, RS2S17, doi:10.1029/2006RS003488.

[131] Bell, T. F., V. P. Pasko, and U. S. Inan (1995), Runaway electrons as a source of red sprites in the mesosphere, *Geophys. Res. Lett.*, 22, 2127.

[132] Milikh, G. M., K. Papadopoulos, and C. L. Chang (1995), On the physics of high altitude lightning. *Geophys. Res. Lett.*, 22, 85.

[133] Roussel-Dupré, R. and A. V. Gurevich (1996), On runaway breakdown and upward propagating lightning, *J. Geophys. Res.*, 101, 2297–2311.

[134] Valdivia, J. A., G. Milikh, and K. Papadopoulos (1997), Red sprites: Lightning as a fractal antenna, *Geophys. Res. Lett.*, 24(24), 3169–3172.

[135] Williams, E. R., V. C. Mushtak, R. Boldi, R. L. Dowden, and Z.-I. Kawasaki (2007b), Sprite lightning heard round the world by Schumann resonance methods, *Radio Sci.*, 42, RS2S20, doi:10.1029/2006RS003498.

[136] Hobara, Y., N. Iwasaki, T. Hayashida, M. Hayakawa, K. Ohta, and H. Fukunishi (2001), Interrelation between ELF transients and ionospheric disturbances in association with sprites and elves, *Geophys. Res. Lett.*, 28, 935–938.

[137] Ganot, M., Y. Yair, C. Price, *et al.* (2007), First detection of transient luminous events associated with winter thunderstorms in the eastern Mediterranean, *Geophys. Res. Lett.*, 34, L12801, doi:10.1029/2007GL029258.

[138] Bell, T. F., S. C. Reising, and U. S. Inan (1996), Continuing currents determined from broadband ELF/VLF magnetic fields radiated by positive cloud-to-ground discharges associated with red sprites, *EOS Supplement*, 77, F61 (abstract).

[139] Cummer, S. A. and U. S. Inan (1997), Measurement of charge transfer in sprite-producing lightning using ELF radio atmospherics, *Geophys. Res. Lett.*, 24, 1731.

[140] Hu, W., S. A. Cummer, W. A. Lyons, and T. E. Nelson (2002), Lightning charge moment changes for the initiation of sprites, *Geophys. Res. Lett.*, 29 (8), 1279, doi:10.1029/2001GL014593.

[141] Greenberg, E., C. Price, Y. Yair, C. Haldoupis, O. Chanrion, and T. Neubert (2009), ELF/VLF signatures of sprite-producing lightning discharges observed during the 2005 *EuroSprite* campaign, *J. Atmos. Sol.-Terr. Phys.*, 71(12), 1254–1266, doi:1016/j.jastp.2009.05.005.

[142] Yair, Y., G. Fischer, F. Simoes, N. Renno, and P. Zarka (2008), Updated review of planetary atmospheric electricity, *Space Sci. Rev.*, 137, 29–49, doi:10.1007/ s11214-008-9349-9.

[143] Ksanfomaliti, L. V. (1979), Lightning in the cloud layer of Venus, *Kosmicheske Issledovaniya*, 17(5), 747–762 (in Russian).

[144] Ksanfomaliti, L. V. (1983a), Electrical activity of the atmosphere of Venus. I. Measurements on descending probes , *Kosmicheske Issledovaniya*, 21(2), 279–296 (in Russian).

[145] Ksanfomaliti, L. V. (1983b), Electrical activity of the atmosphere of Venus. II. Satellite measurements , *Kosmicheske Issledovaniya*, 21(4), 619–633 (in Russian).

[146] Ksanfomaliti, L. V. (1985), *Planet Venus* , Nauka, Moscow *(in Russian)*. .

[147] Scarf, F. L. and C. T. Russell (1983), Lightning measurements from the Pioneer Venus Orbiter, *Geophys. Res. Lett.*, 10(12), 1192–1195.

[148] Taylor, W. W. L., F. L. Scarf, C. T. Russell, and L. H. Brace (1979), Evidence for lightning on Venus, *Nature*, 279, 614–616.

[149] Eden, H. F. and B. Vonnegut (1973), Electrical breakdown caused by dust motion in low-pressure atmospheres: consideration for Mars, *Science*, 180, 962.

[150] Farrell, W. F., M. L. Kaiser, M. D. Desch, *et al.* (1999), Detecting electrical activity from Martian dust storms, *J. Geophys. Res.*, 104, 3795.

[151] Melnik, O. and M. Parrot (1998), Electrostatic discharge in Martian dust storms, *J. Geophys. Res.*, 103(A12), 29107–29117.

[152] Renno, N. O., A. Wong, S. K. Atreya, I. de Pater, and M. Roos-Serote (2003), Electrical discharges and broadband radio emission by Matian dust devils and dust storms, *Geophys. Res. Lett.*, 30(22), 2140.

[153] Bar-Nun, A. (1975), Thunderstorms on Jupiter, *Icarus*, 24, 86–94.

[154] Desch, S. J., W. J. Borucki, C. T. Russell, and A. Bar-Nun (2002), Progress in planetary lighting, *Rep. Prog. Phys.*, 65, 955–997.

[155] Little, B., C. D. Anger, A. P. Ingersoll, *et al.*, The Galileo SSI Team (1999), Galileo images of lightning on Jupiter, *Icarus*, 142, 306–323.

[156] Fischer, G., W. S. Kurth, U. A. Dyudina, *et al.* (2007a), Analysis of a giant lightning storm on Saturn, *Icarus*,190, 528–544.

[157] Fischer, G., W. S. Kurth, D. A. Gurnett, *et al.* (2011), A giant thunderstorm on Saturn, *Nature*, 475, 75–77, doi:10.1038/nature10205.

[158] Lammer, H., T. Tokano, G. Fischer, *et al.* (2001), Lightning activity of Titan: can Cassiny/Huygens detect it?, *Planet. Space Sci.*, 49, 561–574.

[159] Tokano, T., G. J. Molina-Cuberos, H. Lammer, and W. Stumptner (2001) Modeling of thunderclouds and lightning on Titan, *Planer. Space Sci.*, 49, 539–560.

[160] Fischer, G., D. A. Gurnett, W. S. Kurth, W. M. Farrell, M. L. Kaiser, and P. Zarka (2007b), Nondetection of Titan lightning radio emissions with Cassini/RPWS after 35 close Titan flybys, *Geophys. Res. Lett.*, 34, L22104, doi:10.1029/ 2007GL031668.

[161] Fischer, G., and D. A. Gurnett (2011), The search for Titan lightning radio emissions, *Geophys. Res. Lett.*, 38, L08206, doi:10.1029/2011GL047313.

[162] Sentman, D. D. (1990), Electrical conductivity of Jupiter's Shallow interior and the formation of a resonant planetary-ionosphere cavity, *ICARUS*, 88, 73–86.

[163] Nickolaenko, A. P. and L. M. Rabinowicz (1982), On the possibility of existence of global electromagnetic resonances on the planets of Solar system, *Space Res.*, 20, 82–89.

[164] Yang, H., V. P. Pasko, and Y. Yair (2006), Three-dimensional finite difference time domain modeling of the Schumann resonance parameters on Titan, Venus, and Mars, *Radio Sci.*, 41, RS2S03, doi:10.1029/2005RS003431.

[165] Simões, F., M. Hamelin, R. Grard, *et al.* (2008), Electromagnetic wave propagation in the surface-ionosphere cavity of Venus, *J. Geophys. Res.*, 113, E07007, doi:10.1029/ 2007JE003045.

[166] Sukhorukov, A. I. (1991), On the Schumann resonances on Mars, *Planet. Space Sci.*, 39(*12*), 1673–1676.

[167] Molina-Cuberos, G. J., J. A. Morente, B. P. Besser, *et al.* (2006), Schumann resonances as a tool to study the lower ionosphere of Mars, *Radio Sci.*, 41, RS1003, doi:10.1029/2004RS003187.

[168] Besser, B. P., K. Schwingenschuh, I. Jernej, *et al.* (2002), Schumann resonances as indicators for lighting on Titan, in *Proceedings of the Second European Workshop on Exo/Astrobiology, Graz, Australia*, 16–19 Sep. 2002 (ESA SP-518, November 2002).

[169] Morente, J. A., G. J. Molina-Cuberos, J. A. Porti, K. Schwingenschuh, and B. P. Besser (2003), A study of the propagation of electromagnetic waves in Titan's atmosphere with the TLM numerical method, *Icarus*, 162, 374–384.

[170] Navarro, E. A., A. Soriano, J. A. Morente, and J. A. Portí (2007), A finite difference time domain model for the Titan ionosphere Schumann resonances, *Radio Sci.*, 42, RS2S04, doi:10.1029/2006RS003490.

[171] Nickolaenko, A. P., B. P. Besser, and K. Schwingenschuh (2003), Model computations of Schumann resonance on Titan, *Planet. Space Sci.*, 51(13), 853–862.

[172] Morente, J. A., J. A. Porti, A. Salinas, and E. A. Navarro (2008), Evidence of electrical activity on Tital drawn from the Schumann resonances sent by Huygens probe, *Icarus*, 195, 802–811.

[173] Simoes, F., R. Grard, M. Hamelin, *et al.* (2007), *A new numerical model for the simulation of ELF wave propagation and the computation of eigen-modes in the atmosphere of Titan: Did Huygens observe any Schuman resonance?* Planet. Space Sci.¸ 55, 1978–1989.
[174] Bullock, T. H. (2002), Biology of brain waves: natural history and evolution of an informationrich sign of activity, in K. Arikan and N. Moore (eds.), *Advances in Electrophysiology in Clinical Practice and Research*, Wheaton, IL:Kjellberg.
[175] Nunez, P. L. (1981), *Electric Fields of the Brain: The Neurophysics of EEG*, New York, NY: Oxford University Press.
[176] König, H. L. (1974), Behavioural changes in human subjects associated with ELF electric fields, in M.A. Persinger, Ed., *ELF and VLF Electromagnetic Field Effects*, Boston, MA: Springer.
[177] Segal, Y., L. Segal, E. Shohami, *et al.* (2016), The effect of electromagnetic field treatment on recovery from spinal cord injury in a rat model – clinical and imaging findings. *Int J Neurorehabil.*, 3, 1000203, https://doi.org/10.4172/2376-0281.1000203.
[178] Pechony, O., C. Price, and A. P. Nickolaenko (2007), Relative importance of the daynight asymmetry in Schumann resonance amplitude records, *Radio Sci.*, 42, RS2S06, doi:10.1029/2006RS003456.
[179] Satori, G., M. Neska, E. Williams, and J. Szendroi (2007), Signatures of the day-night asymmetry of the Earth-ionosphere cavity in high time resolution Schumann resonance records, *Radio Sci.*, 42, RS2S10, doi:10.1029/2006RS003483.
[180] Price, C. and V. Mushtak (2001), The impact of the August 27, 1998, g-ray burst on the Schumann resonances, *J. Atmos. Sol-Terr. Phys.*, 63, 1043–1047.
[181] Williams, E. R. and G. Satori (2007), Solar radiation-induced changes in ionospheric height and the Schumann resonance waveguide on different timescales, *Radio Sci.*, 42, RS2S11, doi:10.1029/2006RS003494.
[182] Shvets, A. and M. Hayakawa (2011), Global lightning activity on the basis of inversions of natural ELF electromagnetic data observed at multiple stations around the world, *Surv. Geophys.*, 32, 705–732, doi 10.1007/s10712-011-9135-1.

Chapter 11

High energetic radiation from thunderstorms and lightning

Joseph R. Dwyer[1] and Hamid K. Rassoul[2]

When atmospheric electric fields exceed approximately 267 kV/m $\times$ n, where n is the density of the air relative to the International Standard Atmosphere (ISA) sea-level value, energetic particle processes need to be considered. Above this threshold, the rate at which energetic electrons gain energy from the electric field is higher than the rate at which these electrons lose energy through ionization and atomic excitations. As a result, such electrons "run away" and gain significant energies, as described first by [1]. Furthermore, upon the collision of runaway electrons with air atoms, secondary energetic electrons are produced that run away as well, resulting in an avalanche of high-energy electrons that grow exponentially with time and distance. This mechanism is known as relativistic runaway electron avalanche (RREA) multiplication and was first introduced by Gurevich, Milikh, and Roussel-Dupré [2]. When these runaway electrons collide with air atoms or molecules, they produce copious amounts of X-rays and gamma rays, which can be observed at large distances as well. For example, such emissions have been observed from space in the form of terrestrial gamma-ray flashes (TGFs), from within and near thunderclouds and on the ground in association with thunderstorms and lightning. In 2003, a new discharge mechanism was introduced by Dwyer [3] where the backward propagation of energetic positrons and backscattered X-rays produce a positive feedback effect that self-sustains the runaway electron production, resulting in significant fluxes of energetic radiation [4–6]. In order to investigate the production of runaway electrons in our atmosphere, Monte Carlo simulations, which simulate the motion and scattering of individual particles, are very useful. Because the interaction cross sections of energetic electrons, positrons and photons are well known and easy to incorporate into Monte Carlo codes, these simulations help us reproduce the behavior of these runaway electron processes accurately, even for complicated electric and magnetic field geometries. In this chapter, the physics of runaway electron production and the accompanying

[1]Department of Physics and Astronomy and Space Science Center (EOS), University of New Hampshire, USA
[2]Department of Physics and Space Sciences, Florida Institute of Technology, USA

emissions will be presented, along with a discussion of Monte Carlo techniques and their applications in runaway electron physics.

11.1 Introduction

Because air typically has a non-zero conductivity, electrical currents are seen to flow in a manner that discharges the field upon the application of strong electrical fields in our atmosphere. This field will eventually disappear if not supplied additional charges. The discharge of the field depends upon the magnitude and geometry of the applied electric fields, the temperature, the density and humidity of the air, the recombination rate of ions, the flux of ionizing radiation, and the presence of metal electrodes or other objects in the air such as ice or water particles. The most common and familiar form of electrical discharge in the air involves the production of low-energy (few eV) electrons through the collision of electrons with air atoms. Specifically, when the free electrons are moving through an applied electric field, they may gain enough energy to liberate other electrons by the impact ionization of air atoms. These new electrons may then repeat the same, creating even more free electrons and so on. If new free electrons are created at a rate faster than the rate at which electrons are lost through their attachment to air, then the number of free electrons will grow exponentially with time, resulting in an avalanche of electrons that moves opposite to the electric field vector (due to the electron's negative charge). This avalanche growth occurs at a threshold of approximately 3 million V/m at sea level, also known as the conventional breakdown field [7]. Below this field, according to the given scenario, a few free electrons are generated, whereas the number of free electrons and hence the conductivity is seen to grow rapidly above this field. As a result, it is difficult to maintain an electric field above the convention breakdown field. It is often seen that the air responds by generating a spark, a hot channel that carries most of the current. Lightning also involves electrical currents that flow along hot channels. The formation of these hot channels is not completely clear, but it is generally believed that at some place and time the electric field in the air had to exceed the conventional breakdown field [8]. One of the biggest puzzles in atmospheric sciences involves understanding the conditions that occur when lightning initiates/strikes: specifically, how and where are the large electric fields formed that are responsible for the production of lightning.

The reason why lightning initiation is so puzzling is that decades of *in situ* measurements of thundercloud electric fields have failed to find electric fields near the conventional breakdown field [9]. The conventional breakdown field, in fact, decreases with air density and can be written as $E_b \cong 3.0 \times 10^6$ V/m $\times\, n$, where n is the density of air relative to that at sea level. Even when considering the reduced density of air at thundercloud altitudes, the measured thundercloud fields still appear to be too low by about a factor of 10. It is possible that the presence of hydrometeors (e.g., rain or hail) inside thunderclouds assists in the initiation of lightning by locally enhancing the fields near their surfaces, thereby allowing electron avalanches to somehow result in a hot channel. However, how exactly this occurs in reality has not yet been fully deduced and thus is an active area of research.

Another interesting yet perhaps related puzzle are the observed emissions of X-rays and gamma rays from thunderclouds and lightning. Many independent observations show that lightning emits bright bursts of X-rays up to a few MeV in energy, whereas thunderclouds emit gamma rays up to several tens of MeV. However, the conventional breakdown process that we described above comprises only low-energy electrons with a little more than 10 eV or so of kinetic energy. As a result, conventional breakdown should not produce any high-energy radiation, certainly not MeV gamma rays. Even the hot lightning channel only reaches about 30,000 K, which is way too cold to even make soft X-rays, and billions of degrees too cold to produce the observed gamma rays thermally [10]. Therefore, there should be other processes at work to generate high-energy emissions. In this chapter, we will discuss these processes and how they can be applied to thunderstorms, lightning and laboratory sparks.

11.2 Observations

We will begin with a brief overview of the measurements of the energetic radiation produced in our atmosphere. In the next section, we will introduce the mechanisms that explain this radiation. Our purpose here is not to provide a comprehensive review of all observations and experiments in this field. Instead, we aim at providing the reader with some of the key measurements that highlight the basic processes involved. In particular, we will not be discussing many of the early observations made over the last 70 years. Instead, we will be referring the interested reader to Suszcynsky, Roussel-Dupré, and Shaw [11], as they provide a useful review.

It is well known for over 100 years that high-energy radiation exists in our atmosphere [12]. Although some of this radiation comes from radioactive decays from elements such as radon, most of it comes from the galactic cosmic rays impinging on our atmosphere from space [13]. At the sea level, most of these energetic particles are secondary muons and electrons with a flux of a few hundred per m^2 per sec. At thundercloud altitudes, this flux is about 100 times higher and most of the charged particles are electrons and positrons. The flux of these cosmic rays varies somewhat depending on the solar cycle, solar activity, and the geographic location. Furthermore, the natural background rate of energetic radiation may increase when it rains or snows due to the washout of atmospheric radioactive particles produced by radon decays [14]. All these fluctuations are, however, relatively modest and well understood.

In 1994, Burst and Transient Source Experiment (BATSE), onboard the Compton Gamma-Ray observatory (CGRO), observed intense millisecond-long bursts of gamma rays emanating from the Earth's atmosphere, which were named as Terrestrial Gamma-ray Flashes or TGFs [15]. Since then, TGFs have been observed by other instruments in low-Earth orbit, including RHESSI, Fermi/GBM & LAT, AGILE, and ASIM [16–21]. It is now known that TGFs typically have a duration of around 100 μs, and energies extending to many tens of MeV. They are produced by thunderstorms at altitudes near 13 km [22–25]. The only viable mechanism for generating such gamma rays is the bremsstrahlung interaction of energetic electrons with air. Specifically, when an MeV electron collides with an atomic nucleus, the rapid acceleration of the electron in the coulomb field of the nucleus emits an energetic photon. Monte Carlo computer simulations that take into account the propagation of gamma rays through the atmosphere, including Compton scattering, photo-electric absorption and pair production, showed about 10^{17} high-energy (MeV) electrons to be present inside or just above the thundercloud in order to generate the number of gamma rays that were measured > 600 km away in space. This is a very large number of energetic electrons. Considering that they were produced in less than 1 msec, the duration of the TGF, the fluence (number per m^2) of the energetic electrons must have been at least ten billion times higher than the fluence of cosmic-ray particles that were passing through the thundercloud. If an aircraft was struck by such a TGF at the source, the radiation dose received by individuals inside the aircraft could be significant [26]. In fact, TGFs are so bright that the spacecraft designed to measure powerful X-ray and gamma-ray bursts from the sun and astrophysical sources often experience large dead-times and pulse pile-ups (i.e., saturation) during TGFs [27,28].

It was later found that a subset of the observed TGFs were actually high-energy electron beams generated by Compton scattering and pair production from the TGF gamma rays in the upper atmosphere [29–31]. These terrestrial electron beams (TEBs) follow the geomagnetic field line in the inner magnetosphere and can be observed thousands of kilometers away and sometimes contain large numbers of positrons [32], illustrating some of the interesting phenomena associated with TGFs.

Currently, TGFs are thought to be produced inside thunderclouds during the initial stage of upward positive intra-cloud (IC) lightning [33–39]. However, the relation between lightning and TGF production is not clear, nor is it well understood why some lightning produce TGFs while others do not [40,41].

In addition to the spacecraft observations, X-rays and gamma rays have been found to be emitted by thunderclouds using both *in situ* and ground-based observations [42–52]. These emissions often take the form of gamma-ray "glows" that can last from seconds to minutes (also called thunderstorm ground enhancements when observed from the ground). Many of these observations show energy spectra similar to TGFs that extend into the multi-MeV range, indicating similar source mechanisms. However, the flux of these gamma rays is generally many orders of magnitude lower than the flux of gamma rays in TGFs. A few events that did exhibit TGF-like fluxes were ground-based observations of TGF-like bursts seen at sea level in Florida [53,54], an observation of a TGF made by the ADELE instrument onboard an NCAR-NSF Gulfstream-V aircraft [40], and the TGF-like events observed on ground in Japan [55–59]. All these events appear to be very similar to the TGFs observed from space.

Another source of energetic radiation in our atmosphere is associated with lightning leaders [3,60]. Indeed, X-rays are now routinely measured from natural cloud-to-ground and rocket-and-wire-triggered lightning. In a series of experiments beginning in 2002 at the University of Florida/Florida Tech International Center for Lightning Research and Testing (ICLRT) at Camp Blanding, FL, it was found that lightning produces significant X-ray emission, up to about an MeV, during the stepped-leader phase of natural lightning and during the dart-leader phase of natural and rocket-and-wire-triggered lightning, with the most intense emission often detected immediately before the return stroke during the attachment process [53,61–64]. The X-rays are usually observed to arrive in short (<1 μs) bursts in coincidence with the step formation of the stepped-leaders and dart-stepped leaders.

Long laboratory sparks in air at 1 atmosphere pressure have also been found to emit X-rays with characteristics very similar to those found in lightning discharges [30,65–69,112]. However, unlike the TGFs and some thundercloud emissions, the energy spectra of X-rays from lightning and laboratory sparks only extend up to an MeV or so, indicating a different mechanism for the production of the runaway electrons [70].

It can be noted in this chapter and often in literature, the terms "X-ray" and "gamma ray" are used interchangeably. Technically, all the energetic photons under consideration here are X-rays, as they are emitted by energetic electrons via the bremsstrahlung process. The name gamma ray is sometimes reserved for photons that are emitted during nuclear processes. However, we have instead followed the astrophysics convention of calling the photons in the MeV range gamma rays and photons in the keV range X-rays, regardless of the mechanism that produced them. (In astrophysics, this is not always known.) A confusing exception occurs when discussing TGFs: all the TGF photons are usually called gamma rays, even the lower energy ones in the keV range, as the TGFs were first observed by a gamma-ray experiment.

Finally, we wish to point out that there exists a growing body of literature on the production and detection of neutrons and radioactive isotopes from thunderclouds and lightning (e.g., [55,56]). It is possible that these neutrons are a natural by-product of the energetic processes discussed in this chapter. However, more research is needed on this topic. In addition to the references above, we refer the interested reader to the review paper by Milikh and Roussel-Dupré [71] and the references therein regarding neutrons.

11.3 Runaway electrons

In 1925, C.T.R. Wilson showed that for sufficiently strong electric fields, electrons could gain large energies from the field [1]. Specifically, when the rate at which an electron gains energy from the field exceeds the rate at which it loses energy, predominately through ionization energy losses, then the electron will gain energy and is said to "run away." The electric field at which this happens is about

2.67×10^5 V/m $\times$ n, where n is the density of air relative to the International Standard Atmosphere (ISA) sea-level value. Note that throughout this chapter, we have presented results relative to the ISA air density, which is different from the reference density used in previous works (e.g., STP). Therefore, the electric field values presented here are different from the ones seen in earlier papers. This field is about a factor of ten below the conventional breakdown field and is comparable to maximum fields seen inside thunderclouds. This is illustrated in Figure 11.1, showing the rate of energy loss of an energetic electron moving in air (effective frictional force). The plot also shows the rate of energy gain from a strong electric field (horizontal line). It can be seen that when electrons have energies above a threshold, ε_{th}, then they may run away. However, if the field to be below ε_{th}, electrons will lose energy and will not run away. Therefore, in order to generate a runaway electron, an energetic "seed" electron with energy above ε_{th} must be provided from an external source. An exception to this rule occurs when the electric field is increased above the critical field, E_c, above the energy loss curve for all kinetic energies. In this case, all free electrons may run away and, in particular, the thermal population created at low energies may run away as well. This mechanism is usually called "cold runaway" or sometimes "thermal runaway," and does not require any external seed particles [72,73]. We will further argue below that cold runaway may be important for X-ray emissions from lightning leaders.

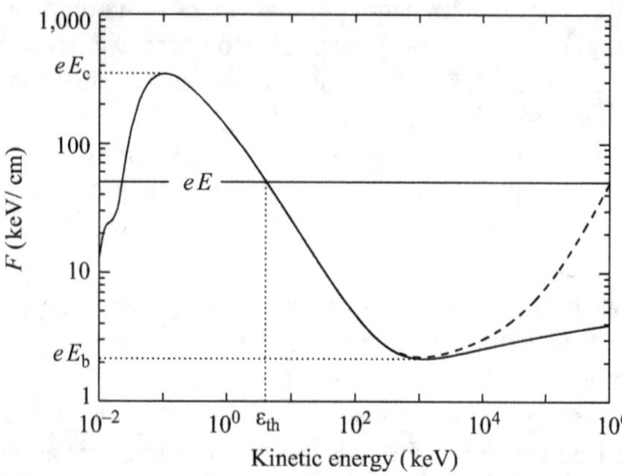

Figure 11.1 *The effective frictional force experienced by a free electron moving through air at STP as a function of kinetic energy. The solid curve is due to inelastic scattering of the electron by air molecules, and the dashed curve indicates the effects of bremsstrahlung emission. The horizontal line shows the electric force from a 5.0 × 10^6 V/m electric field. Runaway electrons occur for kinetic energies greater than the threshold energy, $\varepsilon > \varepsilon_{th}$. In the figure, E_c is the critical electric field strength for which low-energy thermal electrons will run away, and E_b is the so-called breakeven field, the minimum field needed to produce relativistic runaway electrons. Figure from Dwyer [70]*

According to Wilson's original work, the energetic seeds that result in runaway electrons are all provided by external sources such as cosmic rays. Therefore, for each atmospheric cosmic-ray electron that arrives, at most one runaway electron may be generated. Although a runaway electron may gain energy and travel farther than the seed particle that it came from, this mechanism will not produce large fluxes of runaway electrons, especially the large fluxes known to be associated with TGFs. On the other hand, it is possible for cold runaway to provide the energetic seeds, which subsequently experience additional energy gain via Wilson's mechanism. This combination could potentially explain TGFs, as will be discussed below.

In 1992, Gurevich, Milikh, and Roussel-Dupré showed that when Møller scattering (electron–electron elastic scattering) is included, the runaway electrons described by Wilson will also undergo avalanche multiplication, resulting in a large number or relativistic runaway electrons for each energetic seed electron injected into the high-field region [2]. Interestingly, based on Wilson's research notes, Williams [74] argued that Wilson was aware of runaway electron avalanche multiplication, referring to it as a "snowball effect." This avalanche mechanism is commonly referred to as the Relativistic Runaway Electron Avalanche (RREA) mechanism [75,76]. In addition, a hypothesized electrical breakdown of the air generated by RREAs has been termed as "runaway breakdown" [77]. Specifically, several authors have claimed that a RREA acting on cosmic rays results in a conductivity increase large enough to result in an electrical breakdown. Dwyer and Babich [78] challenged these conductivity calculations and argued against the use of the term "runaway breakdown," as RREA is not really an electrical breakdown.

In 2003, Dwyer introduced a new mechanism that involves positive feedback effects from positrons and energetic photons. This mechanism, illustrated in Figure 11.2, comprises avalanches of runaway electrons that emit bremsstrahlung X-rays. These X-rays may either Compton backscatter or pair-produce in air. If the backscattered photons propagate to the start of the avalanche region and produce other runaway electrons, either via Compton scattering or photoelectric absorption, then a secondary avalanche is created. Alternatively, the positrons created by pair-production often turn around in the ambient electric field and run away opposite to electrons. The positrons are relativistic, allowing them to travel for many hundreds of meters before annihilating. If these positrons propagate to the start of the avalanche region, they can produce additional runaway electrons via hard elastic scattering with atomic electrons in the air (i.e., Bhabha scattering), thereby producing secondary avalanches. These secondary avalanches can in turn emit more X-rays that Compton scatter or pair-produce, resulting in more feedback and avalanches. This positive feedback effect allows the creation of RREAs to be self-sustaining, no longer requiring an external source of energetic seed electrons. As a result of this positive feedback, the number of runaway electron avalanches increases exponentially on a timescale measured in microseconds.

The two principal feedback mechanisms that will be referred to as X-ray feedback (also known as gamma-ray or photon feedback) and positron feedback were originally described by Dwyer [79], who also used a Monte Carlo simulation

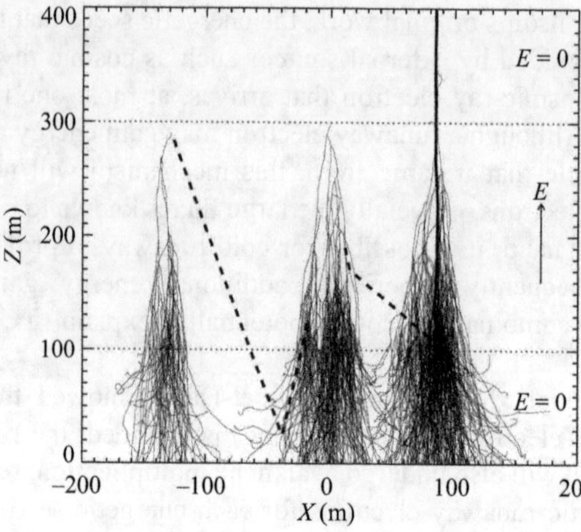

Figure 11.2 The relativistic feedback mechanism. Partial results of the Monte Carlo simulation are shown. The light tracks are the runaway electrons, the dashed lines are the X-rays, and the dark track is a positron. The entire avalanche is initiated by one, 1 MeV, seed electron injected at the top center of the volume. The horizontal dotted lines show the boundaries of the electric field volume (E = 1,000 kV/m). For clarity, only a small fraction of the runaway electrons and X-rays produced by the avalanche are plotted. The avalanches on the left and right illustrate the X-ray feedback and positron feedback mechanisms, respectively. Figure from Dwyer [79]

to calculate the electric field thresholds necessary for feedback to be important (also see [4]). In addition, second-order feedback effects such as feedback from bremsstrahlung X-rays emitted from the backward propagating positrons and feedback from the 511 keV gamma rays emitted by the annihilating positrons can occur as well [80]. To distinguish the feedback mechanisms that involve high-energy particles described here from the low-energy feedback mechanisms occurring in ordinary Townsend gas discharges, these feedback mechanisms, as described above, are jointly known as relativistic feedback [80].

Because the discharge currents generated by relativistic feedback grow exponentially on very short timescales, the electric field will be discharged very quickly, regardless of the charging currents. As a result, relativistic feedback describes a new internal state of the system, a self-sustained discharge that does not rely on externally supplied particles. Once the relativistic feedback starts, the electric field will always discharge, at least partially. Therefore, this can be considered a novel form of electrical breakdown. Because relativistic feedback can generate large electrical currents while emitting little visible light, it is sometimes referred to as "dark lightning" [81]. Indeed, Dwyer [79] showed that large electric fields are highly unstable due to relativistic feedback. Relativistic feedback is important

Table 11.1 Mechanisms for generating relativistic runaway electrons

Name	Reference	New feature	Increase in energetic electrons	Discharge time
Runaway electron	Wilson [1]	Electron energy gain from electric field	$\times 1$	~ hours
Relativistic runaway electron avalanche (RREA)	Gurevich *et al.* [2]	Møller scattering	Up to $\times 10^5$	~10 sec
Relativistic feedback	Dwyer [79]	Backward-propagating runaway positrons and Compton scattered X-rays	Up to $\times 10^{13}$	<1 msec

because it can naturally explain very large fluxes of energetic electrons and gamma rays as well as severely limit the electric field regimes in which alternative mechanisms (i.e. runaway breakdown) may operate.

Table 11.1 summarizes the three mechanisms that produce relativistic runaway electrons, as discussed above. It includes the new feature in each mechanism that is added to the previous mechanism that increases the number of resulting energetic electrons. For Wilson's runaway electrons, the previous mechanism involves the ambient atmospheric cosmic rays. The time for the electric field to discharge due to the increase in conductivity from ionization is also listed. In addition, cold runaway is important, as it provides seed particles for Wilson's runaway electrons and RREA, and is thus treated as a fourth runaway electron generating mechanism.

11.4 Monte Carlo simulations

Monte Carlo simulations are useful in modeling runaway electron physics, including the production and propagation of X-rays and gamma rays and the positron and X-ray feedback mechanisms. Monte Carlo codes are used to extract many fundamental quantities that describe runaway electrons, including avalanche lengths and times, lateral and longitudinal diffusion coefficients for the runaway electrons, and positron production rates and feedback rates. For any given electric and magnetic field configuration, including time-varying fields, Monte Carlo simulations can accurately calculate the production of all the resulting energetic particles. These calculations have been successfully compared with TGF observations to determine their properties. In order to simulate the changes to the field that result from the runaway discharge, the runaway electrons are sometimes modeled using a separate transport code that simulates the propagation of the particles as a whole, not modeling every individual particle. The Monte Carlo simulations, in this case, are useful for providing the parameters used by the transport code, such as diffusion coefficients, runaway avalanche propagation speeds and avalanche times [25,82].

One productive Monte Carlo code was developed in 2002 by J. R. Dwyer [25,64,70,78–80,83–87]. This simulation of relativistic runaway electron avalanche is capable of modeling the development and propagation of runaway electron avalanches in any gaseous medium for both spatially and time-varying electric and magnetic fields. This simulation includes, in an accurate form, all the important interactions involving runaway electrons, positrons, X-rays, and gamma rays. These interactions include energy losses through ionization, atomic excitation, bremsstrahlung, and Møller scattering. The simulation fully models elastic scattering using a shielded-Coulomb potential and includes bremsstrahlung production of X-rays and gamma rays and the subsequent propagation of the photons, including photoelectric absorption, Rayleigh scattering, Compton scattering and pair-production. In addition, it includes positron propagation (and annihilation) and the generation of energetic seed electrons via Bhabha scattering of positrons and via Compton scattering and photoelectric absorption of energetic photons. In recent work, the REAM Monte Carlo simulation also includes low-energy electron and ion currents and self-consistent electric fields changes [88].

In practice, the Monte Carlo code propagates the electrons and positrons in successive small time steps. During each time step, the forces acting on the particles from the electric and magnetic fields plus the effective frictional force from the ionization energy loss, and the atomic excitations are found. The relativistic equations of motion are then solved using the fourth-order Runge–Kutta method. During each time step, the probability of interactions, such as the electron-electron elastic scattering, the elastic scattering with the atomic nuclei, and the bremsstrahlung interactions, are calculated from the relevant cross-sections. A random number generator is then used to determine if the particles have actually undergone each interaction for that time step based on these probabilities. If an interaction does occur, interaction cross-section and relativistic kinematics are used to calculate the momentum and energy changes that result from the interaction and the production of either secondary electrons or bremsstrahlung photons. These secondary particles are then added to the Monte Carlo code and simulated along with the original particles. For electrons and positrons, we are often only interested in the particles that are involved in the runaway electron avalanches, and so electrons below the threshold energy ε_{th} are usually discarded to save computational time.

Because Monte Carlo simulations include all the relevant physics in their calculations, a description of the important runaway electron physics describes the ingredients of the Monte Carlo code as well. For electrons moving through air with kinetic energies above a few hundred eV, most of the energy is lost in ionizing the air, with a smaller amount going into atomic excitations. The ionization/atomic excitation energy losses per unit length along the path of the energetic particles path are well described by the famous Bethe equation [89]

$$F_b(\varepsilon') \equiv -\frac{d\varepsilon'}{dx} = \frac{2\pi N_{air} Z r_e^2 mc^2}{\beta'^2} \times \left[\ln\left(\frac{m^2c^4(\gamma'^2-1)(\gamma'-1)}{I^2} \right) - \left(1 + \frac{2}{\gamma'} - \frac{1}{\gamma'^2} \right) \ln 2 + \frac{1}{\gamma'^2} + \frac{(\gamma'-1)^2}{8\gamma'^2} - \delta_{den}(\gamma') \right] \tag{11.1}$$

where N_{air} is the number density of the gas atoms ($N_{air} = 5.39\times10^{25}$ m^{-3} for air at sea level); Z is the average atomic number of the gas atoms ($Z = 7.26$ for air); $r_e = 2.818\times10^{-15}$ m is the classical electron radius; $mc^2 = 511$ keV is the rest energy of an electron; $\beta' = v'/c$, and $\gamma' = 1/\sqrt{1-\beta'^2}$ is the Lorentz factor of the energetic electron; I is the effective ionization potential (e.g., $I = 85.7$ eV for air) and δ_{den} is a small correction due to the density effect. Equation (11.1) is accurate above a few hundred eV and agrees with the curve in Figure 11.1 for these energies.

As the energetic electrons move through the air, some of the secondary electrons are produced with substantial energies. The production of secondary electrons is described by Møller scattering (electron–electron elastic scattering), given by the cross-section [90]

$$\frac{d\sigma_{Moller}}{d\varepsilon} = \frac{2\pi\, r_e^2 mc^2}{\beta'^2}\left[\frac{(\gamma'-1)^2 m^2c^4}{\varepsilon^2(mc^2(\gamma'-1)-\varepsilon)^2} - \frac{(2\gamma'^2+2\gamma'-1)}{\varepsilon\,(mc^2(\gamma'-1)-\varepsilon)\gamma'^2} + \frac{1}{m^2c^4\gamma'^2}\right], \tag{11.2}$$

where ε is the kinetic energy of the scattered atomic electron and β' and γ' are the speed divided by c and Lorentz factor of the incident electron.

Because the energy losses from Møller scattering are explicitly modeled in the Monte Carlo simulation, these losses should not also be included in the ionization energy loss F_b. Following Lehtinen, Bell and Inan [91], this is remedied by subtracting the quantity

$$\begin{aligned} F_{Moller} &= N_{air} Z \int_{\varepsilon_{\min}}^{mc^2(\gamma'-1)/2} \varepsilon\,\frac{d\sigma_{Moller}}{d\varepsilon}\,d\varepsilon \\ &= \frac{2\pi N_{air} Z r_e^2 mc^2}{\beta'^2}\left[\ln\left(\frac{mc^2(\gamma'-1)}{2\varepsilon_{\min}}\right) + 1 - \frac{\varepsilon_{\min}}{(mc^2(\gamma'-1)-\varepsilon_{\min})}\right. \\ &\quad \left. -\left(1+\frac{2}{\gamma'}-\frac{1}{\gamma'^2}\right)\ln\left(\frac{2\,(mc^2(\gamma'-1)-\varepsilon_{\min})}{mc^2(\gamma'-1)}\right) + \frac{(\gamma'-1)^2}{8\,\gamma'^2} - \frac{\varepsilon_{\min}^2}{2\,m^2c^4\gamma'^2}\right] \end{aligned} \tag{11.3}$$

from the energy loss rate (11.1) when calculating the effective frictional force (energy loss rate):

$$F_{eff} = F_b - F_{Moller} \tag{11.4}$$

In (11.3), the upper limit of the integration comes from the maximum energy for an indistinguishable secondary electron during the scattering process.

When photons are emitted via bremsstrahlung, the photon propagation involves four principle interactions: photoelectric absorption, Compton Scattering, pair-production and Rayleigh scattering. At high energies, Compton scattering is well described by the famous Klein–Nishina formula. At intermediate energies, tabulated values may be used. Tabulated values for the cross-sections for photoelectric absorption and pair-production are also used. Rayleigh scattering is calculated using the standard cross-sections for this process. In all cases, when secondary electrons or positrons are created from the basic interactions described above, their emission angle is calculated using the kinematics of the interaction. The new electrons and positrons are then added to the simulation.

The positrons created via pair production are simulated by the Monte Carlo in a manner similar to the simulation of the runaway electrons. The ionization energy loss equation is modified slightly and the Møller scatter cross-section is replaced with the Bhahba scattering cross-section. In practice, the propagation of the runaway positrons looks very similar to that of electrons, with the exception that avalanches of runaway positions are not produced like the avalanches of electrons. On the other hand, the positions generate energetic secondary electrons that spawn new runaway electron avalanches; this is the positron feedback process. As these positrons are propagated, the annihilation cross-section is used to determine whether they annihilate with atomic electrons. The runaway positrons are very energetic, reaching many tens of MeV and, therefore, the annihilation cross-section for the positrons is small, allowing the positrons to travel on the order of a kilometer at sea level before annihilating. When the positrons do annihilate, the two annihilation gamma rays are added to the simulation.

In the Monte Carlo simulation described above, elastic scattering of the runaway electrons (and runaway positrons) with atoms with an atomic number Z are calculated using the shielded-Coulomb potential

$$V(r) = \frac{Ze}{4\pi\,\varepsilon_o r}\exp(-r/a) \tag{11.5}$$

where r is the spherical radius. This potential is an approximate expression for the potential derived from the Thomas-Fermi model of the atom when $a = 183.8\,\lambda\!\!\!^{-}\,Z^{-1/3}$ [92], where $\lambda\!\!\!^{-}$ is the Compton wavelength. As already discussed, the elastic scattering of the runaway electrons with the individual atomic electrons (Møller scattering) is also included in the simulation.

As derived in Dwyer's work (2007), the elastic scattering differential cross-section per solid angle is given by

$$\frac{d\sigma}{d\Omega} = \frac{1}{4}\left(\frac{Zr_e}{\beta^2\gamma}\right)^2 \cdot \frac{(1-\beta^2\sin^2(\theta/2))}{\left(\sin^2(\theta/2) + \frac{\hbar^2}{4p^2a^2}\right)^2} \tag{11.6}$$

where all the symbols have their usual meaning. The Monte Carlo simulation fully models the runaway electron propagation by allowing electrons and positrons to scatter according to this cross-section. Accurately modeling this elastic scattering is very important because it results in the electrons being scattered off the electric field lines, affecting the rate of energy gain, and causes spatial diffusion of the runaway electrons as the avalanche propagates. The latter reduces the maximum conductivity produced by a runaway electron avalanche, affecting models of lightning initiation [86].

11.5 Energy spectrum

An important characteristic of the runaway electron avalanches is their steady-state energy spectrum. After runaway electron avalanches cover more than a few avalanche lengths, the energy spectrum approaches a steady-state configuration. Although the number of runaway electrons grows exponentially, as the avalanche continues, the overall shape of the energy spectrum does not change. The reason behind this is that as individual electrons gain energy, they always create new lower-energy electrons via Møller scattering, resulting in an exponential energy spectrum.

The runaway electron energy distribution (spectrum) in an avalanche may be described by the approximate equation:

$$\frac{f_{re}}{\tau_{re}} = -\frac{d}{d\varepsilon}(b(\varepsilon)f_{re}) + N_{air}Z\int_{\varepsilon}^{\infty}\frac{d\sigma_{Moller}}{d\varepsilon}v' f_{re}\,d\varepsilon' \tag{11.7}$$

where $b(\varepsilon) = d\varepsilon/dt$ is the average rate of change of the energy of the electrons; $f(\varepsilon, t) = \frac{dN}{d\varepsilon}$ is the number of electrons per unit energy [93]; and τ_{re} is the runaway electron avalanche time. The first term on the right side describes the rate of change in the number of electrons due to energy loss or gain. The second term on the right describes the creation of new electrons due to Møller scattering.

Although runaway electrons may propagate for energies above ε_{th} (see Figure 11.1), in practice, many electrons that have energies only slightly above this threshold do not run away due to the scattering processes. Instead, we can define the boundary energy, ε_b, at which point as many electrons gain energy as lose energy. This boundary energy is usually a few times larger than ε_{th}, as can be found from Figure 11.1 [78]. Equation (11.7) shows that for $\varepsilon \gg \varepsilon_b$, the electrons run away in a direction opposite the applied electric field vector, resulting in a field-aligned beam. Therefore, $b(\varepsilon) \approx v(eE - F_b)$. For relativistic electrons, $F_b(\varepsilon) = F_d \approx 260$ kV/m $\times\, n$, where n is the density of air relative to ISA value and $v = 0.89c$. F_d is the average energy loss rate, which is approximately constant for relativistic particles. If we consider energies that are high enough above ε_b so that most of the new electrons are injected below that energy, then (11.7) becomes

$$\frac{f_{re}}{\lambda} = -(eE - F_d)\frac{df_{re}}{d\varepsilon} \tag{11.8}$$

where $\lambda = v\tau_{re}$. The solution to this equation is

$$f_{re} = f_o \exp\left(\frac{-\varepsilon}{(eE - F_d)\lambda}\right) \tag{11.9}$$

Monte Carlo simulations [79] show that the avalanche length may be written approximately as

$$\lambda \approx \frac{7.3\text{ MeV}}{eE - F_d} \tag{11.10}$$

Substituting this in (11.9) gives

$$f_{re} = f_o \exp\left(\frac{-\varepsilon}{7.3\ \text{MeV}}\right) \tag{11.11}$$

independent of the electric field or the density of air. In contrast, the energy spectrum from the runaway electrons for the cold runaway electron mechanism do not possess any specific shape and may extend up too much smaller energies (e.g., a few 100 keV) depending on the nature of the electric field. As a result, the observation of an X-ray (gamma-ray) spectrum that only extends into the 100 keV range, which is typical for lightning and laboratory sparks, is a strong indication that cold runaway and not RREA is most important in the production of the runaway electrons.

11.6 RREA parameters from Monte Carlo simulations

From Monte Carlo simulations, the avalanche threshold field in air, E_{th}, is 267 kV/m × n [79,94]. The avalanche (e-folding) length is well fit by the empirical relation,

$$\lambda = \frac{7300\ \text{kV}}{[E - (260\ \text{kV/m}) \times n]} \tag{11.12}$$

valid over the range (300–3,000 kV/m) × n , where the electric field E is measured in kV/m [79,83].

The fluence (electrons/m^2) of runaway electrons in an avalanche passing through a plane, perpendicular to the avalanche direction (z-direction), can be written approximately as

$$\Phi_{re}(r,z) = \frac{1}{4\pi\,(D_\perp/v)\,(z - z_o)} \exp\left(\frac{(z - z_o)}{\lambda} - \frac{r^2}{4(D_\perp/v)\,(z - z_o)}\right) S(z - z_o) \tag{11.13}$$

where r in this case is the cylindrical radius $r^2 = (x - x_o)^2 + (y - y_o)^2$, and S is the step function. The sea level values of the lateral diffusion coefficient, $D_\perp/v$, as a function of electric field strength, is found by Monte Carlo simulations and is well described by the following empirical formula [86]:

$$D_\perp/v = (6.22 \times 10^4)E^{-1.79}\ \text{(m)} \tag{11.14}$$

The diffusion of the runaway electrons in the longitudinal direction along the avalanche can be similarly described [86] and is found to be

$$D_{||}/v = (4.04 \times 10^3)E^{-1.57}\ \text{(m)} \tag{11.15}$$

In (11.14) and (11.15), E is in units of kV/m, and $D_{\perp}/v$ and $D_{||}/v$ in meters. To calculate the diffusion coefficients, $D_{\perp}$ and $D_{||}$, $D_{\perp}/v$, and $D_{||}/v$ should be multiplied by the avalanche speed presented in Coleman and Dwyer [83] or by the approximate value $v = 0.89c$. Because runaway electron avalanches follow the similarity rule with reduced atmospheric density, (11.14) and (11.15) may be used to find the values at other air densities by converting the actual E to the sea level equivalent field, and then scaling D/v by $1/n$.

11.7 Relativistic feedback

In the Monte Carlo simulations, the relativistic feedback is recorded by keeping track of the electrons and photons that pass through the plane midway between the top and the bottom of the avalanche region. The initial runaway electrons and photons are labeled as the first-generation particles. Any secondary electrons, photons or positrons, regardless of the production mechanisms, carry the same generation number as the runaway electrons that led to the production of the new particles. Once an electron passes through the mid-plane, moving in the direction opposite to the electric field vector, its generation number is increased by 1. The feedback factor, γ, is defined as the ratio of the number of runaway electrons with generation number $N+1$, passing through the mid-plane, divided by number of runaway electrons, with generation number N, passing through the mid-plane. Note that γ in this case is not the Lorentz factor, which shares the same symbol. The feedback factor is the number of new seed electrons that are created per initial seed electron. The feedback factor is analogous to the second Townsend coefficient for low-energy discharges.

The feedback factor describes the rate at which new avalanches are generated. It is important not to confuse this rate with the rate that positrons or X-rays contribute to the runaway electron growth in a given avalanche. For example, there may be only one electron–positron pair created for every 1,000 runaway electrons in an avalanche, making a negligible contribution to the development of an individual avalanche. However, only one such positron is needed to generate an entirely new avalanche, resulting in a dramatic increase in the number of runaway electrons through the production of an ever-increasing number of new secondary avalanches. As a result, the review paper by Milikh and Roussel-Dupre [71] came to an erroneous conclusion regarding the applicability of the relativistic feedback (see [95]).

Dwyer [80] calculated the flux of runaway electrons from the relativistic feedback and compared this flux with results of the standard RREA model. Let S_o be the flux of energetic seed particles that run away, e.g. the flux due to atmospheric cosmic-ray particles and radioactive decays. Based on the altitude and the geographic location, S_o is in the range 100–10,000 $m^{-2}\ s^{-1}$. If no feedback was occurring, then according to the RREA model, the flux of runaway electrons at the end of the avalanche region would be

$$F_{RREA} = S_o \exp(\xi), \text{ where } \xi = \int_0^L \frac{dz}{\lambda} \tag{11.16}$$

In (11.16), ξ is the number of e-folding lengths and is equal to L/λ for a uniform field.

When relativistic feedback (RF) is under consideration, then the runaway electron flux at time t is the sum of all the feedback generations up until that time. For the case $t >> \tau_{fb}$, where τ_{fb} is the time required for one feedback cycle,

$$F_{RF} = \begin{cases} S_o \exp(\xi)\exp(t/\tau')/(\gamma - 1), & \gamma > 1 \\ S_o (t/\tau_{fb})\exp(\xi), & \gamma = 1 \\ S_o \exp(\xi)/(1-\gamma), & \gamma < 1 \end{cases} \tag{11.17}$$

where

$$\tau' \equiv \tau_{fb}/\ln(\gamma) \tag{11.18}$$

is the e-folding time required to increase the flux of runaway electrons due to feedback (assuming $\gamma > 1$). Note that for $\gamma > 1$ and a uniform electric field, (11.17) is proportional to $\gamma^{t/\tau}\exp(L/\lambda)$, which is the same as (11.2) in Dwyer [79]. Furthermore, for the case $\gamma \ll 1$, i.e. very little feedback, F_{RF} approaches the standard result for the RREA model given by (11.16).

It should be noted that relativistic feedback also operates when $\gamma < 1$, as seen in (11.17). In this case, the discharge does not become self-sustainable. However, the effect of the feedback can dramatically increase the number of runaway electrons produced per seed particle injected as γ approaches 1.

On the other hand, for $\gamma > 1$, the discharge becomes self-sustaining and increases exponentially with time. Comparing (11.16) and (11.17) (for $\gamma > 1$), the ratio of the runaway electron flux from relativistic feedback (RF) to that from the RREA model is then

$$\frac{F_{RF}}{F_{RREA}} \approx \frac{\exp(t/\tau')}{(\gamma - 1)} \tag{11.19}$$

Equation (11.19) denotes the ratio of X-rays emitted by the two mechanisms, as the X-ray emission is proportional to the flux of runaway electrons. Detailed calculations show that the ratio given by (11.19) can reach trillions.

The calculations of the feedback factor, γ, are complicated and depend upon the details of the electric field geometry. The threshold value for feedback to become self-sustainable ($\gamma = 1$) for several shapes of the electric field region has been found and is presented in Figure 11.3.

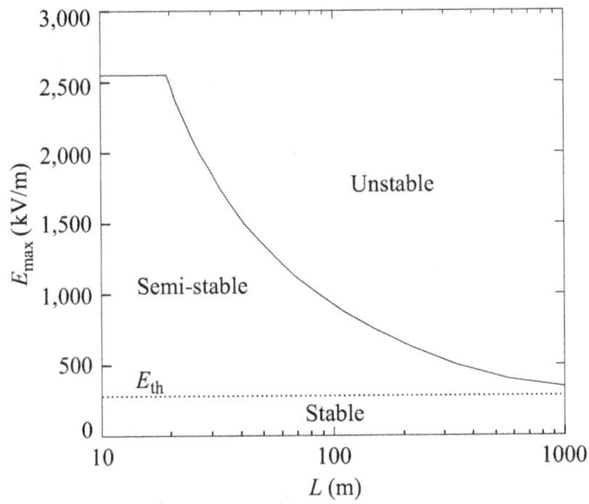

Figure 11.3 The maximum static electric field strength achievable in air versus the length of the electric field region for the case where the lateral width of the avalanche region is much larger than the length of the avalanche region. This curve satisfies the condition where the feedback factor $\gamma = 1$, barring the horizontal part above 2,500 kV/m, which represents the conventional breakdown field. The dotted line shows the runaway avalanche threshold. For the electric field configurations presented on the upper right side of the figure, the field is highly unstable and will quickly discharge until it drops below the feedback threshold, moving to the lower left side of the figure. Figure from Dwyer [79]

11.8 Quantifying TGF source properties

A challenge that occurs when comparing TGF models to observations or when comparing one TGF model to another is agreement regarding consistent quantities for comparison. For example, earlier works have often used the total number of runaway electrons in the TGF source region to describe the size of the TGF, 10^{17} runaway electrons being a typical number reported (e.g., [25]). However, the total number of runaway electrons inferred from the observed gamma-ray fluence is highly model dependent, depending on the electric field strength at the source region, which is not necessarily known. To address this issue, Dwyer *et al.* [87] introduced a more model independent quantity to describe TGFs: the total mass per unit area traversed by all the runaway electrons (i.e., the total grammage) during the TGF, Ξ, which is defined as the total distance traveled by all the runaway electrons along the electric field lines multiplied by the local air mass density along their paths. Several key properties of TGFs, such as the gamma-ray emission rate, the optical power emitted, the electric current moment, and the VLF–LF radio frequency emission may be directly calculated from Ξ and its time derivative, making Ξ a particularly useful quantity to work with. Monte Carlo simulations show that for a typical TGF gamma-ray recorded by spacecraft the corresponding grammage is of the order of $\Xi_0 = 10^{18}$ g/cm^2, which is known as the standard total grammage.

For TGF models, the time derivative of total grammage (in units of g cm^{-2} s^{-1}) traversed by all the runaway electrons can be found as

$$\frac{d\,\Xi}{dt} = v_{re} \int n_{re}\rho \, d^3\vec{x} \tag{11.20}$$

where n_{re} is the number density of runaway electrons in the source region; ρ is the mass density of air as a function of altitude; v_{re} is the average speed of the runaway electrons in the RREA along the local electric field line; and the volume integral is over the entire source region. Equation (11.20) may be integrated to give Ξ as a function of time.

The rate that gamma rays are emitted throughout the entire source region is related to the grammage as follows:

$$\frac{dN_\gamma}{dt} = \frac{1}{\Gamma_\gamma} \frac{d\,\Xi}{dt} \tag{11.21}$$

For energies >1 MeV at the source, Monte Carlo simulations have found that $\Gamma_\gamma = 33.2$ g/cm^2 [87]. The total number of gamma rays emitted during the entire TGF is then

$$N_\gamma = \frac{\Xi}{\Gamma_\gamma} \tag{11.22}$$

Therefore, if we know N_γ, we can calculate Ξ and vice versa. For instance, starting with the gamma-ray fluence observed by a spacecraft in a low-Earth orbit, Monte Carlo simulations can then be used to find the number of gamma rays, N_γ, emitted in the source region, which can then be used to determine Ξ using (11.22). For a given fluence at the spacecraft, N_γ will depend on the depth of the source region, measured as the column depth of atmosphere above the source region (in units of g/cm^2); the grammage will also depend on the atmospheric column depth. Figure 11.4 shows such a calculation of the grammage versus atmospheric column depth for two different spacecraft locations and three different TGF beam geometries. Note that the radio emission from the TGF can be used to infer the source altitude, supporting constrain the grammage in Figures 11.4 and 11.5.

As runaway electrons propagate, they ionize the air, resulting in a large number of free low-energy electrons and low-energy ions. These electrons and ions then drift in the electric field, resulting in sizable electric currents. The total current

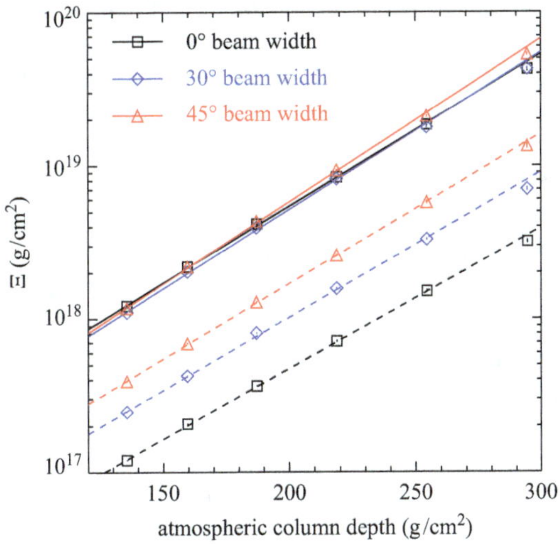

Figure 11.4 *The total grammage versus atmospheric column depth for a standard TGF with a fluence of 0.1 cm^{-2} (> 100 keV) at a spacecraft altitude of 500 km. The top three lines and data are for polar angles (of the spacecraft location relative to the zenith) between 30° and 40°, and the bottom three are for polar angles between 0° and 10°. Figure from Dwyer* et al. *[87]*

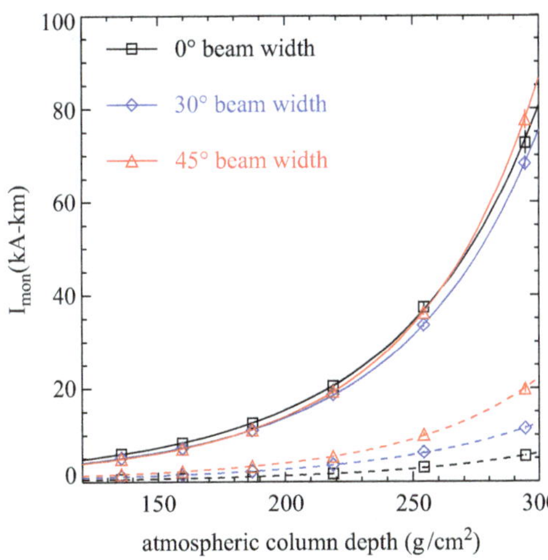

Figure 11.5 *The peak vertical current moment versus atmospheric column depth for a standard TGF with a fluence of 0.1 cm^{-2} (> 100 keV) at a spacecraft altitude of 500 km. For this figure, a TGF duration of σ_{TGF} = 37 μs, corresponding to a T_{50} of 50 μs, is used. The top three lines and data are for polar angles between 30° and 40° and the bottom three for polar angles between 0° and 10°. Figure from Dwyer* et al. *[87]*

moment produced by a TGF from this ionization is found to be

$$I_{mom} = \int \left\{ en_{re}\rho\, v_{re}\alpha\, \tau_a \mu_e E + e\int_{-\infty}^{t} n_{re}\rho\, v_{re}\alpha\, (\mu_+ + \mu_-)Edt' \right\} d^3\vec{x}$$
$$\approx e\alpha\bar{\tau}_a\bar{\mu}_e\bar{E}\frac{d\,\Xi}{dt} + e\alpha\,(\bar{\mu}_+ + \bar{\mu}_-)\bar{E}\,\Xi \qquad (11.23)$$

where the constant α is the ionization rate of runaway electrons in a RREA per g/cm^2 of path length along the electric field line ($\alpha = 6.42\times10^4$ cm^2/g); τ_a is the two and three-body attachment time of the low-energy electrons; μ_e, μ_+, and μ_- are the mobilities of the low-energy electrons and positive and negative ions, respectively; e is the charge of the electron; and E is the electric field strength [87]. In (11.23), the average values are also used for the attachment times, mobilities and the electric field strengths: $\bar{\tau}_a$, $\bar{\mu}_e$, and $\bar{E}$, respectively.

The current moments calculated for the same TGF shown in Figure 11.4 are shown in Figure 11.5 as a function of column depth of the source region. It can be observed that very large current moments, comparable to lightning return strokes, are possible.

The current moment may be used to calculate the VLF–LF radio frequency emissions produced by the TGF. For instance, for large horizontal distances, R, from the source region the radiation electric field is given by

$$E_{rad} = \frac{\sin\theta}{4\pi\varepsilon_0 c^2 R}\frac{d\,I_{mom}}{dt} \qquad (11.24)$$

where all the symbols have their usual meaning [96]. Indeed, a close association between the TGF gamma-ray emissions and radio pulses produced during the TGFs has been found [34,96]. Figure 11.6 shows an example of the radiation field (in this case, the radiation magnetic field B_{rad}) observed on the ground from a TGF observed by the Fermi spacecraft, along with model fits.

For the optical emission directly associated with the runaway electrons, the main emission mechanism is the fluorescence of air (mostly from nitrogen). The optical power from the fluorescence of the air is the rate of energy deposit multiplied by the fluorescent yield Y:

$$P_{optical} = Y\,F_d\frac{d\,\Xi}{dt} \qquad (11.25)$$

where F_d is the average energy loss per runaway electron in a RREA per g/cm^2. Monte Carlo simulation found that $F_d \approx 2{,}110$ keV cm^2/g, and the fluorescent yield, which scales approximately inversely with air density for altitudes below 20 km, $Y = 1.3\times10^{-21}$ J/keV $\times\, n^{-1}$ in the visible range [81].

In summary, the grammage Ξ connects, in a fairly model-independent way, the TGF gamma-ray flux, the radio frequency emissions and the optical emissions, all of which are measurable quantities. The grammage may also be determined from TGF models such as Monte Carlo simulations, allowing the models to be compared in a consistent way with observations.

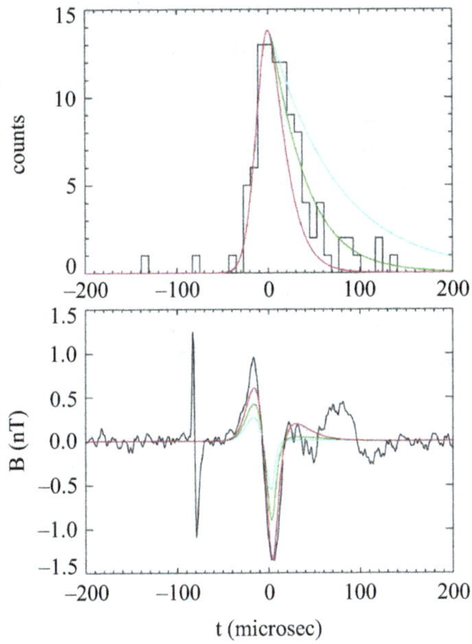

Figure 11.6 *Top panel: Fermi/GBM counts versus time of a TGF (black). The smooth curves show different model fits. Bottom panel: The magnetic field measured by the Duke VLF/LF sensor about 500 km away (black) for the same event. The smooth curves show the predicted RF emissions based on the sources shown in the top panel. Figure from Dwyer and Cummer [97]*

11.9 Theory and observations

Several conclusions can be drawn from the currently available observations of x-rays and gamma rays from thunderclouds and lightning:

1. The energy spectra from lightning and laboratory sparks are too soft, i.e. too low in energy, to be consistent with the relativistic runaway electron avalanche (RREA) production. The fluxes are also too high to be explained by the standard RREA seeded by background radiation [70]. Although relativistic feedback could account for the fluxes, this mechanism does not seem suitable for lightning and is not possible for laboratory sparks given the low voltages involved. This leaves the so-called cold runaway electron production in the strong electric fields associated with the streamer heads or leader tips as the most likely mechanism [70,73]. It is possible that these energetic electrons experience some additional energy gain in the electric fields as described by the Wilson runaway electron mechanism. However, to date, there has been no evidence that RREA multiplication, as described by Gurevich *et al.* [2], plays any role in the energetic radiation from lightning or sparks.

2. The gamma-ray glows seen from thunderclouds are consistent with RREA and perhaps in some cases are just seen as Wilson runaways acting on the ambient cosmic-ray background. As thunderclouds charge and the electric field grows, the feedback threshold may be approached. This will result in a growing discharge current from the runaway electrons and their resulting ionization that may eventually balance the charging current in the thundercloud [98]. In such cases, a near steady-state glow of gamma rays may result. However, it is not clear at this time if feedback effects are important for describing the observed gamma-ray emissions that from last seconds to minutes.
3. Dwyer [84] showed that the flux of RREAs acting on cosmic rays cannot explain TGFs. It was also shown that based on the timing and fluxes, extensive cosmic-ray air showers seeding RREAs cannot explain TGFs. Recent measurements of the lightning associated with TGFs show that they occur during the early stages of normal +IC lightning and that little charge is transferred when the TGF begins [99]. This makes models that involve runaway electron production above the thunderstorms in fields caused by the lightning discharge very unlikely. Dwyer [84] argued that this leaves only two viable hypotheses: TGFs are caused by relativistic feedback or TGFs are caused by cold runaway seeding RREAs. Both explanations may involve lightning. For feedback, the charge moment change from an upward-propagating lightning may drive the system over the feedback threshold, resulting in a self-sustained production of runaway electrons and a very rapid burst of gamma rays [82]. Alternatively, the lightning leaders inside a thundercloud may emit runaway electrons similar to that seen near the ground, seeding RREAs in either the thunderstorm electric field or the electric field produced by the lightning itself [26,73,100–103]. Using the luminosity of runaway electrons measured from triggered lightning of about 10^{17} electrons per second [64], with an additional RREA multiplication of 10^4 (consistent with the feedback limit), 10^{17} runaway electrons would result in 10^{-4} sec (the duration of a TGF) with the correct 7 MeV energy spectrum. More work is required in order to determine which mechanism (if either) is involved in the production of TGFs. At this time, it is not clear if models involving RREA can explain the energy spectrum measured by the AGILE spacecraft, which followed a power law up to 100 MeV [104]. As discussed above, the RREA spectrum is exponential and does not appear to be consistent with these new observations.
4. Lightning initiation: The idea that runaway electron avalanches seeded by extensive cosmic-ray air showers may initiate lightning has gained considerable popularity in recent years [105]. However, using the diffusion coefficients calculated by Dwyer [86] along with the avalanche multiplication limit from X-ray and positron feedback [79,80], it is found that even 10^{17} eV cosmic-ray air showers do not produce conductivities large enough to significantly alter the electric field inside a thundercloud. As a result, at present, no compelling theoretical argument exists to suggest that cosmic-ray extensive air showers initiate lightning.

5. Narrow bipolar events (NBEs): An interesting question is whether radio frequency pulses produced by extensive air showers/runaway electron avalanches can account for some or all narrow bipolar events, as suggested by others [106,107]. Narrow bipolar events are large (~1–10 V/m at 100 km) radio frequency pulses with rise times typically in the order of 1 μs [8]. Dwyer *et al.* [108] calculated the size of the radio frequency pulses produced by air showers and RREAs and found that they were small to explain NBEs. Dwyer and Babich [78] also calculated the conductivity resulting from runaway electron avalanches and found that previous estimates were at least an order of magnitude too large, which raises serious questions about the size of the radio frequency pulses calculated by other authors. Arabshahi *et al.* [109] modeled the current pulses generated by RREAs, acting on extensive cosmic-ray air showers, and found that unreasonably large thundercloud electric fields were required to account for the NBEs. Furthermore, it is not clear if models involving RREAs can explain the strong HF and VHF radio emissions that are characteristics of compact intracloud discharges (CIDs), which are closely associated NBEs. Based on the recent observations using high-speed VHF interferometers, CIDs are now appearing to be a result of a new discharge mechanism known as fast positive breakdown, which may involve a rapidly propagating network of streamers [110]. However, whether this has any connection with the high-energy processes such as the TGFs is not clear.
6. TGFs are associated with some of the largest electrical discharges in our atmosphere, with peak currents sometimes reaching half a million amps (about 100 times larger than the average peak current of in-cloud lightning flashes), resulting in large LF/VLF pulses (sferics) that have been observed from the ground at great distances [24]. Indeed, TGFs have been shown to be associated with the recently discovered energetic in-cloud pulses (EIPs), a new class of high-current discharges produced by thunderclouds [111]. Because the same energetic runaway electrons that produce the gamma rays also create the ionization that is thought to produce the currents and radio pulses, simultaneous observations of the radio and gamma ray emissions have proven fruitful for determining the properties of the TGF sources. However, many challenges remain: We still do not have a clear understanding of the source mechanism(s) responsible for TGFs and the connections between TGFs, EIPs, and normal intracloud lightning.

11.10 Summary

Great progress has been made in the last 20 years regarding the measurement of the energetic radiation from thunderclouds and lightning and developing theory and models to explain these emissions, and with the recent launch of the Atmosphere-Space Interactions Monitor (ASIM) onboard the international space station continued progress is expected in the coming years. To date, four basic mechanisms have been used to describe the production of runaway electrons and the resulting energetic radiation: Wilson runaway electrons; RREA, relativistic feedback, and cold runaway. Although all four share some features and some underlying physics, their behavior and the regimes of applicability are sufficiently different, which is useful in treating them independently when describing the production of energetic radiation in our atmosphere. Discharges involving runaway electrons also behave very differently from conventional discharges that involve only low-energy electrons. Because runaway electrons may be involved in thundercloud and lightning processes, understanding these interesting and novel categories of electrical discharges is an important part of atmospheric physics.

Acknowledgments

This material is based on the work supported by the Air Force Office of Scientific Research under award number FA9550-16-1-0396. This work was also supported in part by the NSF grant ATM 0607885 and by the DARPA grant HR0011-1-10-1-0061. Any opinions, findings, and conclusions or recommendations expressed in this material are those of the author(s) and do not necessarily reflect the views of and are not endorsed by the sponsor.

References

[1] Wilson, C. T. R., The acceleration of beta-particles in strong electric fields such as those of thunder-clouds, *Proc. Cambridge Phil. Soc.*, 22, 534–538, 1925.

[2] Gurevich, A. V., G. M. Milikh, and R. A. Roussel-Dupré, Runaway electron mechanism of air breakdown and preconditioning during a thunderstorm, *Phys. Lett. A*, 165, 463, 1992.

[3] Dwyer, J. R., M. A. Uman, H. K. Rassoul, *et al.*, Energetic radiation produced during rocket-triggered lightning, *Science*, 299, 694–697, 2003.

[4] Babich, L. P., E. N. Donskoy, I. M. Kutsyk, and R. A. Roussel-Dupré, The feedback mechanism of runaway air breakdown, *Geophys. Res. Lett.*, 32, L09809, 2005, doi:10.1029/2004GL021744.

[5] Kutsyk, I. M., L. P. Babich, and E. N. Donskoi, Self-sustained relativistic-runaway-electron avalanches in the transverse field of lightning leader as sources of terrestrial gamma-ray flashes, *Jetp Lett.* 94, 606–609, 2011, https://doi.org/10.1134/S0021364011200094.

[6] Skeltved, A. B., N. Østgaard, B. Carlson, T. Gjesteland, and Celestin, S., Modeling the Relativistic Runaway Electron Avalanche and the feedback mechanism with GEANT4. *J. Geophys. Res.: Space Phys.*, 119, 9174–9191, 2014, https://doi.org/10.1002/2014JA020504.

[7] Raether, H., *Electron Avalanches and Breakdown in Gases*, Butterworth & Co. Ltd., Washington, D.C., 1964.

[8] Rakov, V. A. and M. A. Uman, *Lightning Physics and Effects*, Cambridge University Press, Cambridge, 2003.

[9] MacGorman, D. R. and W. D. Rust, *The Electrical Nature of Storms*, Oxford University Press, New York, NY, 1998.

[10] Uman, M. A., *The Lightning Discharge*, Dover Publications Inc., Mineola, NY, 2001.

[11] Suszcynsky, D. M., R. Roussel-Dupre, and G. Shaw, Ground-based search for X rays generated by thunderstorms and lightning, *J. Geophys. Res.*, 101, 23,505, 1996.

[12] Hess, V.F., Observation of penetrating radiation of seven balloon flights, *Phys. Z.*, 13,1084–1091, 1912.

[13] Hillas, A. M., *Cosmic Rays*, Pergamon, Oxford, UK, 1972.

[14] Foote, R. S. and N. E. Frick, Time variations of natural gamma radiation, *Environ. Geosci.*, 8(2), 130–139, 2001.

[15] Fishman, G. J., P. N. Bhat, R. Mallozzi, *et al.*, Discovery of intense gamma-ray flashes of atmospheric origin, *Science*, 264, 1313, 1994.

[16] Briggs, M. S., G. J. Fishman, V. Connaughton, *et al.*, A First results on terrestrial gamma ray flashes from the Fermi Gamma-ray Burst Monitor, *J. Geophys. Res.*, 115(A7), CiteID A07323, 2010.

[17] Heumesser, M., Chanrion, O., Neubert, T., *et al.*, Spectral observations of optical emissions associated with terrestrial gamma-ray flashes. *Geophys. Res. Lett.*, 48, e2020GL090700, 2021, https://doi.org/10.1029/2020GL090700.

[18] Marisaldi, M., F. Fuschino, C. Labanti, *et al.*, Detection of terrestrial gamma ray flashes up to 40 MeV by the AGILE satellite, *J. Geophys. Res.*, 115, A00E13, 2010, doi:10.1029/2009JA014502.

[19] Neubert, T., N. Østgaard, V. Reglero, *et al.*, A terrestrial gamma-ray flash and ionospheric ultraviolet emissions powered by lightning. *Science*, 367 (6474), 183–186, 2020, https://doi.org/10.1126/science.aax3872.

[20] Østgaard, N., T. Neubert, V. Reglero, *et al.*, First 10 months of TGF observations by ASIM, *J. Geophys. Res.: Atmos.*, 124, 14024–14036, 2019, https://doi.org/10.1029/2019JD031214.

[21] Smith, D. M., L. I. Lopez, R. P. Lin, and C. P. Barrington-Leigh, Terrestrial gamma-ray flashes observed up to 20 MeV, *Science*, 307, 1085–1088, 2005.

[22] Carlson, B., N. Lehtinen, and U. Inan, Constraints on terrestrial gamma-ray flash production derived from satellite observations, *Geophys. Res. Lett.*, 34(8), CiteID L08809, 2007.

[23] Cummer, S. A, Y. Zhai, W. Hu, D. M. Smith, L. I. Lopez, and M. A. Stanley, Measurements and implications of the relationship between lightning and terrestrial gamma ray flashes, *Geophys. Res. Lett.*, 32, L08811, 2005, doi:10.1029/2005GL022778.

[24] Cummer, S. A., M. S. Briggs, J. R. Dwyer, *et al.*, The source altitude, electric current, and intrinsic brightness of terrestrial gamma ray flashes, *Geophys. Res. Lett.*, 41, 8586–8593, 2014, doi:10.1002/2014GL062196.

[25] Dwyer, J. R., The initiation of lightning by runaway air breakdown, *Geophys. Res. Lett.*, 32, L20808, 2005, doi:10.1029/2005GL023975.

[26] Dwyer, J. R., D. M. Smith, M. A. Uman, *et al.*, Estimation of the fluence of high-energy electron bursts produced by thunderclouds and the resulting radiation doses received in aircraft, *J. Geophys. Res.*, 115, D09206, 2010, doi:10.1029/2009JD012039.

[27] Gjesteland, T., N. Østgaard P. H. Connell, J. Stadsnes, and G. J. Fishman, Effects of dead time losses on terrestrial gamma ray flash measurements with the burst and transient source experiment, *J. Geophys. Res*, 115(3), CiteID A00E21, 2010.

[28] Grefenstette, B. W., D. M. Smith, J. R. Dwyer, and G. J. Fishman, Time evolution of terrestrial gamma ray flashes, *Geophys. Res. Lett.*, 35(6), CiteID L06802, 2008.

[29] Cohen, M. B., U. S. Inan, R. K. Said, *et al.*, A lightning discharge producing a beam of relativistic electrons into space, *Geophys. Res. Lett*, 37(18), CiteID L18806, 2010.

[30] Dwyer, J. R., B. W. Grefenstette and D. M. Smith, High-energy electron beams launched into space by thunderstorms, *Geophys. Res. Lett.*, 35, L02815, 2008, doi:10.1029/2007GL032430.

[31] Sarria, D., N. Østgaard, P. Kochkin, *et al.*, Constraining spectral models of a terrestrial gamma-ray flash from a terrestrial electron beam observation by the atmosphere-space interactions monitor, *Geophys. Res. Lett.*, 48, e2021GL093152, 2021, https://doi.org/10.1029/2021GL093152.

[32] Briggs, M. S., V. Connaughton, C. Wilson-Hodge, *et al.*, Electron-positron beams from terrestrial lightning observed with Fermi GBM, *Geophys. Res. Lett.*, 38(2), CiteID L02808, 2011.

[33] Connaughton, V., M. S. Briggs, R. H. Holzworth, *et al.*, Associations between Fermi Gamma-ray Burst Monitor terrestrial gamma ray flashes and sferics from the World Wide Lightning Location Network, *J. Geophys. Res.*, 115(A12), CiteID A12307, 2010.

[34] Cummer, S. A., G. Lu, M. S. Briggs, *et al.*, The lightning-TGF relationship on microsecond timescales, *Geophys. Res. Lett.*, 38, CiteID L14810, 2011.

[35] Inan, S. U., S. C. Reising, G. J. Fishman, and J. M. Horack, On the association of terrestrial gamma-ray bursts with lightning and implications for sprites, *Geophys. Res. Lett.*, 23, 1017–1020, 1996.

[36] Lu, G., R. J. Blakeslee, J. Li, D. M. Smith, *et al.*, Lightning mapping observation of a terrestrial gamma-ray flash, *Geophys. Res. Lett.*, 37(11), CiteID L11806, 2010.

[37] Lu, G., S. A. Cummer, J. Li, F. Han, D. M. Smith, and B. W. Grefenstette, Characteristics of broadband lightning emissions associated with terrestrial gamma ray flashes, *J. Geophys. Res.*, 116(A3), CiteID A03316, 2011.

[38] Shao, X.-M., T. Hamlin, and D. M. Smith, A closer examination of terrestrial gamma-ray flash-related lightning processes, *J. Geophys. Res.*, 115(9), CiteID A00E30, 2010.

[39] Stanley, M. A., X.-M. Shao, D. M. Smith, *et al.*, Link between terrestrial gamma-ray flashes and intra-cloud lightning discharges, *Geophys. Res. Lett*, 33(6), CiteID L06803, 2006.

[40] Smith, D. M., J. R. Dwyer, B. J. Hazelton, *et al.*, The rarity of terrestrial gamma-ray flashes, *Geophys. Res. Lett*, 38(8), CiteID L08807, 2011.

[41] Splitt, M. E., S. M. Lazarus, D. Barnes, *et al.*, Thunderstorm characteristics associated with RHESSI identified terrestrial gamma-ray flashes, *J. Geophys. Res.*, 115, A00E38, 2010, doi:10.1029/2009JA014622.

[42] Brunetti, M., S. Cecchini, M. Galli, G. Giovannini, and A. Pagliarin, Gamma-ray bursts of atmospheric origin in the MeV energy range, *Geophys. Res. Lett.*, 27, 1599–1602, 2000.

[43] Chilingarian, A., A. Daryan, K. Arakelyan, *et al.*, Ground-based observations of thunderstorm-correlated fluxes of high-energy electrons, gamma rays, and neutrons, *Phys. Rev. D*, 82(4), id. 043009, 2010.

[44] Chilingarian, A., G. Hovsepyan, and A. Hovhannisyan, Particle bursts from thunderclouds: natural particle accelerators above our heads, *Phys. Rev. D*, 83(6), id. 062001, 2011.

[45] Eack, K. B., W. H. Beasley, W. D. Rust, T. C. Marshall, and M. Stolzenburg, Initial results from simultaneous observation of X rays and electric fields in a thunderstorm, *J. Geophys. Res.*, 101(D23), 29,637–29,640, 1996.

[46] Eack, K. B., W. H. Beasley, W. D. Rust, T. C. Marshall, and M. Stolzenburg, X-ray pulses observed above a mesoscale convective system, *Geophys. Res. Lett.*, 23(21), 2915–2918, 1996.

[47] McCarthy, M., and G. K. Parks, Further observations of x-rays inside thunderstorms, *Geophys. Res. Lett.*, 12, 393–396, 1985.

[48] Parks, G. K., B. H. Mauk, R. Spiger, and J. Chin, X-ray enhancements detected during thunderstorm and lightning activities. *Geophys. Res. Lett.*, 8, 1176–1179, 1981.

[49] Shaw, G. E., Background cosmic count increases associated with thunderstorms, *J. Geophys. Res.*, 72(18), 4623–4626, 1967.

[50] Tsuchiya, H., T. Enoto, S. Yamada, T. Yuasa, M. Kawaharada, and T. Kitaguchi, Detection of high-energy gamma rays from winter thunderstorms, *Phys. Rev. Lett.*, 99,165002, 2007.

[51] Wada, Y., Enoto, T., Nakamura, Y. *et al.*, Gamma-ray glow preceding downward terrestrial gamma-ray flash, *Commun. Phys.*, 2, 67, https://doi.org/10.1038/s42005-019-0168-y, 2019.

[52] Whitmire, D. P., Search for high-energy radiation near lightning strokes, *Lettere Al Nuovo Cimento*, 26, 16, 497–501, 1979.

[53] Dwyer J. R., M. A. Uman, H. K. Rassoul, *et al.*, Measurements of x-ray emission from rocket-triggered lightning, *Geophys. Res. Lett.*, 31, L05118, 2004, doi:10.1029/2003GL018770.

[54] Hare, B. M., M. A. Uman, J. R. Dwyer, *et al.*, Ground-level observation of a terrestrial gamma ray flash initiated by a triggered lightning, *J. Geophys. Res. Atmos.*, 121, 6511–6533, 2016, doi: 10.1002/2015JD024426.

[55] Bowers, G. S., D. M. Smith, G. F. Martinez-McKinney, *et al.*, Gamma ray signatures of neutrons from a terrestrial gamma ray flash, *Geophys. Res. Lett.*, 44, 10,063–10,070, 2017, https://doi.org/10.1002/2017GL075071.

[56] Enoto, T., Y. Wada, Y. Furuta, *et al.*, Photonuclear reactions in lightning discovered from detection of positrons and neutrons, *Nat. Lett.*, 2017, the issue of November 23, DOI: 10.1038/nature24630.
[57] Smith, D. M., Bowers, G., Kamogawa, M., *et al.*, Characterizing upward lightning with and without a terrestrial gamma ray flash, *J. Geophys. Res.: Atmos.*, 123, 11,321–11,332, 2018, https://doi.org/10.1029/2018JD029105.
[58] Wada, Y., Enoto, T., Nakamura, Y., *et al.*, High peak-current lightning discharges associated with downward terrestrial gamma-ray flashes, *J. Geophys. Res.: Atmos.*, 125, e2019JD031730. https://doi.org/10.1029/2019JD031730, 2020.
[59] Wada, Y., Morimoto, T., Nakamura, Y., *et al.*, Characteristics of low-frequency pulses associated with downward terrestrial gamma-ray flashes, *Geophys. Res. Lett.*, 49, e2021GL097348. https://doi.org/10.1029/2021GL097348, 2022.
[60] Moore, C. B., K. B. Eack, G. D. Aulich, and W. Rison, Energetic radiation associated with lightning stepped-leaders, *Geophys. Res. Lett.*, 28, 2141–2144, 2001.
[61] Dwyer, J. R., M. A. Uman, H. K. Rassoul, *et al.*, X-ray bursts associated with leader steps in cloud-to-ground lightning, *Geophys. Res. Lett.*, 32, L01803, 2005, doi:10.1029/2004GL021782.
[62] Howard, J., M. A. Uman, J. R. Dwyer, *et al.*, Co-location of lightning leader x-ray and electric field change sources, *Geophys. Res. Lett.*, 35(13), CiteID L13817, 2008.
[63] Howard, J., M. Uman, C. Biagi, *et al.*, RF and x-ray source locations during the lightning attachment process, *J. Geophys. Res.*, 115(D6), D06204, 2010, doi:10.1029/2009JD012055.
[64] Saleh, Z., J. Dwyer, J. Howard, *et al.*, Properties of the x-ray emission from rocket-triggered lightning as measured by the Thunderstorm Energetic Radiation Array (TERA), *J. Geophys. Res.*, 114, D17, 2009, doi:10.1029/2008JD011618.
[65] Dwyer, J. R., H. K. Rassoul, Z. Saleh, M. A. Uman, J. Jerauld and J. A. Plumer, X-ray bursts produced by laboratory sparks in air, *Geophys. Res. Lett.*, 32, L20809, 2005, doi:10.1029/2005GL024027.
[66] March, V. and J. Montanyà, Influence of the voltage-time derivative in X-ray emission from laboratory sparks, *Geophys. Res. Lett.*, 37(19), CiteID L19801, 2010.
[67] March, V. and J. Montanyà, X-rays from laboratory sparks in air: the role of the cathode in the production of runaway electrons, *Geophys. Res. Lett.*, 38 (4), CiteID L04803, 2011.
[68] Nguyen, C. V., A. P. J. van Deursen, and U. Ebert, Multiple x-ray bursts from long discharges in air, *J. Phys. D: Appl. Phys.* 41, 234012, 2008.
[69] Rahman, M., V. Cooray, N. A. Ahmad, J. Nyberg, V. A. Rakov, and S. Sharma, X rays from 80-cm long sparks in air, *Geophys. Res. Lett.*, 35, L06805, 2008, doi:10.1029/2007GL032678.

[70] Dwyer, J. R., Implications of x-ray emission from lightning, *Geophys. Res. Lett.*, 31, L12102, 2004, doi:10.1029/2004GL019795.
[71] Milikh, G. and R. Roussel-Dupré, Runaway breakdown and electrical discharges in thunderstorms, *J. Geophys. Res.*, 115, A00E60, 2010, doi:10.1029/2009JA014818.
[72] Gurevich, A. V., On the theory of runaway electrons, *Soviet Physics JETP*, 12(5), 904–912, 1961.
[73] Moss, G. D., V. P. Pasko, N. Liu, and G. Veronis, Monte Carlo model for analysis of thermal runaway electrons in streamer tips in transient luminous events and streamer zones of lightning leaders, *J. Geophys. Res.*, 111(A2), CiteID A02307, 2006.
[74] Williams, E. R., The origin and context of C. T. R. Wilson's ideas on electron runaway in thunderclouds, *J. Geophys. Res*, 115, A00E50, 2010, doi:10.1029/2009JA014581.
[75] Babich, L. P., I. M. Kutsyk, E. N. Donskoy, and A. Yu. Kudryavtsev, New data on space and time scales of a relativistic runaway electron avalanche for thunderstorms environment: Monte Carlo calculations, *Phys. Lett. A.*, 245, 460–470, 1998.
[76] Babich, L.P., E. N. Donskoy, I. M. Kutsyk, *et al.*, Comparison of relativistic runaway electron avalanche rates obtained from Monte Carlo simulations and kinetic equation solution, *IEEE Trans. Plasma Sci.*, 29(3), 430–438, 2001.
[77] Gurevich, A. V., and K. P. Zybin, Runaway breakdown and electric discharges in thunderstorms, *Physics-Uspekhi*, 44, 1119, 2001.
[78] Dwyer, J. R. and L. Babich, Low-energy electron production by relativistic runaway electron avalanches in air, *J. Geophys. Res.*, 116, A09301, 2011, doi: 10.1029/2011JA016494.
[79] Dwyer, J. R., A fundamental limit on electric fields in air, *Geophys. Res. Lett.*, 30, 2055, 2003.
[80] Dwyer, J. R., Relativistic breakdown in planetary atmospheres, *Phys. Plasmas*, 14(4), 042901-042901-17, 2007.
[81] Dwyer, J. R., N. Y. Liu, and H. K. Rassoul, Properties of the thundercloud discharges responsible for terrestrial gamma-ray flashes, *Geophys. Res. Lett.*, 40, 4067–4073, 2013, doi:10.1002/grl.50742.
[82] Dwyer, J. R., The relativistic feedback discharge model of terrestrial gamma ray flashes, *J. Geophys. Res.*, 117, A02308, 2012, doi:10.1029/2011JA017160.
[83] Coleman, L. M., and J. R. Dwyer, The propagation speed of runaway electron avalanches, *Geophys. Res. Lett.*, 33, L11810, 2006, doi:10.1029/2006GL025863.
[84] Dwyer, J. R., The source mechanisms of Terrestrial Gamma-ray Flashes (TGFs), *J. Geophys. Res.*, 113(D10), CiteID D10103, 2008.
[85] Dwyer, J. R., Energetic radiation and lightning, In: H. D. Betz, U. Schumann, P. Laroche (eds.), *Lightning: Principles, Instruments and Applications*, Springer, New York, NY, 2009.

[86] Dwyer, J. R., Diffusion of relativistic runaway electrons and implications for lightning initiation, *J. Geophys. Res*, 115, CiteID A00E14, 2010.

[87] Dwyer, J. R., N. Liu, J. Eric Grove, H. Rassoul, and D. M. Smith, Characterizing the source properties of terrestrial gamma ray flashes, *J. Geophys. Res. Space Phys.*, 122, 8915–8932, 2017, doi:10.1002/2017JA024141

[88] Dwyer, J. R., Terrestrial gamma-ray flashes initiated by positive leaders, *Phys. Rev. D,* 104, 043012, 2021.

[89] Bethe H. and J. Ashkin, Transport of radiation through the matter, In: E. Segre (ed.) *Experimental Nuclear Physics*, vol. 1, John Wiley, New York, NY, Part 2, 1953.

[90] Berestetskii, V. B., E. M. Lifshitz, and L. P. Pitaevskii, *Quantum Electrodynamics*, Pergamon Press, Oxford, 1982.

[91] Lehtinen, N. G., T. F. Bell, and U. S. Inan, Monte Carlo simulation of runaway MeV electron breakdown with application to red sprites and terrestrial gamma ray flashes. *J. Geophys. Res.*, 104, 24699–24712, 1999.

[92] Mott, N. F. and H. S. W. Massey, *The Theory of Atomic Collisions*, 3rd ed., Oxford University Press, London, 1965.

[93] Ginzburg, V. L. and S. I. Syrovatskii, *The Origin of Cosmic Rays*, Pergamon, London, New York, NY, 1964.

[94] Babich, L. P., E. N. Donskoy, R. I. Il'kaev, I. M. Kutsyk, and R. A. Roussel-Dupré, Fundamental parameters of a relativistic runway electron avalanche in air, *Plasma Phys. Rep.*, 30, 616–624, 2004.

[95] Dwyer, J. R. and H. K. Rassoul, Comment on "Runaway breakdown and electrical discharges in thunderstorms," by G. Milikh and R. Roussel-Dupré, *J. Geophys. Res.*, 116, A08312, 2011, doi:10.1029/2011JA016670.

[96] Connaughton, V., M. S. Briggs, S. Xiong, *et al.*, Radio signals from electron beams in terrestrial gamma ray flashes, *J. Geophys. Res. Space Phys.*, 118, 2313–2320, 2013, doi: 10.1029/2012JA018288.

[97] Dwyer, J. R. and S. A. Cummer, Radio emissions from terrestrial gamma-ray flashes, *J. Geophys. Res. Space Phys.*, 118, 3769–3790, 2013, doi:10.1002/jgra.50188.

[98] Kelley, N., D. Smith, J. Dwyer, *et al.*, Relativistic electron avalanches as a thunderstorm discharge competing with lightning, *Nat. Commun.* 6, Article number: 7845, 2015, doi:10.1038/ncomms8845.

[99] Cummer, S. A., F. Lyu, M. S. Briggs, G. Fitzpatrick, O. J. Roberts, and J. R. Dwyer, Lightning leader altitude progression in terrestrial gamma-ray flashes, *Geophys. Res. Lett.*, 42, 7792–7798, 2015, doi:10.1002/2015GL065228.

[100] Carlson, B., N. Lehtinen, and U. Inan, Terrestrial gamma ray flash production by active lightning leader channels, *J. Geophys. Res.* 115(A10), A10324, 2010.

[101] Celestin, S., Xu, W., and V. Pasko, Terrestrial gamma ray flashes with energies up to 100 MeV produced by nonequilibrium acceleration of

electrons in lightning, *J. Geophys. Res.*, 117, A05315, 2012, doi:10.1029/2012JA017535.

[102] Köhn, C., M. Heumesser, O. Chanrion, K. Nishikawa, V. Reglero, and T. Neubert, The emission of terrestrial gamma ray flashes from encountering streamer coronae associated to the breakdown of lightning leaders, *Geophys. Res. Lett.*, 47, e2020GL089749, 2020, https://doi.org/10.1029/2020GL089749.

[103] Pallu, M., S. Celestin, F. Trompier, and Klerlein, M., Estimation of radiation doses delivered by terrestrial gamma ray flashes within leader-based production models, *J. Geophys. Res.: Atmosp.*, 126, e2020JD033907, 2021, https://doi.org/10.1029/2020JD033907.

[104] Tavani, M., M. Marisaldi, C. Labanti, *et al.*, Terrestrial gamma-ray flashes as powerful particle accelerators, *Phys. Rev. Lett.*, 106(1), id. 018501, 2011.

[105] Gurevich, A. V., K. P. Zybin, and R. A. Roussel-Dupré, Lightning initiation by simultaneous effect of runaway breakdown and cosmic ray showers, *Phys. Lett. A*, 254, 79–87, 1999.

[106] Gurevich, A. V., L. M. Duncan, Yu. V. Medvedev, and K. P. Zybin, Radio emission due to simultaneous effect of runaway breakdown and extensive atmospheric showers, *Phys. Lett. A*, 301, 320–326, 2002.

[107] Gurevich, A. V. and K. P. Zybin, High energy cosmic ray particles and the most powerful discharges in thunderstorm atmosphere, *Phys. Lett. A*, 329 (4&5), 341–347, 2004.

[108] Dwyer, J. R., M. A. Uman, and H. K. Rassoul, The remote measurement of thunderstorm electrostatic fields, *J. Geophys. Res.*, 114, D09208, 2009, doi:10.1029/2008JD011386.

[109] Arabshahi, S., J. R. Dwyer, A. Nag, V. A. Rakov, and H. K. Rassoul, Numerical simulations of compact intracloud discharges as the relativistic Runaway Electron Avalanche-Extensive Air Shower process, *J. Geophys. Res. Space Phys.*, 119, 479–489, 2014, doi: 10.1002/2013JA018974.

[110] Rison, W., P. R. Krehbiel, M. G. Stock, *et al.*, Observations of narrow bipolar events reveal how lightning is initiated in thunderstorms, *Nat. Commun.*, 7, Article number: 10721, 2016.

[111] Lyu, F., S. A. Cummer, and L. McTague, Insights into high peak current in-cloud lightning events during thunderstorms, *Geophys. Res. Lett.*, 42, 6836–6843, 2015, doi:10.1002/2015GL065047.

[112] Dwyer, J. R., M. A. Uman, H. K. Rassoul, *et al.*, A study of x-ray emission from laboratory sparks in air at atmospheric pressure, *J. Geophys. Res.*, 113 (D23), CiteID D23207, 2008.

Chapter 12

Excitation of visual sensory experiences by electromagnetic fields of lightning

Vernon Cooray[1] *and Gerald Cooray*[2,3]

12.1 Introduction

Lightning flashes can interact with a human in six different ways [1–3]. These are through direct strikes, side flashes, ground currents, shock waves, connecting leaders, and electromagnetic fields. A direct strike results when the lightning channel terminates on the person's body. In this case, there is a direct current injection from the lightning flash into the body as a result of which the person generally may experience cardiac and/or respiratory arrests; about 20% of the victims die as a result. A side flash happens when the human is located close to an object struck by lightning. The potential gradient created by the current flow along the object may give rise to a discharge between the object and the human. A portion of the lightning current will flow along this discharge path and pass through the body. Depending on the path along which the current travels through the body, the injuries could be similar to that of a direct strike. Someone standing close to the point at which lightning strikes could be injured by ground currents as the current flowing through the ground short-circuits its path by passing into the body from one leg and flowing out from the other. Since the current path is not directly through the brain or the heart, the injuries tend to be less severe than those caused by a direct strike. Injuries can also arise from shock waves created by the lightning channel. During a lightning strike, the channel temperature can rise to about 25,000 K and the channel pressure can increase to several atmospheres. The rapid expansion of heated air creates a shock wave that can injure a person located close to a lightning strike. Someone standing in the vicinity of a lightning strike could also be injured by an aborted connecting leader. Connecting leaders are initiated from grounded objects, including humans, under the influence of the electric field generated by the down-coming stepped leader. The resultant electric field is concentrated on sharp points and on the tips of grounded objects, increasing the electric field at these points to several times the background electric field. In the

[1]Department of Electrical Engineering, Uppsala University, Sweden
[2]Wellcome Trust Centre for Neuroimaging, Institute of Neurology, University College London, UK
[3]Department of Clinical Neuro Science, Karolinska Institute, Stockholm, Sweden

case of a person standing, the maximum field enhancement would take place on the top of the head. The electric field continues to grow as the stepped leader approaches the ground, and when the electric field at the tip of an object exceeds a critical value, a corona discharge (known in the popular literature as St. Elmo's fire) is initiated from it. This corona discharge continues to grow as the stepped leader approaches the ground and when the charge associated with it reaches a critical level, a thermalized electrical discharge travelling from the tip towards the down-coming stepped leader is formed. Several objects, including humans, in the vicinity of a lightning channel may launch connecting leaders, but only one of them will succeed in joining with the stepped leader. The object that initiated the 'successful' connecting leader will receive a direct strike. As the electric field collapses during the lightning strike, the connecting leaders issued by other objects will be aborted, but the current generated by these aborted leaders could be large enough to cause injury especially if it happened to be created from a person's head. When the person is located at a sufficiently large distance where the electric field is not large enough to launch a connecting leader, a corona discharge may still be issued from the person's head.

The electric current passing through the body during a direct strike or during the launch of a connecting leader can be large enough to cause either cardiac arrest or a respiratory arrest, or both. If, however, someone is located at a distance large enough to ensure that they will not be affected by direct strikes, connecting leaders or ground currents, the person can still be affected by the lightning-generated electromagnetic fields and the corona currents generated from the body.

In this chapter, we will consider the possible interactions, either direct or indirect, of the lightning-generated electromagnetic fields with the brain or the visual system of humans to induce visual sensations. Some of these visual sensations are known as phosphenes in the medical literature. Since some of these visual sensations could be misinterpreted as ball lightning, this subject is of interest for lightning researchers due to the still unsolved problem of the origin of ball lightning.

12.2 Features of ball lightning

Ball lightning has been seen and described since antiquity and recorded in many places around the globe. Because ball lightning has not been produced in the laboratory and the authenticity of the available photographs is questionable, the properties of ball lightning have to be extracted from eyewitness records. Observed properties of ball lightning have been summarized in several books [4–6], in which it is described as being spherical in shape, although other shapes such as teardrops or ovals have also been reported. In rare occasions, ball lightning shaped like rods has also been observed. The diameter of ball lightning is usually 10–40 cm but occasionally ball lightning as large as 1 m has been reported. The lifetime of ball lightning is reported to be about a few seconds, but occasionally ball lightning with a duration of as long as one minute has been observed. Ball lightning may manifest in different colours such as red, red-yellow, yellow, white, green, and purple. In some cases, the intensity of ball lightning may increase with time and becomes a dazzling white before it disappears explosively. The structure of the ball lightning may vary from one report to another. Sometimes a solid core surrounded by a translucent envelope is reported; in other cases it is a rotating structure or a structure that emits spark-like phenomena. Most ball lightning phenomena are said to move horizontally; however, descriptions including motionlessness, a zig-zag movement or moving from a cloud towards the ground can also be found. The speed of movement is reported to be walking speed, i.e. about 1–2 m/s, and ball lightning that moves against the wind has also been reported. Some reports describe sharp or acrid odours and 10% of the observers report sounds associated with the manifestation. The ball may disappear silently or explosively. However, in many cases in which the ball exploded in the vicinity of different objects, no movement of objects or damage to them has been reported.

Ball lightning is usually observed during thunderstorms, but a significant number of sightings have also been reported during fine weather without any connection to thunderstorms or lightning. They are also been reported under stressful natural conditions, such as tornadoes, storms and earthquakes. Most of the ball lightning has been sighted indoors, and in one case it was observed inside a commercial aircraft. Ball lightning has also been reported by sailors under stormy conditions but without any lightning. According to observations, balls of lightning can enter or depart from closed rooms, and pass through solid walls and closed windows without any apparent change in its structure and without causing any damage.

12.3 Alternative explanations

There is a wide variety of theories on ball lightning, but so far none of them is able to explain all of the observed features. The great difficulty of encompassing all observed features of ball lightning into a single theory makes it highly probable that many observations and experiences that have no connection to ball lightning are also categorized as ball lightning experiences. If visual sensations are generated by lightning electromagnetic fields, some of these could also be misinterpreted as a ball lightning observation.

So far scientists have identified four pathways that could give rise to visual stimulations, with features similar to ball lightning reports, through the interaction of lightning electromagnetic fields with the visual system or the brain. These are (i) interaction of the magnetic field with the retina or the visual cortex to generate phosphenes, (ii) interaction of the magnetic field or the intermittent light of lightning flashes with the visual system or the brain to induce occipital seizures that accompany visual hallucinations similar to the features documented in ball lightning reports, (iii) interaction of the high energetic radiation with the retina to induce phosphenes, and (iv) stimulation of the brain or the visual system by electric currents of corona discharges caused by the close electric field. Let us consider these alternatives one at a time.

12.3.1 Visual sensations produced by the magnetic fields generated by lightning

A phosphene is a visual sensation that is characterized by perceiving luminous phenomena without light entering the eye. Normal visual perception is created by the conversion of light falling on the retina into electrical signals by photoreceptors and subsequent interpretation of this signal by the occipital brain. Phosphenes are created when such electrical signals are created by other means in the absence of light stimuli. Phosphenes can be induced by direct stimulation of the retina or the optical nerve, either by mechanical, magnetic or electrical means. Phosphenes are created by the interaction of low frequency magnetic fields with the retina or the visual cortex as the time-varying magnetic fields generate currents that disrupt their normal electrical activity.

The occurrence of phosphenes during the interaction of magnetic fields with the brain has been documented in the medical literature related to transcranial magnetic stimulation (TMS) studies. In TMS investigations, time-varying magnetic fields are applied repeatedly to the brain using circular coils located on the head [7]. In clinical studies, TMS has provided a noninvasive means of evaluating distinct excitatory and inhibitory functions of the human cerebral cortex [8,9]. A pulse of duration 0.2–0.5 ms repeated three to five times at a frequency of 25 Hz has been shown to induce phosphenes [10]. A single pulse TMS can create a phosphene of a maximum of 50–100 ms duration [11]. Induction of phosphenes by stimulating the visual cortex would require an induced electric field of magnitude 20–50 V/m [10]. It is of interest to note, e.g. that Bernhardt [12] has estimated that the critical current density in the brain necessary to perturb normal biological functions is about 1 A/m^2. For a brain conductivity of about 0.1 S/m [13], this translates to about 10 V/m. This electric field strength is approximately of the same magnitude as the induced electric field values necessary to stimulate cortical phosphenes in TMS studies.

Peer and Kendl [14] concluded that the magnetic fields generated by lightning within about 100 m of a lightning strike are large enough to generate magnetic phosphenes through interaction with the visual cortex. A revised analysis conducted by Peer *et al.* [15] showed that for stronger-than-average currents cortical phosphenes stimulation can be induced only by lightning striking within a few metres. In the case of lightning flashes at medium distances (about 50 m), only the initial peak, whose width is about 10 to 20 μs, of the magnetic field derivative of rather high current derivatives ($dI/dt > 100$ kA μs^{-1}) could exceed the phosphenes threshold. Taking into account that these pulses are shorter than established axion excitation periods and since the retinal phosphene stimulation threshold is much lower than cortical stimulation they suggested that retinal phosphenes stimulation is the most probable for lightning electromagnetic fields.

12.3.2 *Visual sensations produced by the epileptic seizures of the occipital lobe*

Cooray and Cooray [16] pointed out that the visual hallucinations produced during the partial epileptic seizures of the occipital lobe are similar to the features of ball lightning observations. They suggested the possibility of either the magnetic field or the intermittent light from the lightning flashes could trigger these seizures establishing the connection between them with thunderstorms and lightning. The following description is based on their study.

An epileptic seizure is an event caused by abnormal, excessive, hypersynchronous discharges from an aggregate of neurons in the brain. The results of this activity can vary from dramatic convulsive activity to experiential phenomena. Epilepsy describes a condition in which a person has recurrent seizures owing to a chronic, underlying process. Its prevalence has been estimated at 5–10 persons per 1,000 worldwide [17]. A few per cent of all epilepsies have been described as being of occipital origin. Visual hallucinations sometimes occur as a result of a seizure in the occipital, temporo-occipital or temporal lobes of the brain [18–20]. Elementary visual hallucinations as perceived and drawn by eight patients with occipital epilepsy are shown in Figure 12.1. Visual hallucinations in the form of luminous objects, that are either circular or ball like, are not uncommon [18,19]. In particular luminous balls of different colours, red, yellow, blue and green, moving horizontally from the periphery of the vision to the centre have been described that may appear to be rotating or spinning. Sometimes, the ball may also appear to have a solid structure surrounded by a thin glow or, in other cases, it appears to generate spark-like phenomena. When the ball is moving towards the centre of the vision it may increase in intensity and, when it reaches the centre, it can 'explode' illuminating the whole field of vision [18]. During the hallucinations, the vision is obscured only in the area occupied by the apparent object. The hallucinations may last for 5–30 s and, rarely, up to a minute. Occipital seizures may spread into other regions of the brain giving auditory, olfactory and sensory sensations. These sensations can take the form of buzzing sounds, the smell of burning rubber, pain with thermal sensation, especially in the arms and the face, and numbness and a tingling sensation.

In some cases, people may experience only one or a few seizures during their lifetime and may not be aware of the reason for the experience. This is specially the case with idiopathic occipital epilepsy, a benign condition affecting predominantly children [21]. Children affected with this condition will often experience only one or a few seizures. Otherwise being in good health, those concerned may categorize their experience as ball lightning encounters. If, as described above, the seizure

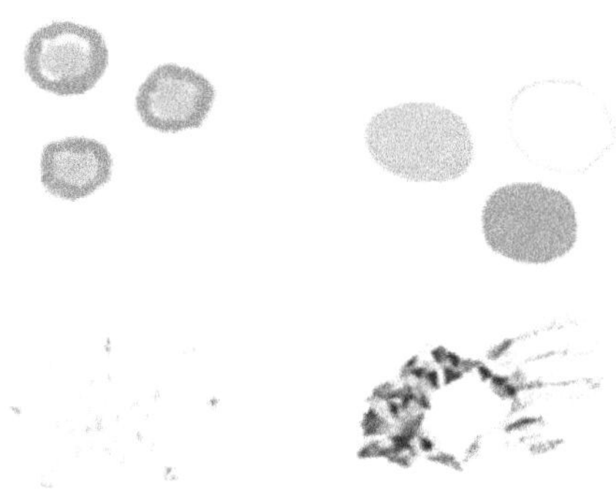

Figure 12.1 Elementary visual hallucinations as perceived and drawn by patients with occipital epilepsy [19]

affected other sensory regions, the resulting experience may appear to have been an electrical effect (the smell of burning, sensation of heat, tingling feeling, etc.) of the ball lightning. Epileptic seizures are a common and important medical problem, with about one in ten persons experiencing at least one seizure at some point in their life. Thus, some of the ball lightning encounters presented in the literature could very well be associated with the experiences of persons who have had an epileptic seizure with visual hallucinations. It is of interest to note that some of the ball lightning observations occur without any association to thunderstorms and such experiences could also be explained by epileptic seizures of the occipital lobe.

12.3.2.1 Possible association with thunderstorms

Cooray and Cooray [16,22] suggested two mechanisms that could trigger partial occipital seizures during thunderstorms. These are the strong magnetic fields produced by nearby lightning flashes and the intermittent light produced by lightning flashes. Let us consider these two possibilities separately.

Magnetic fields of return strokes

Over the last few decades, scientists have been considering the safety issues related to the medical use of time-varying magnetic fields [23,24]. From a combination of theory and experiment, they have come up with threshold levels for the magnetic time derivatives necessary for nerve and cardiac stimulation [24]. Figure 12.2 shows the rate of change of a magnetic field applied in the form of a ramp required

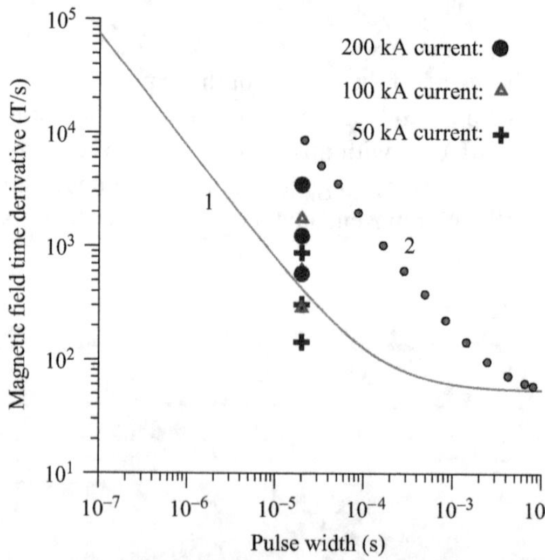

Figure 12.2 The rate of change of magnetic field applied in the form of a ramp required to excite the nerves (curve 1) and induce cardiac stimulation (curve 2) [23,24]. On the same figure, the peak magnetic field derivative for lightning flashes 10, 5, and 2 m from the channel is also depicted for currents of 50, 100, and 200 kA

to excite nerves and to induce cardiac stimulation. In this graph, the horizontal axis gives the duration of the ramp and the vertical one the rate of change of the magnetic field. The curve for the cardiac stimulation corresponds to the most sensitive population percentile, while the nerve stimulation corresponds to the mean population. Usually, the safe level for magnetic fields in MRI is set to be about three times less than the one corresponding to the nerve stimulation [23].

In order to study whether the magnetic field of a close lightning strike could stimulate nerves in the brain, Cooray and Cooray [16] calculated the time derivative of magnetic fields of lightning flashes with different currents located at different distances and compared the results with the threshold magnetic field time derivatives necessary to stimulate nerves in the brain. In the calculations, the transmission line model is used together with a return stroke speed of 1.5×10^8 m/s (see Chapter 3, Volume 1). The current at the channel base of the return stroke was represented by the following waveform that has been adopted in lightning protection standards:

$$i = \frac{I}{k} \cdot \frac{(t/\tau_1)^{10}}{1 + (t/\tau_1)^{10}} \cdot \exp(-t/\tau_2) \tag{12.1}$$

where I gives the peak current and $\tau_1 = 19.0$ μs, $\tau_2 = 485$ μs, and $k = 0.93$. Calculations were conducted for peak currents of 50, 100, and 200 kA. Since the full width of the lightning-generated magnetic field derivative evaluated in the study is about 20 μs, Cooray and Cooray [16] superimposed the peak values of these magnetic field time derivatives in Figure 12.2 at 20 μs pulse width. First note that the magnetic field derivative of a strong lightning flash striking close to a person could not induce cardiac stimulation. This is in agreement with the conclusions made by Andrew's *et al.* [25]. On the other hand, observe that, depending on the distance to the current path, the peak values of magnetic time derivatives exceed the values required for nerve stimulation. Recall, too, that the curve for nerve stimulation corresponds to the mean for the population, whilst the most sensitive percentile may lie about factor two below this [23,24]. The results show that a person located within a few metres of the path of a lightning current could be exposed to a magnetic field derivative that is large enough to stimulate neurons in the brain. This, together with the observed fact that intracranial magnetic stimulation, where the brain is exposed to strong magnetic field derivatives, can cause seizures in epileptic patients [9], makes a strong case for the nonzero probability of a close lightning flash triggering an epileptic seizure of the occipital lobe.

Cooray and Cooray [16] also pointed out that the striking distance, i.e. the distance of attraction, of lightning flashes supporting large currents is larger than several tens of metres and the chances that lightning flashes of this magnitude will strike ground within 10 m of a human standing on open ground is rather small. In such cases, the lightning flash would terminate on the human. However, they pointed out that there are several situations in which a person could be exposed to the magnetic fields generated by strong lightning flashes striking within 10 m. One such example is a person standing within 10 m of a tree or a high object struck by lightning. A similar scenario could also occur, e.g. when lightning strikes a protected building. The current of the lightning flash flows along the down conductors of the lightning protection system and the person could be standing within metres of such a conductor during a lightning strike. Thus, the probability that a human could be exposed to the magnetic field generated by a strong lightning flash within 10 m is not negligible, according to Cooray and Cooray [16].

Effect of intermittent light
In a letter to the editor of the *Journal of Neurology, Neurosurgery and Psychiatry*, Cooray and Cooray [22] suggested that the intermittent light produced by lightning flashes could also be a possible trigger for the epileptic seizures. An epileptic seizure could be triggered in a person having a lower seizure threshold upon exposure to certain visual stimuli. The nature of the stimuli that can trigger a seizure may vary from one person to another. The visual trigger for a seizure is generally intermittent such as flashing lights or rapidly changing images. The clinical data available at present confirm that the frequency of the flashing lights that induce epileptic seizures lie in the range 5–30 pulses/s [26]. Interestingly, lightning flashes also emit light pulses with a frequency that lies in the above range. A lightning flash is a composite event containing many high current events called return strokes. Each of these return strokes generates a strong light pulse since it heats the lightning channel to temperatures higher than 25,000°C. A typical lightning flash may contain up to about three to five return strokes, but in some cases the number of return strokes can exceed 20 [27]. The time interval between each event is about 50 ms on average, and, therefore, the light emitted by a lightning flash would pulsate at a frequency of about 20 pulses/s. Now, in severe thunderstorms, the lightning flashing rate can exceed 60 lightning flashes/min and in extreme cases it may increase to 500 flashes/min [27]. With the light from each lightning flash pulsating at a frequency of 20 pulses/s and with more than 60 such flashes taking place in a minute, the light generated by such thunderstorms might be responsible for provoking a seizure, especially when the ambient light level is low (e.g. night time when the flashes are more obvious against the dark background), in a person who is sensitive to pulsating light.

Interestingly, Ferrie *et al.* [28] show that light pulses can trigger partial epileptic seizures of the occipital lobe. As described previously, the work done by Panayiotopoulos *et al.* [19] shows that such seizures can induce visual hallucinations with the victim remaining in a conscious state throughout the seizure.

12.4 Visual effects produced by energetic radiation

Cooray *et al.* [29] suggested that the energetic radiation generated by thunderstorms and lightning could induce phosphenes in a person located in the vicinity of thunderstorms and lightning. The following description is adapted from their study.

Energetic radiation produced by radium can give rise to phosphenes was first noted by Diesel in 1899 [as referred to in 30]. A resurge of studies related to the creation of phosphenes by energetic radiation took place after the reports of light flashes observed in space by Apollo astronauts. It was first reported by Buzz Aldrin after Apollo 11 flight to the moon in 1969. Actually the possibility of visual effects caused by energetic radiation in space was first predicted by Tobias in 1934 [31]. Based on a 29 questionnaire distributed to 98 astronauts, 47 out of 58 who responded had experienced phosphenes (so-called light flashes at the time) at least in some space flights but not necessarily in all the flights. They were mostly noted just before going to sleep when the eye is adapted to the darkness. However, some have experienced them during daytime. The shapes of the phosphenes were either stripe, comet shaped, single dot, several dots, or blobs. The colours were mostly white, but some had yellow, orange, blue, green and red. Schematic drawings of phosphenes observed by astronauts on Apollo flights are shown in Figure 12.3 [32]. Majority of the astronauts had perceived some kind of motion to the phosphenes. Most of the time, they were moving horizontally and sometimes diagonally, but never vertical. Experiments conducted in

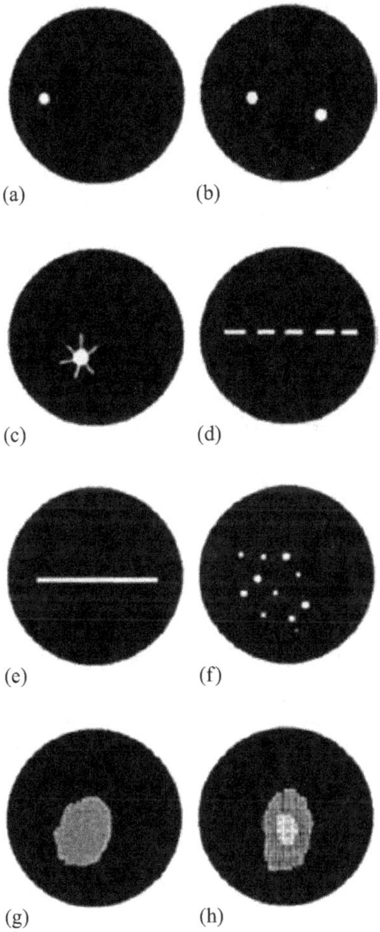

Figure 12.3 Schematic diagrams of the phosphenes observed by astronauts in Apollo missions [adapted from Ref. 32]

subsequent space missions have confirmed that the phosphenes are created by the passage of energetic particles through the visual system. The apparent motion of the phosphenes observed had to be physiological because the flight time of the energetic particles through the optical system was on the order of nanoseconds.

The observations of Apollo astronauts and subsequently the astronauts in the Skylab had motivated some scientists to expose themselves to low-energy particle beams to study the occurrence of phosphenes. Other experiments were conducted using cosmic rays. These experiments managed to reproduce what the astronauts have experienced. Phosphenes were seen by human volunteers from neutron generators at energies of 3, 8, 14, and 300 MeV as well as from Cf-252 at around 1 MeV [as summarized in Reference 33]. Phosphenes have been also observed after exposure to low doses of X-rays. In another study, 10 cancer patients whose eyes were therapeutically irradiated with 6–18 MeV electrons reported phosphenes. Nine reported seeing blue phosphenes while one reported white.

Studies conducted in space show that the probability of phosphene induction by energetic particles depends on their Linear Energy Transfer (LET) in tissue. The threshold LET necessary for the creation of phosphenes is about 10 MeV/cm and the probability increases to about 5% at 50 MeV/cm [34].

Based on this experimental data that confirm the induction of phosphenes in humans by energetic radiation, Cooray *et al.* [29] suggested the possibility of induction of phosphenes in humans located in the vicinity of lightning flashes. They showed that the energy dissipation by energetic radiation in the eye of a person located in the vicinity of lightning strikes exceeds the phosphene threshold. Let us consider their analysis now.

12.4.1 *Induction of phosphenes by the energetic radiation of lightning and thunderstorms*

X-rays and gamma rays have been found to be emitted by thunderclouds using both in situ and ground-based observations [35–42]. X-rays are now routinely measured from natural and rocket-triggered lightning flashes [43–47]. Large bursts of gamma rays, called terrestrial gamma-ray flashes, have been observed to emanate from thunderclouds by several spacecrafts [48–50]. The source of the high-energy radiation is the acceleration of electrons to relativistic energies by the electric fields of thunderclouds or lightning flashes. Cooray *et al.* [29] pointed out that the energetic radiation generated by thunderstorms and lightning is strong enough to generate phosphenes in a person located in the vicinity of lightning flashes. According to Cooray *et al.* [29], there are four scenarios that are of interest in connection with phosphenes stimulations. Following is a description of the four scenarios adapted from the work of Cooray *et al.* [29].

12.4.1.1 Energetic electrons generated by leader steps

X-ray bursts have been observed within several hundred metres of stepped leaders, dart stepped leaders and dart leaders. Dwyer *et al.* [45] measured X-ray energies up to 250 keV, but later Saleh *et al.* [51] measured a triggered lightning flash with X-ray energies extending up to about 1 MeV. From the measured energy spectrum of the X-ray photons, Saleh *et al.* [51] have estimated that a dart leader step should produce about 10^{17} energetic electrons per second with an average energy of about 1 MeV. Assuming that the radius of the electron beam is about 1 m, Dwyer *et al.* [52] have estimated the fluence of the electrons in the vicinity of the source to be 3×10^6 electrons/cm^2. This will decrease to 3×10^4 electrons/cm^2 if the radius is assumed to be 10 m. Dwyer *et al.* [52] considered $r = 10$ m to be a reasonable upper limit because this corresponds to the length of one dart leader step near to the ground.

Cooray *et al.* [29] calculated the fluence of electrons that will be injected through the roof of a building by a subsequent return stroke during a direct lightning strike and showed that for an input fluence (at the roof) of 3×10^6 electrons/cm^2 the energy dissipated in the eye of a person located inside the building is about 3 MeV/cm. This calculation is based on the assumption that the spatial distribution of the electrons is isotropic. In reality, the electrons could be beamed and consequently the energy dissipation could be higher. Cooray *et al.* [29] also pointed out that in the case of first return strokes, the electrons can be further accelerated in the electric field of the final jump region and, consequently, the energy dissipation in the eye tissue by a first return stroke could be higher. These energy dissipation values are beyond the limits necessary for phosphene generation.

12.4.1.2 X-ray bursts from lightning

Saleh *et al.* [51] estimated the average energy of X-ray photons emitted by the dart leader step to be a couple of hundred kiloelectron volts. The peak X-ray luminosity was estimated to be 4×10^{15} photons/s. Since the duration of the last dart leader step is about a microsecond, the total number of photons generated by a leader step is about 4×10^9. Assuming that the radius of the X-ray beam is about 1 m, the fluence of the X-rays in the vicinity of the source is about 10^5 photons/cm^2. This will decrease to 10^3 photons/cm^2 if the radius is assumed to be 10 m.

Conducting an analysis similar to the one conducted for electrons, Cooray *et al.* [29] estimated that for a person located inside a building the energy dissipation in the eye by this radiation would be 9 MeV/cm to 125 MeV/cm depending on whether the incident fluence is 10^3 photons/cm^2 or 10^5 photons/cm^2. These energy dissipation levels are beyond the threshold energies necessary for phosphenes stimulation.

12.4.1.3 Terrestrial gamma rays

Terrestrial gamma ray flashes are intense bursts of X-rays and gamma rays lasting a few milliseconds or less. Until recently, it had been assumed, based on theoretical calculations, that the energy of the flashes follows a power law spectra with a cutoff near 10 MeV [53–55]. Recently, Tavani *et al.* [56] found that the spectrum extends up to 100 MeV without exponential attenuation. The energy content of this extra high-energy component was about 10% of the total energy. In 2004, Dwyer *et al.* [41] observed a gamma ray flash with energies as high as 10 MeV at ground level generated during a lightning flash. The fluence of the photons in the flash which occurred at a height of about 8 km with energies up to about 10 MeV was about 1 photon/cm^2. Analyzing this data, Cooray *et al.* [29] showed that a similar burst from a 5 km height cloud would dissipate about 250 MeV/cm in eye tissue that again is beyond the phosphenes threshold.

12.4.1.4 Possible effects inside airplanes

Cooray *et al.* [29] pointed out that another situation where humans could be exposed directly to bursts of energetic electrons and neutrons could arise if an airplane in flight were to intercept the source region of an ongoing terrestrial gamma ray flash. In such an encounter, the passengers would be exposed to a burst of electrons of fluence 5×10^8 electrons/cm^2 having energies of the order of 10 MeV or more and delivered in less than 1 ms inside an aircraft. These electrons could also dissipate energies much beyond the phosphenes threshold in the eye.

12.4.2 *Concluding remarks concerning the possibility of phosphenes stimulation by energetic radiation of lightning and thunderstorms*

The information presented by Cooray *et al.* [29] shows conclusively that the energetic radiation produced by thunderstorms and lightning flashes could dissipate enough energy inside the human body to stimulate phosphenes in the visual system. Since the appearance of these phosphenes is similar to ball lightning, they suggested that some of the ball lightning observations could indeed be phosphenes generated by the interaction of this radiation with the visual system.

12.5 Stimulation of phosphenes by Corona currents

Recently, Cooray *et al.* [57] demonstrated that the corona current generated from a persons head could be strong enough to excite phosphenes. As mentioned in the introduction, if a person stands within the striking distance of lightning flash, he or she will be exposed to a direct strike. However, as the distance between the person and the lightning flash increases, he would be exposed to an aborted connecting leader rather than receive a direct strike [58]. As the distance between the human and the lightning channel increases further, the background electric field generated by the down-coming stepped leader may decrease to such a level that only a corona burst is issued from the person's head, without a connecting leader. Based on a model identical to the one that was used previously to estimate currents in an aborted leader by Becerra and Cooray [58], Cooray *et al.* [57] estimated the magnitude and duration of the corona current from a head of a person under those circumstances.

Figure 12.4 shows the results obtained by Cooray *et al.* [57] for the growth of the corona current generated from a head as a function of time as the stepped leader approaches the ground. Cooray *et al.* [57] hypothesized that, since the skin resistance and the resistivity of the cranium are rather high, a considerable portion of this current will flow through the brain. Assuming that the conductivity of both the grey and white matter is about 0.1 S/m [13], Cooray *et al.* [57] evaluated the electric field generated by the corona current in the brain using the relationship $J = \sigma E$,

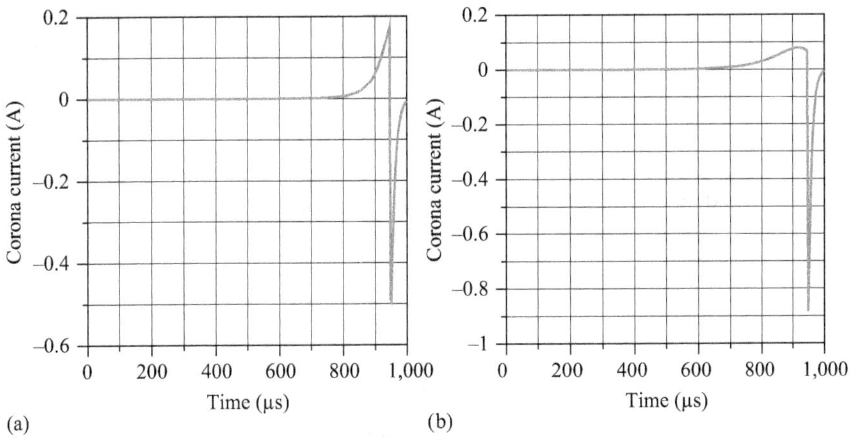

Figure 12.4 Corona current generated from a person's head at certain distance from the lightning channel (a) 30 kA current at a distance of 40 m from the person. (b) 75 kA current at a distance of 90 m from the person. The speed of the down-coming stepped leader is 5×10^5 m/s [adapted from Ref. 57]

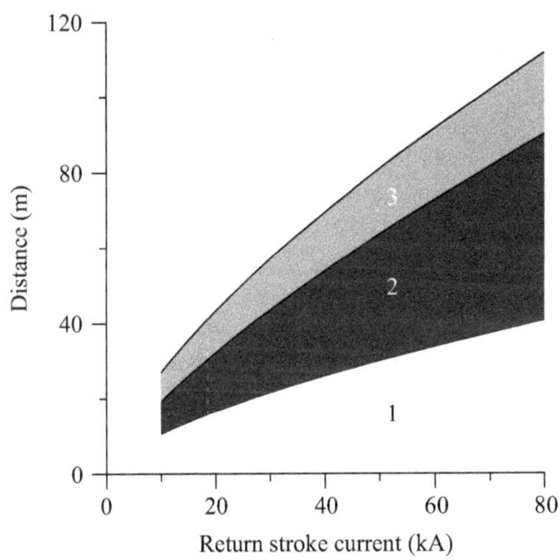

Figure 12.5 The distance from a stepped leader within which a human will experience a direct strike (region 1), an aborted connecting leader (region 2) and a corona discharge large enough to excite cortical phosphenes (region 3) as a function of the prospective return stroke current associated with a stepped leader. Regions 1 and 2 were calculated previously by Becerra and Cooray [58]

where J is the current density, E is the electric field and σ the conductivity. Assuming that the electric field in the brain tissue necessary to generate phosphene is about 20 V/m (based on TMS studies), the distance over which phosphenes could be generated by corona currents is evaluated as a function of return stroke peak current. The results are presented in Figure 12.5. In this figure, the region marked 1 shows the area within which a human will be exposed to a direct strike. In region 2, a human will be exposed to an aborted connecting leader. These regions were previously calculated and published by Becerra and Cooray [58]. In region 3, the amplitude of the corona current generated from a person's head is large enough to generate phosphenes.

Cooray *et al.* [57] also pointed out that electric fields large enough to generate corona discharges may also appear inside houses during direct strikes if the house is made of insulating materials such as wood. A person located in such a house could be exposed to rather high electric fields capable of generating a corona discharge from the head during direct lightning strikes to the house. In such cases, visual perceptions similar to phosphenes could occur indoors.

12.6 Concluding remarks

It is important to stress here that none of the authors of the papers [14–16,22,29] and [57] including the authors of this book chapter claimed that ball lightning is nothing but a visual sensation produced by the direct or indirect interaction of the electromagnetic fields of nearby lightning flashes. However, their work shows the possibility that some of the ball lightning reports could be contaminated by the phosphene experiences caused by lightning electromagnetic fields. Their results may also help to separate the real physical facts of ball lightning from lightning electromagnetic effects on the brain paving the way for the future progress in the ball lightning research.

References

[1] Andrews, C. J., M. A. Cooper, M. Darveniza and D. Mackerras (eds.), *Lightning Injuries: Electrical, Medical and Legal Aspects*, CRC Press, Boca Roton, FL, 1992.

[2] Anderson, R. B., Does a fifth mechanism exist to explain lightning injuries, *IEEE Eng. Med. Biol.*, vol. 20, pp. 105–113, 2001.

[3] Cooray, V., C. Cooray and C.J. Andrews, Lightning caused injuries in humans, *J. Electrostat.*, vol. 65, no. 5–6, pp. 386–394, 2007.

[4] Singer, S., *The Nature of Ball Lightning*, Plenum Press, New York, NY, 1971.

[5] Barry, J. D., *Ball Lightning and Bead Lightning*, Plenum Press, New York, 1980

[6] Stenhoff, M., *Ball Lightning – An Unsolved Problem in Atmospheric Physics*, Plenum Press, New York, NY, 1999.

[7] Walsh, V. and A. Cowey, Nature reviews, *Neuroscience*, vol. 1, pp. 73–80, 2000

[8] Currà, A., N. Modugno, M. Inghilleri, M. Manfredi, M. Hallett and A. Berardelli, Transcranial magnetic stimulation techniques in clinical investigation, *Neurology*, vol. 59, pp. 1851–1859, 2002.

[9] Tassinari, C. A., M. Cincotta, G. Zaccara and R. Michelucci, Transcranial magnetic stimulation and epilepsy, *Clin. Neurophysiol.*, vol. 114, pp. 777–798, 2003.

[10] Marg, E., Magnito-stimulation of vision: direct non invasive stimulation of the retina and the visual brain, *Optom. Vis. Sci.*, vol. 68, pp. 427–440, 1991.

[11] Marg, E. and D. Rudiak, Phosphenes induced by magnetic stimulation over the occipital brain: description and probable site of stimulation, *Optom. Vis. Sci.*, vol. 71, pp. 301–311, 1994.

[12] Bernhardt, J., The direct influence of electromagnetic fields on nerve and muscle cells of man within the frequency range 1 Hz to 30 MHz, *Radiat. Environ. Biophys.*, vol. 16, p. 309, 1979.

[13] Gabriel, C., A. Peyman and E. H. Grant, Electrical conductivity of tissue below 1 MHz, *Phys. Med. Biol.*, vol. 54, pp. 4863–4878, 2009.

[14] Peer, J. and A. Kendl, Transcranial stimulability of phosphenes by long lightning electromagnetic pulse, *Phys. Lett.*, vol. A 374, pp. 2932–2935, 2010.

[15] Peer, J., V. Cooray, G. Cooray and A. Kendl, Erratum and addendum to Transcranial stimulability of phosphenes by long lightning electromagnetic pulse, *Phys. Lett.*, vol. A 374, p. 2932, 2010.

[16] Cooray, G. and V. Cooray, Could some ball lightning observations be optical hallucinations caused by epileptic seizures?, *Open Atmos. Sci. J.*, vol. 2, pp. 101–105,2008.

[17] Harrison, T. R. (ed.), *Principles of Internal Medicine*, 14th ed., McGraw-Hill, New York, 1998.

[18] Blom, S. T. Tomson and C-E Westerberg, Epilepsy, in S-M Aquilonius and J. Fagius (Eds.), *Neurology*, Liber, Stockholm, 2000.

[19] Panayiotopoulos, C. P., Elementary visual hallucinations, blindness, and headache in idiopathic occipital epilepsy: differentiation from migraine, *J. Neurol. Neurosurg. Psychiatry*, vol. 66, pp. 536–540, 1999.

[20] Bien, C. G., F. D. Benninger and H. Urbach, Localizing value of epileptic visual auras, *Brain*, vol. 123, pp. 244–253, 2000.

[21] Taylor, I., I. E. Scheffer and S. F. Berkovic, Occipital epilepsies: identification of specific and newly recognized syndromes, *Brain*, vol. 126, pp. 753–769, 2003.

[22] Cooray, V. and G. Cooray, Could the intermittent light generated by lightning flashes trigger epileptic seizures?, Letter to the Editor, *J. Neurol. Neurosurg. Psychiatry*, http://jnnp.bmj.com/letters?first-index = 31&hits = 10, 2010.

[23] Reilly, J. P., Peripheral nerve stimulation by induced electric currents: exposure to time-varying magnetic fields, *Med. Biol. Eng. Comput.*, vol. 27, pp. 101–110, 1989.

[24] Schaefer, D. J., J. D. Bourland and J. A. Nyenhuis, Review of patient safety in time-varying gradient fields, *J. Magn. Reson. Imaging*, vol. 12, pp. 20–29, 2000

[25] Andrews, C., M. A. Cooper, T. Kotsos, N. Kitagawa and D. Mackerras (eds.), Magnetic effect of lightning strokes on the human heart, *J. Lightning Research*, vol. 1, pp. 158–165, 2007.

[26] Harding, G.F.A. and P. M. Jeavons, *Photosensitive Epilepsy*, 2nd ed., MacKeith Press, London, 1994.

[27] Rakov, A. A. and M. A. Uman, *Lightning: Physics and effects*, Cambridge University Press, Cambridge, 2003.

[28] Ferrie, C. D., P. De Marco, R. A. Grunewald, S Giannakodimos and C P Panayiotopoulos, Video game induced seizures, *J. Neurol. Neurosurg. Psychiatry*, vol. 57, pp. 925–931, 1994.

[29] Cooray, V., G. Cooray and J. Dwyer, On the possibility of phosphenes being generated by the energetic radiation from lightning flashes and thunderstorms, Submitted for *Phys. Lett. A*, 2011 [A summary of the work is also presented at International Symposium of Lightning Protection (SIPDA), Fortaleza, Brazil, 2011, under the title 'Some of the ball lightning observations could be visual sensations produced by energetic radiation of thunderstorms and lightning'].

[30] Lipetz, L. E., The X-ray and radium phosphenes, *Br. J. Ophthalmol.*, vol. 39, pp. 577–598, 1955.

[31] Tobias, C. A., Radiation hazards in high altitude aviation, *Aviation Med.* vol. 23, p. 345, 1952.

[32] Akatov, Yu., V. Arkhangelsky, V. Petrov, K. Truckhanov, S. Adveev, Yu. Ozerov, *et al.*, Biomedical aspects of light flashes observed by astronauts during space fights. In *Proceedings of 6th European Symposium on Life Sciences Research in Space*, Trondheim, Norway, June 16–20, ESA SP390, 1996.

[33] Steidley, K. D., The radiation phosphene, *Vision Res.*, vol. 30, pp. 1139–1143, 1990.

[34] Fuglesang, C., Using the human eye to image space radiation or the history and status of the light flash phenomena, *Nucl. Instr. Meth. Phys. Res. A*, vol. 580, pp. 861–865, 2007.

[35] Shaw, G. E., Background cosmic count increases associated with thunderstorms, *J. Geophys. Res.*, vol. 72, pp. 4623–4626, 1967, doi:10.1029/JZ072i018p04623.

[36] Parks, G. K., B. H. Mauk, R. Spiger and J. Chin, X-ray enhancements detected during thunderstorm and lightning activities, *Geophys. Res. Lett.*, vol. 8, pp. 1176–1179, 1981, doi:10.1029/GL008i011p01176.

[37] McCarthy, M. and G. K. Parks, Further observations of X-rays inside thunderstorms, *Geophys. Res. Lett.*, vol. 12, pp. 393–396, 1985, doi:10.1029/GL012i006p00393.

[38] Eack, K. B., W. H. Beasley, W. D. Rust, T. C. Marshall and M. Stolzenburg, X-ray pulses observed above a mesoscale convection system, *Geophys. Res. Lett.*, vol. 23, pp. 2915–2918, 1996, doi:10.1029/96GL02570.

[39] Brunetti, M., S. Cecchini, M. Galli, G. Giovannini and A. Pagliarin, Gamma-ray bursts of atmospheric origin in the MeV energy range, *Geophys. Res. Lett.*, vol. 27, pp. 1599–1602, 2000, doi:10.1029/2000GL003750.

[40] Chubenko, A. P., V. P. Antonova, S. Y. Kryukov, *et al.*, Intense X-ray emission bursts during thunderstorms, *Phys. Lett. A*, vol. 275, pp. 90–100, 2000, doi:10.1016/S0375-9601(00)00502-8.

[41] Dwyer, J. R., H. K. Rassoul, M. Al-Dayeh, *et al.*, A ground level gamma-ray burst observed in association with rocket-triggered lightning, *Geophys. Res. Lett.*, vol. 31, L05119, 2004, doi: 10.1029/2003GL018771.

[42] Tsuchiya, H., T. Enoto, S. Yamada, *et al.*, Detection of high-energy gamma rays from winter thunderstorms, *Phys. Rev. Lett.*, vol. 99, 165002, 2007, doi:10.1103/PhysRevLett.99.165002.

[43] Moore, C. B., K. B. Eack, G. D. Aulich and W. Rison, Energetic radiation associated with lightning stepped-leaders, *Geophys. Res. Lett.*, vol. 28, pp. 2141–2144, 2001, doi:10.1029/2001GL013140.

[44] Dwyer, J. R., M. A. Uman, H. K. Rassoul, *et al.*, Energetic radiation produced during rocket triggered lightning, *Science*, vol. 299, pp. 694–697, 2003, doi:10.1126/science.1078940.

[45] Dwyer, J. R., H. K. Rassoul, M. Al-Dayeh, *et al.*, Measurements of X-ray emission from rocket triggered lightning, *Geophys. Res. Lett.*, vol. 31, L05118, 2004, doi:10.1029/2003GL018770.

[46] Dwyer, J. R., H. K. Rassoul, M. Al-Dayeh, *et al.*, X-ray bursts associated with leader steps in cloud-to-ground lightning, *Geophys. Res. Lett.*, vol. 32, L01803, 2005, doi:10.1029/2004GL021782.

[47] Howard, J., M. A. Uman, J. R. Dwyer, *et al.*, Co-location of lightning leader X-ray and electric field change sources, *Geophys. Res. Lett.*, vol. 35, L13817, 2008, doi:10.1029/ 2008GL034134, 2008.

[48] Fishman, G. J., P. N. Bhat, R. Mallozzi, *et al.*, Discovery of intense gamma-ray flashes of atmospheric origin, *Science*, vol. 264, pp. 1313–1316, 1994, doi:10.1126/science.264.5163.1313.

[49] Smith, D. M., L. I. Lopez, R. P. Lin and C. P. Barrington-Leigh, Terrestrial gamma-ray flashes observed up to 20 MeV, *Science*, vol. 307, pp. 1085–1088, 2005.

[50] Cohen, M. B., U. S. Inan and G. R. Fishman, Terrestrial gamma ray flashes observed aboard compton gamma ray observatory/burst and transient source experiment and ELF/VLF radio atmospherics, *J. Geophys. Res.*, vol. 111, D24109, 2006, doi:10.1.1029/2005JD006987.

[51] Saleh, Z., J. Dwyer, H. Rassoul, *et al.*, Properties of the X-ray emission from rocket-triggered lightning as measured by the Thunderstorm Energetic Radiation Array, *J. Geophys. Res.*, vol. 114, D17210, 2009, doi:10.1029/ 2008JD011618.

[52] Dwyer, J. R., D. M. Smith, M. A. Uman, Z. Saleh, B. Grefenstette and H. K. Rassoul, Estimation of the fluence of high-energy electron bursts produced by thunderclouds and resulting radiation doses received in aircraft, *J. Geophys. Res.*, vol., 115, D09206, 2010, doi:10.1029/2009JD012039.

[53] Dwyer, J. R. and D. M. Smith, A comparison between Monte Carlo simulations of runaway breakdown and terrestrial gamma-ray flash observations, *Geophys. Res. Lett.*, vol. 32, L22804, 2005, doi:10.1029/2005GL023848.

[54] Lehtinen, N. G., T. F. Bell and U. S. Inan, Monte Carlo simulation of runaway MeV electron breakdown with application to red sprites and terrestrial gamma ray flashes, *J. Geophys. Res.*, vol. 104, pp. 24699, 1999.
[55] Dwyer, J. R., Source mechanisms of terrestrial gamma-ray flashes, *J. Geophys. Res.*, vol. 113, D10103, 2008, doi:10.1029/2007JD009248.
[56] Tavani, M., M. Marisaldi, C. Labanti, *et al.*, Terrestrial gamma-ray flashes as powerful particle accelerators, *Phys. Rev. Lett.*, vol. 106, p. 018501, 2011.
[57] Cooray, V., G. Cooray and M. Becerra, Corona currents generated from a person's head in the vicinity of lightning strikes—stimulation of phosphenes, 7th Asia-Pacific International Conference on Lightning, Chengdu, China, 2011.
[58] Becerra, M. and V. Cooray, On the interaction of lightning upward connecting positive leaders with humans, *IEEE Trans. Electromagn. Compat.*, vol. 51, no. 4, pp. 1001–1008, 2009.

Chapter 13

Lightning location systems

Gerhard Diendorfer[1] and Wolfgang Schulz[1]

13.1 Introduction

In the early days, the risk of lightning strikes was described by the average number of thunderstorm days or thunderstorm hours, where a thunderstorm day is defined as an "*Observational day during which thunder is heard at the station*" [1]. On the basis of long-term records of thunderstorm days by the meteorological services, maps showing the so-called isoceraunic level were produced for the individual countries, from which the regional thunderstorm hazard could be obtained [2]. The first attempts to locate lightning discharges date back to the 1920s. W. Watt in [3] describes the considerations and experimental observations made at that time, which made it possible to identify thunderstorms and lightning discharges as the main cause of the observed electromagnetic disturbances in the early days of long-range radio communication.

In the beginning, the problem of electromagnetic disturbances in radio communication was the driving force for research activities related to lightning electromagnetic fields and their source location. Only since the middle of the 1980s lightning detection has been applied in the field of thunderstorm observation for meteorological applications, in the field of lightning risk assessment and in fault analysis for power utilities. These new applications also placed significantly higher demands on the performance of the lightning location systems (LLS), especially with regard to detection efficiency and location accuracy. Lightning detection has thus evolved from a real-time thunderstorm observation tool to the most precise detection of each individual electrical discharge in lightning both in cloud-to-ground (CG) flashes and discharges within the thundercloud, often referred to as intracloud lightning (IC).

In a typical thunderstorm, the number of IC discharges exceeds the number of CG discharges many times with an average ratio of IC/CG discharges being in the range of 5–10, but this ratio can be much higher in individual thunderstorms. As CG lightning poses the greatest danger to humans and property, at the beginning of the development of LLS, the focus was on the best possible and reliable detection of CG lightning. Only in recent years, the trend is increasingly moving towards the detection of total lightning. New lightning data have recently become available through

[1]Austrian Lightning Detection and Information System (ALDIS), OVE Service GmbH, Austria

the detection of lightning discharges by the Geostationary Lightning Mapper (GLM) on the satellites GOES West and GOES East [4]. Compared to land-based LLS, optical detection from space has the advantage that lightning activity is monitored over continents and oceans with approximately the same detection quality. The main limitation of lightning detection from space is the lack of differentiation between CG and IC discharges and the comparatively limited detection accuracy due to the spatial resolution of the optical sensor in the range of 4–5 km looking to earth surface from a distance of about 36,000 km. In this chapter, we will focus on ground-based systems only which are employing Magnetic Direction Finding (MDF) and/or Time of Arrival (TOA) technique, as data from these systems are used for many applications from lightning risk management to severe storm forecast.

13.2 Methods of lightning detection

Assuming a more or less vertical channel, at least at the lowest few hundred meters above ground, each lightning stroke emits a broadband electromagnetic signal that propagates radially in all directions around the lightning channel at the speed of light. A lightning discharge is either located by a network of ground-based sensors measuring the angle of incidence of the electromagnetic wave (Figure 13.1), or by the so-called time-of-arrival method (TOA), where the location is estimated from the arrival time differences of the lightning radiated field at the LLS sensors (Figure 13.2). A given difference in arrival time at two sensors defines a hyperbola, and when there are multiple sensors, a hyperbola is defined by each possible pair of sensors in whose intersection the source of the signal, the point of impact, is located.

When the angle of incidence is derived from the induced voltages in two orthogonal loop antennas, resulting from the magnetic field components, this method to detect lightning is called magnetic direction finding (MDF).

Interferometry is another method to estimate the angle of field incidence at a sensor site, where the phase difference in the signals of two antennas is proportional to the angel of incidence as shown in Figure 13.3.

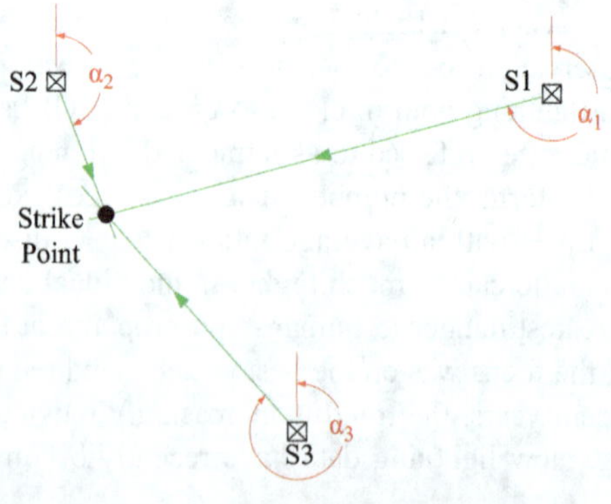

Figure 13.1 Principle of obtaining the angel of field incidence with three sensors employing MDF

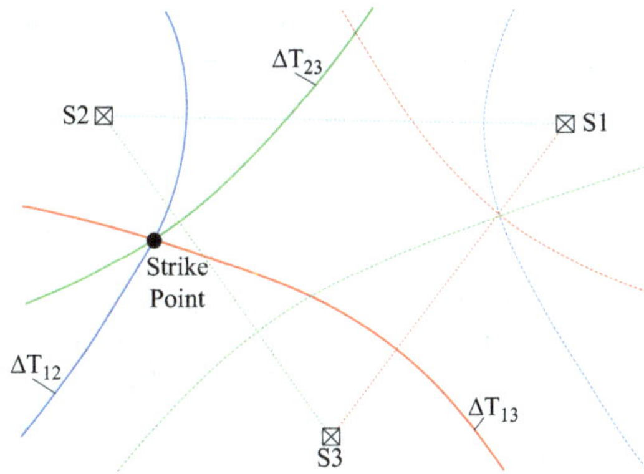

Figure 13.2 Principle of strike point estimation from observed time of arrival differences (TOA) at sensors S1, S2, and S3

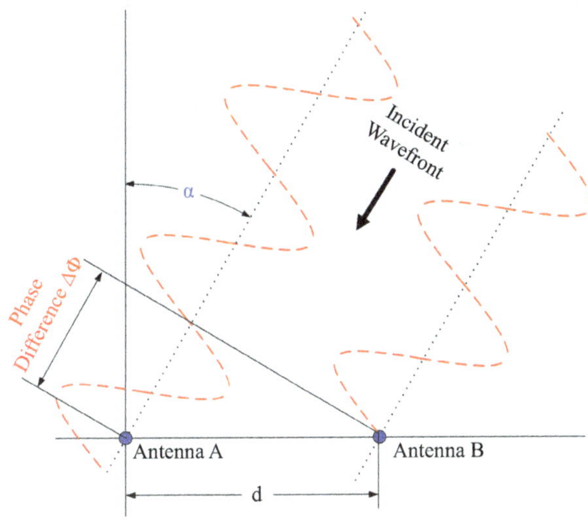

Figure 13.3 Principle of obtaining the angel of field incidence with interferometry, where λ is the wavelength of the signal

In principle, a minimum of two sensors is required for the MDF method and at least three sensors are required for the TOA method to locate a stroke. At least one additional sensor is required for the MDF method if the strike point is located along or near the connecting line of the two sensors (glancing intersection of the two direction lines) or for the TOA method when under certain conditions a three sensor network results in an ambiguous solution. In general, the quality of a LLS increases with an increasing number of available sensors, because, for example, a temporary communication outage of a sensor has hardly any noticeable effect, or a sufficient number of sensors is still available if a very high peak current stroke results in a field pulse exceeding the maximum measurable field of nearby sensors (sensor overrange).

One of the major challenges for early TOA systems was the need for precise time synchronization of multiple sensors located hundreds of kilometers apart. A location accuracy in the range of 100 m requires a maximum permissible absolute time deviation between the sensors of a few 10 nsec. This could only be achieved in the 1990s with the widespread and permanent availability of GPS satellites and led to the final breakthrough of the TOA method. In the mid-1990s, experience demonstrated that the combined use of MDF and TOA techniques offers more robust estimates of lightning locations than either alone. The TOA technique proved to be less dependent on local site errors than direction finding, which is affected by re-radiation from local electrically conducting objects of various dimensions.

Lightning location is an optimization task, where from all the reported sensor data (angle and/or time of arrival) a stroke location is determined, which minimizes the deviations of the measured data (least square method).

13.3 Lightning EM fields and their detection in different frequency ranges

Both CG flashes and IC discharges radiate electromagnetic energy over a wide range of frequencies, where most of this energy is in pulses or high-frequency "bursts" whose rise times and durations vary over a wide range in the time domain. These emissions can be roughly assigned to different frequency ranges as VLF, LF, or VHF known from communications technology (see Table 13.1). The breakdown processes that occur during the formation of a new lightning channel or the very fast processes that lead to the reillumination of previously used lightning channel segments produce emissions predominantly in the VHF range. In the case of return strokes, a large, impulsive current with amplitudes ranging from of a few kA up to several 100 kA propagates from the attachment point at ground level towards the cloud in the long channel previously prepared by the stepped leader or the dart leader. The strongest field emissions of return strokes are observed in the LF and VLF bands. Return strokes are the dominating source of field emissions observed in the VLF band but relatively little activity is produced by the high-current components of return strokes in the VHF band. Stepped leaders in negative CG flashes also produce impulsive emissions associated with each step of a length on the order of tens to hundreds of meters. The emissions of a stepped leader become more continuous in time when the leader approaches ground and the branched structure becomes more complex.

Cloud discharges produce tens to hundreds of small pulses that have most of their energy in the upper LF range and higher frequencies. More details on field waveforms generated by cloud and CG processes can be found in [5,6]. Due to the differences in the rates and peak fields of electromagnetic radiation at different

Table 13.1 Electromagnetic field emissions of lightning processes in different frequency ranges (adapted from [7])

Frequency range	Frequency	Mode of signal propagation	Discharge process(es)/ lightning event(s)	Typical number of emissions per flash (order of magnitude)
VLF	3–30 kHz	Ground wave and Earth-ionosphere waveguide	Return strokes in CG flashes and compact intracloud discharges	1–10
			Large-amplitude cloud pulses including preliminary breakdown	1–100
LF	0.03–3 MHz	Ground wave and Earth-ionosphere waveguide	Return strokes in CG flashes	1–10
			Cloud pulses including preliminary breakdown, compact intracloud discharges, and K-changes	1–100
			CG leader steps	10
HF	3–30 MHz	"Mixed" ground wave and line of sight (affected by blockage due to presence of objects in line of sight)	Various in-cloud and leader processes	10–100
VHF	30–300 MHz	Line of sight (affected by blockage due to presence of objects in line of sight)	Breakdown of "virgin air" during channel formation	10–1,000
			Dart leaders and K-changes, compact intracloud discharges	10–100

frequencies coming from a specific process in the development of a lightning discharge, different techniques are better suited to detect different processes in CG and IC discharges. Vertically polarized transient pulses in the LF and VLF ranges propagate along the Earth's surface over long distances and they are used to detect and locate return strokes and a fraction of IC discharges.

13.4 Peak current estimate

According to the transmission line (TL) model [8], the peak current I_p is related to the far-field peak E_p and to the return-stroke speed (assuming that v_{RS} = const, the ground is perfectly conducting, and the return-stroke front has not reached the top of the channel) by the following expression:

$$I_p = \frac{2.\pi.\epsilon_0.c^2.D}{v_{RS}}.E_p \tag{13.1}$$

where D is the horizontal distance between the lightning channel and the observation point, v_{RS} is the return stroke velocity, and c is the speed of light.

For an assumed typical return stroke velocity of $v_{RS} = 1.10^8$ m/s, we can estimate a 5 V/m peak field per kiloampere peak current at a reference distance of D = 100 km.

Assuming an average value for the returns stroke velocity v_{RS} (e.g., $v_{RS} = 1.2 \times 10^8$ m/s), the linear relation in (13.1) is used to infer lightning peak currents from measured peak fields. The return stroke speed v_{RS} varies in a range of 1/3–2/3 of the speed of light [9] and hence this variability of v_{RS} introduces some uncertainty in the peak current estimate from the remotely measured electric or magnetic field for a given event. It was shown in [10], that a statistical estimation e.g., in terms of a mean value and standard deviation of lightning peak current is possible. This result provides, to some extent, a theoretical justification for the use of lightning location systems (LLS) to infer parameters of statistical lightning current distributions.

It should be mentioned that the linear relationship used to derive the peak current from the peak field is not based on the TL model alone. The existence of such a linear relationship was also confirmed by the simultaneous measurements of currents and fields from triggered lightning [11] and lightning to instrumented towers [12].

When locating IC discharges in the LF range, it should be noted that for IC discharges, the location provided by the LLS has limited significance. In contrast to CG strokes, there is no clearly defined impact point on the ground and a point coordinate for IC discharge would only make sense as the projection of a vertically oriented lightning channel in the cloud on the ground surface. As mentioned above, the LLS determines a fictitious point coordinate on the ground level for which the available sensor data (angle and/or time) show the best agreement in the least square optimization algorithm.

Even more questionable are the LLS estimated peak currents of IC discharges. Today the same conversion from measured peak field to peak current is used as for CG strokes. But it has to be noted that for IC discharges there is no defined attachment point where this amplitude could be theoretically measured. For CG strokes, it is well documented that the current peak decreases with height above ground and the rise time increases [13]. For CG strokes the current waveforms at ground level i(0, t) have been measured at the strike point in triggering lightning or on instrumented towers. It has been shown that the TL model employed for the calculation of I_p from E_p (13.1) results in good agreement with the direct current measurements, at least for negative subsequent strokes up to peak currents of −40 kA. A well-documented validation for LLS estimated higher peak currents of negative strokes and for positive strokes in general is not yet available. It is also not yet clear whether the same conversion from E_p to I_p, as currently used in LLS, is applicable for first and subsequent strokes, or whether different parameters would have to be used. A validation of positive strokes and first negative strokes has not been successful so far, because for these validation process primarily only triggered flashes or flashes to instrumented towers are available and these flashes are almost exclusively upward initiated flashes, where there is no first stroke preceded by a stepped leader.

13.5 CG/IC discrimination

Proper classification of CG and IC discharges by an LLS is important because, for example, CG discharges that are misclassified as IC can lead to errors in correlating transmission line outages with lightning strikes or can lead to unwarranted denials when reviewing insurance claims for equipment damaged by lightning. On the other hand, misclassified IC discharges as CG discharges result in an overestimate of the ground flash density used for lightning risk assessments, or the peak current distribution of CG discharges being shifted towards smaller median values, since the LLS estimated peak currents of IC discharges are usually smaller than those of CG discharges. Correct CG versus IC classification is still one of the biggest challenges in lightning detection. One of the reasons is that validation of the misclassification rate is only possible to a very limited extent and any time when a new or improved algorithm is implemented, the new algorithm needs to be validated again with ground truth data. Data from triggered lightning and instrumented towers, typically used for LLS validation, are CG strokes only and hence can only be used to check the misclassification of CG as IC discharges, but not the other way round.

Two very different methods are used for the classification of IC and CG discharges. One is based on the distinct field waveform characteristics of the different types of discharges and is used e.g. in the NLDN [14,15] and the EUCLID [16] network, the other one is based on the estimation of the height of the field radiation source above ground and is used in LINET type networks [17].

As shown in Figure 13.4, typical fields of CG strokes and IC discharges are quite different in their waveforms. A previously used parameter for IC versus CG discrimination was the peak-to-zero (PTZ) time which is typically greater than 10 μs or so in CG stroke field pulses and is less than 10 μs in radiated IC field pulses.

The limitation of this method is given by the fact that the frequency distributions of the PTZ times of IC and CG field pulses are not completely separated. In individual cases an IC discharge can have a long PTZ and vice versa a CG discharge can have a short PTZ, which can then lead to misclassification. For the final classification of a discharge, someone can also take advantage of the fact that it was located by several sensors and not all sensors report exactly the same field parameters due to different distances and propagation paths. The most significant misclassification rate based on waveform discrimination was observed for small positive CG strokes, which actually were mostly IC discharges [14]. This observation led to the fact that, for example, up to March 2016 in the NLDN all located

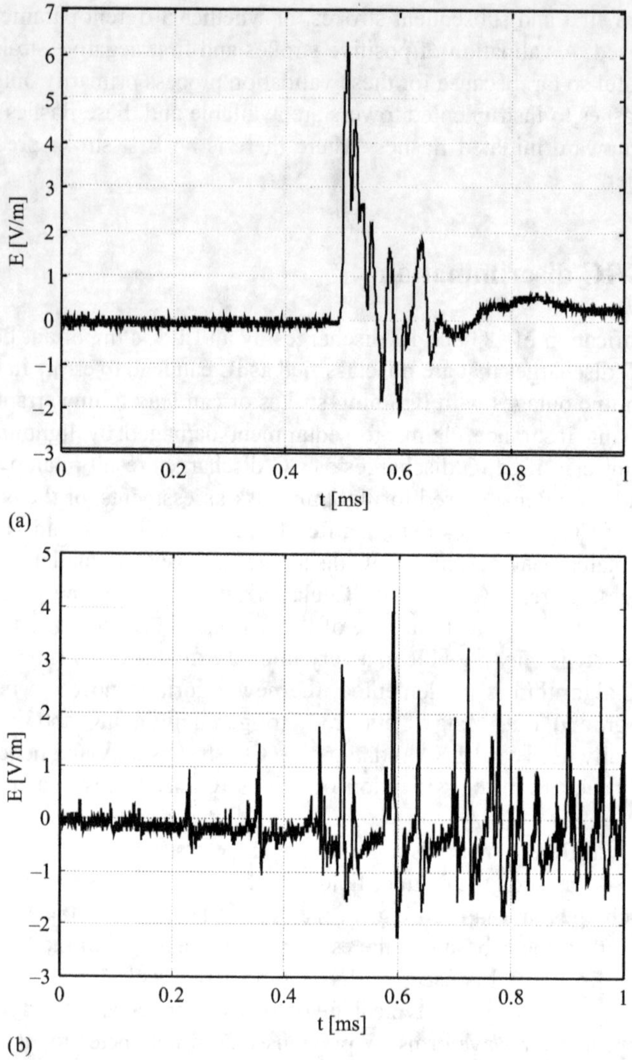

Figure 13.4 Typical vertical E-fields of (a) first stroke in CG and (b) an IC discharge

positive discharges with amplitudes smaller than +15 kA were classified as IC discharges [15]. Nowadays most of the networks use more than one waveform parameter to classify a discharge. By evaluating more than one waveform parameter the above-mentioned drawbacks can be somewhat reduced.

The other method for classifying IC discharges is e.g., used by the LINET network and is based on the assumption that CG strokes emit VLF/LF radiation at low heights near ground level while IC discharges emit from substantial elevations

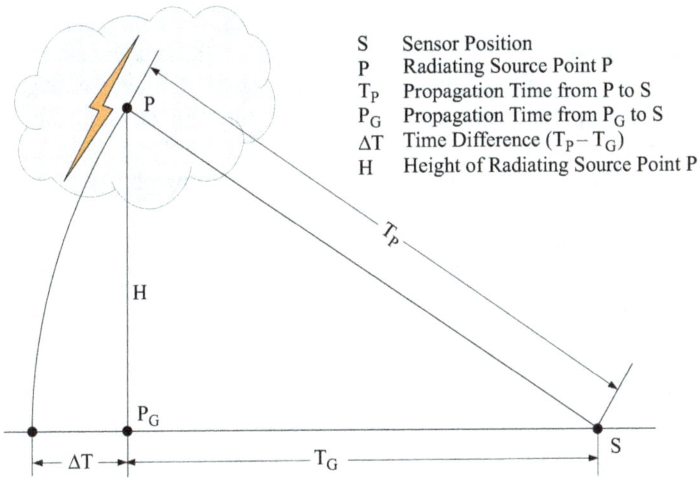

Figure 13.5 Estimate of height of field radiation source from the observed time deviation ΔT at a nearby sensor (adapted from [17])

within the thundercloud [17]. The principle of this 3D discrimination method is shown in Figure 13.5.

When the actual source of radiation is at a height H above ground and a location of the discharge is estimated by the classical TOA method, a time deviation ΔT is observed at the sensor closest to the strike point P_G. This ΔT and the known distance between the estimated strike point P_G and the sensor S allows to estimate the height H of the radiating source. If this estimated height H is above a certain threshold, this discharge is classified as IC, below this height threshold it is classified as CG. This method requires contribution of a nearby sensor within a range of 50–100 km. Some sources for limitation of this method are listed in [17] including statistical time errors, propagation effects due to finite ground conductivity and terrain, systematic time errors and nearfield saturation effects.

In summary, CG/IC classification and its validation in lightning detection is still one of the most challenging issues. On the one hand, the misclassification rate is dependent on a large number of influencing factors and therefore cannot be described by a single number for a given LLS. Consequently, each CG versus IC classification validation of a LLS is strictly speaking only valid for the given network setting and location (sensor baseline, terrain, etc.).

13.6 Grouping of strokes to flashes and ground strike points (GSP)

LLS detect basically pulses of electromagnetic fields which have been radiated by either CG strokes or IC discharges. CG strokes together with IC pulses are called events. IC pulses can be either part of a CG or an IC flash.

The physical entity flash is created from LLS data by grouping located events together (discharges occurring in a certain time period and in a certain region). The limits of this grouping algorithm are so called monster flashes (see e.g. [18,19]), which exceed the typically applied grouping parameters of a maximum spatial extent of 10 km and a maximum discharge duration of 1 s. Furthermore, individual CG strokes of a CG flash may attach ground in one or more ground strike points (GSP).

In former times, when LLS detected mainly CG strokes only, strokes were grouped to CG flashes, meaning a CG flash contained CG strokes only. Nowadays also IC pulses are grouped to flashes and therefore there are 3 different types of flashes detected by the LLS:

- CG flashes containing only CG strokes
- CG flashes containing CG strokes and IC pulses
- IC flashes containing only IC pulses

Figure 13.6 shows as an example a CG flash and an IC flash. The CG flash in this example contains three CG strokes and two IC pulses, the IC flash contains solely IC pulses.

A first algorithm to group CG strokes to flashes was described by Cummins *et al.* [14]. The parameters suggested there to group strokes to flashes were a maximum distance of strokes to the first stroke of 10 km, a maximum flash duration of 1s and a maximum inter-stroke interval of 500 ms. The algorithm was designed to allow grouping of strokes of different polarity to the same flash. Those parameters were finally applied in the majority of the LLS all over the world.

In former times, the ground flash density N_G was used to determine the risk of an object to be hit by lightning assuming that all strokes of a flash hit the same ground strike point. During recent years, it was realized that different strokes in a flash may hit in one or more ground strike points (see [20,21]). LLS data were used for years to determine the ground flash density N_G in a certain region. During the

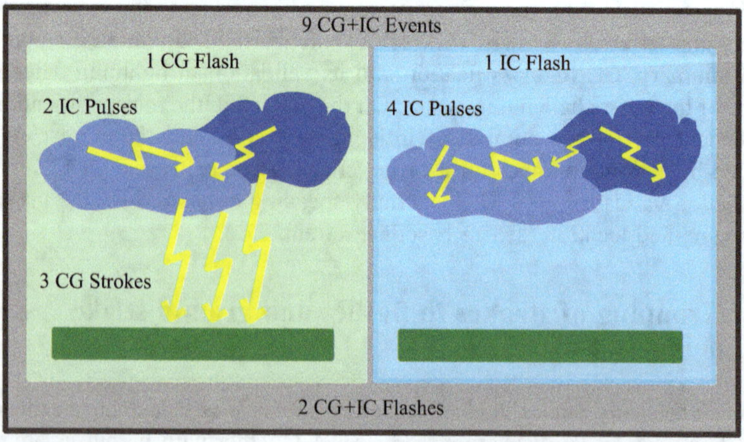

Figure 13.6 Example of a CG flash and an IC flash

recent years the location accuracy of LLS improved to a level where it is nowadays possible to identify individual ground strike points within a flash. Therefore, it is now also possible to determine the regional values of ground strike point densities N_{SG} from LLS data and this parameter describes best the local risk of a lightning strike to objects like buildings or transmission lines. For this reason, N_{SG} has also already been implemented in recently published IEC standards [22] and will replace the ground flash density in lightning risk evaluation procedures.

There are two basic methods to determine which strokes within a flash belong to the same GSP. Those methods are:

(a) based on the distances of strokes to each other [23–25];
(b) based on waveform parameters of individual strokes [26].

Different methods of the type (a) are thoroughly described and compared to each other in [27].

13.7 Measurement errors in LLS

Besides several waveform parameters, LLS sensors measure the arrival time, the angle of incidence of the electromagnetic field, and the peak of the field pulse in order to determine the strike point location and the peak current of the return stroke. All those measured parameters are biased by systematic and random errors.

In this section, we will focus on what causes systematic errors and we will give at the end a description of the so-called confidence ellipse which is determined by the random errors of the sensor measurements.

13.7.1 Systematic angle/amplitude errors (also called site errors)

Systematic angle and amplitude errors of LLS sensors are called site errors. The reason for the existence of site errors is the absorption and re-radiation of the incident lightning electromagnetic field near the sensors antenna by metallic objects such as power lines or fences and features of the terrain [28]. Angle site errors $\beta(\theta)$ caused by a single object such as power lines or a metallic fence have in a first approximation a simple characteristic given by

$$\beta(\theta) = A \sin(2\theta + \phi) \tag{13.2}$$

where θ is the measured angle, ϕ is a constant, and A is the maximum of the site error. The physical background for this two-cycle sinusoidal function is explained in [29]. Also underground cables to a sensor or close to a sensor can generate site errors and cause an angle error with such a two-cycle sinusoidal function [30].

So-called amplitude site errors have the same origin as the angle site errors [31]. They are always positive if they are caused by metallic objects such as power lines or fences, but may be negative because of field reflections by the local terrain (see [28]).

Systematic angle and amplitude site errors are often determined with a statistical method using sensor data collected over a period of time. For this method, a sufficient number of strokes located by an optimization method (if only angle is used) or with usage of time information is necessary. Several papers in literature, e.g. [32–35] describe this method in detail.

13.7.2 Systematic time error

Timing errors in the reported sensor data are the result of a combination of propagation effects due to finite ground conductivity and the elongation of the propagation path between strike point and sensor location due to hilly terrain [36]. Both effects are particularly evident in mountainous terrain, e.g., in the Alps. In such mountainous regions, the timing error can be significant and as an example, the angle and distance dependent time errors of a sensor located in the Alps (Sensor #2 in Schwaz, Austria) are shown in Figure 13.7. This sensor is located in a mountain valley that stretches from west to east and is surrounded by high mountains (up to 3,000 m). The highest mountains are in the south of the sensor site. Compared to sensors located in a more or less flat region, this sensor site shows a relatively complex structure for timing errors. It shows large time errors in the west

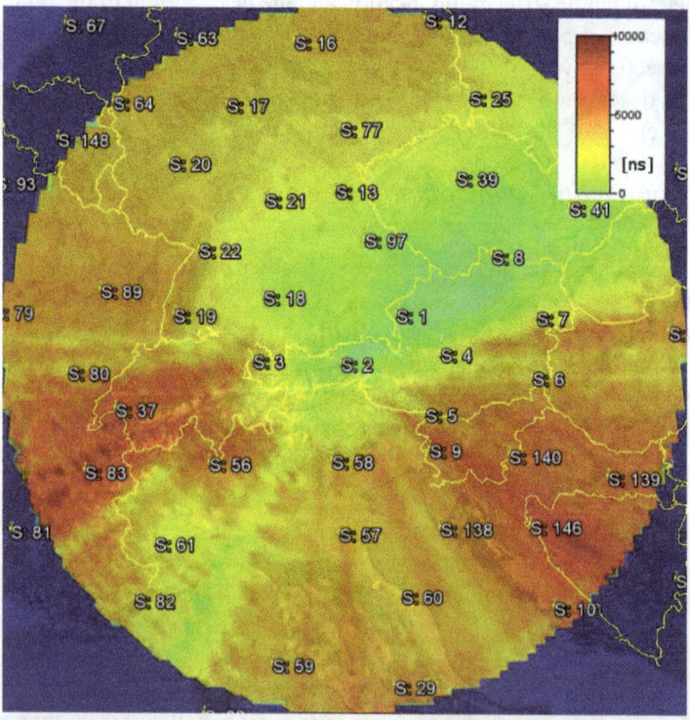

Figure 13.7 Example of a timing error for Austrian Sensor #2 (Schwaz). The sensor is located in the center of the circular area.

and in the south-east of sensor 2. All the regions in blue color are outside the 600 km operational range of this sensor and are therefore not considered in this evaluation of the timing errors.

As done for angle site error correction, the determination of the systematical timing error is also performed by a statistical iterative method using optimized positions.

13.7.3 Confidence ellipse

A first statistical approach of the problem of locating an object of unknown position based on reports of angle of incidence taken from two or more stations (direction finders) whose positions are known, was done in [37]. It was assumed that no systematic error is present in the data, and that the errors in the reported angles from each direction finder can be adequately described by a Gaussian, or "normal," error probability distribution with zero mean as given in (13.3). The probability that a given sensor will report an angle between $\theta + \psi$ and $\theta + \psi + \mathrm{d}\psi$ is

$$P(\psi).d\psi = \frac{1}{\sigma_\psi.\sqrt{2\pi}}.e^{-\frac{\psi^2}{2.\sigma_\psi^2}}.d\psi \tag{13.3}$$

where θ is the true angle to the lightning strike position and ψ is the error in the sensor reported angle. The same basic assumptions apply to the error distribution of the sensor reported times.

It can be shown that for any calculated position the confidence region of the calculated location is an ellipse [38]. We can calculate a 50% confidence (error) ellipse, under the assumption that the distributions of angle and time errors are Gaussian. A detailed discussion of these models, along with the assumptions used, is given in [14,37].

The median (50%) confidence ellipse circumscribes a region centered on the computed (optimum) location, within which there is a 50% probability that the stroke actually occurred. Confidence ellipse is determined by the number and relative position of sensors contributing to a given stroke location as well as by the standard deviation of the time and angle measurements for each of the used sensors. The describing parameters of the confidence ellipse (length of semi-major axis, eccentricity, and orientation) are provided by the LLS for each event. It should be remembered that the confidence ellipse of a given stroke is a statistical measure and says nothing about the actual location accuracy in this particular case.

13.8 Performance characteristics of LLS

Ideally, to verify the performance of an LLS, a larger number of lightning strokes distributed over the entire coverage area of the LLS, with precisely known time, location, and peak current of each stroke, should be available as so-called ground truth data for reference. In reality, such ground truth data are only available to a very limited extent, such as from measurements at high towers or triggered lightning and high-speed video recordings of individual lightning strikes in combination with field measurements. Direct comparison of two different LLSs sharing a common coverage area or detailed statistical analysis of the data from the LLS itself are used in an attempt to overcome the lack of ground truth data.

For the reasons stated above, various methods for evaluating LLS have been developed over the past 20 years. The following sections briefly describe these methods and discuss the advantages and disadvantages of each.

The main performance parameters of an LLS are:

- Flash and/or stroke detection efficiency (DE) is the percentage of flashes or strokes detected as a percentage of the total number of flashes or strokes that occurred in reality. Since it is sufficient for locating a multi-stroke flash, if one of the strokes was located, the flash DE is always larger than the stroke DE.
- Location accuracy (LA) is the median distance between real stroke locations and the stroke locations given by the LLS.
- Peak current estimation accuracy is the difference of the peak current estimate of the LLS from the range normalized peak field and the true peak current of the located stroke.
- Lightning type classification accuracy describes the ability to differentiate between different processes such as in-cloud pulses and cloud-to-ground strokes. Located but misclassified events may include true CG strokes classified as IC discharges or IC discharges misclassified as CG strokes.

The importance of each of the parameters described above may also be determined by the application of the lightning location data. For example, the accuracy of the peak current estimate is much more important for the analysis of transmission line failures by a power grid operator than when the LLS data are used for thunderstorm warning applications by a meteorological service. A comprehensive evaluation of different validation methods of LLS is given in [7].

13.8.1 LLS self-reference

In this method, an evaluation of the LA and DE of an LLS is done based on the statistical analysis of the parameters provided by the LLS itself, such as the standard deviation of the sensor timing error, the semi-major axis length of the 50% confidence ellipse, and the number of sensors reporting the event (NSR). The evaluated LLS data should be collected over a longer period after the LLS is properly calibrated. This method can provide a relatively simple and good estimate of the DE and LA of an LLS. However, it does not provide an absolute measure of DE or LA, or this method cannot provide information on amplitude determination or the goodness of detection of CG and IC strokes. Examples of studies based on this method include [39–41].

13.8.2 Rocket triggered lightning and lightning strikes to tall objects

In this method, data from experiments with rocket-triggered lightning or lightning strikes to tall objects (e.g., instrumented towers for lightning current measurements) are used as the ground truth to evaluate the performance characteristics of an LLS. Ideally, the lightning triggering site or the instrumented tower should be located within the coverage area of the LLS and not at its edge or outside. Since the location of the CG strokes and their peak current are available completely independent of the LLS, it is possible to determine the flash/stroke DE, LA, polarity, and peak current estimate accuracy of the LLS. Since only CG strokes are available as reference, the misclassification rate can only be obtained in the direction of CG strokes being misclassified as IC but not in the other direction (IC events being classified as CG strokes).

For example, rocket-triggered lightning data, acquired at the International Center for Lightning Research and Testing (ICLRT), at Camp Blanding, Florida were used in [42] to evaluate the performance characteristics of the U.S. National Lightning Detection Network (NLDN). For the data acquired in the summers of 2001–2003 flash and stroke DE were estimated to be about 84% and 60%, respectively. Median location error was about 600 m, with larger location errors (greater than 2 km) being associated with strokes having smaller peak currents (5–10 kA). After an upgrade of the NLDN in 2004 a flash and stroke DE of 92% and 76%, respectively, and a median absolute location error of 308 m were obtained for 2004–2009 rocket triggered lightning data [43].

Earth Networks Total Lightning Network (ENTLN) performance characteristics were evaluated in [44] using natural cloud-to-ground lightning data collected at the Lightning Observatory in Gainesville (LOG) and rocket-triggered lightning data acquired in 2014 and 2015 at Camp Blanding, Florida. Based on the rocket-triggered flashes, the values for flash and stroke DE were determined to be 100% and 97%, respectively, with a median location error of 215 m.

In [45], the performance of the LLS of the power grid operator in Guangdong Province, China, was conducted based on triggered lightning flashes obtained in Conghua, Guangzhou, during 2007–2011 and natural lightning flashes to tall structures obtained in Guangzhou during 2009–2011. A flash and stroke DE of about 94% and 60%, respectively, was obtained. The median location error was estimated to be about 489 m.

Evaluation of LLS performance based on data from instrumented towers have been performed using e.g. data recorded at the Gaisberg Tower in Austria [16,46], at the Peissenberg Tower in Germany [47], the Säntis Tower in Switzerland [48], and CN Tower in Toronto, Canada [49].

To validate the accuracy of the LLS peak current estimate, only directly measured lightning peak currents for rocket-triggered lightning and lightning strikes to tall objects are available as ground-truth data. Stroke peak currents are estimated by LLS using (13.1), assuming a constant return stroke speed v_{RS}. In (13.1), infinite ground conductivity is assumed resulting in an inverse distance dependency. In order to correct for some effects of field attenuation due to finite ground conductivity, in some networks (e.g. NLDN, EUCLID) a signal propagation model is applied to the sensor reported peak fields [14].

The results of the comparison of $N = 464$ directly measured return stroke peak currents at the Gaisberg Tower with the correlated peak currents estimated by the EUCLID system are shown in Figure 13.8.

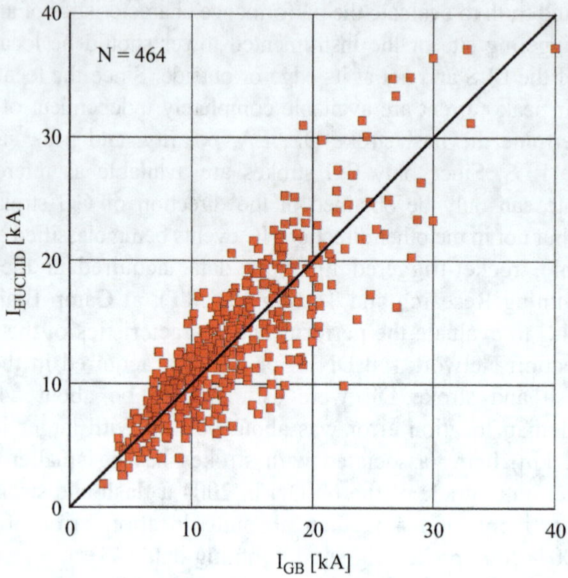

Figure 13.8 EUCLID peak current estimates plotted vs. directly measured stroke peak currents at the Gaisberg Tower (2005–2014). Black line represents the LLS peak current estimate assuming TL model and a return stroke velocity $v_{rs} = 1.2 \times 10^8$ m/s (adapted from [16])

Looking at all the data in Figure 13.8, the EUCLID peak currents are on average somewhat smaller than the actual measured ones, although the differences in individual events may well be over- or underestimated by up to a factor of two. This is largely due to the different return stroke velocities, which can also vary by a comparable order of magnitude, but must be assumed to be constant for estimating the peak current by LLS according to (13.1).

13.8.3 Video and E-field measurements

To avoid some of the limitations of triggered flashes or upward initiated flashes from tall objects, GPS time correlated field and high-speed video recordings are used as ground-truth to validate LLS performance (see e.g. [45,50–52]). Unlike triggered flashes and instrumented towers, video and field recordings can be made at a wide variety of locations within the coverage area of an LLS and are therefore not limited to single points. The challenge with this method is to find suitable locations for video recording even before the onset of lightning activity, from which ideally the lightning channels are visible in their entire extent from cloud base to the ground contact. The video and E-field recording method provides the best and most complete means of evaluating proper separation of CG and IC discharges, since both CG and IC discharges are recorded. On the other hand, with this method the lightning peak currents can only be checked to a limited extent by evaluating the peak fields in case of known distance between field recording site and strike point location. The LLS location accuracy can also only be evaluated indirectly by analyzing the distances between LLS provided impact points of strokes in a flash in which, according to the camera recording, all strokes occurred in the same channel [53].

13.8.4 Intercomparison among LLS that cover a common area

The performance of an LLS under test can be compared to another LLS which is used as a reference. However, in such an approach, the reference LLS must be as well calibrated as possible, its quality must be characterized by independent ground truth data, and the test and reference networks must overlap as much as possible. With this method, some conclusions can be drawn about DE, LA, polarity and peak current estimation accuracy, and lightning classification accuracy of the test LLS compared to the reference LLS. Nevertheless, in the case of observed differences in individual lightning events, such an LLS comparison usually leaves the question unanswered as to which of the two LLSs, the reference LLS or the tested LLS, is now correct or incorrect, as long as also the reference network is not without any limitations, which is the case most of the time. Therefore, using independently measured or observed ground truth data of lightning events as described above should be the preferred method for the validation of any LLS.

13.8.5 Summary

While a single or a combination of the techniques described above can be used to evaluate the performance characteristics of an LLS system, it is important to understand the strengths and limitations of each method used in order to understand what can and cannot be quantified with confidence about the performance characteristics of a given LLS. Also, the user of LLS data should always strive to obtain the most accurate information possible of the performance characteristics of the LLS being used to avoid misinterpretation when using the LLS data. An example is the lightning peak current of a particular lightning strike that has caused damage to a technical facility. For numerical simulations of the transient fault sequences, it should then be noted that the amplitude supplied by the LLS as an input variable is only available with an accuracy of $\pm$ 20% or more, and it therefore makes little sense to regard the results obtained by the simulation programs as highly accurate.

Another often overlooked limitation in the application of LLS data is the significantly reduced DE of upward flashes, which mainly originate from elevated objects such as transmission towers or wind turbines. Long-term observations at the Gaisberg Tower have shown that only 43% of upward flashes could be located by the LLS [54], compared to 98% flash DE in Austria for downward flashes [16]. More than half of the upward initiated flashes were of the so-called ICC_{Only} type [55] and had no fast rising current components, such as return strokes, that are required for proper field radiation and LLS detection with sensors at distances up to several hundred kilometers.

References

[1] American Meteorological Society (AMS). Glossary of Meteorology [Internet]. 2021 [cited 2021 Jul 24]. Available from: https://glossary.ametsoc.org/wiki/Thunderstorm_day

[2] World Distribution of Thunderstorm Days. Part 1: Tables. WMO Report. Geneva, Switzerland; 1953.

[3] Watson WRA. Weather and wireless. *Quarterly Journal of the Royal Meteorological Society*. 1929;(3):273–301.

[4] Goodman SJ, Blakeslee RJ, Koshak WJ, *et al.* The GOES-R geostationary lightning mapper (GLM). *Atmos Res.* 2013;125–126(0):34–49. Available from: http://linkinghub.elsevier.com/retrieve/pii/S0169809513000434

[5] Weidman CD and Krider EP. The fine structure of lightning return stroke wave forms. *J Geophys Res Ocean.* 1978;83(C12):6239–47. Available from: http://dx.doi.org/10.1029/JC083iC12p06239

[6] Weidman CD and Krider EP. The radiation field wave forms produced by intracloud lightning discharge processes. *J Geophys Res Ocean.* 1979;84(C6):3159–64. Available from: http://dx.doi.org/10.1029/JC084iC06p03159

[7] Nag A, Murphy MJ, Schulz W, and Cummins KL. Lightning locating systems: insights on characteristics and validation techniques. *Earth Sp Sci.* 2015;2:65–93. Available from: http://doi.wiley.com/10.1002/2014EA000051

[8] Uman MA, McLain DK, and Krider EP. The electromagnetic radiation from a finite antenna. *Am J Phys.* 1975;48(January).

[9] Rakov VA. Lightning return stroke speed. *J Light Res.* 2007;1(1935):80–9.

[10] Rachidi F, Bermúdez J-L, Rubinstein M, and Rakov VA. On the estimation of lightning peak currents from measured fields using lightning location systems. *J Electrostat.* 2004;60(2–4):121–9. Available from: https://www.sciencedirect.com/science/article/abs/pii/S0304388604000245

[11] Willett JC, Idone VP, Orville RE, *et al.* An experimental test of the "transmission-line model" of electromagnetic radiation from triggered lightning return strokes. *J Geophys Res Atmos.* 1988;93(D4):3867–78. Available from: http://dx.doi.org/10.1029/JD093iD04p03867

[12] Pichler H, Mair M, and Diendorfer G. Correlated current and far field records from lightning discharges to the gaisberg tower. In: 2007 International Symposium on Lightning Protection (IX SIPDA), 2007. Available from: https://www.aldis.at/fileadmin/user_upload/aldis/publication/2007/2_SIPDA2007_Pichler.pdf

[13] Jordan DM, Rakov VA, Beasley WH, and Uman MA. Luminosity characteristics of dart leaders and return strokes in natural lightning. *J Geophys Res Atmos.* 1997;102(D18):22025–32. Available from: http://dx.doi.org/10.1029/97JD01838

[14] Cummins KL, Murphy MJ, Brdo EA, Hiscox WL, Pyle RB, and Pifer AE. A combined TOA/MDF technology upgrade of the U.S. National Lightning

Detection Network. *J Geophys Res.* 1998;103(D8):9035. Available from: http://dx.doi.org/10.1029/98JD00153

[15] Murphy MJ, Cramer JA, and Said RK. Recent history of upgrades to the U.S. National Lightning Detection Network. *J Atmos Ocean Technol.* 2020; 38(3):573–85. Available from: http://journals.ametsoc.org/doi/10.1175/JTECH-D-19-0215.1

[16] Schulz W, Diendorfer G, Pedeboy S, and Poelman D. The European lightning location system EUCLID – Part 1: performance analysis and validation. *Nat Hazards Earth Syst Sci.* 2016;16(2):595–605. Available from: https://nhess.copernicus.org/articles/16/595/2016/

[17] Betz H-D, Schmidt K, Oettinger WP, and Wirz M. Lightning detection with 3-D discrimination of intracloud and cloud-to-ground discharges. *Geophys Res Lett.* 2004;31(11):L11108. Available from: http://dx.doi.org/10.1029/2004GL019821

[18] Lang TJ, Pédeboy S, Rison W, *et al.* WMO world record lightning extremes: longest reported flash distance and longest reported flash duration. *Bull Am Meteorol Soc.* 2016, BAMS-D-16-0061.1. Available from: https://journals.ametsoc.org/view/journals/bams/98/6/bams-d-16-0061.1.xml

[19] Peterson M, Lang TJ, Bruning EC, *et al.* New world meteorological organization certified megaflash lightning extremes for flash distance (709 km) and duration (16.73 s) recorded from space. *Geophys Res Lett.* 2020;47(16):1–9. Available from: https://agupubs.onlinelibrary.wiley.com/doi/full/10.1029/2020GL088888

[20] Thottappillil R, Rakov VA, Uman MA, Beasley WH, Master MJ, and Shelukhin D V. Lightning subsequent-stroke electric field peak greater than the first stroke peak and multiple ground terminations. *J Geophys Res Atmos.* 1992;97(D7):7503–9. Available from: http://dx.doi.org/10.1029/92JD00557

[21] Rakov VA, Uman MA, and Thottappillil R. Review of lightning properties from electric field and TV observations. *J Geophys Res Atmos.* 1994;99 (D5):10745–50. Available from: http://dx.doi.org/10.1029/93JD01205

[22] IEC 62858. Lightning density based on lightning location systems (LLS) – General Principles. 2019.

[23] Pedeboy S. Identification of the multiple ground contacts flashes with lightning location systems. In: 22nd International Lightning Detection Conference & 4th International Lightning Meteorology Conference (ILDC/ILMC). Broomfield, CO, USA; 2012.

[24] Campos LZS, Cummins KL, and Pinto OJ. An algorithm for identifying ground strike points from return stroke data provided by Lightning Location Systems. In: 2015 9th Asia-Pacific International Conference on Lightning (APL). Nagoya, Japan; 2015. p. 475–8.

[25] Matsui M, Michishita K, and Yokoyama S. Characteristics of negative flashes with multiple ground strike points located by the japanese lightning detection network. *IEEE Trans Electromagn Compat.* 2019;61(3):1–8.

[26] Cummins KL. Analysis of multiple ground contacts in cloud-to-ground flashes using LLS data: the impact of complex terrain. In: 22nd International

Lightning Detection Conference & 4th International Lightning Meteorology Conference (ILDC/ILMC). Broomfield, CO, USA; 2012.

[27] Poelman DR, Schulz W, Pedeboy S, *et al.* Global ground strike point characteristics in negative downward lightning flashes – Part 2: algorithm validation. *Nat Hazards Earth Syst Sci.* 2021;21(6):1921–33. Available from: https://nhess.copernicus.org/articles/21/1921/2021/

[28] Kohlmann H, Schulz W, and Rachidi F. Evaluation of site errors in LLS magnetic direction finding caused by large hills using the 3D-FDTD technique. *Earth Sp Sci.* 2021. *Accepted for Publication*.

[29] Schulz W. *Performance Evaluation of Lightning Location Systems. PhD Thesis*, Technical University of Vienna; 1997.

[30] Schulz W, Diendorfer G, Hofbauer F, Stimmer A, and Mair M. Site errors in magnetic direction finding due to buried cables. In: 1998 24th International Conference on Lightning Protection (ICLP). Birmingham, England; 1998.

[31] Schulz W and Diendorfer G. Amplitude site error of magnetic direction finder. In: 2002 26th International Conference on Lightning Protection (ICLP). Cracow, Poland; 2002. pp. 4–7.

[32] Hiscox WL, Krider EP, Pifer AE, and Uman MA. A systematic method for identifying and correcting "site errors" in a network of magnetic direction finders. In: International Conference on Lightning and Static Electricity (ICOLSE). 1984.

[33] Passi RM and López RE. A parametric estimation of systematic errors in networks of magnetic direction finders. *J Geophys Res.* 1989;94 (D11):13319. Available from: https://agupubs.onlinelibrary.wiley.com/doi/abs/10.1029/JD094iD11p13319

[34] López RE and Passi RM. Simulations in site error estimation for direction finders. *J Geophys Res.* 1991;96(D8):15287. Available from: https://agupubs.onlinelibrary.wiley.com/doi/abs/10.1029/91JD01398

[35] Schulz W and Diendorfer G. Site-Error-Korrektur bei einem Blitzortungssystem mit Magnetic-Direction-Findern. *Elektrotechnik und Informationstechnik.* 110 (5), 1993, Springer Wien New York 1993.

[36] Schulz W and Diendorfer G. Evaluation of a lightning location algorithm using an elevation model. In: 2000 25th International Conference on Lightning Protection (ICLP). Rhodes, Greece; 2000.

[37] Stansfield RG. Statistical theory of D.F. Fixing. *J Inst Electr Eng IIIA Radiocommun.* 1947;94(15):762–70.

[38] Gotthardt E. *Einführung in die Ausgleichsrechnung*. Herbert Wichmann Verlag. Karlsruhe; 1968.

[39] Cummins KL, Murphy MJ, Cramer JA, Scheftic WD, Demetriades NWS, and Nag A. Location accuracy improvements using propagation corrections: a case study of the U.S. National Lightning Detection Network. In: 21st International Lightning Detection Conference & 3rd International Lightning Meteorology Conference (ILDC/ILMC). Orlando, FL; 2010.

[40] Honma N, Murphy MJ, Cummins KL, Pifer AE, and Rogers T. Improved lightning locations in the Tohoku Region of Japan using propagation and waveform onset corrections. *IEEJ Trans Power Energy*. 2012;133(2):195–202.

[41] CIGRE Report 376. Cloud-to-ground lightning parameters derived from lightning location systems – the effects of system performance. CIGRE Report 376. *CIGRE Report* 376; 2009.

[42] Jerauld JE, Rakov VA, Uman MA, *et al.* An evaluation of the performance characteristics of the U.S. National Lightning Detection Network in Florida using rocket-triggered lightning. *J Geophys Res Atmos*. 2005;110(D19): D19106. Available from: http://dx.doi.org/10.1029/2005JD005924

[43] Nag A, Mallick S, Rakov VA, *et al.* Evaluation of U.S. National Lightning Detection Network performance characteristics using rocket-triggered lightning data acquired in 2004–2009. *J Geophys Res Atmos*. 2011;116 (D2):D02123. Available from: http://dx.doi.org/10.1029/2010JD014929

[44] Zhu Y, Rakov VA, Tran MD, *et al.* Evaluation of ENTLN performance characteristics based on the ground truth natural and rocket-triggered lightning data acquired in Florida. *J Geophys Res Atmos*. 2017;122(18):9858–66. Available from: http://dx.doi.org/10.1002/2017JD027270

[45] Chen L, Zhang Y, Lu W, *et al.* Performance evaluation for a lightning location system based on observations of artificially triggered lightning and natural lightning flashes. *J Atmos Ocean Technol*. 2012;29(12):1835–44. Available from: http://journals.ametsoc.org/doi/abs/10.1175/JTECH-D-12-00028.1

[46] Diendorfer G, Schulz W, and Mair M. Evaluation of a LLS based on lightning strikes to an instrumented tower. In: 16th International Lightning Detection Conference (ILDC). Tucson, Arizona, USA; 2000.

[47] Paul C and Heidler F. Properties of three types of M-components and ICC-pulses from currents of negative upward lightning measured at the peissenberg tower. *IEEE Trans Electromagn Compat*. 2018;99:1–8.

[48] Azadifar M, Rachidi F, Rubinstein M, *et al.* Evaluation of the performance characteristics of the European Lightning Detection Network EUCLID in the Alps region for upward negative flashes using direct measurements at the instrumented Säntis Tower. *J Geophys Res Atmos*. 2016;121(2):595–606. Available from: http://doi.wiley.com/10.1002/2015JD024259

[49] Kazazi S, Mutlag AH, and Liatos P. Evaluation of NALDN performance characteristics in the vicinity of the CN Tower based on tall-structure lightning. *Electr Power Syst Res*. 2017;153:19–31. Available from: https://doi.org/10.1016/j.epsr.2016.12.005

[50] Schwalt L, Pack S, and Schulz W. Ground truth data of atmospheric discharges in correlation with LLS detections. *Electr Power Syst Res*. 2020;180(2):106065.

[51] Mata CT, Hill JD, Mata AG, and Cummins KL. Evaluation of the performance characteristics of the Cglss and Nldn systems based on two years of ground-truth data from Launch Complex 39B, Kennedy Space Center, *24rd Int Light Detect Conf 6th Int Light Meteorol Conf*. 2014.

[52] Poelman D, Schulz W, and Vergeiner C. Performance characteristics of distinct lightning detection networks covering Belgium. *J Atmos Ocean*

Technol. 2013;30(5):942–51. Available from: https://journals.ametsoc.org/view/journals/atot/30/5/jtech-d-12-00162_1.xml

[53] Biagi CJ, Cummins KL, Kehoe KE, and Krider EP. National Lightning Detection Network (NLDN) performance in southern Arizona, Texas, and Oklahoma in 2003–2004. *J Geophys Res Atmos.* 2007;112(D5):D05208. Available from: http://dx.doi.org/10.1029/2006JD007341

[54] Diendorfer G, Pichler H, and Schulz W. LLS Detection of Upward Initiated Lightning Flashes. In: 2015 9th Asia-Pacific International Conference on Lightning (APL). Nagoya, Japan; 2015. pp. 1–5.

[55] Diendorfer G, Pichler H, and Mair M. Some parameters of negative upward-initiated lightning to the gaisberg tower (2000–2007). *Electromagn Compat IEEE Trans.* 2009;51(3):443–52. https://ieeexplore.ieee.org/abstract/document/5089467

Index

www.ingramcontent.com/pod-product-compliance
Lightning Source LLC
LaVergne TN
LVHW011555200926
848484LV00001B/318